Youth Physical Activity and Sedentary Behavior

Challenges and Solutions

Alan L. Smith, PhD
Purdue University

Stuart J.H. Biddle, PhD
Loughborough University

Editors

Human Kinetics

Library of Congress Cataloging-in-Publication Data

Youth physical activity and sedentary behavior : challenges and
solutions / Alan L. Smith and Stuart J.H. Biddle, editors.
p. ; cm.
Includes bibliographical references and index.
ISBN-13: 978-0-7360-6509-2 (hard cover)
ISBN-10: 0-7360-6509-1 (hard cover)
1. Obesity in adolescence--Prevention. 2. Health behavior in adolescence. 3. Exercise--Psychological aspects. I. Smith, Alan L., 1968- II. Biddle, Stuart.
[DNLM: 1. Obesity--prevention & control. 2. Adolescent. 3. Attitude to Health. 4. Child. 5. Exercise. WD 210 Y83 2008]
RJ399.C6Y68 2008
618.92'398--dc22

2008009922

ISBN-10: 0-7360-6509-1
ISBN-13: 978-0-7360-6509-2

The Web addresses cited in this text were current as of January 2008, unless otherwise noted.

Acquisitions Editor: Myles Schrag; **Developmental Editor:** Elaine H. Mustain; **Managing Editor:** Melissa J. Zavala; **Copyeditor:** Joyce Sexton; **Proofreader:** Joanna Hatzopoulos Portman; **Indexer:** Betty Frizzéll; **Permission Manager:** Dalene Reeder; **Graphic Designer:** Fred Starbird; **Graphic Artist:** Dawn Sills; **Cover Designer:** Nancy Rasmus; **Photographer (interior):** © Human Kinetics, unless otherwise noted; **Photo Asset Manager:** Laura Fitch; **Photo Office Assistant:** Jason Allen; **Art Manager:** Kelly Hendren; **Associate Art Manager:** Alan L. Wilborn; **Illustrator:** Alan L. Wilborn; **Printer:** Thomson-Shore, Inc.

Printed in the United States of America 10 9 8 7 6 5 4 3 2 1

Human Kinetics
Web site: www.HumanKinetics.com

United States: Human Kinetics
P.O. Box 5076
Champaign, IL 61825-5076
800-747-4457
e-mail: humank@hkusa.com

Canada: Human Kinetics
475 Devonshire Road Unit 100
Windsor, ON N8Y 2L5
800-465-7301 (in Canada only)
e-mail: orders@hkcanada.com

Europe: Human Kinetics
107 Bradford Road
Stanningley
Leeds LS28 6AT, United Kingdom
+44 (0) 113 255 5665
e-mail: hk@hkeurope.com

Australia: Human Kinetics
57A Price Avenue
Lower Mitcham, South Australia 5062
08 8372 0999
e-mail: info@hkaustralia.com

New Zealand: Human Kinetics
Division of Sports Distributors NZ Ltd.
P.O. Box 300 226 Albany
North Shore City
Auckland
0064 9 448 1207
e-mail: info@humankinetics.co.nz

To Austin, Emma, and Sarah, my support network and inspiration
A.L.S.

To Fiona, Jack, and Greg, and my mother, Jean
S.J.H.B.

Contents

Foreword

Young children love to play, enjoy all kinds of motion, would rather hop and run than walk, and wiggle and squirm when they are supposed to sit still. Teenagers seem like a different species. They love to play—computer games. They enjoy the motion of riding in a car. Their need to be cool and look good often leads them to avoid sweating. They are content to stare at screens for hours at a time.

This book answers three big questions about youth physical activity:

- How does this change with age happen?
- Does it matter?
- What can be done about it?

Let's start with the second question because the answer is perfectly clear—the decline in physical activity *probably* matters. It is surprisingly difficult to document the health effects of low physical activity and high levels of sedentary behavior among youth. Some have argued that teenagers don't get heart attacks, osteoporosis, colon cancer, or other serious consequences of inactive lifestyles that afflict adults. This argument is heard less often now that childhood obesity is perceived as a worldwide health crisis and type 2 diabetes is a common diagnosis among young people. Many health professionals predict that complications of childhood obesity and diabetes will eventually cause great suffering and huge burdens for health care systems around the world. The youth obesity and diabetes epidemics are bringing unprecedented attention to physical activity from scientists, health officials, and policy makers for the first time.

What about the first question—what accounts for the decline in physical activity, and possibly its enjoyment, from childhood to adolescence? Though several chapters in this volume provide detailed answers, I can summarize the reasons for the decline as "them" and "us." There is a biological basis for the decline in physical activity with age, seen in many species from flies and fish to rats and monkeys. A major mechanism appears to be the disappearance of dopamine receptors in a part of the brain that links movement and reward centers (Ingram, 2000). As we age, we lose part of our ability to derive pleasure from movement.

We can't blame teen sloth only on "their" biology. "We" adults are part of the problem. Children's jumping and shouting during play is often annoying to parents. One of the phrases children hear most often is some variation of "sit down and be quiet." Schools reduce physical education and recess time so that more information can be crammed into immobile brains. Governments have been reluctant to invest in physical education, after-school, and community programs that would make safe, affordable, and enjoyable activity opportunities available to all youth.

Though low-cost activity programs are scarce in many countries, options for sedentary entertainment are constantly expanding. The average American child who has difficulty finding time to be active for one hour a day watches an average of three hours of television, plus time in front of other screens. Marketers for movies and video games advertise aggressively for teens' time and money. The appeal of cars as symbols of independence is reinforced by the world's largest advertising budgets. The suburban design of communities that separates homes from places where people want to go began in the United States, but it has spread to Australia, Europe, Asia, and South America. In the suburbs, most children are unable to walk to school or anywhere else, so an entire category of physical activity can disappear.

This brings us to the third question—what can be done about the epidemic of inactivity? This is where Drs. Smith and Biddle's book makes its most important contributions. The book promises solutions, and many solutions are needed. How can we convince parents that daily physical activity is essential for their child's health? How can the lure of heavily promoted electronic entertainment be countered? What will it take for all schools to provide high-quality physical education? What will it take for society to commit to building communities where children can travel by bike or foot to most of the places they need to go? These are some of the big questions about youth activity levels and sedentary behavior that are being taken seriously for the first time.

Thus, this book is well timed and is likely to be read carefully for guidance about how to get kids more active. This book is the first comprehensive look at the science of youth physical activity that covers the range from cells to psyche to schools to society. The international authors bring a needed global perspective to global problems. Let's hope this book leads to improved science and finds its way into the hands of decision makers who can put into practice many of the solutions identified by some of the world's leading experts.

James F. Sallis, PhD
San Diego State University
June 2007

Ingram, D.K. (2000). Age-related decline in physical activity: Generalization to non-humans. *Medicine and Science in Sports and Exercise, 32,* 1623-1629.

List of Contributors

Elva M. Arredondo
Center for Behavioral
and Community Health Studies
San Diego State University
San Diego, California, United States of America

Guadalupe X. Ayala
Center for Behavioral
and Community Health Studies
San Diego State University
San Diego, California, United States of America

Robert J. Brustad
School of Sport & Exercise Science
University of Northern Colorado
Greeley, Colorado, United States of America

Janice Causgrove Dunn
Faculty of Physical Education and Recreation
University of Alberta
Edmonton, Alberta, Canada

Nikos L.D. Chatzisarantis
School of Psychology
University of Plymouth
Plymouth, United Kingdom

Peter R.E. Crocker
School of Human Kinetics
University of British Columbia
Vancouver, British Columbia, Canada

Nicoleta Cutumisu
Faculty of Physical Education & Recreation
University of Alberta
Edmonton, Alberta, Canada

David A. Dzewaltowski
Department of Kinesiology
Kansas State University
Manhattan, Kansas, United States of America

John P. Elder
Center for Behavioral
and Community Health Studies
San Diego State University
San Diego, California, United States of America

Stuart J. Fairclough
Faculty of Education, Community and Leisure
Liverpool John Moores University
Liverpool, United Kingdom

Antonio Manuel Fonseca
Faculty of Sport
University of Porto
Porto, Portugal

Michael Gard
School of Human Movement Studies
Charles Sturt University
Bathurst, Australia

Donna L. Goodwin
Faculty of Physical Education and Recreation
University of Alberta
Edmonton, Alberta, Canada

Trish Gorely
School of Sport and Exercise Sciences
Loughborough University
Loughborough, United Kingdom

Valerie Hadd
School of Human Kinetics
University of British Columbia
Vancouver, British Columbia, Canada

Martin S. Hagger
Risk Analysis, Social Processes,
and Health Group, School of Psychology
University of Nottingham
Nottingham, United Kingdom

Jacqueline Kerr
Departments of Psychology
and Public Health
San Diego State University
San Diego, California, United States of America

Kent C. Kowalski
College of Kinesiology
University of Saskatchewan
Saskatoon, Saskatchewan, Canada

Robert M. Malina
Tarleton State University
Stephenville, Texas, United States of America
and Department of Kinesiology
and Health Education
University of Texas
Austin, Texas, United States of America

Simon J. Marshall
Department of Exercise & Nutritional Sciences
San Diego State University
San Diego, California, United States of America

Suzanna M. Martinez
Center for Behavioral
and Community Health Studies
San Diego State University
San Diego, California, United States of America

Meghan H. McDonough
Department of Health and Kinesiology
Purdue University
West Lafayette, Indiana, United States of America

Mary McElroy
Department of Kinesiology
Kansas State University
Manhattan, Kansas, United States of America

Chad D. Meyerhoefer
Agency for Healthcare Research and Quality
Rockville, Maryland, United States of America

Claudio R. Nigg
Department of Public Health Sciences
University of Hawaii at Manoa
Honolulu, Hawaii, United States of America

Raheem J. Paxton
Cancer Research Center of Hawaii
Honolulu, Hawaii, United States of America

Nicola D. Ridgers
Research Institute for Sport and Exercise Sciences
Liverpool John Moores University
Liverpool, United Kingdom

Brian E. Saelens
Departments of Pediatrics
and Psychiatry & Behavioral Sciences
Seattle Children's Hospital Research
Institute, University of Washington
Seattle, Washington, United States of America

Jo Salmon
School of Exercise & Nutrition Sciences
Deakin University
Burwood, Victoria, Australia

John C. Spence
Faculty of Physical Education & Recreation
University of Alberta
Edmonton, Alberta, Canada

David J. Stensel
School of Sport and Exercise Sciences
Loughborough University
Loughborough, United Kingdom

Gareth Stratton
Research Institute for Sport and Exercise Sciences
Liverpool John Moores University
Liverpool, United Kingdom

Anna Timperio
School of Exercise & Nutrition Sciences
Deakin University
Burwood, Victoria, Australia

Darren C. Treasure
Competitive Advantage
International Performance Systems
Scottsdale, Arizona, United States of America

Runar Vilhjalmsson
Faculty of Nursing
University of Iceland
Eirbergi, Eiriksgotu, Iceland

C.K. John Wang
National Institute of Education
Nanyang Technological University
Singapore

Gregory J. Welk
Department of Kinesiology
Iowa State University
Ames, Iowa, United States of America

Preface

Over the past two decades there has been significant interest in the health implications of physical activity and sedentary behavior, as well as in how to promote active living. This interest has intensified among the research, professional, and policy-making communities in recent years as obesity rates have climbed and disease states linked with sedentary living have been identified as the main causes of death in developed countries. Of tremendous concern is the increased incidence of obesity, diabetes, and other disease states among children (treated by authors in this text as youth aged 12 years or younger) and adolescents (youth over age 12 years through at least the late teen years). Such conditions are believed to be largely preventable with consistent and appropriate participation in physical activity, independently from or alongside other health behaviors. Physical activity appears to offer a host of physical, psychological, and social benefits to youth. Yet, we appear to be facing a growing physical *inactivity* problem in this population. The present state of affairs may represent only the proverbial tip of the iceberg, and it is essential that we comprehensively address the youth physical inactivity challenge to find a workable solution to it. Notwithstanding this, because it is sometimes difficult to identify clear health outcomes in youth and because measuring physical activity and sedentary behavior accurately can be problematic, this area of study is complex. This stimulated us to bring together a wide-ranging set of chapters to address the topic of youth physical activity and sedentary behavior.

Although a great deal of research from a host of disciplinary and theoretical perspectives has been conducted on physical activity levels, and to a lesser extent on sedentary behavior, there is no resource to our knowledge that brings together the corpus of work that addresses the unique developmental experiences of children and adolescents in these behavioral contexts. In our research careers we have devoted our energies to the study of psychosocial processes within general physical activity, physical education, and organized sport contexts and have come to recognize the value of adopting a youth-centered vantage when conceptualizing, executing, and interpreting our work. Therefore, our goals in assembling this book were to consider the fundamental questions and assumptions framing research on youth physical activity and sedentary behavior; communicate the knowledge base on how developmental, economic, psychological, and social factors link with the youth physical inactivity problem; and overview youth-specific approaches to the reduction of sedentary living and the promotion of active living. Despite our primary allegiance in the past to psychosocial issues, we recognize that complex behaviors, such as youth physical activity, can be properly understood only if we adopt a multidisciplinary and multifaceted approach, as advocated in ecological models.

In pursuing these goals we sought contributions from internationally recognized experts on youth physical activity behavior. As youth physical inactivity is a multifaceted worldwide health challenge, we enlisted a group of authors representing Australia, Europe, and North America from diverse scholarly backgrounds. In addressing their respective topics, the authors were charged with sharing contemporary thought and knowledge on *both* sedentary behavior and physical activity behavior of youth. Although sedentary behavior and physical activity behavior are often conceptualized as falling on a single continuum, recent thinking is that this perspective may limit our progress in addressing youth physical inactivity. Inactivity may not be sufficiently understood as stemming simply from the absence of conditions linked to physical activity behavior. That is, sedentary behaviors can coexist alongside physically active pursuits, and the two behaviors may have different determinants. At present, however, the knowledge base on sedentary behavior is much less well developed than that on physical activity, and it remains to be seen if understanding sedentary behavior will offer assistance in addressing the youth physical inactivity problem. Thus, we requested that authors carefully distinguish physical activity levels, ranging from high activity to inactivity, from specific sedentary behaviors.

This book will appeal to those with an interest in youth health behavior and physical activity, especially, but not exclusively, as approached from psychosocial viewpoints. The text is mainly constructed for students and scholars and is appropriate for use in upper-level undergraduate and graduate courses in developmental sport and exercise psychology, physical activity and health, behavioral medicine, health promotion, physical education teaching, and youth physical activity. Depending on the structure of the course, the book may be used as a comprehensive stand-alone resource, or, alternatively, chapters may be selected that offer important background for discussions on specific issues. Scholars will find the book to be a comprehensive, cutting-edge resource on youth physical activity and sedentary behavior.

The book is organized into three parts. Part I addresses the conceptualization of physical activity and sedentary behavior. Fundamental to the practice of science and solving challenging problems is asking good questions and making sound assumptions. What is physical inactivity? How do we measure it? Why is physical activity behavior important? Should youth inactivity be framed as a social problem? Is it ethical to promote physical activity behavior or discourage sedentary behavior in youth? In this part of the book, the authors address these issues by overviewing the theoretical, historical, and ethical underpinnings of youth physical activity and sedentary behavior research.

The remaining two sections of the book build on this foundation. Part II contains chapters with specific emphasis on developmental and individual-level factors associated with physical activity levels and sedentary behavior in youth. Here the authors communicate that to effectively address youth physical inactivity it is necessary to consider physical maturation, genetics, attitudes, motivational characteristics, and self-perceptions as well as challenges that stem from physical disability. Of course, physical activity and sedentary behavior are expressed within a host of social contexts, and therefore we must consider the world that youth live in if we are to better understand these behaviors. In part III the authors overview the knowledge base on how social agents such as parents and peers, social agencies such as schools and organized sport, living environments, and broader factors (community, economic, and cultural) link to youth physical activity and sedentary behavior.

We are hopeful that readers will find this book both comprehensive and thought provoking. In bringing together this exceptional and diverse group of contributors, we seek to provide a platform for the next wave of youth physical activity and sedentary behavior research and to catalyze new thinking that moves us closer to solving the public health challenge posed by youth inactivity.

Acknowledgments

To complete a project of this magnitude requires the support and inspiration of many. We feel extremely fortunate to have been surrounded by talented and passionate mentors, colleagues, and students over our careers. Whatever contributions we make or successes we enjoy as a result of this project should be attributed to them.

We specifically would like to thank the authors for their enthusiastic participation in the project, their thoughtful and competent work, and their tolerance of our nagging queries. This top-notch group of scholars taught us much over the course of the project. We also thank Drs. John Coveney, David Lavallee, and James F. Sallis for their guidance as we drafted and refined the book proposal and sought suggestions for authors. Very helpful reviews of selected chapters were provided by Drs. George H. Avery, Robert C. Eklund, William A. Harper, and Sarah Ullrich-French. We are grateful for their assistance. We also thank Dr. Thomas W. Rowland and an anonymous reviewer for their thoughtful, exhaustive feedback on the book manuscript. The final product is significantly enhanced as a result of their efforts. We also wish to acknowledge the expert support of the staff at Human Kinetics. In particular we thank Myles Schrag for shepherding the project from the proposal stage through the finishing touches with patience and great humor. His deliberate efforts, along with those of Elaine Mustain (developmental editor), Melissa Zavala (managing editor), Joyce Sexton (copyeditor), Dalene Reeder (permissions), and Fred Starbird and Dawn Sills (graphics), were essential in bringing the manuscript to final form.

We owe a debt of gratitude to our respective institutions, Purdue University (United States) and Loughborough University (United Kingdom), for the provision of research leaves, resources, and dynamic intellectual environments. And, finally, we thank our families for the endless supply of supportive words, tolerance, and laughter over the course of this project.

PART

I

Conceptualization of Youth Physical Activity and Sedentary Behavior

Part I addresses fundamental issues and assumptions pertaining to youth physical activity and sedentary behavior across both childhood (through about age 12 years) and adolescence (delimited to the teen years in the present text). In the study of any health behavior, we need to know how to define key terms and measure the behavior in question (chapter 1) and whether the behavior is associated with specific health outcomes (chapter 2). In addition, it is important to understand physical activity trends in a public health context through a sociohistorical analysis (chapter 3).

Also presented are conceptual perspectives used to frame the key issues (chapter 4) and philosophical questions pertaining to the characterization of active and inactive young people as well as the promotion of physical activity (chapter 5). A closer look at key conceptual perspectives is important for a better understanding of the need for theoretically driven work in this field. Moreover, it is important to critically analyze our beliefs and views about youth activity at a time when many assumptions are taken for granted. In short, this section of the book provides vital foundational knowledge on youth physical activity and sedentary behavior.

CHAPTER

1

Definitions and Measurement

Simon J. Marshall, PhD ▪ Gregory J. Welk, PhD

There has been tremendous interest in assessing and promoting physical activity among children, due in large part to the highly publicized increase in the prevalence of pediatric obesity. The purpose of this chapter is to describe the issues associated with measuring physical activity and sedentary behavior in youth. Definitions and key concepts related to physical activity levels and sedentary behavior are first described. The unique nature of children and children's movement patterns is then discussed, followed by a review of the specific techniques that may be effective for measuring physical activity and sedentary behavior. The chapter concludes with some implications and recommendations for researchers and professionals. This chapter is intended to serve as a broad introduction to the measurement issues that affect our ability to understand activity and sedentary behavior patterns during childhood and adolescence.

Recently, behavioral scientists have advocated the use of a behavioral epidemiology framework to help guide the sequence and spectrum of descriptive, analytic, and intervention studies involving physical activity and sedentary behavior (Sallis & Owen, 1999). Behavioral epidemiology is the scientific study of the etiology and distribution of behaviors that affect health and disease. More importantly, behavioral epidemiology has the explicit purpose of understanding health behaviors so that they can be influenced as part of a population-wide effort to prevent disease and promote health. The behavioral epidemiology framework applied to physical activity, sedentary behavior, and health specifies five main research phases in a rationally ordered sequence. Table 1.1 summarizes these research phases. Each phase of research is intended to build upon previous phases, and for this reason the framework provides a nice "road map" for how best to prioritize and sequence our research efforts in both physical activity and sedentary behavior. With a behavioral epidemiology framework it is clear why the development of valid and reliable measures of physical activity and sedentary behavior is an important research priority. Accurate assessment tools (Phase II) help us better understand the determinants or correlates of physical activity levels (Phase III), which in turn helps us focus our intervention efforts

Table 1.1 The Behavioral Epidemiology Framework Applied to the Study of TV Viewing as a Sedentary Behavior

Phase	Purpose	Example
I	Establish the links between physical (in)activity and health	Determine if a dose–response relationship exists between television viewing (TV) and excess adiposity among adolescents
II	Develop methods for accurately assessing physical (in)activity	Establish the validity and reliability of a self-report measure of TV viewing
III	Identify factors that influence levels of physical (in)activity	Identify the descriptive epidemiology of TV viewing among adolescents Identify modifiable determinants of excessive TV viewing
IV	Evaluate interventions to increase physical activity and reduce physical inactivity	Conduct a randomized controlled trial to reduce TV viewing among adolescents
V	Translate research into practice	Incorporate a new program into a school health promotion curriculum to help adolescents reduce their TV viewing between 3 p.m. and 6 p.m.

Adapted from Sallis and Owen 1999.

on factors most likely to bring about behavior change (Phase IV). There is also the possibility for reverse sequencing in the research phases. For example, as more valid and reliable measures of physical activity and sedentary behavior are developed, we are able to better clarify relationships between physical activity levels and health (Phase I).

Understanding Physical Activity and Inactivity

A basic issue that has confounded our understanding of the relationship between human movement and health is ambiguity regarding what physical activity and inactivity actually are. Thus, we begin our discussion with definitions and a brief exploration of the nature of physical activity and inactivity.

Physical Activity

Physical activity is generally defined as "any bodily movement produced by skeletal muscles that results in energy expenditure" (Caspersen, Powell, & Christenson, 1985, p. 234). This definition has become widely accepted by the scientific community, and it acknowledges that physical activity is a behavior that can be conceptualized on a continuum from minimal to maximal movement. For example, fidgeting, using a computer, walking to school, playing soccer, and the 100 m sprint are all types of physical activity. However, some experts argue that this definition is too broad because it downplays the importance of how much bodily movement (or energy expenditure) is needed to improve health. Therefore some experts and organizations, such as the American College of Sports Medicine, define physical activity as bodily movement that causes a *substantial increase* in energy expenditure. For this text, the broader definition of physical activity is preferred because it is difficult to agree on what constitutes a *substantial* increase and there is evidence of a dose–response relationship between total energy expenditure and all-cause mortality (Lee & Skerrett, 2001).

Another term used to describe the physical activity habits of young people is "exercise." Exercise is viewed as "a subset of physical activity that is planned, structured, and repetitive bodily movement done to improve or maintain one or more components of physical fitness" (American College of Sports Medicine, 2006, p. 3). Exercise is therefore a health behavior that can influence physical fitness. For younger children the concept of exercise is of limited use because they are unlikely to engage in activity that is consciously planned and structured (by them) for the specific purpose of improving or maintaining their fitness. However, when we talk about the physical activity habits of young people, it is important to distinguish between these terms because the problem of youth inactivity may not necessarily be a direct consequence of a "lack of exercise" or "poor fitness." The relationships between physical activity and fitness are discussed in more detail in chapter 2.

Physical Inactivity and Sedentary Behavior

The concept of **physical *in*activity** is less simple to define because the term implies an absence of physical activity, which, using the definition presented earlier, can really occur only during sleep! Approaches to defining youth inactivity can be classified as either norm referenced or criterion referenced. With use of a norm-referenced approach, an individual is compared to his or her peers on how much (or how little) physical activity he or she engages in. An example of a norm-referenced definition is "a physical activity level that is lower than in healthy individuals of similar age, gender, cultural, and socioeconomic background" (Bar-Or & Rowland, 2004, p. 388). The advantage of a norm-referenced approach is that individuals are compared only to those who are from a similar demographic subgroup. This helps to "level the playing field" when one is making comparisons between individuals from different populations, because factors that affect physical activity can be taken into account. However, norm-referenced approaches for defining physical inactivity are limited. First, they do not consider the *dose* of physical activity thought to be important for good health. Young persons who engage in 1 h of moderate-to-vigorous physical activity (MVPA) on five days of the week (considered an adequate dose by health experts) might still be classified as inactive if 50% of their peers engage in more activity than this. Second, a norm-referenced approach may mask true changes in an individual's level of physical activity if all members of a group change to the same degree.

From a criterion-referenced perspective, an individual would be described as physically inactive if he or she did not meet a specific threshold or level of physical activity. For example, the American College of Sports Medicine (2006) defines having a sedentary lifestyle as "not participating in a regular exercise program or not meeting the minimal physical activity recommendations from the U.S. Surgeon General" (p. 22). Criterion-referenced approaches are also used by epidemiologic surveillance systems because they assess the prevalence of inactivity by measuring against a minimum criterion for physical activity or energy expenditure thought necessary to obtain health benefits (Bernstein, Morabia, & Sloutskis, 1999). For example, the Youth Risk Behavior Survey (YRBS), a national self-report survey of high-priority health-risk behaviors among U.S. youth and young adults, was used to classify individuals as "insufficiently active" if they participated in no moderate or vigorous physical activity during the seven days preceding the survey (Centers for Disease Control and Prevention, 2004). Similarly, the 2002 Health Survey for England (Sproston & Primatesta, 2003) classified young people as inactive if they achieved less than 30 min of at least moderate-intensity physical activity a day on five or more days of the week over the seven days before the survey.

However, concluding that young people are *insufficiently active* is different from concluding that young people are *physically inactive*, though this distinction often gets overlooked in both the professional and lay literature. While physical inactivity is an increasingly common term, a more appropriate label for the concept of inactivity is *sedentary behavior* because this term reflects the fact that a diverse range of behaviors can be considered "inactive." The word *sedentary* derives from the Latin verb *sedere*, meaning *to sit*. This is helpful because it reminds us that sedentary behaviors usually involve sitting. Different types of sedentary behavior, as well as the challenges of measuring the time spent in different activities, are discussed in more detail later in the chapter.

Understanding Active and Sedentary Behavior in Youth

In order to understand the recommendations for, and measures of, physical activity, we first need to understand how and why young people differ from adults with regard to their physical activity habits, as well as the characteristics of physical activity that are measurable.

Unique Features

Children differ from adults in a number of ways, and one must consider these distinctions when assessing, evaluating, or promoting physical activity. Children's activity patterns are influenced by the greater biological need for physical activity (Rowland, 1998), as well as by the inherent drive for exploration and the inability to delay gratification. Children tend to move more spontaneously and accumulate physical activity in small bursts throughout the day (Welk, Corbin, & Dale, 2000). This intermittent (lifestyle) activity pattern in children is consistent with the public health goals but is often difficult to assess and study.

A full understanding of children's activity patterns also necessitates an evaluation of sedentary behavior. The combination of low levels of health-enhancing physical activity and high levels of sedentary behavior underlies the true problem of "youth inactivity." A way to understand and link physical activity and sedentary behavior is through the unifying framework of energy expenditure. The term *total energy expenditure* (TEE) refers to an individual's entire energy output, measured in kilocalories (kcal) or kilojoules (kJ). Total energy expenditure consists of basal metabolic rate (BMR) (the energy required to maintain basic physiologic processes at rest), diet-induced thermogenesis (the energy required to transport, digest, and absorb food), and physical activity thermogenesis (the energy required for bodily movement). For most people, physical activity composes only 15% to 30% of their TEE.

This physical activity portion can be further divided into different intensity categories based on multiples of the metabolic rate when seated at rest (which approximates 3.5ml O_2 per kg of body weight per minute). These multiples are referred to as *metabolic equivalents* (METs). For example, a 3.5 MET activity requires 3.5 times the energy expenditure of sitting at rest (i.e., 12.25ml O_2 per kg of body weight per minute). Because there are an infinite number of MET values, we group activities into the intensity categories of Light (1.5-3.0 METs), Moderate (3.0-6.0 METs), and Vigorous (>6 METs) activity.

An important message is that all forms of movement contribute to TEE. Participation in vigorous physical activity requires a higher rate of energy expenditure but occurs over shorter periods of time. Light activity, in contrast, has a lower rate of energy expenditure but can be done for longer periods of time.

A concept that helps to clarify some of the ambiguity in defining physical activity and sedentary behavior is non-exercise activity thermogenesis (NEAT) (Levine et al., 2001). NEAT includes the energy expenditure associated with posture allocation (e.g., sitting, standing, and lying), fidgeting, and routine daily movements such as walking, performing house chores, and playing. Thus, NEAT can also be thought of as the energy expenditure associated with sedentary behavior and light activity. Because NEAT is likely to be the largest component of activity thermogenesis—we spend the majority of our life in this state—it reminds us that energy expended during activities of daily living can be extremely important for maintaining caloric balance. Figure 1.1 summarizes

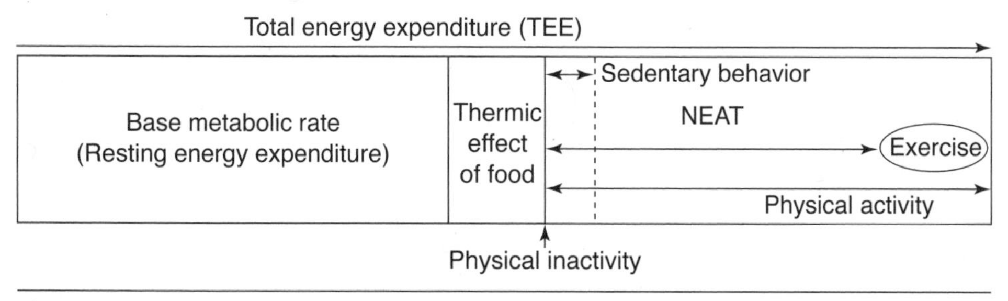

Figure 1.1 Relationships among the different concepts used to describe physical activity. NEAT = non-exercise activity thermogenesis.

the relationships between the different physical activity and inactivity concepts discussed so far.

Characteristics of Physical Activity and Sedentary Behavior

The five principal characteristics of physical activity are frequency, intensity, duration, type, and domain. Because sedentary behavior is a subdomain of physical activity, at least from an energy expenditure perspective, these characteristics can also be used to describe sedentary behavior:

- **Frequency** refers to number of times the physical activity is performed within a specific time period (e.g., bouts per week, month, or year).
- **Intensity** refers to the magnitude of the physiologic response to physical activity and is often quantified by the amount of metabolic work performed (e.g., kilocalories expended). Because it is difficult to measure metabolic work directly, intensity is often captured using physiologic surrogates (e.g., heart rate) or perceptual categories (e.g., very light, light, moderate, hard, very hard). Physical activity intensity may also be expressed in relative or absolute terms. Relative intensity is defined by a workload expressed as a percentage of an individual's maximum capacity (e.g., 60% of maximal heart rate), whereas absolute intensity refers to the workload expressed in units that are independent of an individual's capacity or tolerance (e.g., a heart rate of >155 beats/min).
- **Duration** refers to the length of time (usually in minutes) the activity is performed.
- **Type** refers to the main physiologic systems used (e.g., aerobic, anaerobic) during the activity, though it can also refer to features of the behavior itself (e.g., walking, jumping, running).
- **Domain** refers to the context or setting in which physical activity occurs (e.g., at school, during leisure time, for transportation) and can be useful for understanding the purpose or intent behind the activity.

Outcome measures of physical activity are often defined by one or more of these characteristics (e.g., number of hours of brisk walking per week; average number of minutes spent above 60% of maximum heart rate during a physical education lesson). However, because different scientific disciplines use different measurement instruments and assess different characteristics of physical activity, it is often difficult to compare the findings across studies. We discuss the common outcome measures of physical activity and sedentary behavior in more detail later in the chapter.

Recommended Levels of Physical Activity and Sedentary Behavior

This section focuses on national and international recommendations for physical activity and sedentary behavior. It is also important to discuss how these recommendations were derived and why some countries have different recommendations than others.

Physical Activity

For years it was assumed that young people should be encouraged to do the same amount of physical activity as adults. However, numerous scientific reports by expert panels and health organizations have led many countries to adopt national guidelines for physical activity levels that are specific to young people. National recommendations from the United States (Corbin & Pangrazi, 2004; U.S. Department of Health and Human Services & U.S. Department of Agriculture, 2005), United Kingdom (Department of Health, 2004), and Australia (Australian Government, 2005b) now suggest that young people should engage in at least 60 min of moderate-intensity physical activity each day. (Because space limitations prevent a full discussion of the rationale and basis for the American, Australian, and British physical activity guidelines for youth, readers are encouraged to consult the original documents to fully understand the basis for the 60 min guideline, as well as the specific recommendations related to the type and intensity of physical activity.)

Moderate intensity usually refers to movements that make you breathe hard, requiring at least as much effort as brisk walking (Bar-Or & Rowland, 2004). The 60 min recommendation, which is double that for adults, is now widely accepted by the scientific community. However, experts agree that direct evidence to support this 60 min recommendation is limited, in part because of a lack of epidemiological and experimental data showing a dose–response relationship between physical activity and health outcomes during childhood (Strong et al., 2005; Twisk, 2001). There are also few data to suggest that physical activity habits carry over, or "track," significantly from childhood and adolescence to adulthood (Malina, 1996).

In addition to recommendations for moderate-intensity physical activity, some countries have also adopted recommendations for the amount of vigorous physical activity young people should do. Refer to table 1.2 for a summary of recommendations for vigorous-intensity activity.

The lack of data to support a specific recommendation has led some countries (New Zealand, for example) to purposefully avoid making recommendations for young people altogether. However, available evidence indicates that the

Table 1.2 Summary of Various National Recommendations for Physical Activity and Sedentary Behavior

Country	Physical activity	Sedentary behavior
Australia	All children and youth should be physically active daily, or nearly every day, as part of play, games, sports, work, transportation, recreation, physical education, or planned exercise, in the context of family, school, and community activities. All children and youth should engage in physical activity of at least moderate intensity for 60 min or more on a daily basis. Children and youth who currently do little activity should participate in physical activity of at least moderate intensity for at least 30 min daily, building up to undertaking 60 min daily.	Children and youth should avoid extended periods of inactivity involving participation in sedentary activities such television watching, video, computer games, and surfing the Internet. Children and young people should not spend more than 2 hours a day using electronic media for entertainment (e.g., computer games, Internet, TV), particularly during daylight hours.
Canada	Children ages 6 to 14 should engage in 60 min of moderate-intensity and 30 min of vigorous-intensity physical activity each day. Inactive youth should increase their activity gradually over a period of 5 months: • For moderate activity: Start with 20 min each day for the first month, adding 10 extra daily minutes each month for the next 4 months. • For vigorous activity: Start with 10 min each day for the first month, adding 5 extra daily minutes for each month for the next 4 months.	Physically inactive children should decrease the time they spend on TV, playing computer games, and surfing the Internet by at least 30 min per day. Over several months, children and youth should decrease by at least 90 min per day the amount of time spent on nonactive pursuits such as watching videos and sitting at a computer.
United Kingdom	Children and young people should achieve a total of at least 60 min of at least moderate-intensity physical activity each day. At least twice a week, this should include activities to improve bone health (activities that produce high physical stresses on the bones), muscle strength, and flexibility.	None
United States	Elementary school–aged children should accumulate at least 30 to 60 min of age-appropriate and developmentally appropriate physical activity from a variety of activities on all, or most, days of the week. An accumulation of more than 60 min, and up to several hours per day, of age-appropriate and developmentally appropriate activity is encouraged. Some of the child's activity each day should be in periods lasting 10 to 15 min or more and include moderate to vigorous activity. This activity will typically be intermittent, involving alternating moderate to vigorous activity with brief periods of rest and recovery. All adolescents should be physically active daily, or nearly every day, as part of play, games, sports, work, transportation, recreation, physical education, or planned exercise, in the context of family, school, and community activities. Adolescents should engage in three or more sessions per week of activities that last 20 min or more at a time and that require moderate to vigorous levels of exertion.	Children should not have extended periods of inactivity.

vast majority of children and adolescents would meet the adult recommendation of 30 min per day, and it is reasonable to assume that young people need more than this in order to develop the movement skills required for a physically active lifestyle during adulthood. Therefore, despite the lack of strong evidence to support the recommendation, the 60 min message does appear a reasonable and logically derived threshold that will encourage the promotion of physical activity, help classify young people as sufficiently or insufficiently active, and enable better planning and implementation of organized programs designed to enhance physical activity participation (Australian Government, 2005b).

Because a minimum of 60 min of moderate-intensity physical activity may be unrealistic for some young people, Canada has developed national recommendations that are incremental and based on current physical activity status (Health Canada, 2002a, 2002b). These include the recommendation that inactive youth increase the amount of time they currently spend being physically active by at least 30 min per day. Because 60 min of daily physical activity should be seen as a minimum amount, Canada's recommendations also suggest that after several months children should try to accumulate at least 90 min (60 min of moderate-intensity plus 30 min of vigorous-intensity activity) of physical activity per day. Although this is a higher recommendation than the 60 min message, new evidence (Andersen et al., 2006) suggests that 90 min of daily physical activity may be needed to prevent insulin resistance in children, an important cardiovascular disease risk factor. Incremental recommendations may be especially useful for inactive children because they encourage the setting of attainable goals, a factor that is important for motivation.

Sedentary Behavior

Very few published reports present specific guidelines for the amount of sedentary behavior that is recommended for young people. However, most experts agree that young people, especially those in elementary or primary school, should avoid prolonged periods of inactivity altogether, especially during daylight hours (Corbin & Pangrazi, 2004). Yet it is difficult to know what constitutes a *prolonged period;* and many sedentary behaviors, such as talking with friends and doing homework, are important for healthy social and cognitive development. Existing recommendations usually target TV viewing or general categories (e.g., using media) of how young people spend their time. Perhaps the most widely cited and scientifically accepted recommendation comes from the American Academy of Pediatrics (AAP, 2001) and focuses mainly on TV viewing. The AAP recommends that parents limit children's total entertainment media time to no more than 2 h of quality programming per day; discourage television viewing for children younger than 2 years; and encourage more interactive activities that will promote proper brain development, such as talking, playing, singing, and reading together.

To date, only Australia and Canada have formally adopted and published guidelines about the time young people should spend engaged in sedentary behavior. The Australian government (2005a) recently recommended that "Children and young people should not spend more than 2 hours a day using electronic media for entertainment (e.g., computer games, Internet, TV), particularly during daylight hours" (para. 1). Health Canada (2002a, 2002b) has published guidelines for reducing the time that some young people spend being sedentary. Specifically, it is recommended that physically inactive children decrease the time they spend watching TV, playing computer games, and surfing the Internet by at least 30 min per day. Over several months, children and youth should decrease by at least 90 min per day the amount of time spent on non-active pursuits such as watching videos and sitting at a computer. These recommendations are particularly important because they are the first to acknowledge evidence that physical activity and sedentary behavior are not two sides of the same coin and may carry independent health risks (Dietz & Gortmaker, 1985; Marshall et al., 2002; Owen et al., 2000). High levels of physical activity and sedentary behavior are able to coexist within the lifestyle of a young person. Thus, efforts to reduce sedentary behavior may be independent of efforts to increase physical activity.

Table 1.2 presents more detailed recommendations for physical activity and sedentary behavior for different countries.

Measuring Physical Activity and Sedentary Behavior

The terminology used by researchers and practitioners to describe and quantify human movement can be confusing. One reason is that there are a variety of subdisciplines in the physical activity sciences that each focus on different issues and use different measurement tools. For example, an exercise physiologist is often interested in the physiologic responses to movement and would likely measure the ventilatory and metabolic features of physical activity, such as oxygen consumption and energy expenditure. An exercise psychologist might be more interested in *what* people are doing in order to better understand the purpose or motivation behind different activities. In contrast, a biomechanist may focus on the kinetic features of movement such as the force or torque on body segments to help advance understanding of how physical movements can be made more mechanically efficient or effective.

The topic of physical activity assessment has been of increasing interest, and a number of excellent reviews have described the relative merits of the various approaches for use with children (e.g., Trost, 2001; Welk et al., 2000). Methodological issues to consider when one is collecting, processing, and interpreting different types of physical activity data have also been described elsewhere (Welk, 2002a). A unique contribution of the present review (and this book) is the systematic exploration of activity and inactivity as related but independent constructs. Because the measurement of inactivity (and sedentary behaviors in particular) has received so little attention in the scientific literature, we have devoted more of our discussion to this topic. The types of instruments used to capture activity and inactivity in adults and young people are essentially the same, but some unique measurement considerations are involved in applying these instruments to study youth physical activity behavior. As previously described, children have unique behavior patterns, and these complicate the assessment of physical activity and sedentary behavior. Children also have less highly developed cognitive skills and more concrete thinking patterns than adults. These do not have direct implications for physical activity or sedentary behavior but are important considerations because self-report instruments are often used to assess these behaviors. Dramatic developmental changes in cognition occur as children age, and this makes comparisons across children of different ages problematic. Differences that appear in the data may be due to actual differences in behavior, but they may also in some part be due to differences in cognitions or abstract reasoning. In this section, we describe the relative advantages and disadvantages of different assessment techniques for studying both physical activity and physical inactivity.

Sources of Measurement Error

To advance research on physical activity and inactivity in youth, it is important to obtain valid and reliable measurements of typical behavior. Measuring physical activity is an inherently complex task. Activity patterns vary from day to day, and from season to season. Sedentary behavior may be somewhat less variable, but the same issues apply. Researchers are generally interested in capturing habitual or usual behavior, either because they want to test whether these are associated with health-related variables or because they are trying to influence physical activity levels through intervention. Physical activity or inactivity is typically assessed over multiple days and then averaged to reflect habitual patterns. A major consideration is determining the number of days needed to obtain reliable indicators of behavior. In general, the number of days needed to get reliable data depends on the amount of variability that exists in the data. More days are needed if there is a great deal of variability whereas fewer days are needed if there is less variability.

It is important to recognize that measurements of physical activity and inactivity provide only estimates of the actual or "true" behavior. These estimates are influenced by both systematic and nonsystematic (random) error. **Systematic error,** or **bias,** refers to a predictable form of error, whereas **random error** is an unpredictable form that can lead to either overestimation or underestimation of the actual behavior. Random error is typically divided into biological variation (naturally occurring, intraindividual variation) and analytical variation (variation due to the way in which the measurements were performed or analyzed). Minimizing random error in measurements is important to ensure that there is sufficient power to identify the true relationship between measures of activity and specific outcomes of interest. A review of the implications of measurement error on physical activity assessments is available in Dale, Welk, and Matthews, 2002.

Techniques for Measuring Physical Activity Levels

A number of measurement techniques are available for assessing physical activity. The selection of an appropriate instrument for a particular application depends largely on the relative importance of accuracy and practicality. This section reviews the advantages and disadvantages of the most common techniques used for assessing physical activity in youth and describes measurement issues that complicate assessments of physical activity. Sirard and Pate (2001) have created a useful characterization of the different categories of assessments, and this system is also employed here. We have sought to characterize the sources of error that may influence each of the various assessment techniques. Special emphasis is placed on computing appropriate outcome measures.

Criterion Measures

The most valid and reliable measures of physical activity are referred to as criterion measures. Although we sometimes refer to criterion measures as "gold standard" measures, this is often misleading because even these measures have limitations and contain some error.

- **Indirect calorimetry and doubly labeled water.** Indirect calorimetry and doubly labeled water (DLW) techniques are viewed as criterion measures of energy expenditure. They are frequently used to validate other measures of physical activity but have less utility for actually measuring physical activity behavior. Indirect calorimetry involves measuring O_2 consumption as a proxy of energy expenditure. A limitation of indirect calorimetry is that it confines a participant to an unnatural laboratory setting. Some researchers have used metabolic chambers to allow for some freedom, but this is a highly constrained environment for studying youth activity patterns. The development of portable metabolic analyzers offers promise for improving the utility of this technique, but it is still highly invasive. Doubly labeled water is a simple, noninvasive approach that requires a participant to ingest harmless isotopic tracers (deuterium and oxygen-18) that become gradually eliminated from the body. This elimination is measured in subsequent urine samples and enables total carbon dioxide production to be measured and energy expenditure to be estimated. Doubly labeled water is very expensive and not useful for studying activity patterns in youth since it is not possible to accurately partition energy expenditure into the components of basal metabolism, diet-induced thermogenesis, and physical activity.
- **Direct observation.** Direct observation techniques offer a number of advantages for studying physical activity in youth. They provide detailed objective information about levels of activity, but the main advantage is that they

provide valuable information about the context of physical activity. Numerous direct observation systems have been developed for measuring the physical activity of young people during physical education lessons, during school recess, and in open spaces such as public parks (McKenzie, 2002). Considerable time and effort are required to collect and process direct observation data, but the additional work is needed for certain studies. Direct observation can provide highly accurate information for controlled studies and serve as a useful criterion measure for other instruments.

Objective Measures

Measures of physical activity are considered "objective" if the data being collected do not need to be cognitively and perceptually processed by the participant; the data exist independently of what a participant thinks about them. The "best" available objective measures are also often criterion measures.

- **Pedometers.** Pedometers are an increasingly popular way to collect information about physical activity in youth. They provide an objective indicator of step counts, and new models not only allow estimations of time spent moving but also possess the ability to recall data over a span of time. A well-known limitation of pedometers is that they are suited only for capturing locomotor behavior. Another challenge is that step counts are inversely proportional to leg (stride) length. Because youth vary widely in biological and chronological age, there may be considerable variability in pedometer values due simply to body size (factors that few studies control for). Still, pedometers have been shown to have good utility for studying youth activity behavior.
- **Accelerometers.** Accelerometers have become the accepted standard for most field-based studies of physical activity in youth. These devices work by measuring the acceleration of body segments or limbs during movement. Small piezoelectric or resistive elements contained within the device translate forces or changes in resistance to electrical signals, which are then filtered and stored as movement counts. While more research has been conducted with the Actigraph monitor (Actigraph LLC, Fort Walton Beach, FL, formerly known as the CSA/MTI), a number of other instruments provide similar information (e.g., Biotrainer, Actical, R3D). A well-known limitation of waist-worn activity accelerometers is the inability to detect physical activities that involve upper body movement, and the inability to capture the increased energy cost of walking up a grade or carrying a load (Welk, 2002b). A number of equations have been developed to characterize the relationship between accelerometer counts and movement, but accurate estimations of individual energy expenditure remain elusive. Equations based on locomotor activities tend to underestimate the energy cost of free-living or lifestyle activities, but equations based on a diverse set of locomotor and lifestyle tasks tend to overestimate the energy cost of locomotor activities (Welk, 2005). At present, accelerometers provide useful estimates for group comparisons but not for individual estimates of activity or energy expenditure.

- **Heart rate monitors.** Heart rate provides an indicator of activity that reflects the true physiological stress on the body. While this circumvents the calibration issue just described for the accelerometers, it introduces a different form of error due to inherent individual differences in heart rate response to activity. More highly fit individuals have a lower heart rate response to activity than less fit individuals, so the absolute differences in heart rate cannot be used to capture differences in the activity levels. Corrections based on resting heart rate or individual calibration (Janz, 2002) can overcome this limitation but add complications to data collection that may limit utility for field-based research.

- **Multichannel activity monitors.** Recent developments in technology have led to the development of combination sensors that integrate data from heart rate monitors and accelerometers. Studies with the new Actiheart monitor (Mini Mitter, Bend, OR) show that this combination device may provide more accurate estimates than either measure used alone. New pattern recognition monitors such as the Sensewear Pro II armband monitor (BodyMedia, Inc., Pittsburgh, PA) may also help to overcome limitations with current accelerometer technology. These devices use multiple sensors to detect the predominant activity being performed and then apply activity-specific algorithms for estimations of energy expenditure.

Subjective Measures

In contrast to objective measures, subjective measures require some level of cognitive or perceptual processing by the participant to create the data. Subjective measures require participants to think about, and record, information regarding their physical activity levels.

- **Self-report.** Self-report techniques are the most widely used, and practical, method for collecting information on physical activity in youth. The category of "self-report" measures actually encompasses a variety of assessment methods (e.g., diaries, logs, interviews, and questionnaires), but these share a common reliance on people to estimate or recall their own activity level. Because of children's limited cognitive capabilities, many have questioned the overall utility of self-report tools. Children may misinterpret the questions being asked; they may have problems recalling the time or intensity of activity (or both). While there is clearly considerable error due to cognitive capabilities, the intermittent nature of children's activity is equally problematic. Self-report instruments can be divided into two basic categories: recall-based measures and general measures.

    - **Recall-based approaches** seek to obtain actual information about a child's activity on a specific day or series of days, while general measures tend to emphasize typical activity behavior. An advantage of recall-based approaches is that they can be processed to obtain relevant outcome measures of interest (time spent in activity, or estimates of energy expenditure). A limitation of recall-based formats is that data must be obtained on multiple days in order to take into account the normal intraindividual variability in activity patterns.

- If the intent is to characterize general activity habits, **general measures** of typical activity may be more useful. The Physical Activity Questionnaire for Children (Crocker et al., 1997), for example, uses a series of questions about general activity patterns to calculate an overall activity score. While the score does not allow for estimations of frequency, intensity, or duration, it may be useful in discriminating between active and inactive children.

Although self-report instruments have clear limitations, in many cases they may provide information that cannot be obtained with other instruments. Moreover, a promising approach for improving the accuracy of self-report measures is to calibrate them against other more objective instruments.

Outcome Measures of Physical Activity

The primary outcome measures of interest in most research studies is the amount of time spent in MVPA or the energy expenditure attributable to physical activity. For self-report and observation tools, time spent in activity can be obtained directly. With electronic monitoring devices, estimates of time are typically computed through converting the raw data into the number of minutes. The usual way to do this is to determine the amount of time that is spent below or above different activity thresholds or cut points. Because electronic monitors take repeated samples during movement and then average them over a predefined interval (also referred to as an "epoch"), it is necessary to use care in selecting the length of this interval. Most experts recommend the use of shorter epochs (<1 min) to more effectively capture children's intermittent activity.

From a behavioral perspective, it is often interesting to evaluate the number of bouts of activity or the frequency of physical activity behavior. The frequency of activity bouts can be directly determined with self-report and observation tools but has to be estimated with most electronic monitoring devices. In either case, obtaining these values requires an appropriate criterion to define what constitutes "regular activity" and what counts as a "bout" of activity. This is often problematic because the results are heavily influenced by the way a bout is operationalized. An appropriate criterion could be the percentage of youth that report two to three bouts of short, intermittent activity totaling 30 to 60 min on at least five days a week. Conclusions about typical activity patterns should not be drawn from one or two isolated days of measurement.

If frequency, intensity, and duration are known, it is possible to estimate energy expenditure. This is a useful outcome measure because it has direct public health significance and is easily interpreted. The limitation is that many assumptions and estimations are involved in the calculation of energy expenditure. Hence, it is likely that researchers will lose statistical power when making comparisons based on energy expenditure estimates. For observation and self-report measures, estimates are typically made using multiples of resting metabolic rate (METs). Because the MET values for various activities are not well established for children (Ainsworth et al., 1993), the estimates using these calculations may not be highly accurate. However, there are now age- and puberty-adjusted MET correction factors (Harrell et al., 2005) that can be applied for some activities, and this represents an important contribution to the literature. For heart rate and motion sensors, a calibration equation is needed to convert the raw unit of measurement into energy expenditure values. These equations are typically developed under lab situations using structured activities and may not generalize to field-based activities. Thus, these estimations must also be interpreted with caution.

Techniques for Measuring Sedentary Behavior

Because there are conceptual differences between physical inactivity and sedentary behavior, the methods for assessing these constructs are different. When the goal is to measure physical inactivity (i.e., an absence of health-enhancing physical activity), physical activity is assessed using one of the measures described previously and a criterion- or normative-based threshold is used to classify a subset of individuals as "inactive." However, when the goal is to assess sedentary behavior, assessment methods usually focus on describing *what* people are doing when they are inactive. Some researchers (e.g., Reilly et al., 2003) have tried to measure sedentary behavior by using accelerometers to create a series of cut points for movement counts that best discriminate between sedentary behavior and physical activity. However, this does not permit an assessment of what young people are actually doing, only the amount of movement that occurs while they are sedentary. Other researchers have attempted to measure energy expenditure while the person is sedentary (Puyau et al., 2002). These methods are not described here because they use the physical activity measures discussed previously. The energy costs of different sedentary behaviors are also very similar because most of these behaviors involve sitting.

This section focuses on the measurement of sedentary behaviors that involve the use of screen-based media (e.g., TV viewing, computer or Internet use, playing video games). It is acknowledged that these behaviors do not entirely represent the many ways young people can be sedentary; however, they are highly prevalent among young people (Marshall, Gorely, & Biddle, 2006) and are most commonly measured and reported in the scientific literature.

Criterion Measures

When the goal is to measure the use of screen-based media, the criterion assessment is **direct observation.** For many studies, directly observing the time that young people spend engaged in sedentary activity is impractical because it is expensive, places a high burden on the researcher, and may cause participants to change their behavior (referred to as "behavioral reactivity") because they know they are being observed. Even video-recorded observations may influence findings because participants are aware that they are being observed and may have to operate the recording equipment themselves. Few data exist that quantify people's reactivity to being observed while they are sedentary, though evidence suggests that fewer people agree to participate in research studies about TV viewing after they learn that they will be observed (Anderson et al., 1985).

We found only six studies that have measured youth sedentary behavior using direct observation, and these have focused exclusively on TV viewing (Anderson et al., 1985; Bechtel, Achelpohl, & Akers, 1972; DuRant et al., 1994, 1996; McKenzie et al., 1992) or TV viewing and video gaming (Borzekowski & Robinson, 1999). Test-retest reliability of observations is generally high (r = 0.8-0.9) regardless of whether observations are made in person or by videotape. There is an urgent need for the continued development of low-burden direct observational tools for youth sedentary behavior, especially those that go beyond measuring TV viewing. This presents a research challenge that, at present, appears to be dictated largely by the sophistication of the technology used for observations.

Subjective Measures

Because the majority of sedentary behavior measures are subjective and there are currently few resources available to guide researchers and practitioners to select appropriate measures, we have devoted more discussion to these assessments here.

Time Use Diaries One method of collecting detailed information about sedentary behavior is through a time use diary (figure 1.2). Generally, diaries are completed by the children themselves or their caregivers (Vandewater, Bickham, & Lee, 2006) and contain detailed records of what the children do and for how long. Because they require ongoing and detailed information to be recorded (every minute of the day must be accounted for), diaries are intrusive to the participant and are usually kept only for short periods (typically one to seven days). Despite these limitations, diary-based methods for reporting sedentary behavior generally have stronger validity data than retrospective recall (Anderson et al., 1985) and appear to provide valid and reliable estimates of behavior when compared to direct observation (Juster, 1985).

Only two studies (Anderson et al., 1985; Bechtel et al., 1972) have systematically evaluated the psychometric properties of a time use diary for collecting information about TV viewing. Anderson and colleagues reported that the test-retest reliability of viewing estimates from two 10-day diaries spaced one month apart was good (r = 0.72), and correlations of viewing time with direct (video) observation were highest when the diary was kept concurrently with actual viewing (r = 0.84). However, when diaries were recorded as "typical" viewing time, associations were weaker (r = 0.48) and viewing time was overestimated by up to 25%, a finding also reported in the second study (Bechtel et al., 1972). Interestingly, Anderson and colleagues noted that when directly observed, children were seen looking at the TV only 67% of the time they were in the

Time	Start/stop indicator	**Description of what you are doing** (e.g., sleeping, eating, doing homework, talking with friends, watching TV, listening to music, on telephone, walking to school, etc.)
AFTER SCHOOL		
3:00		End of school
:01		
:02		
:03		
:04		
:05		
:06		
:07		Walk home from school
:08		
:09		
:10		
:11		
:12	Arrive home	
:13		
:14		
3:15		
:16	Turn on TV	Make sandwich
:17		
:18		
:19		
:20		
:21		
:22		
:23		
:24		
:25		
:26		
:27		
:28		
:29		Watch TV
3:30		
:31		
:32		Read magazine and eat sandwich
:33		
:34		
:35		
:36		
:37		

Figure 1.2 Time use diary.

room with the TV turned on. This may be one reason why proxy self-reports of children's TV viewing may overestimate true viewing time.

A number of other studies (Katzmarzyk et al., 1998; Schmitz et al., 2001; Shannon, Peacock, & Brown, 1991) have used quasi- or "blocked" diaries in which youth are asked to recall sedentary activities they have participated in during "blocks" of time (typically 30 min) during the day. Compared to minute-by-minute diaries, block diaries are easier and less time-consuming to complete because participants simply place an "X" in the blocks in which they watched TV, played video games, or participated in other sedentary behavior. Although these measures are often valid and reliable, it is not possible with this method to compute the number of hours that young people spend engaged in specific behaviors. This remains a major limitation of the block diary approach.

Diaries have been used successfully with children as young as age 6 years, and there is evidence that many young people are able to give as much diary information as adults (Meeks & Mauldin, 1990). Extensive compliance data for diary-based methods are lacking, but common sense suggests that efforts to increase the attractiveness and simplicity of the diary and reduce the time required for each entry should reduce respondent burden, decrease the number of missing data points, and improve compliance rates.

Self-Report Questionnaires and Checklists Measuring sedentary behavior using self-reports of young people is a cost-effective alternative to direct observation and time use diaries. Self-reports are usually obtained from paper-and-pencil questionnaires or in-person interviews in which young people are asked to recall the frequency and duration of their sedentary behavior over a specific time frame (e.g., yesterday) or in general (e.g., on a "typical" weekday). For questionnaires, a retrospective recall of the previous day or a "typical day" is recommended. If the "previous day" format is used, then repeated administrations are required to adequately capture typical behavior. If the focus of assessment is screen-based media, researchers should collect data during weekdays and weekend days because of known differences across these periods (Vandewater et al., 2006). Questions in self-report surveys usually refer to the "time spent sitting" or the time engaged in specific behaviors. In the case of multiple behaviors, activities are usually presented in a checklist. The response format of questionnaires and checklists is usually free recall (i.e., open-ended) or on an ordinal or interval scale anchored by units of time (e.g., 0-1 h/day, 1-2 h/day). Compared to questions about the time spent sitting, checklists about specific behaviors are probably easier for most children and adolescents to understand. Checklists of sedentary behavior (e.g., Marshall et al., 2002; Norman et al., 2005; Utter et al., 2003) usually demonstrate acceptable ($r > 0.7$) levels of test-retest reliability, though most studies have been limited to samples of adolescents. The internal consistency among sedentary behaviors in a checklist is, however, typically low (Cronbach alphas = 0.4 to 0.5). This suggests that interrelationships among different sedentary behaviors are complex and that outcome measures based on aggregates of time may mask unique behavior-specific patterns.

Unfortunately, few published data are available to support the validity and reliability of self-reports of sedentary behavior, whether they are based on specific activities (e.g., TV viewing) or groups of activities (e.g., screen-based media). Of the measures with published validity and reliability data, Schmitz and colleagues' (2004) five-item recall of TV viewing and video game and computer use appears particularly promising because it incorporates most of the desirable elements already discussed. Youth typically overestimate the frequency, duration, and intensity of physical activity, and it might be reasonable to assume that recall of sedentary behavior is plagued by the same problem. However, the magnitude and direction of estimation error in recall of sedentary behavior may be different from that of physical activity because an estimate of intensity is not required and social desirability may lead to the systematic underreporting of sedentary time. The reasons for recall estimation errors are complex, but have to do partly with the cognitive complexity involved with reconstructing events from memory and the tendency of youth to respond in a socially desirable manner (Welk et al., 2000).

Only criterion validity studies can examine the extent to which estimation errors are influencing self-reports of sedentary behavior, and these studies must involve comparisons with concurrent direct observation, the "gold standard." To the best of our knowledge, no studies have actually done this, and this remains a major limitation of the current literature. Most validity studies of questionnaire-based self-reports of sedentary behavior involve comparisons with similar measures or other constructs known to be related to sedentary behavior. However, due to the infancy of theoretical models for understanding and predicting sedentary behavior, the "related constructs" are often taken from the physical activity domain, which may not be appropriate for understanding patterns of sedentary behavior (Owen et al., 2000). In other studies the researchers have attempted to validate questionnaire-based measures of sedentary behavior by correlating sedentary time with physical activity scores (Norman et al., 2005; Vilhjalmsson & Thorlindsson, 1998). Although this may provide convergent validity evidence for aggregates of sedentary time, it is an inappropriate method for validating self-reports of TV viewing, video game playing, or computer use because evidence suggests that these activities are largely uncorrelated with physical activity (Marshall et al., 2004). A more appropriate method would be to correlate questionnaire-based self-reports of individual behaviors with data from time use diaries or parent reports of the young person's behavior.

Recently, Bryant and colleagues (2007) published a systematic review of the self-report methods used to assess TV viewing in samples of children and adolescents. Of the 88 published studies that have used self-reports of TV viewing by young people, the vast majority relied on paper-and-pencil questionnaires (81%) completed by the child (80%) about his or her "typical" viewing habits (66%), usually in hours per day (78%). In 5% of the studies, respondents were asked to estimate their viewing time by using the length or number of TV programs watched. Only 15% of published studies presented data on the psychometric properties of the instruments. Asking young people under 14 years of age to recall the number of minutes or hours per day spent watching television may not yield reliable estimates of behavior. Evidence suggests that younger children (<14 years) think about their viewing time not in terms of minutes or hours, but according to the programs watched and the time of day when viewing occurred, as well as the time spent completing homework and participating in sports (Schmitz et al., 2004). This is an important finding because it helps us understand the cognitive strategies that young people use to recall the sedentary activities they have participated in. It also suggests that self-report questions about TV viewing may benefit from memory probes and recall strategies that refer to when young people watch TV and what they watch. Recent advances in research methodology and technology have led to the development of a number of novel self-report assessment tools of sedentary behavior that could assist with recall. For example, handheld computers have been used to collect real-time in situ data about sedentary behavior in adolescents (Marshall & Readdy, 2005), and desktop computers have been used to assist one-day recalls (Ridley, Olds, & Hill, 2001) by incorporating devices that help with time estimation.

Several issues are worth noting about the measurement of sedentary behaviors using retrospective self-reports. The majority of measures that are currently being used have not been adequately evaluated for validity and reliability. Many single behaviors were assessed using single-item self-reports that are unlikely to provide valid or stable estimates of typical behavior. The recall frame also varied greatly, with studies reporting TV viewing time across different periods such as the past evening, day, week, or month. In some studies, the parent provided a proxy report of a child's TV viewing time. However, previous research (Armstrong et al., 1998; Gortmaker et al., 1996; Guillaume et al., 1997) demonstrates that parent and child estimates of behavior are often poorly correlated (typically r = 0.2-0.3), suggesting either that considerable error exists or that different constructs are being measured. For children younger than 10 years it is recommended that proxy reports from a parent or older sibling be used in place of the child's self-report. However, further research is needed to establish reporting guidelines for using such proxy reports.

One strategy that has been shown to increase the validity of self-reports from children and adolescents is to shorten the recall period over which they are asked to remember information, preferably to the previous day (Welk et al., 2000). This reduces the cognitive demand of recall; but the trade-off is that these estimates may not represent habitual behavior, and thus multiple days of assessment are necessary. It is also reasonable to assume that patterns of sedentary behavior exhibit seasonal variation, perhaps similar to that of physical activity (Snel & Twisk, 2001). However, few studies have measured sedentary behavior at different times of the year or attempted to statistically control for seasonal artifacts. For frequent behaviors, such as television viewing and computer use, fewer days of assessment are necessary. However, for infrequent or sporadic behaviors such as "hanging out" at a shopping mall or going to the movies, more days of assessment will be required.

Outcome Measures

The primary outcome measure of interest in most research studies is the amount of time spent engaged in sedentary behavior. When it is of interest to estimate the total time that young people spend being sedentary, simply summing the time spent in individual behaviors is not recommended. Aggregating duration estimates of individual sedentary behaviors is likely to cause total sedentary time to be overestimated because many young people engage in multiple sedentary activities simultaneously. Using measures of the time spent in specific behaviors as a proxy for the total time spent being sedentary is also problematic because of the many ways in which young people spend their free time. Even the most prevalent sedentary behavior, TV viewing, usually accounts for only 15% to 20% of the total time most young people spend awake.

For studies using direct observation, outcome measures are usually presented as the number of minutes the child or adolescent spends sitting or is engaged with sedentary technology during a predefined observation period. For self-report measures, the number of minutes (or hours) per day or week is usually used. When the response options are based on ordinal categories, the outcome measure should be presented as the frequency of responses or the number of children responding in each category. When the frequency of behavior is used as an outcome measure, it is usually expressed as the number of bouts or days per week that a young person reports participating in specific activities. In assessment of TV viewing, however, frequency is a poor outcome measure because most children watch at least some television each day. Nevertheless, for intervention studies that encourage children to turn off the television set for entire days, this may be an appropriate measure.

For diary-based measures, duration-based outcomes are appropriate unless a block diary has been used. In the latter case, the outcome should be presented as the number of blocks for which a particular behavior was recorded over a specified period. Other measures have asked children to report the number of TV programs watched in the past day or week, and this is promising because it assists with recall accuracy. However, program-based outcomes are generally limited because children's viewing is rarely continuous and may be accompanied by other activities. Outcome measures of sedentary behavior that are based on a proxy or direct measure of energy expenditure or intensity are limited for several reasons. For example, most sedentary activities involve sitting and therefore are similar in terms of energy cost. This reduces the variability in the data, making it difficult to identify factors that discriminate between sedentary behaviors. Also, energy expenditure or intensity-based outcomes tell us nothing about what young people are actually doing.

APPLICATIONS FOR RESEARCHERS

The selection of a measure of physical activity or sedentary behavior should be based most directly on the research question. Deciding between different tools that capture the same dimension may depend on the relative importance of accuracy and practicality. Sometimes it is important to obtain an estimate of energy expenditure, while other studies may require only an indicator of overall activity level or the time spent doing a specific behavior. Efforts to estimate energy expenditure from raw activity data may add unnecessary error if only total activity levels are needed. Researchers should also become familiar with potential sources of error that may influence their measurements and should take active steps to reduce these errors. Measurement error reduces power for statistical analyses and therefore makes drawing valid conclusions difficult. One should use a standardized script when administering self-report surveys, and detailed rules should govern how data are cleaned and processed.

APPLICATIONS FOR PROFESSIONALS

Encourage children and adolescents to get at least 60 min of moderate-intensity physical activity every day. Moderate-intensity activity makes you breathe hard and is similar to fast walking. For inactive children, encourage them to meet this target incrementally, starting with 20 min each day and building up to 60 min after several months. Try to promote all kinds of enjoyable movement, not just exercise and organized sports, especially among younger children. Encourage activities that involve intermittent but frequent bursts of fun activity rather than continuous bouts. Television viewing is the most prevalent sedentary behavior in young people but is not a good marker of a sedentary lifestyle. For physically inactive children, target reductions in multiple sedentary behaviors. One strategy is to decrease by 30 min each day the time they spend in activities such as watching TV, playing computer games, and surfing the Internet.

References

Ainsworth, B.E., Haskell, W.L., Leon, A.S., Jacobs, D.R., Montoye, H.J., Sallis, J.F., et al. (1993). Compendium of physical activities: Classification of energy costs of human physical activities. *Medicine and Science in Sports and Exercise*, *25*, 71-80.

American Academy of Pediatrics. (2001). Policy statement: Children, adolescents and television (RE0043). *Pediatrics*, *107*(2), 423-426.

American College of Sports Medicine. (2006). *ACSM's guidelines for exercise testing and prescription* (7th ed.). Philadelphia: Lippincott Williams & Wilkins.

Andersen, L.B., Harro, M., Sardinha, L.B., Froberg, K., Ekelund, U., Brage, S., et al. (2006). Physical activity and clustered cardiovascular risk in children: A cross-sectional study (The European Youth Heart Study). *Lancet, 368*(July 22), 299-304.

Anderson, D., Field, D., Collins, P., Lorch, E., & Nathan, J. (1985). Estimates of young children's time with television: A methodological comparison of parent reports with time-lapse video home observation. *Child Development, 56,* 1345-1357.

Armstrong, C.A., Sallis, J.F., Alcaraz, J.E., Kolody, B., McKenzie, T.L., & Hovell, M.F. (1998). Children's television viewing, body fat, and physical fitness. *American Journal of Health Promotion, 12*(6), 363-368.

Australian Government. (2005a). *Australia's physical activity recommendations for children and young people.* Retrieved November 6, 2006, from www.aodgp.gov.au/internet/wcms/publishing.nsf/Content/health-pubhlth-strateg-active-recommend.htm.

Australian Government. (2005b). *Discussion paper for the development of recommendations for children's and youths' participation in health promoting physical activity* (No. 3704). Australian Government, Department of Health and Ageing. The Commonwealth of Australia, Canberra.

Bar-Or, O., & Rowland, T.W. (2004). *Pediatric exercise science: From physiologic principles to health care application.* Champaign, IL: Human Kinetics.

Bechtel, R.B., Achelpohl, C., & Akers, R. (1972). Correlates between observed behavior and questionnaire responses on television viewing. In E.A. Ruhinstein, C.A. Comstock, & I.P. Murray (Eds.), *Television and social behavior: Television in day-to-day life: Patterns of use* (Vol. 4, pp. 274-344). Washington, DC: Government Printing Office.

Bernstein, M.S., Morabia, A., & Sloutskis, D. (1999). Definition and prevalence of sedentarism in an urban population. *American Journal of Public Health, 89*(6), 862-867.

Borzekowski, D.L., & Robinson, T.N. (1999). Viewing the viewers: Ten video cases of children's television viewing behaviors. *Journal of Broadcasting and Electronic Media, 43,* 506-528.

Bryant, M.J., Lucove, J.C., Evenson, K.R., & Marshall, S.J. (2007). Measurement of television viewing in children and adolescents: A systematic review. *Obesity Reviews, 8,* 197-209.

Caspersen, C.J., Powell, K.E., & Christenson, G.M. (1985). Physical activity, exercise, and physical fitness: Definitions and distinctions for health-related research. *Public Health Reports, 100*(2), 126-130.

Centers for Disease Control and Prevention. (2004). Surveillance summaries. *Morbidity and Mortality Weekly Report, 53*(May 21, SS-2).

Corbin, C.B., & Pangrazi, R.P. (2004). *Physical activity for children: A statement of guidelines for children ages 5-12* (2nd ed.). Reston, VA: National Association for Sport and Physical Education.

Crocker, P.R.E., Bailey, D.A., Faulkner, R.A., Kowalski, K.C., & McGrath, R. (1997). Measuring general levels of physical activity: Preliminary evidence for the Physical Activity Questionnaire for Older Children. *Medicine and Science in Sports and Exercise, 29*(10), 1344-1349.

Dale, D., Welk, G.W., & Matthews, C.E. (2002). Methods for assessing physical activity and challenges for research. In G. Welk (Ed.), *Physical activity assessments for health-related research* (pp. 19-34). Champaign, IL: Human Kinetics.

Department of Health. (2004). *At least five a week: Evidence on the impact of physical activity and its relationship to health. A report from the Chief Medical Officer* (No. 2389). London, UK: Department of Health.

Dietz, W.H., & Gortmaker, S.L. (1985). Do we fatten our children at the television set? Obesity and television viewing in children and adolescents. *Pediatrics, 75*(5), 807-812.

DuRant, R.H., Baranowski, T., Johnson, M., & Thompson, W.O. (1994). The relationship among television watching, physical activity, and body composition of young children. *Pediatrics, 94*(4), 449-455.

DuRant, R.H., Thompson, W.O., Johnson, M., & Baranowski, T. (1996). The relationship among television watching, physical activity, and body composition of 5- or 6-year-old children. *Pediatric Exercise Science, 8*, 15-26.

Gortmaker, S.L., Must, A., Sobol, A.M., Peterson, K., Colditz, G.A., & Dietz, W.H. (1996). Television viewing as a cause of increasing obesity among children in the United States, 1986-1990. *Archives of Pediatric and Adolescent Medicine, 150*, 356-362.

Guillaume, M., Lapidus, L., Bjorntorp, P., & Lambert, A. (1997). Physical activity, obesity, and cardiovascular risk factors in children: The Belgian Luxembourg Child Study II. *Obesity Research, 5*(6), 549-556.

Harrell, S., McMurray, R.G., Baggett, C.D., Pennell, M.L., Pearce, P.F., & Bangdiwala, S.I. (2005). Energy costs of physical activities in children and adolescents. *Medicine and Science in Sports and Exercise, 37*(2), 329-336.

Health Canada. (2002a). *Canada's physical activity guide for children* (No. 39-611/2002-2E). Ottawa: Minister of Public Works and Government Services Canada.

Health Canada. (2002b). *Canada's physical activity guide for youth* (No. H39-611/2002-1E). Ottawa: Minister of Public Works and Government Services Canada.

Janz, K.F. (2002). Use of heart rate monitors to assess physical activity. In G. Welk (Ed.), *Physical activity assessments for health-related research* (pp. 143-161). Champaign, IL: Human Kinetics.

Juster, F.T. (1985). The validity and quality of time use estimates obtained from recall diaries. In F.T. Juster & F.P. Stafford (Eds.), *Time, goods, and well-being* (pp. 63-91). Ann Arbor, MI: Institute for Social Research.

Katzmarzyk, P.T., Malina, R.M., Song, T.M.K., & Bouchard, C. (1998). Television viewing, physical activity, and health-related fitness of youth in the Québec Family Study. *Journal of Adolescent Health, 23*(5), 318-325.

Lee, I.M., & Skerrett, P.J. (2001). Physical activity and all-cause mortality: What is the dose-response relation? *Medicine and Science in Sports and Exercise, 33* Suppl.(9), S459-S471.

Levine, J., Melanson, E.L., Klaas, R., Westerterp, K.R., & Hill, J.O. (2001). Measurement of the components of nonexercise activity thermogenesis. *American Journal of Physiology: Endocrinology and Metabolism, 281*, 670-675.

Malina, R.M. (1996). Tracking of physical activity and physical fitness across the lifespan. *Research Quarterly for Exercise and Sport, 67*(Suppl. 3), 48-57.

Marshall, S.J., Biddle, S.J.H., Gorely, T., Cameron, N., & Murdey, I.M. (2004). Relationships between media use, body fatness and physical activity among children and youth: A meta-analysis. *International Journal of Obesity, 28*, 1238-1246.

Marshall, S.J., Biddle, S.J.H., Sallis, J.F., McKenzie, T.L., & Conway, T.L. (2002). Clustering of sedentary behaviors and physical activity among youth: A cross-national study. *Pediatric Exercise Science, 14*(4), 401-417.

Marshall, S.J., Gorely, T., & Biddle, S.J.H. (2006). A descriptive epidemiology of youth sedentary behavior: A review and critique. *Journal of Adolescence, 29*, 333-349.

Marshall, S.J., & Readdy, R.T. (2005). *The development of an electronic diary to collect in-situ data about sedentary behavior and eating*. Paper presented at the fourth annual conference of the International Society of Behavioral Nutrition and Physical Activity, Amsterdam, The Netherlands.

McKenzie, T. (2002). Use of direct observation to assess physical activity. In G. Welk (Ed.), *Physical activity assessments for health-related research* (pp. 179-195). Champaign, IL: Human Kinetics.

McKenzie, T.L., Sallis, J.F., Nader, P.R., Broyles, S.L., & Nelson, J.A. (1992). Anglo- and Mexican-American preschoolers at home and at recess: Activity patterns and environmental influences. *Journal of Developmental and Behavioral Pediatrics, 13*(3), 173-180.

Meeks, C.B., & Mauldin, T. (1990). Children's time in structured and unstructured leisure activities. *Lifestyles: Family and Economic Issues, 11*(3), 257-281.

Norman, G.J., Schmid, B.A., Sallis, J.F., Calfas, K.J., & Patrick, K. (2005). Psychosocial and environmental correlates of adolescent sedentary behaviors. *Pediatrics, 116*(4), 908-916.

Owen, N., Leslie, E., Salmon, J., & Fotheringham, M.J. (2000). Environmental determinants of physical activity and sedentary behavior. *Exercise and Sport Sciences Reviews, 28*(4), 165-170.

Puyau, M.R., Adolph, A.L., Vohra, F.A., & Butte, N.F. (2002). Validation and calibration of physical activity monitors in children. *Obesity Research, 10*(3), 150-157.

Reilly, J.J., Coyle, J., Kelly, L., Burke, G., Grant, S., & Paton, J.Y. (2003). An objective method for measurement of sedentary behavior in 3- to 4-year olds. *Obesity Research, 11*(10), 1155-1158.

Ridley, K., Olds, T., & Hill, A. (2001). A multimedia activity recall for children and adolescents (MARCA): Description and validation. *Journal of Science and Medicine in Sport, 5*(4), s113.

Rowland, T.W. (1998). The biological basis of physical activity. *Medicine and Science in Sports and Exercise, 30*(3), 392-399.

Sallis, J.F., & Owen, N. (1999). *Physical activity and behavioral medicine*. Thousand Oaks, CA: Sage.

Schmitz, K.H., Harnack, L., Fulton, J.E., Jacobs, D.R., Gao, S., Lytle, L.A., et al. (2004). Reliability and validity of a brief questionnaire to assess television viewing and computer use by middle school children. *Journal of School Health, 74*(9), 370-377.

Schmitz, K.H., Lytle, L.A., Phillips, G.A., Murray, D.M., Birnbaum, A.S., & Kubik, M.Y. (2001). Psychosocial correlates of physical activity and sedentary leisure habits in young adolescents: The Teens Eating for Energy at School Study. *Preventive Medicine, 34*(2), 266-278.

Shannon, B., Peacock, J., & Brown, M.J. (1991). Body fatness, television viewing and calorie-intake of a sample of Pennsylvania sixth grade children. *Journal of Nutrition Education, 23*(6), 262-268.

Sirard, J.R., & Pate, R.R. (2001). Physical activity assessment in children and adolescents. *Sports Medicine, 31*, 439-454.

Snel, J., & Twisk, J. (2001). Assessment of lifestyle. In A. Vingerhoets (Ed.), *Assessment in behavioral medicine* (pp. 245-276). Hove, UK: Brunner-Routledge.

Sproston, K., & Primatesta, P. (2003). *Health Survey for England 2002. The health of children and young people.* London: The Stationery Office.

Strong, W.B., Malina, R.M., Blimkie, C.J.R., Daniels, S.R., Dishman, R.K., Gutin, B., et al. (2005). Evidence based physical activity for school-age youth. *Journal of Pediatrics, 146*, 732-737.

Trost, S.G. (2001). Objective measurement of physical activity in youth: Current issues, future directions. *Exercise and Sport Sciences Reviews, 29*(1), 32-36.

Twisk, J. (2001). Physical activity guidelines for children and adolescents: A critical review. *Sports Medicine, 31*, 617-627.

U.S. Department of Health and Human Services & U.S. Department of Agriculture. (2005). *Dietary guidelines for Americans* (6th ed., No. HHS-ODPHP-2005-01-DGA-A). Washington, DC: Government Printing Office.

Utter, J., Neumark-Sztainer, D., Jeffery, R., & Story, M. (2003). Couch potatoes or french fries: Are sedentary behaviors associated with body mass index, physical activity, and dietary behaviors among adolescents? *Journal of the American Dietetic Association, 103*(10), 1298-1305.

Vandewater, E.A., Bickham, D.S., & Lee, J.H. (2006). Time well spent? Relating television use to children's free-time activities. *Pediatrics, 117*(2), 181-191.

Vilhjalmsson, R., & Thorlindsson, T. (1998). Factors related to physical activity: A study of adolescents. *Social Science and Medicine, 47*(5), 665-675.

Welk, G. (Ed.). (2002a). *Physical activity assessments for health-related research.* Champaign, IL: Human Kinetics.

Welk, G. (2002b). Use of accelerometry-based activity monitors to assess physical activity. In G. Welk (Ed.), *Physical activity assessments for health-related research* (pp. 125-141). Champaign, IL: Human Kinetics.

Welk, G.W. (2005). Principles of design and analyses for the calibration of accelerometry-based activity monitors. *Medicine and Science in Sports and Exercise, 37*(11 Suppl.), S501-511.

Welk, G., Corbin, C.B., & Dale, D. (2000). Measurement issues in the assessment of physical activity in children. *Research Quarterly for Exercise and Sport, 71*(2), 59-73.

CHAPTER

2

Youth Health Outcomes

David J. Stensel, PhD ▪ Trish Gorely, PhD ▪ Stuart J.H. Biddle, PhD

It is a common assumption that physical activity in childhood and adolescence is beneficial to health and that, conversely, physical inactivity in these periods is detrimental to health. The aim of this chapter is to examine the evidence underlying these assumptions. The predominant focus of the chapter is on evidence linking physical activity levels in youth with health outcomes in youth. Where possible, evidence linking activity levels in youth with health outcomes in adulthood is also presented. It should be appreciated at the outset, however, that there is limited evidence linking activity levels in youth with health outcomes in adulthood.

Another factor that one should bear in mind while reading this chapter is the nature of the evidence linking physical activity levels with health outcomes. Much of this evidence comes from uncontrolled or nonrandomized trials or from observational studies. This is considered category C evidence by the U.S. National Institutes of Health (Kesaniemi et al., 2001); category C evidence is inferior to category B evidence (i.e., evidence from a limited number of well-designed randomized controlled trials) and category A evidence (i.e., evidence from many randomized controlled trials involving large numbers of subjects). Thus, it is important for the reader to appreciate that while there are many observational studies linking physical activity levels with health outcomes, these studies do not provide proof of cause and effect, and this is a limitation of much of the evidence presented here.

Although there are many health issues that are important to young people, the focus in this chapter is on overweight and obesity, type 2 diabetes, cardiovascular disease risk, skeletal health, and mental health. These are considered major health issues for young people and also issues for which there is a reasonable body of evidence concerning potential links with physical activity levels. In particular, the issue of obesity in youth and the link between this condition and type 2 diabetes is topical and therefore deserves special attention. This chapter focuses on recent evidence related to the areas just outlined. Additional information can be found on these topics in our previous review (Biddle, Gorely, & Stensel, 2004).

Overweight and Obesity

The issue of obesity continues to attract great interest both within the research community and within the media. Consistent with trends in adults, obesity prevalence in children and adolescents appears to be increasing in the United Kingdom (Bundred, Kitchiner, & Buchan, 2001; Chinn & Rona, 2001; Stamatakis et al., 2005), the United States (Ogden et al., 2006; Strauss & Pollack, 2001), and other developed and developing countries (Speiser et al., 2005).

Measuring Obesity

In the majority of studies, obesity is defined according to age- and sex-specific percentile guidelines for body mass index (BMI: calculated by dividing weight in kilograms by height in meters squared—kg/m^2), such as those developed by Cole and colleagues (2000) (see table 2.1). A clear limitation here is the inability of BMI to distinguish between fat and lean tissue. That is, a high BMI score may indicate high levels of adiposity but, alternatively, could indicate a large muscle mass. Despite this limitation, many still endorse the use of the BMI because it correlates well with more direct fatness measures, it is simple and inexpensive to use, and the likelihood of misclassification is small because most individuals with high BMI values have excess body fat (Must & Anderson, 2006; Reilly, 2006; Steinberger et al., 2005).

Table 2.1 International BMI Cutoff Points for Overweight and Obesity by Sex for Ages 2 through 18 Years[1]

	BMI 25 KG/M^2		BMI 30 KG/M^2	
Age (years)	Males	Females	Males	Females
2	18.4	18.0	20.1	20.1
2.5	18.1	17.8	19.8	19.5
3	17.9	17.6	19.6	19.4
3.5	17.7	17.4	19.4	19.2
4	17.6	17.3	19.3	19.1
4.5	17.5	17.2	19.3	19.1
5	17.4	17.1	19.3	19.2
5.5	17.5	17.2	19.5	19.3
6	17.6	17.3	19.8	19.7
6.5	17.7	17.5	20.2	20.1
7	17.9	17.8	20.6	20.5
7.5	18.2	18.0	21.1	21.0

	BMI 25 KG/M²		BMI 30 KG/M²	
Age (years)	**Males**	**Females**	**Males**	**Females**
8	18.4	18.3	21.6	21.6
8.5	18.8	18.7	22.2	22.2
9	19.1	19.1	22.8	22.8
9.5	19.5	19.5	23.4	23.5
10	19.8	19.9	24.0	24.1
10.5	20.2	20.3	24.6	24.8
11	20.6	20.7	25.1	25.4
11.5	20.9	21.2	25.6	26.1
12	21.2	21.7	26.0	26.7
12.5	21.6	22.1	26.4	27.2
13	21.9	22.6	26.8	27.8
13.5	22.3	23.0	27.2	28.2
14	22.6	23.3	27.6	28.6
14.5	23.0	23.7	28.0	28.9
15	23.3	23.9	28.3	29.1
15.5	23.6	24.2	28.6	29.3
16	23.9	24.4	28.9	29.4
16.5	24.2	24.5	29.1	29.6
17	24.5	24.7	29.4	29.7
17.5	24.7	24.8	29.7	29.8
18	25.0	25.0	30.0	30.0

[1]For those aged 18 or older the definition of overweight is a body mass index (BMI) of 25 up to 30, and the definition of obese is a BMI of 30 or higher. The values presented in this table are predictive of a BMI of 25 and 30 kg/m² at age 18. Obtained by averaging data from Brazil, Great Britain, Hong Kong, the Netherlands, Singapore, and the United States.

From T.J. Cole et al., 2000, "Establishing a standard definition for child overweight and obesity worldwide: International survey", *British Medical Journal* 320: 1242. With permission of the BMJ Publishing Group.

Findings from recent studies examining the health risks of elevated BMI values in adults are inconsistent. For example, recent studies of men and women in the United States (Adams et al., 2006) and Korea (Jee et al., 2006) conclude that elevations in BMI are associated with excess mortality, whereas a recent systematic review involving 40 studies and over 250,000 participants showed that overweight (BMI 25.0 to 29.9 kg/m²) is associated with a lower risk of total and cardiovascular mortality than normal weight (BMI 20.0 to 24.9 kg/m²).

Moreover, the review concludes that obese individuals (BMI 30.0 to 35.0 kg/m^2) are not at increased risk of total mortality or cardiovascular mortality compared with normal-weight individuals (Romero-Corral et al., 2006). This has led to the suggestion that the BMI should be left aside as a clinical and epidemiological measure of cardiovascular risk (Franzosi, 2006). This lack of clarity regarding the relationship between elevations in BMI and ill health in adulthood makes the link between elevated BMI and ill health in youth more tenuous, because the rationale for determining BMI in young people is to predict future risk of ill health.

Although the extent to which elevations in BMI are predictive of future risks of ill health in young people is unclear, this does not mean that excess adiposity in youth is not detrimental to health. Many studies indicate that overweight and obesity (regardless of the definitions used) are associated with a variety of health risks both in youth and in adulthood (British Medical Association, 2005). Moreover, overweight and obesity in youth is a strong predictor of overweight and obesity in adulthood (Matton et al., 2006; Whitaker et al., 1997). A challenge at present is to develop a more precise marker than BMI for the health risks associated with excess adiposity. Waist circumference may be one way forward in this respect. Waist circumference is a marker of central body fat accumulation and has been linked to an increased risk of metabolic complications such as type 2 diabetes (Wang et al., 2005). One recent study concluded that trends of increasing waist circumference in British children and adolescents in the last 10 to 20 years have greatly exceeded those in BMI, suggesting that BMI is a poor proxy for central fatness and that BMI has systematically underestimated the prevalence of obesity in young people (McCarthy, Ellis, & Cole, 2003). Another alternative to BMI is body fat percentage and body fat reference curves for children that have recently been published (McCarthy et al., 2006). Issues remain, however, regarding the best way to measure body fat in children and adolescents and also the point at which excess body fat becomes detrimental to health in young people.

Activity Levels and Adiposity

Notwithstanding the difficulties in defining obesity in young people, many researchers believe that young people are getting fatter and that inactivity is a major cause of this trend (Gard & Wright [2005] are a notable exception). Definitive evidence linking inactivity with childhood and adolescent obesity is hard to come by, however. The strongest and most consistent determinant of obesity in childhood and adolescence appears to be obesity early in life (Baird et al., 2005; Reilly et al., 2005; Vogels et al., 2006; Wardle et al., 2006).

Although some cross-sectional studies have shown an inverse association between physical activity levels and BMI or body fatness in children and adolescents, the associations are often weak (Ekelund et al., 2005; Ruiz et al., 2006; Sulemana, Smolensky, & Lai, 2006). For example, Ekelund and colleagues observed that physical activity (assessed using self-report questionnaires) explained less than 4% of the variance in fat mass (measured using air-displacement plethysmography) in a group of 190 male adolescents. Moreover, there was no relationship between physical activity and fat mass in a group of 255 female adolescents in this study. Lack of an association between physical activity and obesity in cross-sectional studies is often attributed to the imprecise nature of self-report physical activity measures. However, another recent cross-sectional study (Rennie et al., 2005) that examined physical activity levels in prepubescent children, using doubly labeled water and heart rate monitoring, showed no difference in activity levels of boys and girls at high risk of obesity (by virtue of obesity in one or both of their parents) compared with boys and girls not at high risk of obesity.

Prospective (longitudinal) studies, while still observational in nature, provide stronger indications of cause and effect than cross-sectional studies. Many prospective studies have investigated the association between activity levels and body fatness or obesity in children. Television viewing has often been used as a marker for inactivity. In a notable prospective study linking television viewing with overweight, Hancox, Milne, and Poulton (2004) examined the association between child and adolescent television viewing and adult health. They followed 1000 individuals in New Zealand from birth to age 26 years. They found that the prevalence of overweight (BMI $\geq$ 25 kg/m^2) was approximately 45% in adults who watched 2 to 3 h of television per weekday between the ages of 5 and 15 years compared with around 25% in those who watched less than 1 h of television per weekday during the same period. Interestingly, this association remained significant after adjustment for BMI at age 5 years and physical activity at age 15 years. This suggests that depressed metabolic rate during television viewing or adverse effects on diet (or both), rather than displacement of physical activity per se, may explain the association between television viewing and overweight or obesity (Ludwig & Gortmaker, 2004).

Another recent longitudinal study provides convincing support for the hypothesis that reductions in physical activity during childhood and adolescence are related to elevations in adiposity in young adulthood (Kimm et al., 2005). Changes in self-reported habitual physical activity were examined in relation to changes in BMI and skinfold thickness in 1152 Black and 1135 White females from the United States who were followed from ages 9 or 10 to ages 18 or 19 years. There was a pronounced decline in physical activity during the transition from childhood to young adulthood, and BMI and sum of skinfolds increased during this period. However, those classified as active during adolescence experienced smaller gains in BMI and sum of skinfolds than those classified as inactive, even after adjustment for energy intake (assessed using three-day food records). At ages 18 or 19 years, BMI differences between active and inactive participants were 2.98 kg/m^2 for Black females and 2.10 kg/m^2 for White females. For comparison, BMI differences between active and inactive participants at baseline (ages 9 or 10 years) were 1.12 and 0.72 kg/m^2 for Black and White females, respectively. Similar findings were reported for sum of skinfolds.

To demonstrate that physical activity level is causally related to overweight and obesity, intervention trials are required, and these preferably are randomized controlled trials. Two recent reviews have addressed the effectiveness of interventions to prevent overweight and obesity in children and adolescents (Doak et al., 2006; Flodmark, Marcus, & Britton, 2006):

- **Doak and colleagues** (2006) restricted their review to school-based interventions. Twenty-five studies met their inclusion criteria, and many, though not all, of these studies were randomized controlled trials. Nineteen of the 25 studies focused on both activity and diet. Three focused on activity alone, one focused on television viewing alone, and two focused on diet alone. Seventeen of the 25 studies (68%) were judged to be effective based on significant reductions in BMI or skinfolds for the intervention group. Four interventions were effective in reducing both BMI and skinfold measures. Two of these targeted reductions in television viewing, and two targeted increased physical activity combined with diet. Importantly, one study that successfully reduced overweight also increased prevalence of underweight. Doak and colleagues commented that few studies indicate outcomes for underweight and that this negative side effect of obesity intervention programs warrants further study.
- **Flodmark and colleagues** (2006) included in their review only studies with at least 12 months of follow-up. They identified 15 overweight or obesity prevention programs with positive outcomes, 24 with neutral outcomes, and none with negative results, though the reason for this may be that negative results were not reported. Thus, 41% of the studies examined, including 40% of the 33,852 youth studied, showed a positive effect from prevention. Flodmark and colleagues concluded that it is possible to prevent obesity in children and adolescents through limited, school-based programs that combine healthy dietary habits and physical activity.

Two recent randomized controlled trials have produced conflicting findings in the area of overweight/obesity prevention in children:

- **The STRIP (Special Turku Coronary Risk Factor Intervention Project for Children) study,** conducted in Finland, monitored the prevalence of overweight in children from age 7 months through age 10 years (Hakanen et al., 2006). Children were placed into either an intervention group (n = 540) or a control group (n = 522) at 7 months of age. Biannually for 10 years, parents and children in the intervention group received individualized advice on diet and physical activity. At the age of 10 years, 10.2% of the intervention girls were overweight (weight for height >20% of the mean weight for height of healthy Finnish children) compared with 18.8% of the control girls. No group differences were observed in overweight prevalence in the boys. Only three children in the intervention group were obese at some point in the study, whereas 14 control children were classified as obese at some point in the study. It was concluded that dietary and activity guidance provided twice a year from infancy decreases the prevalence of overweight in school-aged girls.

- In contrast, a **Glasgow, Scotland, cluster randomized trial** conducted in 36 nurseries in that city showed no effect of physical activity (three 30 min sessions per week over 24 weeks) on BMI of preschool children (mean age 4.2 years) (Reilly et al., 2006). Differences in the nature of the intervention and the duration of follow-up are possible explanations for the divergent findings in these two studies.

Overall, the majority of recent evidence appears to support a role for physical activity in overweight and obesity prevention in youth. This suggests that increasing physical activity and decreasing sedentary behaviors may be an effective strategy in the management of overweight and obesity in young people. This issue has been addressed in two recent review papers (Atlantis, Barnes, & Fiatarone Singh, 2006; Watts et al., 2005):

- **Watts and colleagues** highlight that there are very few randomized controlled studies investigating the efficacy of exercise training in obese children or adolescents, and state that many existing studies have been poorly controlled and have not specifically stratified the independent effects of exercise versus dietary modification. This last point is also true of many of the studies that have focused on prevention. The authors conclude that although exercise training does not consistently decrease body weight or BMI, it is associated with beneficial changes in fat and lean body mass, stressing the importance of comprehensive assessment of body composition in future studies.
- **Atlantis and coworkers** come to essentially the same conclusions. They state that 155 to 180 min of aerobic exercise per week at moderate-to-high intensity is effective for reducing body fat in overweight youth but that effects on body weight and central obesity are inconclusive.

Modifiability of Physical Activity Levels

Even though some intervention studies suggest that physical activity is effective to some degree in preventing or managing obesity, there is still a question mark regarding the extent to which changes in physical activity levels will be maintained in the long term. Moreover, some studies have indicated that deliberate attempts to increase the physical activity levels of children at certain points in a day may be countered by compensatory reductions in physical activity levels at other times in the day. For example, Wilkin and colleagues (2006) observed that a fivefold variation in time-tabled physical education (i.e., physical education taught in school time) explained less than 1% of the total variation in physical activity, assessed using accelerometers, in young children. The authors' interpretation of this finding is that physical activity in children is under central biological regulation. The implications are that alterations in the environment that make it more conducive to physical activity may have little impact on overall activity levels in a child who is "programmed" to be inactive. Support for this notion comes from another recent study showing that physical activity levels at age 4, based on maternal report, inversely relate to a sedentary lifestyle at age 10 to 12 years (Hallal et al., 2006). However, other research suggests that familial resemblance in physical activity is explained predominantly by shared environmental factors and not by genetic factors (Franks et al., 2005), conflicting with the notion that some children are programmed to be inactive.

Type 2 Diabetes

The concern over the increasing prevalence of overweight and obesity in young people arises due to the associated health risks. In adults the greatest health risk of obesity is type 2 diabetes, which in turn increases the risk of cardiovascular disease. Studies have consistently demonstrated a dose–response association between obesity, assessed using BMI, waist circumference, or waist-to-hip ratio, and risk of type 2 diabetes (Colditz et al., 1990; Wang et al., 2005). In the Nurses' Health Study, women whose BMI was ≥35 kg/m^2 were 60.9 times more likely to develop type 2 diabetes during eight years of follow-up than women whose BMI was <22 kg/m^2 (Colditz et al., 1990). This is an extreme example; more usually, the risk of type 2 diabetes associated with obesity is in the region of 5 to 10 times higher than for normal-weight individuals.

Rising Incidence Among Young People

Until relatively recently, type 2 diabetes was also known as "adult-onset diabetes," but this term is now outdated due to the emergence of type 2 diabetes in children and adolescents. According to Pinhas-Hamiel and Zeitler (2005), the rising prevalence of type 2 diabetes in children and adolescents was initially recognized in the United States in the 1990s and is now noted in many countries around the globe. The increase in pediatric type 2 diabetes has primarily affected minority populations such as the Pima Indians in America and those of Pakistani, Indian, or Arabic origin in the United Kingdom (McKnight-Menci, Sababu, & Kelly, 2005; Pinhas-Hamiel & Zeitler, 2005); however, type 2 diabetes has also been identified in obese White children (Drake et al., 2002).

Many believe that the emergence and increasing prevalence of type 2 diabetes in young people are linked with the increased prevalence of obesity in children and adolescents. Aylin, Williams, and Bottle (2005) report that among patients aged 18 years and younger, the obesity admission rate in English hospitals increased 63.5%, from 52 per million to 85 per million, from 1996-1997 to 2003-2004, and the diabetes admission rate rose 44.4%, from 18 per million to 26 per million, during the same period. The authors point out that one possible explanation for these increases is improved screening. They go on to caution, however, that if these results "... are due to an actual increase in cases, then the patients seen in hospitals will represent only a small fraction of an increasing population of young people affected by obesity and type 2 diabetes" (Aylin, Williams, and Bottle, 2005, p. 1167).

Convincing evidence for a link between obesity and type 2 diabetes in young people is provided by a study examining the prevalence of impaired glucose tolerance, a marker for the risk of diabetes, in a group of severely obese (BMI > 95th percentile for age and sex) children and adolescents (Sinha et al., 2002). Impaired glucose tolerance was detected in 25% of 55 obese children (ages 4 to 10 years) and 21% of 112 obese adolescents (ages 11 to 18 years). More recent studies, albeit cross-sectional in nature, confirm an association between excess adiposity and insulin resistance in children and adolescents (Khan et al., 2003; Viner et al., 2005).

Physical Activity and Type 2 Diabetes

To the authors' knowledge, intervention trials have not been conducted to assess the effectiveness of physical activity in preventing type 2 diabetes in young people. However, physical activity and physical fitness are inversely related to insulin resistance (Kasa-Vubu et al., 2005) and positively associated with insulin sensitivity (Imperatore et al., 2006; Wennlof et al., 2005) in young people. Moreover, there is extensive literature in adults demonstrating that exercise is effective in the prevention and management of type 2 diabetes (for reviews see Bassuk & Manson, 2005; LaMonte, Blair, & Church, 2005). This evidence includes two randomized controlled trials (U.S. Diabetes Prevention Program; Finnish Diabetes Prevention Study) demonstrating that lifestyle intervention is effective in reducing the incidence of type 2 diabetes in adults at high risk of developing the disease (Knowler et al., 2002; Tuomilehto et al., 2001). In the U.S. Diabetes Prevention Program, lifestyle intervention was found to be more effective than the drug metformin in preventing type 2 diabetes. Moreover, though new drugs are becoming available for the treatment of type 2 diabetes, these are associated with serious side effects including heart failure (Heneghan, Thompson, & Perera, 2006). Therefore, based on the available evidence it would appear prudent to recommend physical activity to children and adolescents as one component of an overall lifestyle package aimed at preventing type 2 diabetes.

Cardiovascular Disease Risk

Unlike the situation with type 2 diabetes, there is no indication that clinical signs and symptoms of cardiovascular disease (CVD) are occurring regularly in children and adolescents. Nevertheless, there is considerable evidence to show that CVD has its origins in childhood and that obesity in childhood is associated with traditional CVD risk factors such as hypertension and dyslipidemia (British Medical Association, 2005), as well as emerging risk factors such as elevated C-reactive protein concentration (Moran et al., 2005). One issue with studies examining the relationship between physical activity and CVD risk is the extent to which this relationship is independent of adiposity. Another issue is the extent to which physical activity can influence CVD risk in children who have normal (i.e., not elevated) risk factor profiles. These issues are examined in this section.

In adults there is substantial evidence that high levels of physical activity or physical fitness protect against CVD (Hardman & Stensel, 2003). The relationship between physical activity or physical fitness and CVD is more difficult to establish in young people because CVD does not typically occur until adulthood. Very long follow-up periods would be required to provide definitive evidence of a link between physical activity or physical fitness in childhood and adolescence and subsequent risk of CVD. Therefore, most studies of young people have examined the relationship between physical activity or physical fitness and risk factors for CVD rather than disease endpoints. A limitation here is that CVD risk factors are not always predictive of disease endpoints. Nevertheless, studies of physical activity or physical fitness and CVD risk in young people are important in identifying the strength and direction of any likely association.

Much research in this area suggests that there is no clear relationship between physical activity and CVD risk in nonobese children and adolescents who have normal (not elevated) levels of CVD risk markers. For example, Tolfrey, Jones, and Campbell (2000) concluded in their review paper that "imposed regular exercise has little, if any, influence on the lipoprotein levels of children and adolescents" (p. 100). Armstrong and Simons-Morton (1994) came to similar conclusions in their review, stating that the results of longitudinal studies on physical activity and blood lipids in adolescents are "unimpressive." The Northern Ireland Young Hearts Project involving 1015 schoolchildren did show a relationship between cardiorespiratory fitness and CVD risk, but this was mediated by fatness. That is, after statistical adjustment for adiposity, the relationship between fitness and CVD risk was no longer significant (Boreham et al., 2001). Similarly, data from the Aerobics Centre Longitudinal Study involving an average of 11 years of follow-up showed no association between adolescent fitness (treadmill time) and CVD risk factors (cholesterol concentration, blood pressure, glucose concentration) in adulthood (Eisenmann et al., 2005b). This having been said, some recent evidence suggests that there is a relationship between physical activity and CVD risk in young people, and there is some evidence that this is independent of adiposity.

A notable example of a cross-sectional study linking physical activity with CVD risk in children is The European Youth Heart Study (Andersen et al., 2006). This study involved 1732 randomly selected 9-year-old and 15-year-old schoolchildren from Denmark, Estonia, and Portugal. Physical activity was measured objectively using accelerometers, and CVD risk was assessed using a composite risk factor score compiled from measurements of systolic blood pressure, triglyceride concentration, total cholesterol/high-density lipoprotein

(HDL) cholesterol ratio, insulin resistance, sum of four skinfolds, and aerobic fitness (maximum power output in a cycle test).

Subjects were divided into quintiles according to physical activity levels (accelerometer counts per minute), and those in the lowest three quintiles for physical activity were found to have a higher CVD risk factor score than those in quintile five, that is, the most active quintile (figure 2.1). Furthermore, the mean time spent above 2000 counts per minute, indicating moderate to vigorous physical activity corresponding to a walking speed of 4 km/h or more, in the fourth quintile was 116 min per day in 9-year-olds and 88 min per day in 15-year-olds (see table 2.2). On the basis of these findings the authors concluded that physical activity guidelines for children and adolescents should be higher than the current international guidelines of 1 h per day or more of at least moderate-intensity physical activity in order to prevent a clustering of CVD risk factors. In an associated commentary, Weiss and Raz (2006) highlight that the inverse association between physical activity and CVD risk observed in The European Youth Heart Study was independent of the degree of adiposity and was similar for lean and overweight children, suggesting that physical inactivity is an independent CVD risk factor in young people.

Sedentary behavior also appears to be a risk factor for CVD in young people. An interesting example is seen in the longitudinal study conducted by Hancox and colleagues (2004), which we mentioned in the section on overweight and obesity. Mean hours of television viewed per weekday from the ages of 5 to 15 years was positively associated not only with prevalence of overweight at

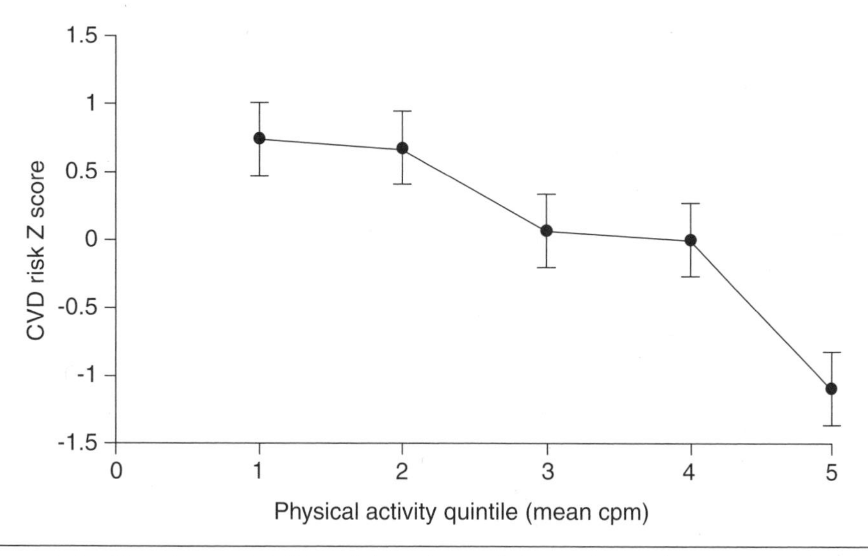

Figure 2.1 Mean Z score (±95% confidence interval) for cardiovascular disease risk by physical activity quintile in 9- and 15-year-old children participating in The European Youth Heart Study.

Reprinted from *The Lancet,* Vol. 368, L.B. Andersen et al., "Physical activity and clustered cardiovascular risk in children: A cross-sectional study," Pages No. 302, Copyright 2006, with permission from Elsevier.

Table 2.2 Daily Time Spent at Physical Activity Intensities Above 2000 Counts per Minute (cpm) in 9- and 15-Year-Old Children Participating in The European Youth Heart Study

	Time > 2000 cpm (min/day [SD])	Mean intensity of minutes spent > 2000 cpm (cpm [SD])
9-YEAR-OLD CHILDREN		
Least active quintile	38 (20)	2869 (1286)
Second quintile	69 (20)	3487 (786)
Third quintile	92 (26)	3649 (746)
Fourth quintile	116 (32)	3728 (651)
Most active quintile	167 (49)	4125 (1117)
15-YEAR-OLD CHILDREN		
Least active quintile	34 (15)	3253 (1080)
Second quintile	53 (24)	3684 (850)
Third quintile	70 (24)	3744 (754)
Fourth quintile	88 (32)	3941 (956)
Most active quintile	131 (47)	4119 (820)

Reprinted from *The Lancet*, Vol. 368, L.B. Andersen et al., "Physical activity and clustered cardiovascular risk in children: A cross-sectional study," Pages No. 303, Copyright 2006, with permission from Elsevier.

age 26 years, but also with CVD risk at age 26 years as indicated by elevated total cholesterol concentration, smoking, and poor fitness levels. This was not the first study to demonstrate an association between television viewing and CVD risk in young people. Wong and colleagues (1992) reported that the risk of hypercholesterolemia is 4.8 times higher in children who watch more than 4 h of television per day compared with those watching less than 2 h per day. Television viewing is also associated in a dose–response fashion with the initiation of youth smoking (Gidwani et al., 2002).

In addition to studies linking sedentary behavior with CVD risk, it has also been demonstrated that aerobic fitness (physical work capacity at a heart rate of 150 beats/min) is independently associated with CVD risk factors in young people (Eisenmann et al., 2005a). Moreover, a prospective cohort study involving 15 years of follow-up has demonstrated that poor fitness in young adulthood is associated with a three- to sixfold increase in the risk of developing diabetes, hypertension, and the metabolic syndrome in middle age, although these associations were diminished to twofold when adjustments were made for BMI. The authors concluded that poor fitness in young adulthood is associated with the development of CVD risk factors and that these associations involve obesity and may be modified by improvement in fitness (Carnethon et al., 2003). Furthermore, a recent study has demonstrated that multifactor intervention (involving exercise, nutrition education, and behavior therapy) is effective in improving CVD risk factors in obese children (Reinehr et al., 2006).

However, there remains a lack of firm evidence from randomized controlled trials to show that physical activity is effective in modifying CVD risk factors in young people. Indeed, Watts and colleagues (2005) recently concluded that exercise training seems to have little effect on blood lipid profile or blood pressure in obese young people, though they noted that studies have demonstrated improved endothelial function in obese children and adolescents after a period of exercise training, and this alone would indicate reduced risk of atherosclerosis provided that exercise habits were maintained.

Skeletal Health

Childhood and adolescence is a crucial period for bone development. According to Vicente-Rodriguez (2006), the prepubertal human skeleton is sensitive to mechanical stimulation elicited by physical activity; therefore, to attain peak bone mass, children should be physically active prior to and throughout puberty. This will result in a greater peak bone mass than would occur in non–physically active children. In our previous review (Biddle et al., 2004), we highlighted several intervention trials demonstrating that weight-bearing activity is effective in increasing bone mineral density in children and adolescents. However, we also noted that there were no longitudinal studies to convincingly demonstrate that gains in bone mineral density achieved in childhood and adolescence are maintained into adulthood or that such gains reduce fracture risk in later life. To the best of our knowledge this remains true.

Several recent studies have confirmed that physical activity is effective in enhancing bone health in children. A notable example involved 51 children (ages 8 to 10 years at baseline) participating in a program named "Bounce at the Bell" (McKay et al., 2005). The program consisted of 10 countermovement jumps three times each day, amounting to a total of 3 min of activity per day over a period of eight months. Seventy-one "matched" children served as controls. At the end of the intervention period, the children in the "Bounce at the Bell" program had gained significantly more bone mineral content at the total proximal femur and the intertrochanteric region than the control children (see figure 2.2). These findings are important because they indicate that a relatively low dose of activity performed daily is effective in increasing bone health in prepubertal/early pubertal children.

Another recent study demonstrating the benefits of exercise for bone health in children is the randomized controlled trial conducted in Hong Kong by Yu and colleagues (2005). In this study, 82 obese children (ages 8 to 11 years) were randomly assigned either to a balanced low-energy diet plus strength

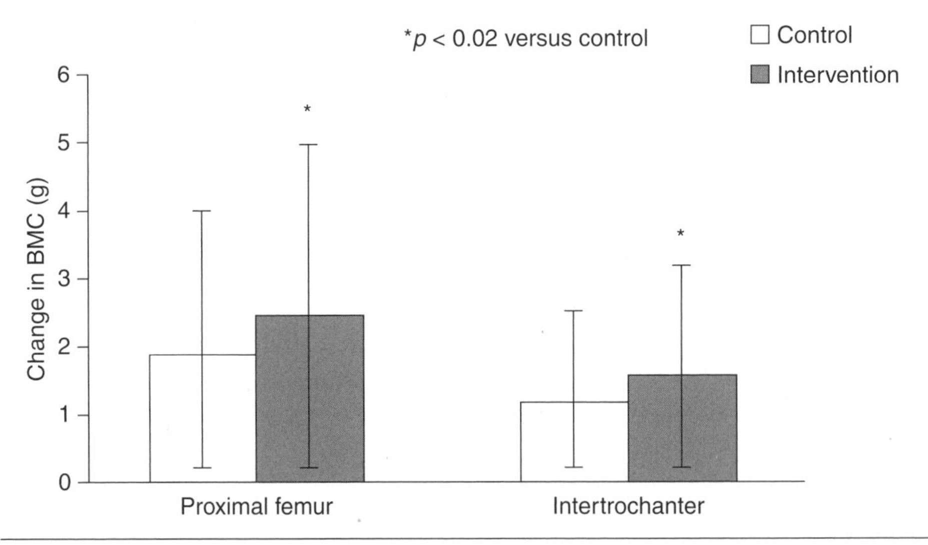

Figure 2.2 Mean (95% confidence interval) change in bone mineral content (BMC) following an eight-month school-based exercise intervention.

Adapted from McKay et al. 2005.

training group or to a diet-alone group. The diets consumed by the children were low in fat (20% to 25%) and total calories (900 and 1200 kcal/day) but contained sufficient protein to support growth. The training group attended a 75 min strength exercise program three times a week for six weeks. At the end of this period, increases in bone mineral content were noted in both groups, but the increase in the training group was significantly larger than that in the diet-alone group.

A final exercise study to mention is the trial conducted over a one-year period by Courteix and colleagues (2005). Premenarcheal girls were assigned to one of four conditions: (1) exercise combined with calcium supplementation, (2) exercise and placebo, (3) sedentary lifestyle and calcium supplementation, and (4) sedentary lifestyle and placebo (allocation to the exercise and sedentary lifestyle groups was based on analysis of physical activity questionnaires). At the end of the intervention, significant gains in bone mineral density were noted in the exercise combined with calcium supplementation group at the total body (6.3% increase), lumbar spine (11% increase), femoral neck (8.2% increase), and Ward's triangle (9.3%). There were no significant changes in bone mineral density in any of the other groups. It was concluded that calcium supplementation increases the effect of exercise on bone mineral acquisition in the period preceding puberty and that calcium supplementation without physical activity does not improve bone mineral density acquisition during this period.

Courteix and coworkers' (2005) conclusion is consistent with the finding of a recent meta-analysis that calcium supplementation in healthy children has no effect on bone density at the hip and lumbar spine (Winzenberg et al., 2006). Moreover, though the meta-analysis showed calcium supplementation to have a small effect at the upper limb, the resulting increase in bone density is unlikely to lead to a decrease in the risk of fracture. Taken together, these findings suggest that physical activity should be central when one is making bone health recommendations for children (Lanou, 2006). It remains to be determined, however, if there are carryover effects from childhood to adulthood with respect to bone health.

In 2004 the American College of Sports Medicine published a position stand on physical activity and bone health (Kohrt et al., 2004). This position stand provides specific guidelines for children and adolescents relative to mode, intensity, frequency, and duration of physical activity. Guidelines for mode of activity target impact activities such as gymnastics, plyometrics, and jumping, as well as moderate-intensity resistance training. Participation in sports that involve running and jumping (soccer, basketball) is believed likely to be of benefit; however, scientific evidence to support this belief is lacking. Intensity should be high in terms of bone-loading forces, but for safety reasons it is recommended that resistance training be at less than 60% of 1-repetition maximum (1RM). The recommended frequency of such activities is at least three days a week with a duration of 10 to 20 min per bout (two times per day or more may be more effective).

Mental Health

There is widespread belief that physical activity is inherently "good" for young people with respect to varied psychosocial outcomes such as self-esteem, mental health, and cognitive functioning. Despite such claims, the evidence is not always clear (Biddle et al., 2004); and there is a lack of prospective population cohort studies and randomized controlled trials (Strong et al., 2005). The majority of studies in this area are cross-sectional; therefore, causality cannot be established because the temporal relationship between exposure and outcome has not been demonstrated. Thus, though there is evidence that physical activity can enhance psychological well-being, such an outcome is not inevitable (Lagerberg, 2005). Nevertheless, the potential psychological benefits of exercise in children and adolescents should not be ignored. This is particularly true in the case of obesity. Obesity in youth may have serious psychosocial consequences, and exercise interventions in obese children may have immediate salutary effects.

The effect of physical activity on mental health in children and adolescents has received significantly less attention than among adult populations (Paluska & Schwenk, 2000). Where this has been investigated, the work has primarily focused on depression, anxiety, and self-concept or self-esteem, and relatively little work has addressed other important aspects of mental health such as stress and emotional distress (Strong et al., 2005). As a consequence, this section is limited to discussions of the relationship between physical activity participation and these three aspects of mental health plus the further psychosocial outcome of cognitive performance. All the work reviewed is premised on a definition of inactivity as "activity absence" (see chapter 1), as there is little systematic work relating individual sedentary behaviors, such as computer and Internet use, to these specific psychosocial outcomes (Subrahmanyam et al., 2000), and much of what exists is small scale and cross-sectional in design (e.g., Colwell & Kato, 2003; Kraut et al., 1998; Roe & Muijs, 1998).

Depressive Disorders

Depressive disorders have been identified throughout the life span, with an estimated 1% of 5- to 16-year-olds in the United Kingdom (Green et al., 2005) and 2.5% of U.S. children and 8.3% of U.S. adolescents (National Institute of Mental Health, 2000) suffering depression. Those experiencing depression can face significant disability as a result (Mutrie, 2000).

Evidence from adult studies demonstrates that physical activity is inversely associated with symptoms of depression (Craft & Landers, 1998; North, McCullagh, & Tran, 1990), and there is some evidence that this relationship is causal (Mutrie, 2000). There is, however, much less evidence for this relationship in children and adolescents (Biddle et al., 2004). In two meta-analyses, a moderate effect, similar to that found in adults, was reported for physical activity on depression in adolescents (Calfas & Taylor, 1994; North et al., 1990). However, this should be viewed cautiously as each meta-analysis included only five studies. Cross-sectional surveys also show an inverse relationship between physical activity participation and depression (see brief review by Strong et al., 2005). For example, Steptoe and Butler (1996) report, in a sample of over 5000 British adolescents, that emotional well-being is positively associated with participation in vigorous recreation and sport. Moreover, Tomson and colleagues (2003) collected data from 933 children 8 to 12 years of age and reported an elevated risk of depressive symptoms among children classified as inactive (2.8 to 3.4 times higher risk) and among children not meeting criterion-referenced standards for health-related fitness (1.5 to 4 times higher risk) compared with their contemporaries who were considered active or fit.

The majority of work in this area is limited because it is cross-sectional in nature, fails to adjust for confounding factors (e.g., sex, socioeconomic status), and employs measures of depression that do not have established validity for adolescent populations (Motl et al., 2004). Motl and colleagues, in a recent two-year longitudinal study in which they attempted to overcome some of these limitations, demonstrated that changes in self-reported frequency of physical activity were inversely associated with changes in depressive symptoms even after controlling for confounding factors (i.e., sex, socioeconomic status; smoking behavior; alcohol consumption; and the value participants placed on their health, appearance, and achievement). Although this observational study does not establish the direction of causality, it does provide stronger evidence for a relationship between physical activity and depressive symptoms in adolescents than do cross-sectional studies.

Anxiety

Active adults report fewer symptoms of anxiety than do inactive adults (Department of Health, 2004; Taylor, 2000), and there is some evidence that this is also true for young people. For example, a meta-analysis of exercise and anxiety reduction (Petruzzello et al., 1991) showed an effect size for those under 18 years of age of 0.20 for state anxiety (K = 21; K = number of effect sizes), 0.47 for trait anxiety (K = 3), and 0.38 for psychophysiological indices of anxiety (K = 8). These compare with effect sizes of 0.24, 0.34, and 0.56 for all ages, respectively, suggesting that anxiety-reducing effects of exercise for younger people are slightly less for state anxiety and psychophysiological indices of anxiety. We can be less certain about the effect of exercise on trait anxiety in young people because of the small number of studies conducted. In a recent narrative review on the outcomes of physical activity in school-age youth, Strong and colleagues (2005) concluded that cross-sectional studies show a weak negative association between physical activity and anxiety, and that quasi-experimental studies show strong negative associations between physical activity and measures of anxiety. However, in both these reviews there were few studies in some categories, and further testing is required before clear conclusions can be drawn.

As with other areas, the obvious limitation here is that these studies do not reveal whether exercise reduces anxiety or simply that less anxious children are more likely to exercise. There are some proposed mechanisms to explain how physical activity may reduce anxiety. Exercise may, for example, act as a distraction from anxiety-provoking situations and thoughts. Biochemical changes (e.g., in circulating neurotransmitters) induced by exercise may also lead to reductions in anxiety, as might the increase in core temperature and sense of well-being that occur during and after exercise. However, direct evidence to support these mechanisms is limited.

Self-Esteem

Self-esteem reflects the degree to which individuals value themselves and is widely viewed as a key indicator of positive mental health and well-being (Fox, 2000). Early reviews concluded that physical activity is associated with the development of self-esteem in young people (Calfas & Taylor, 1994; Gruber, 1986; Mutrie & Parfitt, 1998). A more recent meta-analysis (Ekeland et al., 2004) examined whether exercise interventions improved global self-esteem among children and young people (aged 3-20 years). The results showed that in the eight trials of an exercise-alone intervention versus a no-intervention control there was a moderate effect in favor of the intervention group (overall standardized mean difference [SMD] random effect model = 0.49, 95% confidence interval [CI] 0.16-0.81). There did not appear to be an effect for type of exercise, though all but one of these studies were based on aerobic exercise or duration of the intervention (<10 weeks vs. >10 weeks).

Comparisons were also made between interventions with healthy children and interventions with children who were at risk or who had defined problems. Significant findings were obtained only for the latter group (SMD = 0.49, 95% CI 0.17-0.82). Most of the trials were of small scale and short duration. No follow-up results were given, so the sustainability of changes could not be assessed. Only one of the trials was considered to be of high methodological quality, but it demonstrated the strongest effects. Four further trials compared the effects of exercise as part of a comprehensive intervention package against no-intervention control groups and showed a moderate positive effect on self-esteem in favor of the intervention (overall SMD = 0.51, 95% CI 0.15-0.88). The size of this effect increased when the only study with healthy participants was excluded (SMD = 0.64, 95% CI 0.22-1.06). There appeared to be an effect for study quality, with studies of moderate bias yielding nonsignificant effects but those of high bias yielding significant results (there were no comprehensive intervention studies with low bias). The authors also noted that the effects of exercise may be underestimated, as across both sections of the review all the studies used "usual activity" as the control treatment and therefore comparisons are not between exercise and complete physical inactivity. Despite the paucity of good-quality research in the area, Ekelund and colleagues concluded that exercise can lead to improvements in self-esteem, at least in the short term and among at-risk youth.

Cognitive Performance

There has been significant interest in the relationship between physical activity, cognitive performance, and academic achievement in young people. The authors of a recent meta-analysis (Sibley & Etnier, 2003) attempted to bring consensus to a conflicting and previously inconclusive body of literature (Mutrie & Parfitt, 1998). This meta-analysis of 44 studies concluded that there is a significant positive relationship (effect size = 0.32, $p < 0.05$) between physical activity and cognitive functioning in young people (ages 4-18 years). Moderator analyses showed that this relationship held for all participants (healthy participants, those with mental impairments, and those with physical disabilities), of all ages (though largest effects were seen in middle school–aged students, 11-13 years old), and across types of activity (e.g., resistance and circuit training, physical education programs, aerobic exercise). The findings of the meta-analysis were qualified by the authors because of lack of control for confounding variables; the relatively small number of peer-reviewed true experimental designs (n = 9, 20%); and measurement issues, particularly of the dependent measure of cognitive functioning. Despite this qualification, Sibley and Etnier conclude that, from a conservative standpoint, time spent in physical activity will not hurt cognitive performance or academic achievement and may actually improve one or both of these.

In summary, physical activity can have positive psychosocial outcomes for young people. The effects appear strongest for self-esteem (at least in the short term), and those who are physically active appear less likely to suffer from mental health problems and may have enhanced cognitive functioning. Although all participants are likely to gain significant benefits, the beneficial effects are likely to be greater in those who have poorer mental health at baseline. However, the evidence is not extensive. Studies are largely cross-sectional, small scale, and lacking in measurement consistency. In addition, while physical activity may enhance psychological well-being, it is possible that the prevailing psychological climate and social interactions inherent in such settings will also be crucial. Unfortunately, such factors are rarely accounted for.

Conclusion

There is strong evidence that the prevalence of overweight and obesity is increasing among children and adolescents and that this is linked to an emergence of type 2 diabetes in young people. There is some evidence implicating physical inactivity in the development of overweight and obesity in young people. Much of this evidence is observational, but recent observational studies are more convincing than previous ones due to the relatively large sample sizes involved, the use of objective measures of physical activity such as accelerometry, or both.

A limited number of well-designed intervention studies also support a role for physical activity in the prevention and management of overweight and obesity in children and adolescents. Recent studies provide good evidence of an association between physical activity levels and risk of type 2 diabetes and CVD in children and adolescents. It must be stressed that this evidence is observational in nature and that risk factors rather than disease endpoints have been used in these studies. However, it is unrealistic at present to contemplate randomized controlled trials involving type 2 diabetes and CVD endpoints in children and adolescents. The same is true for bone health; there is good evidence that physical activity can enhance bone accrual in children and adolescents but limited evidence that this leads to a reduced risk of fractures in later life.

For mental health, available evidence indicates that physical activity can enhance self-esteem and cognitive function and reduce symptoms of depression and anxiety; but as with the other health outcomes mentioned here, much of this evidence comes from observational studies. Thus, the challenge for future research is to provide more concrete evidence that many of the associations between activity levels and health outcomes described here are causal in nature. The most realistic solution to this challenge is to pursue a combination of large-scale observational studies, using recent developments in technology to assess activity levels, together with a limited number of randomized controlled trials employing large subject numbers and long follow-up periods.

Finally, it should be noted that the term physical activity denotes a wide variety of behaviors and that each of these may relate to different aspects of health. For example, endurance-type activities (walking, running, swimming, cycling) will be best for promoting cardiovascular health, whereas weight-bearing activity will be most effective for bone health and resistance exercise may be best from the point of view of injury prevention. If the goal is weight loss, then overall energy expenditure should be emphasized. One should bear these differing relationships in mind when recommending physical activity, and a well-rounded physical activity lifestyle is probably best for optimal health in children and adolescents.

APPLICATIONS FOR RESEARCHERS

The prevalence of overweight and obesity is increasing among children and adolescents, and this is linked to the emergence of type 2 diabetes in young people. There is evidence implicating physical inactivity in the development of overweight and obesity in young people and supporting a role for increased physical activity and reduced sedentary behavior in the management of overweight and obesity. However, this evidence is not strong, as few randomized controlled trials have been conducted. The same is true of evidence linking physical inactivity with type 2 diabetes, CVD, bone health, self-esteem, cognitive function, anxiety, and depression. In all of these areas, randomized controlled trials are required to strengthen the evidence base. Ideally, these trials would involve large sample sizes and long follow-up periods. Despite the limitations, however, there is sufficient evidence from robust observational studies to support a role for physical activity in optimizing physical and mental health in young people.

APPLICATIONS FOR PROFESSIONALS

Available evidence supports a role for physical activity in enhancing many aspects of physical and mental health in young people. Physical activity may reduce the likelihood of overweight and obesity, type 2 diabetes, CVD, anxiety, and depression in children and adolescents as well as enhance their cognitive function, self-esteem, and bone health. It should be recognized that there are limitations to the evidence linking physical activity to health in young people, and it does not always follow that physical activity will be beneficial. Moreover, an overemphasis on physical activity in young people may have detrimental effects, such as an increased prevalence of underweight in obesity prevention programs. Despite these caveats, young people should engage in 1 h or more of physical activity daily for optimal physical and mental development. This activity may take a variety of forms, including unsupervised play, active transportation, and structured activities (e.g., exercise, physical education, sport).

References

Adams, K.F., Schatzkin, A., Harris, T.B., Kipnis, V., Mouw, T., Ballard-Barbash, R., et al. (2006). Overweight, obesity, and mortality in a large prospective cohort of persons 50 to 71 years old. *New England Journal of Medicine, 355*, 763-778.

Andersen, L.B., Harro, M., Sardinha, L.B., Froberg, K., Ekelund, U., Brage, S., et al. (2006). Physical activity and clustered cardiovascular risk in children: A cross-sectional study (The European Youth Heart Study). *Lancet, 368*, 299-304.

Armstrong, N., & Simons-Morton, B. (1994). Physical activity and blood lipids in adolescents. *Pediatric Exercise Science, 6*, 381-405.

Atlantis, E., Barnes, E.H., & Fiatarone Singh, M.A. (2006). Efficacy of exercise for treating overweight in children and adolescents: A systematic review. *International Journal of Obesity, 30*, 1027-1040.

Aylin, P., Williams, S., & Bottle, A. (2005). Obesity and type 2 diabetes in children, 1996-7 to 2003-4. *British Medical Journal, 331*, 1167.

Baird, J., Fisher, D., Lucas, P., Kleijnen, J., Roberts, H., & Law, C. (2005). Being big or growing fast: Systematic review of size and growth in infancy and later obesity. *British Medical Journal, 331*, 929.

Bassuk, S.S., & Manson, J.E. (2005). Epidemiological evidence for the role of physical activity in reducing the risk of type 2 diabetes and cardiovascular disease. *Journal of Applied Physiology, 99*, 1193-1204.

Biddle, S.J.H., Gorely, T., & Stensel, D.J. (2004). Health-enhancing physical activity and sedentary behaviour in children and adolescents. *Journal of Sports Sciences, 22*, 679-701.

Boreham, C., Twisk, J., Murray, L., Savage, M., Strain, J.J., & Cran, G. (2001). Fitness, fatness, and coronary heart disease risk in adolescents: The Northern Ireland Young Hearts Project. *Medicine and Science in Sports and Exercise, 33*, 270-274.

British Medical Association. (2005). *Preventing childhood obesity.* London: British Medical Association Board of Science.

Bundred, P., Kitchiner, D., & Buchan, I. (2001). Prevalence of overweight and obese children between 1989 and 1998: Population based series of cross sectional studies. *British Medical Journal, 322*, 326-328.

Calfas, K.J., & Taylor, C. (1994). Effects of physical activity on psychological variables in adolescents. *Pediatric Exercise Science, 6*, 406-423.

Carnethon, M.R., Gidding, S.S., Nehgme, R., Sidney, S., Jacobs, D.R., & Liu, K. (2003). Cardiorespiratory fitness in young adulthood and the development of cardiovascular disease risk factors. *Journal of the American Medical Association, 290*, 3092-3100.

Chinn, S., & Rona, R.J. (2001). Prevalence and trends in overweight and obesity in three cross sectional studies of British children, 1974-94. *British Medical Journal, 322*, 24-26.

Colditz, G.A., Willett, W.C., Stampfer, M.J., Manson, J.E., Hennekens, C.H., Arky, R.A., et al. (1990). Weight as a risk factor for clinical diabetes in women. *American Journal of Epidemiology, 132*, 501-513.

Cole, T.J., Bellizzi, M.C., Flegal, K.M., & Dietz, W.H. (2000). Establishing a standard definition for childhood overweight and obesity worldwide: International survey. *British Medical Journal, 320*, 1240-1243.

Colwell, J., & Kato, M. (2003). Investigation of the relationship between social isolation, self-esteem, aggression and computer game play in Japanese adolescents. *Asian Journal of Social Psychology, 6*, 149-158.

Courteix, D., Jaffré, C., Lespessailles, E., & Benhamou, L. (2005). Cumulative effects of calcium supplementation and physical activity on bone accretion in premenarchal children: A double-blind randomised placebo-controlled trial. *International Journal of Sports Medicine, 26*, 332-338.

Craft, L.L., & Landers, D.M. (1998). The effect of exercise on clinical depression and depression resulting from mental illness: A meta-analysis. *Journal of Sport and Exercise Psychology, 20*, 339-357.

Department of Health. (2004). *At least five a week: Evidence on the impact of physical activity and its relationship to health.* London: Department of Health.

Doak, C.M., Visscher, T.L.S., Renders, C.M., & Seidell, J.C. (2006). The prevention of overweight and obesity in children and adolescents: A review of intervention programmes. *Obesity Reviews, 7*, 111-136.

Drake, A.J., Smith, A., Betts, P.R., Crowne, E.C., & Shield, J.P.H. (2002). Type 2 diabetes in obese white children. *Archives of Disease in Childhood, 86*, 207-208.

Eisenmann, J.C., Katzmarzyk, P.T., Perusse, L., Tremblay, A., Després, J-P., & Bouchard, C. (2005a). Aerobic fitness, body mass index, and CVD risk factors among adolescents: The Québec family study. *International Journal of Obesity, 29*, 1077-1083.

Eisenmann, J.C., Wickel, E.E., Welk, G.J., & Blair, S.N. (2005b). Relationship between adolescent fitness and fatness and cardiovascular disease risk factors in adulthood: The Aerobics Centre Longitudinal Study (ACLS). *American Heart Journal, 149*, 46-53.

Ekeland, E., Heian, F., Hagen, K.B., Abbott, J., & Nordheim, L. (2004). Exercise to improve self-esteem in children and young people. *The Cochrane Database of Systematic Reviews.* Issue 1, CD003683.

Ekelund, U., Neovius, M., Linné, Y., Brage, S., Wareham, N.J., & Rössner, S. (2005). Associations between physical activity and fat mass in adolescents: The Stockholm Weight Development Study. *American Journal of Clinical Nutrition, 81*, 355-360.

Flodmark, C.E., Marcus, C., & Britton, M. (2006). Interventions to prevent obesity in children and adolescents: A systematic literature review. *International Journal of Obesity, 30*, 579-589.

Fox, K.R. (2000). The effects of exercise on self-perceptions and self-esteem. In S.J.H. Biddle, K.R. Fox, & S.H. Boutcher (Eds.), *Physical activity and psychological well-being* (pp. 88-117). London: Routledge.

Franks, P.W., Ravussin, E., Hanson, R.L., Harper, I.T., Allison, D.B., Knowler, W.C., et al. (2005). Habitual physical activity in children: The role of genes and the environment. *American Journal of Clinical Nutrition, 82*, 901-908.

Franzosi, M.G. (2006). Should we continue to use BMI as a cardiovascular risk factor? *Lancet, 368*, 624-625.

Gard, M., & Wright, J. (2005). *The obesity epidemic: Science, morality and ideology.* London: Routledge.

Gidwani, P.P., Sobol, A., DeJong, W., Perrin, J.M., & Gortmaker, S.L. (2002). Television viewing and initiation of smoking among youth. *Pediatrics, 110*, 505-508.

Green, H., McGinnity, A., Meltzer, H., Ford, T., & Goodman, R. (2005). *Mental health of children and young people in Great Britain 2004*. Basingstoke, UK: Palgrave Macmillan.

Gruber, J. (1986). Physical activity and self-esteem development in children: A meta-analysis. In G. Stull & H. Eckert (Eds.), *Effects of physical activity on children* (pp. 330-348). Champaign, IL: Human Kinetics.

Hakanen, M., Lagström, H., Kaitosaari, T., Niinikoski, H., Näntö-Salonen, K., Jokinen, E., et al. (2006). Development of overweight in an atherosclerosis prevention trial starting in early childhood: The STRIP study. *International Journal of Obesity, 30*, 618-626.

Hallal, P.C., Wells, J.C.K., Reichert, F.F., Anselmi, L., & Victora, C.G. (2006). Early determinants of physical activity in adolescence: Prospective birth cohort study. *British Medical Journal, 332*, 1002-1007.

Hancox, R.J., Milne, B.J., & Poulton, R. (2004). Association between child and adolescent television viewing and adult health: A longitudinal birth cohort study. *Lancet, 364*, 257-262.

Hardman, A.E., & Stensel, D.J. (2003). *Physical activity and health: The evidence explained.* London: Routledge.

Heneghan, C., Thompson, M., & Perera, R. (2006). Prevention of diabetes: Drug trials show promising results, but have limitations. *British Medical Journal, 333*, 764-765. [Editorial]

Imperatore, G., Cheng, Y.J., Williams, D.E., Fulton, J., & Gregg, E.W. (2006). Physical activity, cardiovascular fitness, and insulin sensitivity among U.S. adolescents: The National Health and Nutrition Examination Survey, 1999-2002. *Diabetes Care, 29*, 1567-1572.

Jee, S.H., Sull, J.W., Park, J., Lee, S.Y., Ohrr, H., Guallar, E., et al. (2006). Body-mass index and mortality in Korean men and women. *New England Journal of Medicine, 355*, 779-787.

Kasa-Vubu, J.Z., Lee, C.C., Rosenthal, A., Singer, K., & Halter, J.B. (2005). Cardiovascular fitness and exercise as determinants of insulin resistance in postpubertal adolescent females. *Journal of Clinical Endocrinology and Metabolism, 90*, 849-854.

Kesaniemi, Y.A., Danforth, E., Jensen, M.D., Kopelman, P.G., Lefebvre, P., & Reeder, B.A. (2001). Dose-response issues concerning physical activity and health: An evidence-based symposium. *Medicine and Science in Sports and Exercise, 33*(Suppl.), S351-S358.

Khan, F., Green, F.C., Forsyth, J.S., Greene, S.A., Morris, A.D., & Belch, J.J.F. (2003). Impaired microvascular function in normal children: Effects of adiposity and poor glucose handling. *Journal of Physiology, 551*, 705-711.

Kimm, S.Y.S., Glynn, N.W., Obarzanek, E., Kriska, A.M., Daniels, S.R., Barton, B.A., et al. (2005). Relation between changes in physical activity and body-mass index during adolescence: A multicentre longitudinal study. *Lancet, 366*, 301-307.

Knowler, W.C., Barrett-Connor, E., Fowler, S.E., Hamman, R.F., Lachin, J.M., Walker, E.A., et al. (2002). Reduction in the incidence of type 2 diabetes with lifestyle intervention or metformin. *New England Journal of Medicine, 346*, 393-403.

Kohrt, W.M., Bloomfield, S.A., Little, K.D., Nelson, M.E., & Yingling, V.R. (2004). American College of Sports Medicine position stand: Physical activity and bone health. *Medicine and Science in Sports and Exercise, 36*, 1985-1996.

Kraut, R., Patterson, M., Lundmark, V., Kiesler, S., Mukopadhyay, T., & Scherlis, W. (1998). Internet paradox: A social technology that reduces social involvement and psychological well-being? *American Psychologist, 53*, 1017-1031.

Lagerberg, D. (2005). Physical activity and mental health in schoolchildren: A complicated relationship. *Acta Paediatrica, 94*, 1699-1701.

LaMonte, M.J., Blair, S.N., & Church, T.S. (2005). Physical activity and diabetes prevention. *Journal of Applied Physiology, 99*, 1205-1213.

Lanou, A.J. (2006). Bone health in children: Guidelines for calcium intake should be revised. *British Medical Journal, 333*, 763-764. [Editorial]

Ludwig, D.S., & Gortmaker, S.L. (2004). Programming obesity in childhood. *Lancet, 364*, 226-227.

Matton, L., Thomis, M., Wijndaele, K., Duvigneaud, N., Beunen, G., Claessens, A.L., et al. (2006). Tracking of physical fitness and physical activity from youth to adulthood in females. *Medicine and Science in Sports and Exercise, 38*, 1114-1120.

McCarthy, H.D., Cole, T.J., Fry, T., Jebb, S.A., & Prentice, A.M. (2006). Body fat reference curves for children. *International Journal of Obesity, 30*, 598-602.

McCarthy, H.D., Ellis, S.M., & Cole, T.J. (2003). Central overweight and obesity in British youth aged 11-16 years: Cross sectional surveys of waist circumference. *British Medical Journal, 326*, 624-626.

McKay, H.A., MacLean, L., Petit, M., MacKelvie-O'Brien, K., Janssen, P., Beck, T., et al. (2005). "Bounce at the Bell": A novel program of short bouts of exercise improves proximal femur bone mass in early pubertal children. *British Journal of Sports Medicine, 39*, 521-526.

McKnight-Menci, H., Sababu, S., & Kelly, S.D. (2005). The care of children and adolescents with type 2 diabetes. *Journal of Pediatric Nursing, 20*, 96-106.

Moran, A., Steffen, L.M., Jacobs, D.R., Steinberger, J., Pankow, J.S., Hong, C.P., et al. (2005). Relation of C-reactive protein to insulin resistance and cardiovascular risk factors in youth. *Diabetes Care, 28*, 1763-1768.

Motl, R.W., Birnbaum, A.S., Kubik, M.Y., & Dishman, R.K. (2004). Naturally occurring changes in physical activity are inversely related to depressive symptoms during early adolescence. *Psychosomatic Medicine, 66*, 336-342.

Must, A., & Anderson, S.E. (2006). Body mass index in children and adolescents: Considerations for population-based applications. *International Journal of Obesity, 30*, 590-594.

Mutrie, N. (2000). The relationship between physical activity and clinically defined depression. In S.J.H. Biddle, K.R. Fox, & S.H. Boutcher (Eds.), *Physical activity and psychological well-being* (pp. 46-62). London: Routledge.

Mutrie, N., & Parfitt, G. (1998). Physical activity and its link with mental, social and moral health in young people. In S.J.H. Biddle, J.F. Sallis, & N. Cavill (Eds.), *Young and active? Young people and health-enhancing physical activity: Evidence and implications* (pp. 49-68). London: Health Education Authority.

National Institute of Mental Health. (2000). *Depression in children and adolescents: A fact sheet for physicians* (NIH Pub. No. 00-4744). Bethesda, MD: National Institute of Mental Health.

North, T.C., McCullagh, P., & Tran, Z.V. (1990). Effect of exercise on depression. *Exercise and Sport Sciences Reviews, 18*, 379-415.

Ogden, C.L., Carroll, M.D., Curtin, L.R., McDowell, M.A., Tabak, C.J., & Flegal, K.M. (2006). Prevalence of overweight and obesity in the United States, 1999-2004. *Journal of the American Medical Association, 295*, 1549-1555.

Paluska, S., & Schwenk, T. (2000). Physical activity and mental health: Current concepts. *Sports Medicine, 29*, 167-180.

Petruzzello, S.J., Landers, D.M., Hatfield, B.D., Kubitz, K.A., & Salazar, W. (1991). A meta-analysis on the anxiety-reducing effects of acute and chronic exercise: Outcomes and mechanisms. *Sports Medicine, 11*, 143-182.

Pinhas-Hamiel, O., & Zeitler, P. (2005). The global spread of type 2 diabetes mellitus in children and adolescents. *Journal of Pediatrics, 146*, 693-700.

Reilly, J.J. (2006). Diagnostic accuracy of the BMI for age in paediatrics. *International Journal of Obesity, 30*, 595-597.

Reilly, J.J., Armstrong, J., Dorosty, A.R., Emmett, P.M., Ness, A., Rogers, I., et al. (2005). Early life risk factors for obesity in childhood: Cohort study. *British Medical Journal, 330*, 1357-1359.

Reilly, J.J., Kelly, L., Montgomery, C., Williamson, A., Fisher, A., McColl, J.H., et al. (2006). Physical activity to prevent obesity in young children: A cluster randomised controlled trial. *British Medical Journal, 333*, 1041.

Reinehr, T., de Sousa, G., Toschke, A.M., & Andler, W. (2006). Long-term follow-up of cardiovascular disease risk factors in children after an obesity intervention. *American Journal of Clinical Nutrition, 84*, 490-496.

Rennie, K.L., Livingstone, M.B.E., Wells, J.C.K., McGloin, A., Coward, W.A., Prentice, A.M., et al. (2005). Association of physical activity with body-composition indexes in children aged 6-8 y at varied risk of obesity. *American Journal of Clinical Nutrition, 82*, 13-20.

Roe, K., & Muijs, D. (1998). Children and computer games: A profile of heavy users. *European Journal of Communication, 13*, 181-200.

Romero-Corral, A., Montori, V.M., Somers, V.K., Korinek, J., Thomas, R.J., Allison, T.G., et al. (2006). Association of bodyweight with total mortality and with cardiovascular events in coronary artery disease: A systematic review of cohort studies. *Lancet, 368*, 666-678.

Ruiz, J.R., Rizzo, N.S., Hurtig-Wennlof, A., Ortega, F.B., Warnberg, J., & Sjostrom, M. (2006). Relations of total physical activity and intensity to fitness and fatness in children: The European Youth Heart Study. *American Journal of Clinical Nutrition, 84*, 299-303.

Sibley, B., & Etnier, J. (2003). The relationship between physical activity and cognition in children: A meta-analysis. *Pediatric Exercise Science, 15*, 243-256.

Sinha, R., Fisch, G., Teague, B., Tamborlane, W.V., Banyas, B., Allen, K., et al. (2002). Prevalence of impaired glucose tolerance among children and adolescents with marked obesity. *New England Journal of Medicine, 346*, 802-810.

Speiser, P.W., Rudolf, M.C.J., Anhalt, H., Camacho-Hubner, C., Chiarelli, F., Eliakim, A., et al. on behalf of the Obesity Consensus Working Group. (2005). Consensus statement: Childhood obesity. *Journal of Clinical Endocrinology and Metabolism, 60,* 1871-1887.

Stamatakis, E., Primatesta, P., Chinn, S., Rona, R., & Falascheti, E. (2005). Overweight and obesity trends from 1974 to 2003 in English children: What is the role of socio-economic factors? *Archives of Disease in Childhood, 90,* 999-1004.

Steinberger, J., Jacobs, D.R., Raatz, S., Moran, A., Hong, C-P., & Sinaiko, A.R. (2005). Comparison of body fatness measurements by BMI and skinfolds vs dual energy X-ray absorptiometry and their relation to cardiovascular risk factors in adolescents. *International Journal of Obesity, 29,* 1346-1352.

Steptoe, A., & Butler, N. (1996). Sports participation and emotional wellbeing in adolescents. *Lancet, 347,* 1789-1792.

Strauss, R.S., & Pollack, H.A. (2001). Epidemic increase in childhood overweight, 1986-1998. *Journal of the American Medical Association, 286,* 2845-2848.

Strong, W.B., Malina, R.M., Blimkie, C.J., Daniels, S.R., Dishman, R.K., Gutin, B., et al. (2005). Evidence based physical activity for school-age youth. *Journal of Pediatrics, 146,* 732-737.

Subrahmanyam, K., Kraut, R.E., Greenfield, P.M., & Gross, E.F. (2000). The impact of home computer use on children's activities and development. *Children and Computer Technology, 10,* 123-144.

Sulemana, H., Smolensky, M.H., & Lai, D. (2006). Relationship between physical activity and body mass index in adolescents. *Medicine and Science in Sports and Exercise, 38,* 1182-1186.

Taylor, A. (2000). Physical activity, anxiety, and stress. In S.J.H. Biddle, K.R. Fox, & S.H. Boutcher (Eds.), *Physical activity and psychological well-being* (pp. 10-46). London: Routledge.

Tolfrey, K., Jones, A.M., & Campbell, I.G. (2000). The effect of aerobic exercise training on the lipid-lipoprotein profile of children and adolescents. *Sports Medicine, 29,* 99-112.

Tomson, L., Pangrazi, R., Friedman, G., & Hutchison, N. (2003). Childhood depressive symptoms, physical activity and health related fitness. *Journal of Sport and Exercise Psychology, 25,* 419-439.

Tuomilehto, J., Lindström, J., Eriksson, J.G., Valle, T.T., Hämäläinen, H., Ilanne-Parikka, P., et al. (2001). Prevention of type 2 diabetes mellitus by changes in lifestyle among subjects with impaired glucose tolerance. *New England Journal of Medicine, 344,* 1343-1350.

Vicente-Rodriguez, G. (2006). How does exercise affect bone development during growth? *Sports Medicine, 36,* 561-569.

Viner, R.M., Segal, T.Y., Lichtarowicz-Krynska, E., & Hindmarsh, P. (2005). Prevalence of the insulin resistance syndrome in obesity. *Archives of Disease in Childhood, 90,* 10-14.

Vogels, N., Posthumus, D.L.A., Mariman, E.C.M., Bouwman, F., Kester, A.D.M., Rump, P., et al. (2006). Determinants of overweight in a cohort of Dutch children. *American Journal of Clinical Nutrition, 84,* 717-724.

Wang, Y., Rimm, E.B., Stampfer, M.J., Willett, W.C., & Hu, F.B. (2005). Comparison of abdominal adiposity and overall obesity in predicting risk of type 2 diabetes among men. *American Journal of Clinical Nutrition, 81*, 555-563.

Wardle, J., Brodersen, N.H., Cole, T.J., Jarvis, M.J., & Boniface, D.R. (2006). Development of adiposity in adolescence: Five year longitudinal study of an ethnically and socioeconomically diverse sample of young people in Britain. *British Medical Journal, 332*, 1130-1135.

Watts, K., Jones, T.W., Davis, E.A., & Green, D. (2005). Exercise training in obese children and adolescents. *Sports Medicine, 35*, 375-392.

Weiss, R., & Raz, I. (2006). Focus on childhood fitness, not just fatness. *Lancet, 368*, 261-262.

Wennlof, A.H., Yngve, A., Nilsson, T.K., & Sjostrom, M. (2005). Serum lipids, glucose and insulin levels in healthy schoolchildren aged 9 and 15 years from Central Sweden: Reference values in relation to biological, social and lifestyle factors. *Scandinavian Journal of Clinical and Laboratory Investigation, 65*, 65-76.

Whitaker, R.C., Wright, J.A., Pepe, M.S., Seidel, K.D., & Dietz, W.H. (1997). Predicting obesity in young adulthood from childhood and parental obesity. *New England Journal of Medicine, 337*, 869-873.

Wilkin, T.J., Mallam, K.M., Metcalf, B.S., Jeffery, A.N., & Voss, L.D. (2006). Variation in physical activity lies with the child, not his environment: Evidence for an "activitystat" in young children (EarlyBird 16). *International Journal of Obesity, 30*, 1050-1055.

Winzenberg, T., Shaw, K., Fryer, J., & Jones, G. (2006). Effects of calcium supplementation on bone density in healthy children: Meta-analysis of randomized controlled trials. *British Medical Journal, 333*, 775-778.

Wong, N.D., Hei, T.K., Qaqundah, P.Y., Davidson, D.M., Bassin, S.L., & Gold, K.V. (1992).. Television viewing and pediatric hypercholesterolemia. *Pediatrics, 90*, 75-79.

Yu, C.C., Sung, R.Y., So, R.C., Lui, K.C., Lau, W., Lam, P.K., et al. (2005). Effects of strength training on body composition and bone mineral content in children who are obese. *Journal of Strength and Conditioning Research, 19*, 667-672.

CHAPTER

3

A Sociohistorical Analysis of U.S. Youth Physical Activity and Sedentary Behavior

Mary McElroy, PhD

Health professionals, government officials, and parents are increasingly troubled by the sedentary lifestyles of young people. A 2002 survey in the United States revealed that 61.5% of youth age 9 to 13 years did not participate in any organized physical activity during their nonschool hours and that 22.6% did not participate in any form of physical activity during their free time (Centers for Disease Control and Prevention, 2003). Increased participation in passive leisure activities such as television viewing and playing video games, coupled with a decline in school-sponsored physical education and community sport programs, has also been linked to the escalating number of overweight and obese youth (Koplan, Liverman, & Kraak, 2005). A recent report released by the Institute of Medicine estimates that one-third of American children and adolescents are either obese or at risk of becoming obese (Koplan et al., 2007). Over the past 30 years, according to the report, obesity rates have nearly tripled for children age 2 to 5 years (from 5% to 14%) and adolescents age 12 to 19 years (from 5% to 17%). For children age 6 to 11 years, the obesity rates have nearly quintupled (from 4% to 19%). These alarming trends have prompted some to claim that the current generation of children will grow into the most obese generation of adults in U.S. history (Hill & Trowbridge, 1998).

The increasing numbers of obese and overweight youth have led policy makers to rank the problem as one of the most critical public health threats of the 21st century (Koplan et al., 2007). Although the health consequences of youth physical inactivity and obesity preoccupy today's health officials, sedentary activity and obesity among youth were far from the minds of health planners a century ago. At the turn of the 20th century, health officials focused their attention on infectious diseases that proved fatal to many children (King, 1993).

Due in large part to the inadequate and unsanitary methods of child and infant care and the unavailability of adequate vaccines and other medicines, more than one-third of all children died before reaching 5 years of age, and one-half did not survive through adolescence (Meckel, 1999). The Spanish flu epidemic in 1918, which claimed the lives of half a million Americans (20 million people worldwide)—many of them young people—reinforced the precarious health status of children and adolescents.

Although today the specific health threats faced by youth differ from those experienced by their predecessors, many of the social and political forces germane to these threats have their roots firmly planted in what Swedish author and feminist, Ellen Key (1900), proclaimed as "the century of the child." The early years of the 20th century marked a period when the U.S. government became involved with child welfare issues that led to the passing of child labor laws and the creation of innovative health agencies and programs such as the Children's Bureau and the Shepard Towner Child Welfare Act (Helfand, Lazarus, & Theerman, 2000).

Insight into the challenges of youth physical inactivity and obesity facing health officials and politicians today is fixed firmly in the century-long struggle to deal with the health of young people. What follows is an exploration of several of the key themes that composed the 20th century children's health movement in the United States. Particular attention is given to the different attitudes regarding the role and importance of physical activity and inactivity in young people's health and the associated reaction of health officials, government agencies, and the general public.

Early 20th Century Views of Children's Health

At the beginning of the 20th century, most Americans lived on farms or in small towns. Few homes had a telephone, electricity, or indoor plumbing. Because medical technology was primitive by contemporary standards, and because there was no national insurance system to provide for older persons, families typically had lots of children in the hope that a few would survive to provide for them in their old age (King, 1993).

During the early years of the 20th century, pediatrics emerged as a distinct branch of medicine. Prior to this time, children's medicine was an under-recognized area, and few physicians identified themselves as interested in the health of children. Perceiving the need for a medical organization independent of the American Medical Association, Abraham Jacobi formed the American Pediatric Society (Viner, 2002). Jacobi championed two principles of children's health that remained central to the children's health movement throughout the 20th century:

- The child is a complex being, and health promotion is needed to address the needs of the whole child, including the cures of some diseases and prevention of others.

- Pediatricians should become politically active, as many of the needs of children are connected to their social and economic surroundings (Markel & Golden, 2005).

By the second decade of the 20th century, the children's health movement had gained momentum. Public health advocates and politicians called for greater protection of mothers and infants as the "year of the child" was declared in 1918. The most important welfare organization, the American Association for the Study and Prevention of Infant Mortality, changed its name to the American Child Hygiene Association (Stern & Markel, 2002). A healthy populace, the organization argued, was important to military and economic well-being. In response, governments provided supplemental income for new mothers, free medical care, and a range of preventive health services aimed at reducing infant mortality rates, which at that time exceeded 300 deaths per thousand live births.

The physical activity habits of early 20th century youth were also very different from those of their modern counterparts. Physical activity was an integral part of the daily routines of children and adolescents. In 1910, 12% of children age 10 to 13 years worked, as did 31% of those age 14 to 15 years (Markel & Golden, 2005). Many young people engaged in hard labor as they worked alongside parents harvesting crops, while others were exposed to serious injury

Factory work, such as that completed by these boys on the cotton loom, exposed youth to significant injury risks.

and in some cases death in industries such as mining or in factory jobs (Gulick & Ayers, 1913). In the prospering urban areas it was common to see young children walking the streets for hours selling newspapers and other items.

Eliminating child labor was a vexing issue. Few states passed laws to keep children out of the labor force; an exception was New York, which passed a law in 1903 banning child labor in factories, on farms, in sweatshops, and in the streets. The federal government later passed the Keating Owen Act of 1916, which banned articles produced by children from being sold as interstate commerce. Unfortunately, the ruling was overturned by the Supreme Court in 1918. Not until the passage of the Fair Labor Standards Act of 1938 were children under 16 years old excluded from full-time labor and those under 18 kept from hazardous occupations. Many children and adolescents would continue to labor on family farms with little or no legislation to protect them. Health officials and parents in the early years of the 20th century were preoccupied with the safety of their children. They had little interest in their children's physical activity levels or in childhood overweight and obesity, because these issues were far less problematic in the United States at that time than they are today.

Early 20th century health reformers were more concerned with the impact of urban-industrial life upon the moral, intellectual, and physical welfare of children. For some, physical activity as one form of the urban youth's leisure-time activities was seen as counterproductive to a child's development (Lewis, 1977). For others, the benefits of physical activity for children, particularly in the form of organized sport, were viewed as positive contributors to character development and citizenship training (McCloy, 1930). Seen as a solution to many of the evils that threatened society, physical activity through physical education and sport programs was enthusiastically promoted by school administrators, social workers, public officials, and religious leaders. Some of the period's most important social and political reformers, such as Jane Adams, Lillian Wald, and Theodore Roosevelt, endeavored to convince the public of the need for government-sponsored physical activity facilities. Their efforts led to the construction of thousands of playgrounds and parks, notably in the congested city environments, and the inclusion of physical education courses in the school curriculum.

Coinciding with a burgeoning child social welfare movement in the United States were tremendous advances in medicine and public health. The development of preventive vaccines against childhood diseases, such as tetanus, poliomyelitis, and a host of other now-forgotten childhood diseases, formed the foundation of children's health care. Life expectancy has increased by more than 30 years since 1900, and much of this improvement is due to the reduction of infant and early childhood mortality. Improved public health measures, along with the emergence of antibacterial medications, would allow health professionals to slowly shift their focus to disease prevention measures such as good nutrition and physical activity.

It was not until the early years of the 20th century that people were urged to visit their doctors periodically to evaluate the general state of their health. The health screening movement began with the examination of school-age children and was fueled by studies reporting that one-fifth to one-third of them had disabilities requiring medical attention. Most disturbing were findings from the mass examination of army draftees in the United States for World War I. Of some 3.7 million males, about 550,000 were rejected, and almost half of the remainder had some physical impairment (Reiser, 1978). The health status of potential draftees for World War II was equally distressing. More than 45% of the first 2 million people examined for military service were rejected for mental or physical reasons. These realities forced the public health community to look more closely at national issues of health, including both diet and exercise.

The First Weight Crisis and American Youth

Just after World War I, health experts became increasingly concerned about the weight of children, but their distress was not focused on the problem of youth overweight and obesity. Between one-quarter and one-third of young people were malnourished, and up to half suffered physical defects that put them at risk for malnutrition (Brosco, 2002). Physicians believed that weight gain was a measure of growth, nutritional status, and overall health. Weight had been the standard for judging the nutrition of infants, and it was natural to carry the idea over to older children. Over the next decade, public health workers weighed and measured millions of children in an effort to identify the malnourished. In addition to their value in managing illness, weight charts were useful in monitoring general health and in screening for disease. Pierre Budin, a Paris physician, was one of the first to use weighing scales in the effort to save infant lives. He instituted infant consultations with mothers during which he encouraged breast-feeding and taught safe child-rearing practices. Using Budin's work as a model, infant welfare workers in England and the United States urged parents to weigh their babies at festivals and clinics and during home visits. In 1918, the U.S. Children's Bureau proposed that the weighing and measuring test should be the first activity for the "year of the child." Under their plan, local committees would organize efforts to weigh every child in a community. By 1919, more than 16,000 communities across the country had conducted weight tests in schools, meeting halls, and town offices. The campaign to weigh every child implied a new use of weight as a screening for health.

The weighing stations that were commonplace in large cities, small towns, and rural communities also served to confirm popular notions that associated poor health with class, race, and ethnicity. St. Louis pediatrician Borden Veeder, for example, reported in 1926 that children from poor families were two and one-half times more likely than others to be underweight for their height. Physicians in the early 20th century already knew that inadequate health care,

© Fox Photos/Hulton Archive/Getty Images

Weigh stations throughout the United States helped provide data the government used to develop standards for healthy youth height and weight after World War I.

limited economic resources, and dirty city environments contributed to the health status of children from poor or immigrant families (Viner, 2002).

By the 1930s, substantial amounts of data had been accumulated on heights and weights of youth in the United States. This information would be incorporated into standardized height and weight charts used by pediatricians to evaluate the developmental progress of children and adolescents. Henry P. Bowditch, a professor of physiology at Harvard Medical School, is generally acknowledged as the first to stimulate interest among American physicians in the field of child growth and development. He recognized the importance of the potential uses of height and weight for medical practice. Physicians believed that weight gain was a measure of growth, nutritional status, and overall health. The impetus for measuring height and weight was the belief that poorly grown and undernourished children could be identified and aided before they became sick. Without such assessments, and the interventions they catalyzed, young people would later be unfit for work or for military service in the case of boys and for motherhood in the case of girls.

Pediatricians did not view the condition of "overnutrition" as a serious health issue in children. During the early decades of the 20th century, fat was considered harmless and mainly hereditary (Schwartz, 1986). Continued concern about underweight patients might have combined with endocrine or heredity arguments to produce considerable skepticism regarding treatment of youth obesity. Even among the children who might be considered heavier than the norm, the idea of restricting children's food in a period when poverty so obviously was connected to malnutrition was not acceptable medical practice (King, 1993).

Youth Physical Activity and National Defense

During World War II, the U.S. government formed a series of organizations to work to improve the health and fitness levels of citizens. A division of Physical Fitness was formed that worked first under the Office of Civilian Defense and later under the Office of Defense. A national committee on physical fitness was created by the president in 1943 to help indoctrinate the public and social institutions on the need for physical conditioning (Reiser, 1978).

The dismal status of young people's health piqued interest in physical education as the realities of World War I and World War II placed special responsibilities upon school programs. Schools seemed likely places to develop the prowess of the soldiers of tomorrow, and many programs focused on preparing young people for the military. At first, military drill formed the foundation of the physical education class; however, Newton Baker, the secretary of war, warned that military drill made little contribution to immediate military strength and that vigorous disciplined physical training of youth furnished better preparation (Van Dalen & Bennett, 1971). Both civilian and military leaders soon agreed that the schools and colleges would contribute most to the war effort by developing youth physical fitness.

Concerns for the physical fitness level of American youth did not end with the conclusion of World War II. In 1953, Hans Kraus and Ruth Hirschland (1954) published data indicating that American youth were less fit than European youth. Nearly 60% of American youth failed at least one item on the Kraus-Weber tests for muscular fitness, whereas less than 10% of children from three European countries (Austria, Italy, and Switzerland) failed at least one item. Because the report was published during the Cold War, concern was reignited about the readiness of American youth for military service. In response, in 1956 U.S. President Dwight D. Eisenhower established the President's Council on Youth Fitness to promote youth fitness nationwide. Over John F. Kennedy's subsequent presidency from 1961 to 1963, considerable energy was devoted to addressing youth fitness, and the council was reorganized and later renamed the President's Council on Physical Fitness. Kennedy was a sport enthusiast who was very concerned with the fitness of the country's youth (see Kennedy, 1960, 1962).

The effects of national concerns for physical fitness were felt in school programs throughout the country and at all grade levels. A report in 1963 estimated that more than 20 million public school children in grades 4 to 12 took part in regular physical fitness programs, a gain of 2.1 million over the previous year (Van Dalen & Bennett, 1971). Thirty states had a fitness council or commission. The physical fitness levels of youth would remain the focal point of many physical education programs for decades to come.

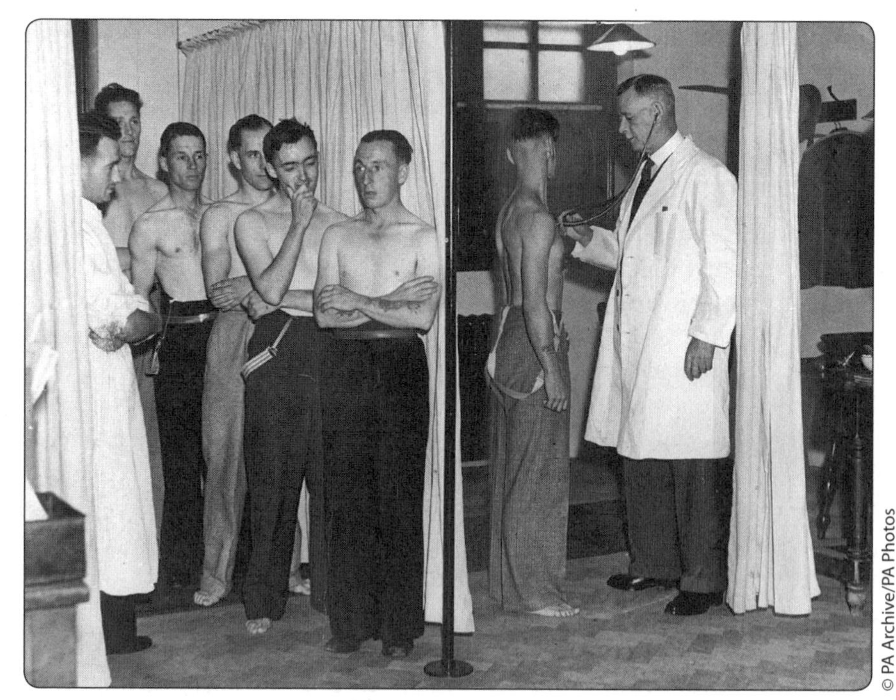

© PA Archive/PA Photos

Medical examinations during World War II showed many recruits to be unfit for military service.

The Intractability of Youth Obesity

The abundance of food in the United States following World War II was a symbol of the country's wealth. Americans of all classes ate more meat and more fruit than their counterparts in Europe and Africa. Postwar prosperity meant that Americans were less concerned with the problems of undernutrition of youth. By the midpoint of the 20th century, "overnutrition" was viewed as America's new health problem. At first attention focused on adults, but before long children and adolescents would be included.

Prior to the 1950s, pediatricians and parents did not view excess weight and fatness as serious problems in young people. Only a handful of articles concerning obese children and youth appeared in the American medical press (Schwartz, 1986). Many physicians viewed overweight as a temporary condition that young people would "outgrow." Even noted pediatrician Dr. Benjamin Spock (1957) declared that mild overweight is common between the ages of 7 and 12 and likely to vanish by the age of 15 without great effort. Prior to the 1960s, most of the specific health warnings attached to campaigns against fat focused on the degenerative process for the heart, arteries, and the digestive organs—problems of later adulthood and not relevant to children.

Research studies published in the 1960s, introducing the concept of the "intractability" of childhood weight, ignited a growing concern over childhood obesity. For example, much attention was given to the study by Abraham and Nordsieck (1960), which followed overweight boys and girls over a 20-year period and showed that 84% of the boys and 87% of the girls were still overweight as adults. Likewise, according to Stunkard (1976), the author of a popular children's health manual, the odds of an overweight child becoming an overweight adult significantly increased if there was no attention to specific weight reduction strategies during the adolescent years. Knittle (1972) offered a biological explanation, that childhood obesity was a problem in adipose tissue cellular development and was set in early childhood.

By the second half of the 20th century, however, obesity and overweight became a psychological and social issue and not just a question of genetic inheritance. Overeating, not lack of physical activity, was considered the culprit. Dismayed health critics complained that somewhere in transition we went from the child who did not have enough to eat to the child who could not be satisfied, no matter how much was offered. The public health community increasingly accepted the widespread hostility to fat and gave it new medical justification. However, the methods to treat youth obesity were largely dietetic. Physical activity for children and adolescents remained important, but the crucial emphasis rested on restraint and good sense in eating. Physical activity was encouraged, but greater attention was given to what young people ate.

Prior to the 1980s, health promoters recognized that behavioral risk factors, such as physical inactivity, played a crucial role in the development of chronic disease. However, the failure of health promotion efforts to be directed at young people reflected two pervasive beliefs: (1) that though physical inactivity was widespread in adults, youth are naturally and spontaneously active, and (2) that the health risks associated with sedentary living, such as diabetes and heart disease, are far more pressing in adults. Interest in measuring physical activity in youth would be the response in the late 1970s to the growing belief that youth health behaviors carry into adulthood. Further, consistent with evolving views on overweight and obesity, dietary habits, food intake, physical activity, and exercise were increasingly viewed as acquired from young people's social environments and critical to their health (Bruch, 1973).

Focusing on Physical Activity

Beginning in the 1980s, physical educators and health professionals slowly shifted their focus from physical fitness toward physical activity of young people. The federal government commissioned two studies of representative population samples of school-aged children known as the National Children and Youth Fitness Studies (NCYFS). NCYFS I examined nearly 9000 children (Ross & Gilbert, 1985) in grades 5 through 12, and NCYFS II more than 4000 children in grades 1 through 4 (Ross & Pate, 1987). These studies provided much information regarding the physical activity behaviors of youth. For example, information regarding physical education enrollment, physical fitness assessments, and physical activity levels was collected from the students participating in these two studies.

Systematic studies also confirmed the prevalence of overweight and obesity among children and adolescents. Comparing data from the National Health Examination Survey collected in 1963 to 1965 to data from the National Health and Nutrition Examination Survey conducted from 1976 to 1980, Gortmaker and colleagues found a 54% increase in the prevalence of obesity among children 6 to 11 years old and a 98% increase in the prevalence of superobesity. Increases in obesity occurred among children of all ages, both sexes, and both Blacks and Whites (Gortmaker et al., 1987). Obese children were defined as those with triceps skinfolds greater than or equal to those of the 85th percentile of children in the earlier data set, while superobese children were defined as those with skinfolds greater than or equal to the 95th percentile. These studies taken together demonstrated that U.S. youth were not participating in adequate physical activity and were experiencing high rates of weight gain.

The 1990s witnessed a series of national reports aimed at promoting specific recommendations for physical activity necessary for the health of young people. *Physical Activity and Health: A Report of the Surgeon General*, while acknowledging that young people are more active than adults, stated that many did not engage in enough physical activity for health (U.S. Department of Health and Human Services [USDHHS], 1996). *The International Consensus Conference on Physical Activity Guidelines for Adolescents* recommended that young people be active daily and engage in three or more sessions per week of activity lasting 20 min or more and requiring moderate to vigorous levels of exertion (Sallis & Patrick, 1994). The Health Education Authority in Great Britain increased the recommendation to 60 min of moderate and vigorous activities per day (Biddle, Sallis, & Cavill, 1998). The National Association for Sport and Physical Education (NASPE) later revised its recommendation, suggesting that children accumulate at least 60 min, and up to several hours, of age-appropriate physical activity on all or most days of the week (NASPE, 2004). Specific guidelines regarding physical activity were also issued for infants, toddlers, and preschool children (NASPE, 2002). It was recommended that toddlers have at least 30 min of daily structured physical activity and that preschoolers accumulate 60 min of daily structured physical activity. Finally, the Healthy People Goals for 2000 and 2010 include explicit mentions of the need for America's youth to increase daily physical activity (USDHHS, 1991, 2000). Though some criticized the lack of good "scientific evidence" on which physical activity guidelines were based (e.g., Twisk, 2001), the value of publishing guidelines was indisputable. These goals were developed under the direction of the federal government at a time when children were being exposed to the new technologies that would encourage sedentary leisure pursuits.

The Growth of the Media Generation

As the United States' wealth increased in the decades after World War II, so did home ownership, rising to nearly 66.3% in 1998 (Sterling & Kittross,

2002). Television was introduced in the United States in the 1950s and shortly thereafter became a prominent fixture in the living room. During the early years of television, there were only three television channels, no videocassette recorders, no home computers, and, for many homes, no air-conditioning. As a result, there was no particular reason to stay inside.

Today, televisions are present in virtually every household in the United States. In 1998, 98% of households had a television (Sterling & Kittross, 2002). Further, the percentage of households with two or more televisions has increased dramatically in recent years. By 1980, 50% of households had more than one television set; and by 2000, 76% of households had more than one set. The proportion of households with three or more sets also rapidly grew from 15% in 1980 to 41% in 2000. Videocassette recorder ownership also increased, from about 66% of households in 1990 to 85% in 2000 (Sterling & Kittross, 2002). The VCR added versatility because it was the first video communication medium that allowed individuals to take control of when, where, how, and what to watch on television or via a television monitor (French, Story, & Jeffery, 2001). Digital video discs brought crystal-clear digitized video images into the home. As television sets, VCRs, and DVD players transformed into "everywhere" appliances and air-conditioning became almost ubiquitous, children and adolescents found more and more reasons to stay inside.

Emerging in the 1980s were several studies pointing to the concern that children and youth were spending time watching television rather than engaging in physical activity. For example, Dietz and Gortmaker (1985) studied nearly 7000 children aged 6 to 11 years and more than 6600 adolescents aged 12 to 17 years and documented their concern in language that piqued public interest. Using an epidemiological framework, they described the prevalence of youth obesity in terms of increases for each additional hour of television viewed daily. Specifically, they found significant associations between television viewing and obesity prevalence for children and adolescents. For adolescents, the prevalence of obesity increased by 2% for each additional hour of average daily television watching. Even after other factors believed to be related to obesity such as race, socioeconomic class, and a variety of family variables were controlled for, the relationship between obesity and television remained. Another study, sponsored by the Kaiser Family Foundation, showed that daily time spent watching television, taped shows, or commercial videos averaged 2.5 h for children aged 2 to 7 years, 4.5 h for children aged 8 to 13 years, and 3.3 h for children aged 14 to 18 years (Roberts et al., 1999). The concerns prompted the Committee on Public Education as part of the American Academy of Pediatrics to recommend that television sets be removed from bedrooms and that viewing be limited to no more than 2 h a day (American Academy of Pediatrics, 2001).

By the 1990s, television in particular was perceived as the cause of the youth obesity epidemic. Television viewing was believed to promote obesity in two ways: (1) by displacing participation in physical activity that would expend more energy and (2) by increasing dietary energy intake either via snacking during viewing or as a result of food advertising that promoted high-calorie products. Other forms of screen time were also blamed for contributing to the promotion of obesity in children and adolescents. Video game playing and computer surfing were also called into question as poor substitutes for outdoor play and other active pursuits. An editorial in the *Journal of the American Medical Association* warned that television viewing likely leads to obesity due to decreases in physical activity (Robinson, 1998).

The impact of what has come to be known as "screen time" on leisure-time activity, however, remains unclear (see chapters 1 and 2). It is uncertain whether time spent using the computer or watching television is substituted for other sedentary activities or whether it takes the place of more active pursuits such as participating in physical activity. Although young people likely have greater access to mediated leisure than in previous generations, a number of studies have suggested caution in blaming technology as a cause for the rising obesity rates.

Several studies, for example, have indicated that the amount of television watching has not really changed significantly during the last 40 years, while others suggest that there is time in the day for both television and more physically active pursuits. Biddle and colleagues (2004), for example, concluded that television viewing and video game playing were largely uncorrelated with physical activity. On the basis of a study of 11- to 15-year-old British youth, known as Project STIL (Sedentary Teenagers and Inactive Lifestyles), they caution us not to assume a cause-and-effect relationship between these popular sedentary behaviors and more active pursuits. Even if more youth are watching television than in previous generations, the researchers maintained, one cannot assume that television viewing is actually replacing time spent in physical activity. They suggested the increased use of motorized transportation instead of walking or biking, or the replacement of physical education classes with more sedentary learning experiences, as potentially more salient factors in promoting physically inactive living. By the closing decades of the 20th century, the search for such factors that predispose young people to inactive living resulted in attention to social, economic, and environmental matters.

Challenges Facing Contemporary Families

Children are strongly influenced by the food and physical activity decisions made by their families (see chapter 11). Contemporary U.S. children and adolescents live in a society where family functioning has changed dramatically in the last several decades. Many of these changes, such as both parents working outside the home and for longer hours, combined with a decline in school physical education programs and disappearing community resources for physical activity, dramatically affect what children eat and how much they participate in physical activity (McElroy, 2002).

One of the major challenges facing families is the requirement that adults juggle employment and child care roles. In 1977, fewer than half of employed males with children under 18 had employed spouses; but by 1997, two-thirds of female spouses were employed. In 75% of dual-career couples in 1997, both partners worked full-time, an increase from the 66% of 20 years earlier (Ventura et al., 1999). Families can find themselves caught in a "time bind" between work and family life as they devote increasing amounts of parental time to paid work (Hochschild, 1989).

The time bind is also responsible for another contemporary phenomenon, that is, the growth of child care centers. In 2005, more than half of 3- to 5-year-old children were enrolled in early childhood programs, which include day care centers, nursery schools, preschools, Head Start programs, and prekindergarten programs (Federal Interagency Forum on Child and Family Statistics, 2006). Dowda and colleagues (2004) found that children attending preschools with more resources and better educated teachers demonstrated significantly higher levels of physical activity than those attending preschools with fewer resources and less educated teachers. Unfortunately, they found that many day care centers did not promote physical activity among the children.

Children in families headed by single mothers or living below the poverty line are also more likely to be in poor or fair health than children in two-parent or more affluent families (Montgomery, Kiely, & Pappas, 1996). In one study, children raised in a family lacking money for basic needs during the preceding year were more likely to be reported by their mothers as presenting acute health problems, a growth delay, and hospitalizations compared to children living in a family not lacking money for basic needs (Séguin et al., 2005). Recent estimates place nearly one in six children living in poverty. Twenty-five percent of children live in families in which only one parent has full-time employment, and 28% of families with children are headed by a single parent.

Despite the abundant food availability and high food wastage in the United States, recent evidence suggests that a large number of U.S. children live in families that have insufficient food. The Community Childhood Hunger Identification Project indicated that nearly one in three children younger than age 12 years in low-income families often went hungry or was at risk of hunger (Wehler et al., 1995). The Third National Health and Nutrition Examination Survey showed that between 1988 and 1994, approximately 2.4 to 3.2 million children younger than 12 years lived in food-insufficient families. An additional 0.7 to 1.3 million teenagers lived in food-insufficient families (Alaimo et al., 1998). Economic and time constraints, as well as the stresses and challenges of daily living, may make healthful eating and increased physical activity a difficult reality on a day-to-day basis for many families.

The term "underserved" first emerged in the 1970s, with a behavioral connotation that labeled poor people as victims of the lack of support services and opportunities in their neighborhoods. William Julius Wilson's (1987) *The*

Truly Disadvantaged: The Inner City, the Underclass, and Public Policy described the consequences of neighborhoods with high-poverty conditions. Wilson used the term *underclass* with reference to how the structural characteristics and economic changes that had occurred in inner cities adversely affected the lives of their residents. This impact stems from a variety of factors, including poverty, substandard housing, limited educational opportunities, and low levels of social integration, all of which lead to poor health and adverse social and economic circumstances (Gostin & Powers, 2006). Although the determinants of childhood obesity are generally of the same nature among all population groups, special consideration is likely needed to assess whether the pathways to preventing childhood obesity in the population will reach subgroups that are most affected (Gostin & Powers, 2006).

A number of studies have pointed to the fundamental barriers to participation in physical activity that exist in underserved communities. Beyond challenges related to low income, youth from poor neighborhoods have less access to recreational facilities and after-school sport than do other people their age. For example, one study showed that the opportunities for after-school physical activities for youth living in urban areas were only one-third of those available for youth in nearby suburban communities. As part of the Play Across Boston Core Project, Cradock and colleagues (2002) also found disparities in access and participation by sex and by race or ethnicity. Boys were found to be twice as likely as girls to participate in sport and physical activity programs. The opportunities for participation are fewer and the level of participation is lower for African American and Hispanic youths than for White youths. The study also indicated that playground quality and the number of sport and recreation facilities varied significantly across the city's neighborhoods.

Children living in poverty are more likely to live in environments with inadequate support for health-promoting behaviors. Specific attention must be given to young people who are at high risk for becoming obese, and this includes children in populations with higher obesity prevalence rates and long-standing health disparities such as African Americans, Hispanic Americans, and Native Americans.

As a result of the growing concern about health disparities, the U.S. Congress in 2000 called for the establishment of the National Institutes of Health National Center on Minority Health and Health Disparities and a strategic plan to address the continuing poor health status of minorities, those with low income, and people living in rural areas (Thompson, Mitchell, & Williams, 2006). Private and public efforts to eliminate health disparities, as outlined in the *Healthy People 2010* objectives (USDHHS, 2000), are critical to the development of community-based programs to address social, economic, and environmental barriers that contribute to the health outcomes of many children and adolescents. An important aspect of these efforts is to vigorously address the barriers to adequate physical activity that exist among these minority, low-income, and rural populations.

Children living in underserved communities spend much of their time indoors in environments that are not conducive to being physically active. They often live in cramped quarters, and activities at home are typically sedentary, involving little or no parental input. Television and video game use is a consistently planned strategy employed by parents to control the chaotic nature of their home environments (Goodway & Smith, 2005). Thus, a comprehensive community commitment is vital to improving early childhood physical activity in urban and poor communities. Underserved communities must create environments with safe playgrounds and parks, walking and bike trails, and neighborhood recreation centers with sport facilities and supervised activities for youth. Initiatives to promote physical activities need to start with very young children and to carefully structure safe environments in which to be active.

Healthy communities contain what Putnam (2000) terms "social capital," which can be defined as the resources available to individuals and groups within communities as a result of their connection to other people. Important to social capital is the concept of collective efficacy, which refers to the ability of community residents to undertake collective action for mutual benefit (McElroy, 2002). Communities that lack strong civic connections find it harder to mobilize to achieve the outcomes necessary to benefit the whole community. Helping at-risk children and adolescents requires an understanding of the complex and interacting influences of the social, economic, and physical environments and ways to improve the community environment in order to (among other things) promote physical activity among youth.

Lessons Learned From the Past for the Future

Many of the physical inactivity problems facing children and adolescents today require broad-based social and environmental responses, similar to those applied a century or more ago when health officials interested in preserving the lives of infants and children had little choice but to approach health issues from such a perspective. The early years represented a period when the U.S. government became involved with children's social welfare issues and took actions ranging from the passing of child labor laws to the creation of innovative health agencies and programs like the U.S. Children's Bureau and the Shepard Towner Child Welfare Act. As we look back at the children's health movement during the 20th century, we can see that the numerous health achievements were a direct result of significant social transformation. As first promoted by Abraham Jacobi in the early part of the 20th century, pediatricians, legislators, and parents today need to be focused on a sociocultural model of pediatrics in order to address the health of the whole child. This approach includes attention to social institutions such as schools and families as well as the creation of healthy community environments. A healthy future for U.S. youth also depends on the ability of all young people to have access to resources and opportunities (McElroy, 2002).

A second important lesson learned is the finding that the significant successes achieved by the children's health movement in the United States bring a set of new challenges. For example, improved public health measures, in combination with the emergence of antibacterial medications, allowed health professionals to shift their attention to disease prevention measures such as good nutrition and physical activity. At mid-20th century, when health officials began to realize that chronic illnesses required the same kind of attention previously given to contagious diseases, it was necessary to shift to a view that the chronic diseases once associated with adulthood are also pertinent to the health of children and adolescents. The Institute of Medicine Committee on Prevention of Obesity in Children and Youth was charged with developing a prevention-focused action plan to decrease the prevalence of obesity in children and adolescents in the United States (Koplan, Liverman, & Kraak, 2005). The primary emphasis of the committee's task was to examine the behavioral, cultural, and social factors associated with today's problem of childhood obesity. Today, in contrast to two decades ago, the children's health message focuses on the importance of physically active lifestyles as a precursor to healthy behaviors in adulthood.

Finally, history teaches us a third lesson: The health challenges facing youth today will require a rigorous national strategy that addresses the problems at multiple points of contact, ranging from the individual to the institutional, community, and public policy levels (McLeroy et al., 1988). And yet, while there is still much to be learned from the public health efforts directed at

APPLICATIONS FOR RESEARCHERS

Serious interest in youth physical activity and inactivity as health issues did not emerge until the 1980s. Initial recommendations regarding youth physical activity were based on the amount of exercise needed to enhance fitness; and, as pointed out by Sallis and Owen (1999), recommendations for youth physical activity have generally been based on studies of physical activity and health among adults. As physical activity is adopted as a public health priority it will be important to identify how much physical activity is needed to maintain good health. Popularly held beliefs, such as the notion that youth physical activity "tracks" into adulthood or that television viewing is causing the youth obesity epidemic, are without much empirical verification. Additional research is necessary to determine if and how time spent in sedentary pursuits, such as watching television, surfing the Internet, or even reading a book, is correlated with youth obesity. Researchers also need to continue to evaluate the importance of social conditions such as changing workforce demographics, poverty, and high-risk communities that have likely affected the physical activity levels of children and adolescents.

APPLICATIONS FOR PROFESSIONALS

Many of the physical inactivity problems facing youth today are those that require broad-based social environmental responses similar to those applied a century or more ago when health officials were interested in saving the lives of infants and children from infectious diseases. Throughout the 20th century, coinciding with the child welfare movement in the United States, health officials, politicians, and the general public slowly recognized the importance of physical activity. Toward the end of the 20th century, public health officials came full circle regarding the importance of addressing social conditions such as changing employment demographics, impoverished living conditions, health disparities, and technological advances (e.g., television, video games, computers) as significant impediments to the health of the next generation. Not until we fully embrace health interventions that address social and physical environments will young people benefit from the more than 100 years of experience associated with the U.S. children's health movement.

young people, only with sufficient political resolve from health professionals, politicians, and other child advocates will future generations benefit. Likely only when the public at large recognizes that improved social conditions (such as socioeconomic status, family, and community conditions) are as important to the improved health of youth as are strategies to change health behaviors of children and adolescents, will we witness the positive health impact on future generations. The challenge before us compels us to build on past experiences so that young people can enter adulthood with the benefits public health advocates have promoted for more than a century.

References

Abraham, S., & Nordsieck, I.M. (1960). Relationship of excess weight in children and adults. *Public Health Reports*, *75*, 263-273.

Alaimo, K., Briefel, R.R., Frongillo, E.A., & Olson, C.M. (1998). Food insufficiency exists in the United States: Results from the Third National Health and Nutrition Examination Survey. *American Journal of Public Health*, *88*, 419-426.

American Academy of Pediatrics. (2001). Children, adolescents, and television. *Pediatrics*, *107*, 423-426.

Biddle, S.J.H., Gorely, T., Marshall, S.J., Murdey, I., & Cameron, N. (2004). Physical activity and sedentary behaviours in youth: Issues and controversies. *Journal of the Royal Society for the Promotion of Health*, *124*, 29-33.

Biddle, S.J.H., Sallis, J.F., & Cavill, N.A. (Eds.). (1998). *Young and active? Young people and health enhancing physical activity: Evidence and implications.* London: Health Education Authority.

Brosco, J.P. (2002). Weight charts and well child care: When the pediatrician became the expert in child health. In A.M. Stern & H. Markel (Eds.), *Formative years: Children's health in the United States, 1880-2000* (pp. 91-120). Ann Arbor, MI: University of Michigan Press.

Bruch, H. (1973). *Eating disorders: Obesity, anorexia nervosa, and the person within.* New York: Basic Books.

Centers for Disease Control and Prevention. (2003). Physical activity levels among children aged 9-13 years—United States, 2002. *Morbidity and Mortality Weekly Report, 52,* 785-788.

Cradock, A., Elayadi, A., Gortmaker, S., Hannow, C., Sobol, A., & Wiecha, J. (2002). *Play across Boston: Summary report.* Boston: Harvard Prevention Research Center.

Dietz, W.H., & Gortmaker, S.L. (1985). Do we fatten our children at the television set? Obesity and television viewing in children and adolescents. *Pediatrics, 75,* 807-812.

Dowda, M., Pate, R.R., Trost, S.G., Almeida, M.J.C.A., & Sirard, J.R. (2004). Influences of preschool policies and practices on children's physical activity. *Journal of Community Health, 29,* 183-196.

Federal Interagency Forum on Child and Family Statistics. (2006). *America's children: Key national indicators of well-being.* Retrieved February 3, 2008, from www.childstats.gov/americaschildren06/tables/pop8a.asp.

French, S.A., Story, M., & Jeffery, R.W. (2001). Environmental influences on eating and physical activity. *Annual Review of Public Health, 22,* 309-335.

Goodway, J.D., & Smith, D.W. (2005). Keeping all children healthy: Challenges to leading an active lifestyle for preschool children qualifying for at-risk programs. *Family and Community Health, 28,* 142-155.

Gortmaker, S.L., Dietz, W.H., Sobol, A.M., & Wehler, C.A. (1987). Increasing pediatric obesity in the United States. *American Journal of Diseases of Children, 141,* 535-540.

Gostin, L.O., & Powers, M. (2006). What does social justice require for the public's health? Public health ethics and policy imperatives. *Health Affairs, 25,* 1053-1060.

Gulick, L.H., & Ayers, L.P. (1913). *Medical inspection of the schools* (2nd ed.). New York: Russell Sage Foundation.

Helfand, W.H., Lazarus, J., & Theerman, P. (2000). The children's bureau and public health at midcentury. *American Journal of Public Health, 90,* 1703.

Hill, J.O., & Trowbridge, F.L. (1998). Childhood obesity: Future directions and research priorities. *Pediatrics, 101,* 570-574.

Hochschild, A. (1989). *The second shift.* New York: Avon Books.

Kennedy, J.F. (1960). The soft American. *Sports Illustrated, 13*(26), 15-17.

Kennedy, J.F. (1962). The vigor we need. *Sports Illustrated, 17*(3), 12-14.

Key, E. (1900). *The century of the child.* New York: Putnam and Sons.

King, C.R. (1993). *Children's health in America: A history.* New York: Twayne.

Knittle, J.L. (1972). Obesity in childhood: A problem in adipose tissue cellular development. *Journal of Pediatrics, 81,* 1048-1059.

Koplan, J.P., Liverman, C.T., & Kraak, V.I. (Eds.). (2005). *Preventing childhood obesity: Health in the balance.* Washington, DC: Institute of Medicine, National Academies Press.

Koplan, J.P., Liverman, C.T., Kraak, V.I., & Wisham, S.L. (Eds.). (2007). *Progress in preventing childhood obesity: How do we measure up?* Washington, DC: Institute of Medicine, National Academies Press.

Kraus, H., & Hirschland, R.P. (1954). Minimum muscular fitness tests in school children. *Research Quarterly, 25*, 178-188.

Lewis, G. (1977). Sport, youth culture and conventionality, 1920-1970. *Journal of Sport History, 4*, 129-150.

Markel, H., & Golden, J. (2005). Successes and missed opportunities in protecting our children's health: Critical junctures in the history of children's health policy in the United States. *Pediatrics, 115*, 1129-1133.

McCloy, C.H. (1930). Character building through physical education. *Research Quarterly, 1*, 41-59.

McElroy, M. (2002). *Resistance to exercise: A social analysis of inactivity.* Champaign, IL: Human Kinetics.

McLeroy, K.R., Bibeau, D., Steckler, A., & Glanz, K. (1988). An ecological perspective on health promotion programs. *Health Education Quarterly, 15*, 351-377.

Meckel, R.A. (1999). *Save the babies: American public health reform and the prevention of infant mortality, 1850-1929.* Ann Arbor, MI: University of Michigan Press.

Montgomery, L.E., Kiely, J.L., & Pappas, G. (1996). The effects of poverty, race, and family structure on US children's health: Data from the NHIS, 1978 through 1980 and 1989 through 1991. *American Journal of Public Health, 86*, 1401-1405.

National Association for Sport and Physical Education. (2002). *Active start: A statement of physical activity guidelines for children birth to five years.* Reston, VA: AAHPERD.

National Association for Sport and Physical Education. (2004). *Physical activity for children: A statement of guidelines for children ages 5-12* (2nd ed.). Reston, VA: AAHPERD.

Putnam, R. (2000). *Bowling alone: The collapse and revival of American community.* New York: Simon & Schuster.

Reiser, S. (1978). The emergence of the concept of screening for disease. *Milbank Memorial Fund Quarterly, 56*, 403-424.

Roberts, D.F., Foehr, U.G., Rideout, V.J., & Brodie, M. (1999). *Kids and media at the new millennium.* Menlo Park, CA: Kaiser Family Foundation.

Robinson, T.N. (1998). Does television cause childhood obesity? *Journal of the American Medical Association, 279*, 959-960. [Editorial]

Ross, J.G., & Gilbert, G.G. (1985). The National Children and Youth Fitness Study: A summary of findings. *Journal of Physical Education, Recreation and Dance, 56*(1), 45-50.

Ross, J.G., & Pate, R.R. (1987). The National Children and Youth Fitness Study II: A summary of findings. *Journal of Physical Education, Recreation and Dance, 58*(9), 51-56.

Sallis, J.F., & Owen, N. (1999). *Physical activity and behavioral medicine.* Thousand Oaks, CA: Sage.

Sallis, J.F., & Patrick, K. (1994). Physical activity guidelines for adolescents: Consensus statement. *Pediatric Exercise Science, 6*, 302-314.

Schwartz, H. (1986). *Never satisfied: A cultural history of diets, fantasies and fat.* New York: Free Press.

Séguin, L., Xu, Q., Gauvin, L., Zunzunegui, M., Potvin, L., & Frohlich, K.L. (2005). Understanding the dimensions of socioeconomic status that influence toddlers' health: Unique impact of lack of money for basic needs in Quebec's birth cohort. *Journal of Epidemiology and Community Health, 59,* 42-48.

Spock, B. (1957). *Commonsense book of baby and childcare* (2nd ed.). New York: Hawthorn Books.

Sterling, C.H., & Kittross, J.M. (2002). *Stay tuned: A history of American broadcasting* (3rd ed.). Mahwah, NJ: Erlbaum.

Stern, A.M., & Markel, H. (Eds.). (2002). *Formative years: Children's health in the United States, 1880-2000.* Ann Arbor, MI: University of Michigan Press.

Stunkard, A.J. (1976). *The pain of obesity.* Palo Alto, CA: Bull.

Thompson, G.E., Mitchell, F., & Williams, M.B. (2006). *Examining the health disparities research plan of the National Institutes of Health: Unfinished business.* Washington, DC: Institute of Medicine, National Academies Press.

Twisk, J.W.R. (2001). Physical activity guidelines for children and adolescents: A critical review. *Sports Medicine, 31,* 617-627.

U.S. Department of Health and Human Services. (1991). *Healthy People 2000* (DHHHS Pub. No. PHS91-50212). Washington, DC: U.S. Government Printing Office.

U.S. Department of Health and Human Services. (1996). *Physical activity and health: A report of the Surgeon General.* Atlanta: Centers for Disease Control and Prevention, National Center for Chronic Disease Prevention and Health Promotion.

U.S. Department of Health and Human Services. (2000). *Healthy people 2010.* Washington, DC: U.S. Government Printing Office.

Van Dalen, D.B., & Bennett, B.L. (1971). *A world history of physical education: Cultural, philosophical, comparative* (2nd ed.). Englewood Cliffs, NJ: Prentice Hall.

Ventura, S.J., Martin, J.A., Curtin, S.C., & Matthews, T.J. (1999). Births: Final data for 1997. *National vital statistics reports* (Vol. 47, No. 18, pp. 1-96). Hyattsville, MD: National Center for Health Statistics.

Viner, R. (2002). Abraham Jacobi and the origins of scientific pediatrics in America. In A.M. Stern & H. Markel (Eds.), *Formative years: Children's health in the United States, 1880-2000* (pp. 23-46). Ann Arbor, MI: University of Michigan Press.

Wehler, C.A., Scott, R.I., Anderson, J.J., Summer, L., & Parker, L. (1995). *The Community Childhood Hunger Identification Project: A survey of childhood hunger in the U.S.* Washington, DC: Food Research and Action Center.

Wilson, W.J. (1987). *The truly disadvantaged: The inner city, the underclass, and public policy.* Chicago: University of Chicago Press.

CHAPTER

Conceptual Perspectives

Claudio R. Nigg, PhD ▪ Raheem J. Paxton, PhD

The rationale for increasing physical activity levels and reducing sedentary behaviors is well outlined in other chapters of this book (e.g., see chapter 2). Children benefit physically, mentally, emotionally, and socially from participating in physical activity (U.S. Department of Health and Human Services, 1996; SPARK, 2004). Physically, these outcomes in youth consist of increased bone mass, aerobic fitness, and high-density lipoproteins as well as reduced risk for hypertension, obesity, and diabetes (Myers et al., 1996). Studies have also shown that students who participate in physical activity programs experience improved psychological health, concentration, and academic achievement (Sallis et al., 1999; SPARK, 2004). Emotionally and socially, youth who participate in sport have higher levels of confidence, stronger self-images, reduced disruptive behavior, and lower levels of depression (Dowling, 2000).

Despite these benefits, less than one-half of young adults are sufficiently active; and the prevalence of overweight is at an all-time high among children and adolescents, reaching epidemic proportions (Centers for Disease Control and Prevention [CDC], 2004). Decreasing physical activity levels and increasing sedentary behavior levels have been declared a public health burden for our society (CDC, 2001; Dishman et al., 2002). There is a strong need for research that identifies correlates of physical activity levels and sedentary behavior and for theories and frameworks that logically combine the correlates to increase our understanding of, and our ability to effectively and efficiently augment, physical activity levels and reduce sedentary behaviors.

Inactivity is an independent contributor to chronic health issues (such as obesity and diabetes); however, we know little about how and why people decrease their inactivity levels. Theorists should devote time and effort to

---

At the time of writing this chapter, the authors were supported by grants R01 CA109941 (PI Nigg); Hawaii Medical Services Association (PI Nigg); R01 HL0799505 (PI Vogt); and R25 CA090956 (PI Maskarinec).

conceptualizing a theory or framework for inactivity, specifically focusing on reducing factors related to its prevalence. Despite the absence of theory or a conceptual framework to examine the nature of inactivity, there is evidence that interventions placing an emphasis on decreasing inactivity can result in positive health behavior change and may reduce obesity (DeMattia, Lemont, & Muerer, 2006). Because interventions that focus on reducing inactivity often stress increasing knowledge, physical activity, or healthy eating patterns, it becomes difficult to tease out the true magnitude of the impact that inactivity interventions have on weight (DeMattia et al., 2006). We need more research on these topics, including follow-up data on study participants six months or more beyond an intervention, in an effort to truly identify the causal relationships among inactivity messages and outcomes.

Necessity and Status of Theory

Although recently there has been much focus on child-centered health education, less attention has been paid to the theoretical underpinnings of interventions (Theunissen & Tates, 2004). Theory-based interventions have the potential to help identify factors related to specific populations and health behaviors, enabling the design of more effective interventions (Biddle & Nigg, 2000; Theunissen & Tates, 2004). Also, in spite of the potential usefulness of models applied in health behaviors, many of them have primarily been developed for adults (Theunissen & Tates, 2004). Caution is warranted as developmental issues like biological maturation, changing social influences (e.g., parents vs. peers), and life responsibilities may differentially influence behavior change. Therefore, it is recommended that researchers incorporate developmental factors when adapting an existing theory for adults to apply to children.

Definition of Theory

A *theory* is a proposed description or model that explains natural phenomena (what is known) and that can be used to make testable predictions of future occurrences or observations (Kerlinger, 1973; Wikipedia, 2006). In other words, a good theory of physical activity or inactivity should be able to organize facts into meaningful wholes and increase clarity about what is known. A good theory predicts relationships, mechanisms, or outcomes. The expected relationships also must be testable through experimentation or be able to be falsified through some empirical observation. Support for a specific theory should come from many strands of evidence rather than a single foundation (Wikipedia, 2006). Ultimately, a theory provides guidance for systematically collecting facts, formulating hypotheses, and extending knowledge (King, 1978). Although in psychology of physical activity and sport, distinctions are sometimes made between the terms theory and model, these distinctions are not consistently observed in the field. We have chosen to follow that convention by using the two terms interchangeably in this chapter.

Further components of theory quality are the following:

- **Parsimony**—the ability to explain things in as simple a fashion as possible while maintaining completeness
- **Generalizability** (a.k.a. **transferability**)—applicability from one situation to the next and from one population to the next
- **Productivity**—the ability to drive experimentation and produce knowledge

These three characteristics determine how useful a theory is; a theory or framework that is not characterized by these qualities is not useful. Physical activity and inactivity research should be grounded in theory, and each theory should be tested and revised through research. Use of theories in our field helps us to verify knowledge about decision making and provides a rationale for gathering reliable and valid data that are essential for effective decision making and implementation (King, 1978). In creating theories or conceptual models for physical activity, it is important to identify particular correlates of behavior change in the population being studied.

Correlates of Youth Activity and Inactivity

Sallis, Prochaska, and Taylor (2000) reviewed the literature and identified a set of consistent correlates of youth physical activity. In children ages 3 to 12 years, positive correlates were

- contextual variables (sex, ethnicity, parental overweight status, program and facility access, and time spent outdoors),
- psychological variables (physical activity preference, intention to be active, enjoyment, attitudes, confidence), and
- behavioral variables (previous physical activity behavior, diet).

In an example of a study using objectively measured physical activity, preadolescent youth (6th graders) with a mean age of 11.4 years were sampled (Trost et al., 1999). This study showed that for boys, physical activity self-efficacy, social norms related to physical activity, and involvement in community physical activity organizations were significant predictors of moderate-to-vigorous physical activity (MVPA). For girls, only physical activity self-efficacy was found to be a predictor of objectively measured physical activity. In contrast to the abundance of literature concerning the determinants of physical activity, less is known about the determinants of inactivity. This is an important, emerging area of research that is necessary for developing theory addressing inactivity.

The Developing Theoretical Foundation

Over the last three decades, the literature applying theory to the field of physical activity has grown substantially (Dishman, 1994; King et al., 2002). The most popular theories driving the field of physical activity to date have been the theory of reasoned action/planned behavior (Ajzen, 1988, 1991; Hausenblas, Carron, & Mack, 1997), the transtheoretical model of behavior change (Prochaska & Marcus, 1994; Prochaska & DiClemente, 1983), social cognitive theory (Bandura, 1986), and more recently self-determination theory (Chatzisarantis & Biddle, 1998; Deci & Ryan, 1985) and the health action process approach (Schwarzer, 1999).

These theories can be placed in larger conceptual categories of belief-attitude, competence-based, control-based, and decision-making approaches (Biddle & Nigg, 2000). This categorization has evolved to belief-attitude, competence-based, control-based, stage-based, and hybrid models (Biddle et al., 2007; chapter 8). The conceptual categories make it evident that these approaches stem from understanding an individual's psychology, targeting motivation, intentions, and behavior. This is undertaken either within the psychological context by itself or within the social psychological context, the immediate or microlevel social environment (Biddle & Nigg, 2000; King et al., 2002). Each approach has directed the attention of the field to some very useful determinants and has provided guidance on how to develop programs and interventions (see preceding section on correlates of youth activity and inactivity).

Specific models and theories translated into the field of physical activity are described in various chapters of this book. For example, although the theory of planned behavior (TPB; Ajzen, 1991) is popular in adult physical activity work, it has been underutilized in examinations of youth physical activity and has not been used to address youth sedentary behavior at all (see chapter 7). Existing results (Craig, Goldberg, & Dietz, 1996; Trost, Saunders, & Ward, 2002; Trost et al., 2002) provide promising evidence for use of the TPB as a model for child and adolescent physical activity interventions, but additions to the TPB or other approaches seem to be necessary with this population (e.g., enjoyment and environmental considerations). Only a small amount of variance in physical activity is accounted for by TPB, and similar results have been shown with other theories. Although there is some evidence that these theories account for a meaningful proportion of variance, there is much more to be explained.

A Health Behavior Change Model

In recent years only one conceptual model has been proposed to increase physical activity among youth: Welk's (1999) Youth Physical Activity Promotion Model. Based on an ecological framework (Green and Kreuter's 1991 Precede-Proceed Model), Welk's model suggests that multiple levels of the environment (e.g., institutional, physical, cultural, social) can directly and indirectly influence behavior. In addition, Welk's model provides a conceptual framework for determining how youth become predisposed to physical activity and how physical activity is enabled and reinforced.

Predisposing factors, which increase the likelihood that youth will be regularly active, include factors such as self-efficacy, perceived competence, enjoyment, beliefs, and attitudes (Welk, 1999). Enabling factors include both biological and environmental elements that allow youth to be physically active. Examples of enabling factors include fitness, access to facilities, skill, and environmental supports for physical activity. Reinforcing factors increase the likelihood that youth will increase and maintain physical activity, and they influence behavior directly and indirectly. Such factors include family, peer, teacher, and coach influence (Kimiecik & Horn, 1998). Welk's model provides a bottom-up framework according to which demographics (e.g., age, gender, race, social-economic status) are considered prior to the establishment of a program. This framework takes into account a given population's specific characteristics and needs. Despite the holistic nature of Welk's model, there are no studies to date that apply its principles. Although the study of sedentary behavior has recently been brought to the forefront (Robinson, 1999), there are no theories or conceptual frameworks specifically designed to address sedentary behaviors. As noted earlier, further work on determinants of inactivity is also needed in order to generate such theory.

A Review of Theory-Based Interventions

We identified several physical activity interventions among children or adolescents. Synopses and results of the identified interventions are reported in table 4.1. We used two criteria for selecting an intervention study: The study had to have been published following the review by Stone and colleagues (1998) and to have been theory based. The majority of the studies were conducted in the school environment, with the remaining conducted in low-income areas, summer camps, subjects' households, and the primary care setting. Each intervention was designed either to increase MVPA, decrease inactivity, or reduce cardiovascular risk factors such as obesity. The most commonly cited theories driving the interventions were the social cognitive theory (Bandura, 1986) and the social ecological model (Stokols, 1996).

The school or after-school environment is particularly popular as a setting for the implementation of interventions because schools are cost-effective and efficient vehicles for providing physical activity instruction and programs that reach a large number of children and adolescents (CDC, 1997; Faucette et al., 1995). Youth who are not engaging in these activities in the school environment may also miss out on these opportunities during their leisure time or at home (Dale, Corbin, & Dale, 2000). Such activities and programs facilitate the skills and knowledge necessary to support an active and healthy lifestyle. Interventions within the school environment have been effective in enhancing students' physical activity–related knowledge (Arbeit et al., 1992; Bush et al., 1989a, 1989b), attitudes (Prokhorov et al., 1993), and level of physical fitness (Kelder, Perry, & Klepp, 1993; Kelder et al., 1995).

Table 4.1 Theory-Based Physical Activity Interventions Among Children and Adolescents

Author and year Experimental design Setting	Intervention name Theoretical components Measures Follow-up time	Sample size Demographic
Baranowski et al. (2003a) Randomized controlled trial Summer camp and Internet programs	Name: The Fun, Food, and Fitness project (Baylor GEMS pilot study)—designed to increase fruit and vegetable (FV) consumption by replacing dietary fat with FV, increase water intake, and increase moderate or vigorous physical activity (PA) to 60 min per day. Girls were trained to set goals, social support was advised, and fun activities were encouraged. Theory: social cognitive theory (SCT) Measures: body mass index (BMI), PA, sexual maturation, diet Follow-up: 12 weeks	73 8-year-old Black girls and their parents (n = 82)
Caballero et al. (2003); Stevens et al. (2003) For others see *Preventive Medicine, 37,* Suppl. I Randomized controlled trial Elementary schools	Name: Pathways—a multicomponent intervention for reducing body fat in American Indian children. Intervention components consisted of classroom curriculum, food service, physical education, and family involvement. Theory: social learning theory and relevant cultural practices (e.g., storytelling) Measures: body composition; PA; PA-related behaviors, attitudes, and knowledge; dietary intake Follow-up: 3 years	1704 American Indian children
Fitzgibbon et al. (2005) Randomized controlled trial Elementary schools	Name: Hip-Hop to Health Jr.—designed to promote healthy eating and PA of children enrolled in Head Start. Theory: SCT, self-determination theory, and the transtheoretical model that incorporates the stages of change Measures: dietary intake, BMI, and PA Follow-up: 2 years	420 primarily African American children (mean age = 4 years; parent mean age = 30)
Ford et al. (2002) Randomized controlled trial Low-income urban community clinic	Name: No name; an intervention to reduce television viewing among low-income urban African American children. Theory: SCT Measures: BMI ≥85th percentile; number of televisions in the home; number of families with television in child's bedroom; number of VCRs in home; number with video game player hooked to a television; hours children spent watching television or videotape and in video game play; overall household television use; days having breakfast with television on; days having dinner with television on; hours playing outside; hours of organized PA Follow-up: 4 weeks	28 families with 7- to 12-year-old children

RESULTS		
Evaluation presented (Yes/No)	**Summary**	**Litmus test to evaluate theory**
Yes, but reported elsewhere	BMI at the end of the summer camp did not vary between conditions; lower calories from fat and beverages reported by intervention group. No changes in PA were reported.	1. Yes 2. No, not reported 3. Yes 4. Yes 5. No, not reported 6. Yes 7. No, not reported 8. No
Yes	There were no significant differences between groups for body composition. Dietary intake was lower among intervention students. No significant differences were observed between the two groups for the PA motion sensor, but self-report revealed higher reported PA among intervention students. Knowledge and self-efficacy for PA increased in the intervention group.	1. Yes 2. No 3. No 4. No 5. No 6. Yes 7. No 8. No
No	At year 1 and year 2 postintervention, the increase in BMI was greater in control students relative to intervention students. No significant differences were observed postintervention for TV viewing or for exercise frequency or intensity.	1. Yes 2. No 3. No 4. No 5. No 6. No, not measured 7. No, not reported 8. Yes
No	Both groups reported differences in television/video watching, playing video games, and total household television use. The intervention group reported a statistically significant increase in organized game play and a nearly significant increase in outside activity. There was also evidence for a decrease in overall family television use and meals eaten in front of the television in the intervention group, although nonsignificant.	1. Yes 2. No 3. No, not reported 4. No, not reported 5. No, not reported 6. No, no independent variables 7. No mediators measured 8. Yes, but for both groups

(continued)

Table 4.1 *(continued)*

Author and year Experimental design Setting	Intervention name Theoretical components Measures Follow-up time	Sample size Demographic
Gortmaker et al. (1999a) Quasi-experimental field trial with matched control Public elementary schools	Name: Eat Well and Keep Moving—an intervention designed to enhance cognitive and behavioral skills and focused on four behavioral outcomes: reduced fat intake and TV viewing, increased PA and FV consumption. Theory: SCT and school-level change theories Measures: dietary intake, PA, TV viewing, FV intake, and dietary and PA knowledge Follow-up: 2 years	479 predominantly African American students
Gortmaker et al. (1999b) Randomized controlled trial Public middle schools	Name: Planet Health—designed to reduce obesity by reducing TV viewing and high-fat food and increasing moderate and vigorous PA and FV consumption. Theory: behavior choice and SCT Measures: BMI, TV viewing, FV intake, and PA Follow-up: 2 years	1560 middle school students
Jamner et al. (2004) Quasi-experimental trial with matched control High school physical education (PE) classes	Name: Project FAB—an intervention designed to modify variables related to PA such as enjoyment, self-efficacy, benefits, barriers, and social support. Theory: behavioral modification (e.g., goal setting, self-monitoring, problem solving) Measures: $\dot{V}O_2$max, body composition, BMI, PA, lifestyle activities, self-efficacy, barriers, social support, enjoyment Follow-up: 4 months	58 sedentary adolescent females in grades 10 through 11
Kelder et al. (2003) Randomized controlled trial with delayed intervention for former control and recruitment of new control PE classes	Name: CATCH-PE and CATCH-ON—took place 5 years after the completion of the original CATCH intervention; 56 former intervention (FI), 20 former control (FC), and 12 unexposed control (UC) schools participated. Theory: SCT and organizational change (OC) Measures: SOFIT and in-depth interviews Follow-up: 1 year	645 3rd- to 5th-grade classes

RESULTS		
Evaluation presented (Yes/No)	**Summary**	**Litmus test to evaluate theory**
No	Dietary fat was reduced and FV consumption was increased among students within the intervention group relative to the control group. The reduction in TV viewing was lower in the intervention group but the difference was not statistically significant. No differences were observed for PA.	1. Yes 2. No 3. No 4. No 5. No 6. Yes 7. No 8. Yes for dietary fat and FV intake
Yes	The prevalence of obesity in girls increased for control schools but decreased for intervention schools. Among boys, no significant difference was observed for obesity; hence, both declined at the same rate in the two groups. The intervention effect for obesity was larger for Black girls. TV viewing decreased in intervention girls and boys compared to control.	1. Yes 2. No 3. No 4. No 5. No 6. No psychosocial variables to compare 7. No, not reported 8. Yes, but only for girls
No	*Physical fitness:* $\dot{V}O_2max$ remained constant in the intervention group but declined in the control group. *Physical activity recall:* The intervention had a significant effect on light, moderate, and total activity. Intervention group increased total energy expenditure while the control group showed a decline. Those in the intervention group were seven times more likely to report hard activity compared to the control group. *Lifestyle activity:* Lifestyle activity was significantly increased in the intervention group but not the control group. *Psychosocial variables:* No effect was observed.	1. Yes 2. No 3. No 4. No 5. No 6. Yes 7. No 8. Yes
Yes	Students in FI schools spent more time in moderate-to vigorous PA and vigorous PA, but these values were not significantly different from those in the FC or UC schools. FI and FC schools spent more time on general knowledge and skills, while UC schools spent more time on game and free play. More FI school teachers reported having the CATCH-PE materials and curriculum compared to FC teachers.	1. Yes 2. No for SCT, but yes for OC 3. No, not reported 4. No, not reported 5. No, not reported 6. Yes 7. No, not reported 8. Yes

(continued)

Table 4.1 *(continued)*

Author and year Experimental design Setting	Intervention name Theoretical components Measures Follow-up time	Sample size Demographic
Nader et al. (1999) Randomized controlled trial 5th- to 8th-grade students	Name: CATCH III—designed to determine whether the 5th-grade intervention resulted in changes in eating and activity attitudes and behaviors at grade 8 of the CATCH II cohort (56 intervention and 40 control schools). Theory: SCT and OC Measures: 24 h diet recall, food checklist, PA checklist, health behavior, blood pressure, lipid and cholesterol levels Follow-up: 3-year follow-up of the original CATCH II cohort	3714 8th-grade students
Neumark-Sztainer et al. (2003) Randomized controlled trial School based	Name: New Moves—a school-based obesity prevention intervention for adolescent girls. Theory: SCT Measures: stage of change, PA, sedentary behavior, BMI, diet-related behaviors (soda intake, breakfast, fast food, weight control behaviors, binge eating), self-acceptance, athletic competence, physical appearance, self-worth, media internalization, exercise benefits, eating benefits, exercise enjoyment, self-efficacy, parental support, peer support, staff support Follow-up: 16 weeks	201 high school adolescent girls
Nigg et al. (2006); Battista et al. (2005) Quasi-experimental study Elementary after-school programs	Name: Fun 5—a PA and FV intervention implemented in elementary after-school programs in Hawaii. Fun 5 offered a variety of organized, noncompetitive, non-gender–specific and fun activities in which children of all skill levels can participate and experience success. Theory: structural ecological model, SCT, theory of planned behavior, stages of change Measures: school-based and leisure-time PA, FV consumption, enjoyment, self-efficacy, intentions, social norms, attitudes, PA stage, perceived behavioral control Follow-up: 1 year	Pilot: n = 533 (48% female) Year 1: n = 453 (54% female). All participants were public elementary school students
Pate et al. (2003) Quasi-experimental trial with matched control Rural communities	Name: Active Winners—designed to increase PA and increase the hypothesized determinants of PA. Active Winners consisted of four parts: Active Home, Active School, Active Kids, and Active Community. Theory: SCT and Pender's health promotion model Measures: PA, self-efficacy, beliefs regarding PA, social influences on PA, intentions to be physically active Follow-up: 18 months	558 predominantly Black middle school students

RESULTS		
Evaluation presented (Yes/No)	**Summary**	**Litmus test to evaluate theory**
Yes	At the end of the trial, intervention students reported significantly lower energy intakes compared to control students. Intervention students reported more minutes of PA compared to control students (30.2 vs. 22.1 min). Health knowledge and healthy food choices were higher in the intervention group. No differences were observed for physiological measures.	1. Yes 2. No, not reported 3. No, not reported 4. No, not reported 5. No, not reported 6. Yes 7. No, not reported 8. Yes
Yes	Program participants rated the study favorably. The majority of the outcome variables did not significantly differ between groups. At postintervention, 31% of intervention students progressed in stage compared to 20% of control group participants, while 19% and 24% regressed in stage by postintervention. At follow-up no change in stage increase was observed among control students, while 38% overall in the intervention group progressed.	1. Yes 2. No 3. No 4. No 5. No 6. Yes 7. No 8. Yes, but only PA stage of change
Yes	Overall, there was a 21% decrease in time spent standing, sitting, and lying down and a 140% increase in moderate and vigorous PA during the after-school program. Results on self-reported leisure-time activity revealed a significant increase in moderate PA over the course of the program. During its pilot phase, Fun 5 did not appear to affect FV consumption. One year after the initial pilot study, Fun 5 resulted in an increase in moderate and vigorous PA and in FV consumption. No changes in psychosocial variables were observed 1 year after implementation (unpublished).	1. No, not reported 2. No 3. No 4. No 5. No 6. No 7. No 8. Yes
Yes	No significant differences were observed in moderate to vigorous PA or psychosocial variables between intervention and control conditions over time. The control group usually reported higher mean scores compared to the intervention group in both boys and girls for psychosocial variables.	1. Yes 2. No 3. No, not reported 4. No 5. No 6. Yes 7. No 8. No

(continued)

Table 4.1 *(continued)*

Author and year Experimental design Setting	Intervention name Theoretical components Measures Follow-up time	Sample size Demographic
Pate et al. (2005); Felton et al. (2005) Randomized cohort design High schools	Name: LEAP—designed to promote PA by changing the instructional practices in the school environment. LEAP, designed to enhance self-efficacy and enjoyment, consisted of two components: LEAP PE and LEAP education. Theory: social ecological model and the coordinated school health approach Measures: PA, BMI, self-efficacy, enjoyment Follow-up: 2 years, but this study provides only 1-year results.	2111 Black and White girls
Resnicow et al. (2000) Quasi-experimental (no control) Public housing units	Name: GO GIRLS—designed to increase FV consumption and PA behavior, reduce TV viewing, and decrease dietary fat, as well as to enhance skills, efficacy, and outcome expectations. Theory: SCT Measures: BMI, outcome expectation, social support, self-efficacy, health knowledge, perceived weight, PA Follow-up: 6 months	57 Black adolescent females
Robinson (1999) Randomized controlled trial Elementary schools	Name: no name; designed to reduce television, videotape, and video game use, changes in adiposity, PA, and dietary intake. Theory: SCT Measures: BMI; tricep skinfold thickness; waist hip circumferences; waist-to-hip ratio; time watching TV, watching movies or videos on a VCR, and playing video games (before and after school, yesterday, and last Saturday). (Children's television/video viewing and game playing were validated by estimated parental reports. Children and parents estimated time spent in sedentary activity [homework, reading, computer use, listening to music, playing instruments, etc.].) Previous day out-of-school activity, organized and nonorganized game play, 1-day food frequency recalls, 20 m shuttle run test, and meals in front of the TV. Follow-up: 7 months	198 4th-grade public elementary students

RESULTS		
Evaluation presented (Yes/No)	**Summary**	**Litmus test to evaluate theory**
Yes	The prevalence of regular vigorous PA was greater in intervention schools than control schools. Eighty percent of girls in both conditions were enrolled in PE classes as 9th-grade students. Previous studies provided evidence that the LEAP intervention influenced psychosocial variables (Dishman et al., 2004).	1. Yes 2. No 3. No 4. No 5. No 6. Yes 7. Not known 8. Yes
No	High attendees reported more social support for diet and exercise changes. No significant differences in physiological outcomes or behavioral measures.	1. Yes 2. No, not reported 3. No 4. No 5. No 6. Yes 7. No 8. No
Yes	BMI, skinfold thickness, waist circumference, and waist-to-hip ratio increased in both groups as expected, but this increase was significantly lower for those in the intervention. A significant difference in TV viewing and video game playing was observed among intervention students compared to controls. A significant reduction in eating in front of the TV was observed. No significant differences were observed for PA levels or 20 m shuttle run test. No significant sex differences were observed.	1. Yes 2. No 3. No 4. No 5. No 6. No, not reported 7. No, not reported 8. Yes

(continued)

Table 4.1 *(continued)*

Author and year Experimental design Setting	Intervention name Theoretical components Measures Follow-up time	Sample size Demographic
Robinson et al. (2003) Two-arm parallel group randomized trial Low-income neighborhoods	Name: Stanford GEMS Pilot Study—an after-school dance and family-based intervention to reduce obesity and sedentary behavior. Theory: SCT (i.e., attention, retention, production, and motivation) Measures: BMI, waist circumstance, sexual maturation, blood plasma, reported media use, TV viewing, eating while watching TV, 24 h dietary recalls, PA accelerometer, overconcerns with weight and shape, self-esteem Follow-up: 12 weeks	61 8- to 10-year-old African American girls
Roemmich et al. (2004) Randomized controlled trial Family based	Name: no name; evaluated the effect of open-loop feedback and reinforcement on PA and TV time in a sample of sedentary youth. Theory: Premack's theory of reinforcement Measures: BMI, objective daily PA, and a habit book used to record time spent in sedentary behaviors (TV time, recreational computer use, handheld video game play, reading, and telephone time) Follow-up: 6 weeks	21 families with children ages 8 to 12 years
Saelens et al. (2002) Randomized controlled trial Primary care setting	Name: Healthy Habits—a behaviorally based weight control intervention for overweight adolescents initiated in primary care. Theory: behavioral modification Measures: BMI, 2-day dietary recalls, 7-day PA recall, 7-day sedentary behavior self-report, problematic eating and weight-related behaviors and beliefs, physician counseling, behavioral skills use, and participant satisfaction Follow-up: 4 months	44 overweight adolescents aged 12 to 16 years

RESULTS		
Evaluation presented (Yes/No)	**Summary**	**Litmus test to evaluate theory**
Yes	Most of the results of the study were nonsignificant; however, evidence of a 7% (91 counts/min) increase in PA counts was observed in the intervention group relative to the control group. Significant differences were observed in total household TV use (20%) and eating dinner with TV on (10%). The treatment group reported 20% fewer hours of TV, videotape, and video game use and 10% fewer meals eaten with the TV on. A statistically significant decrease in the treatment group was observed for overconcerns with weight and shape.	1. Yes 2. No, but the four components are explained 3. No 4. No 5. No 6. Yes 7. No 8. Yes, but only household TV use
No	There were no group differences for BMI over the course of the study. The open-loop feedback group increased PA by 24%, which was a greater increase than in the control group. Although TV time was not significantly different, subjects in the open-loop feedback group reduced TV time by 20 min per day while the control group increased TV time by 13 min.	1. Yes 2. No 3. No 4. No, not reported 5. No 6. No independent variables measured 7. No mediators measured 8. Yes
No	At posttreatment, there was a statistically significant difference in BMI over time between the healthy habits group and the typical care or control group. More healthy habits adolescents reduced their BMI when compared to the typical care adolescents (40% vs. 10.5%). No significant differences between conditions were observed at posttreatment for dietary fat intake, PA, sedentary behavior, or problematic eating- and weight-related behaviors or beliefs. At follow-up, the treatment-by-time interaction for BMI remained significant; however, linear contrasts for BMI from posttreatment to follow-up revealed no differential change in BMI scores. From baseline to follow-up, more healthy habits adolescents had decreased BMI scores from baseline values than typical care adolescents (55.6% vs. 15.8%). No significant interactions by conditions were observed among the remaining variables. Healthy habits participants reported higher rates of eating- and PA-specific behavioral skills compared to typical care students.	1. Yes 2. Yes 3. Yes 4. Yes 5. No 6. Yes, measured 7. Yes, use of skills changed 8. Yes, but only BMI

(continued)

Table 4.1 *(continued)*

Author and year Experimental design Setting	Intervention name Theoretical components Measures Follow-up time	Sample size Demographic
Sallis et al. (2003) Randomized controlled trial Public middle schools	Name: M-SPAN—a PA and nutrition intervention implemented in middle schools with the intent to influence the school environment and policies. Theory: structural ecologic model Measures: SOFIT, System for Observing Play and Leisure Activity in Youth (SOPLAY), PA, dietary fat Follow-up: 2 years	24 public middle schools, 1109 participants
Warren et al. (2003) Four-arm randomized controlled trial Oxford primary schools	Name: Be Smart—a school- and family-based intervention to prevent obesity of children in three primary schools in Oxford, U.K., randomized into four conditions (Be Smart [control], Eat Smart, Play Smart, and Eat Smart Play Smart). Theory: SCT Measures: anthropometry (BMI, skinfold thickness, and circumference from the waist, hip, upper arm, and head), nutrition knowledge, PA (transportation to school, playground activity, and lunchtime activities), 24 h recall, parental questionnaires (PA, nutrition knowledge, and social and medical history) Follow-up: 14 months	213 children ages 5 to 7 years selected from three primary schools

School Based

The interventions we identified that were conducted in schools often infused science-based health education within normal physical education classes and encouraged teachers to reduce management time (e.g., SPARK, CATCH, LEAP, Project FAB). Additional strategies included providing parents with newsletters regarding the program to build support for the students involved and encouraging the reduction of television viewing (e.g., HIP-HOP to Health Jr.). M-SPAN in particular promoted activity through media messages on bulletin boards throughout the school and encouraged structured activity before, during, and after school (Sallis et al., 2003).

Despite the level and magnitude of the interventions, several resulted in little or no change in physical activity. For example, no intervention effects were observed for activity or inactivity among the Baylor GEMS, Eat Well

RESULTS		
Evaluation presented (Yes/No)	Summary	Litmus test to evaluate theory
No	Intervention schools increased PA at a greater rate than control schools. Gender-specific results revealed that the increase was significant for boys but not girls. Boys increased PA in both PE and leisure time, while girls increased PA only during PE. No significant differences were observed for fat intake between groups.	1. Yes 2. No 3. Yes, but not described fully 4. Yes, but not described fully 5. No 6. Yes 7. No, not reported 8. Yes
Yes	There were no significant differences in overweight from baseline to follow-up. Nutrition knowledge increased in all groups. There was a small increase in number of children walking to school. An increase in activity in playground activities at morning break was reported in all groups, but was higher in all intervention groups. Overall there was a significant increase in FV consumption; however, this increase was even higher for the Be Smart and Eat Smart groups. No significant differences were observed in 24 h recalls between groups. Outcome evaluation suggested that children in the Eat Smart group scored significantly higher in nutrition knowledge compared to children in the Be Smart and Play Smart groups. No significant sex differences were observed.	1. Yes 2. No 3. No 4. No 5. No 6. Yes 7. No, no PA/inactivity mediators measured 8. Yes

and Keep Moving, and HIP-HOP to Health Jr. interventions. Conversely, interventions such as CATCH-PE, CATCH-ON, Project FAB, M-SPAN, Pathways, Be Smart, and Fun 5 resulted in positive behavioral outcomes. It is important to note that few studies reported changes in psychosocial variables as a result of the intervention. Interventions that successfully influence proposed mediators are more likely to have successful behavioral outcomes (Baranowski et al., 2003b). More research is needed in this area to determine the efficacy of school-based interventions.

Clinically Based

Clinically based physical activity and weight control interventions among children and adolescents have provided some evidence of long-term efficacy (Epstein et al., 1998). Pediatric primary care settings are ideal environments for delivering theory-based research approaches to target manageable health risk behaviors, given the number of children and adolescents who see a doctor over the course of any given year (Epstein et al., 1998). Despite the practicality of the avenue, data are lacking on interventions in this area. Interventions in the primary care setting (Ford et al., 2002; Saelens et al., 2002) were plagued by small sample sizes and suffered from a lack of statistical power.

The Healthy Habits intervention was a multicomponent behavioral intervention for weight control among overweight adolescents in the primary care setting. The Healthy Habits intervention was based on concepts of behavioral modification (e.g., goal setting, problem solving, self-monitoring). Behavioral skills use among adolescents in the experimental condition was higher than in the typical care condition. In addition, the intervention resulted in an overall decrease in body mass index (BMI) among all participants (Saelens et al., 2002). Ford and colleagues (2002) conducted an intervention in a low-income urban community clinic in Atlanta, Georgia. The intervention was based on the social cognitive theory, with families receiving a brief counseling session. Decreases in children's television, videotape, and video game use were observed in the intervention condition. In addition, an increase in physical activity was observed among those in the behavioral intervention group.

A number of studies that were not reviewed because they were not theory based (e.g., Dennison et al., 2004; Faith et al., 2001; Simon et al., 2004) resulted in significant changes in physical activity, sedentary behavior, or obesity. Recent reviews have focused on interventions that target sedentary behavior and obesity among children and adolescents (DeMattia et al., 2006; Sharma, 2006). Although some of the approaches varied, there was evidence that programs targeting sedentary behaviors and obesity are quite effective (DeMattia et al., 2006). Sharma noted that television watching seems to be the most modifiable behavior, followed by physical activity. Most of the interventions we identified focused only on individual behavior change approaches that emphasize short-term changes. More longitudinal interventions are needed that include long-term follow-up data.

Critical Evaluation of Applied Theory

Current interventions often focus on psychosocial models within schools or community settings and have had modest success. However, limitations of these efforts are apparent. Few researchers have measured changes in the constructs of the theory inspiring their intervention; or, when they have done so, they measured only one construct to capture the totality of the theory. For example, researchers examining the effectiveness of school-based interventions grounded in the social cognitive theory commonly measure only the construct of self-efficacy, neglecting other components such as outcome expectations, reinforcement, and goal setting. In order to improve both theory and intervention, we must consider all concepts and constructs within a theory to find out what components are really working (Sharma, 2006). In addition, some consensus is needed in the field regarding appropriate, psychometrically sound instruments for specific populations or environments.

Studies often produce weak results, if any at all, or may affect only one subset of the intended population. These problems often arise because researchers are not conducting preliminary research (e.g., qualitative inquiry, elicitation studies) in an effort to understand perspectives of the target population; this results in poor selection of the procedures used to influence potential mediators (e.g., psychosocial variables) in desired directions (Baranowski, Anderson, & Carmack, 1998). Similarly, selecting a theory for an intervention that is irrelevant to behavior change in a certain population will lead to nonsignificant findings. Hence, it is important to know relevant moderators (e.g., age, gender, and ethnicity) of one's intervention as well as the key mediators that influence behavioral change; this knowledge has a direct impact on the efficacy of an intervention (Baranowski & Jago, 2005).

In one example of evaluation of a theory-based intervention, investigators explored the intervening effects of the LEAP program on proposed social cognitive mediators (i.e., self-efficacy, goal setting, satisfaction, and outcome expectancy) in an effort to increase self-reported physical activity (Dishman et al., 2004; Pate et al., 2005). The intervention successfully influenced self-efficacy, goal setting, and self-reported physical activity. In addition, self-efficacy partially mediated the effects of the intervention on physical activity, although no mediating effects were observed for goal setting, satisfaction, or outcome expectancy (Dishman et al., 2004). LEAP may have been effective because the mediating variables were related to physical activity and effective intervention strategies were in place to influence the variables in the desired direction.

Extensive process evaluations are needed to clearly assess how intervention components influence mediators. Baranowski and Stables (2000) outline a number of components relevant to successful process evaluations:

> *Recruitment* and *maintenance* of participants, *context* within which the program functions, *resources* available to the program and the participants, *implementation* of the program, *reach* of materials into (or receipt by) the target group, *barriers* to implementing the program, *initial use* of program activities, *continued use* of program-specified activities, and *contamination* of treatment and control groups. (p. 158)

Evaluating these components provides clear inferences about the effectiveness of the intervention, allowing one to determine successes and barriers to implementation. The implementation of process evaluation within youth physical activity research helps bridge the gap between research and practice, resulting in interventions that are transferable and that can produce significant effects on behavioral outcomes.

Studies employing process or formative evaluations that are described in table 4.1 include Active Winners, Pathways, CATCH, Fun 5, New Moves, LEAP,

Planet Health, Stanford GEMS, and Be Smart. The major limitation identified by Active Winners (Pate et al., 2003) was a lack of full implementation of all program elements. Relative to Pathways (Caballero et al., 2003; Stevens et al., 2003), though a positive intervention effect for physical activity was identified, the authors failed to relate changes in the psychosocial variables to elements of the evaluation. Baranowski and Jago (2005) also suggested that the family component of Pathways (Caballero et al., 2003; Stevens et al., 2003) might have had a limited reach. One of the more successful interventions implemented in recent years was CATCH (Kelder et al., 2003; Nader et al., 1999). Process measures indicated that the study was implemented as planned, but no associations among outcomes and psychosocial variables were ever reported. One year after its initial pilot study, Fun 5 (Battista et al., 2005; Nigg et al., 2006) replicated an increase in moderate and vigorous physical activity. Evaluation of Fun 5 suggested that the majority of the site coordinators and group leaders enjoyed implementing Fun 5 and perceived that enjoyment and enthusiasm among the kids increased with intervention implementation. Such belief in and support for a program usually result in successful implementation, which increases the probability of desired outcomes. For process evaluation information on the remaining interventions, see the related articles.

A Litmus Test for Evaluating the Use of a Theory

As we begin to evaluate theory as applied in youth physical activity interventions, it would be valuable to generate an evaluation protocol that outlines key considerations. Thus we offer the following "litmus test" comprising the factors to consider when one is evaluating the use of theory in this research area:

1. Is a theory identified? A theory may be explicitly stated or presented via a logic model. The chances are that if no theory is identified, the approach is not theory based or is based only on one or more parts of theories.

2. Is the theory described? If so, this suggests that the individuals doing the intervention understand how and why behavior change should come about. Lip service is provided too commonly in the literature: That is, a theory is named (e.g., health belief model, protection motivation theory), but no description or a limited description is given. This usually leads to problems with the next issue.

3. Are all components of the theory translated into the intervention or the components thereof? One should be able to discern what part of the intervention maps to the posited theoretical mediators. For illustrations of translating theory to intervention, we guide the reader to a special issue of *Health Education Research* (Nigg, Allegrante, & Ory, 2002).

4. Is there evidence that all of the intervention components were implemented? Such evidence is usually presented as part of the process evaluation. The process evaluation should provide information on the fidelity of the treatment or intervention for all the intervention components.

5. Are the components of the theory assessed? Frequently the major variable of a theory is assessed and then conclusions are made about its effectiveness or usefulness (e.g., self-efficacy—social cognitive theory; stages of change—transtheoretical model). Two major categories of variables should usually be evident:

- Moderators, which are preexisting conditions that influence the effectiveness of an intervention—by definition, moderator variables either exist prior to the intervention or program or are quantified at baseline
- Mediators, which are the mechanisms through which an intervention is expected to work—for a variable to qualify as a mediator, changes must occur *during* an intervention (Kraemer et al., 2002)

6. Are the theory variables and the outcome congruent? An intervention addressing the theory of planned behavior variables for overall physical activity may not be effective (may be too broad) to increase one specific behavior (e.g., swimming).

7. Did the mediators change during the intervention? This must occur in order to allow the claim that any change in the outcome is due to the mediator. If the mediators do not change, the reason may be faulty theory, and the validity of the theory being used needs to be questioned. For example, an intervention that uses dramatic portrayals of heart attacks to motivate adolescents to be physically active but does not elicit a change in perceived severity, which is thought to lead to protective behavior (physical activity), might not be appropriate for this population. As another example, a participant may not interpret the intervention as expected. The intervention may have targeted social support (getting one's parents to go for a walk), but the participant may have interpreted it as removing a barrier (the barrier being not to go for a walk because the family wants to spend time together). A process evaluation usually does not address this issue. One way to obtain this information is to have qualitative or structured interviews with the participants once the program or intervention is over.

We caution, however, that if the mediators do not change during the intervention, the lack of change may also be due to a failure in the research. There may be several reasons for this. For one, the intervention may not have addressed the mediators appropriately. For example, the interventionist is trained to talk about obtaining social support, but does not talk about the sources of such support, how to ask for it, or how to capitalize on it. Another possible reason is that the participant did not attend to or understand what to do. The researcher may have created excellent expert reports that addressed all the theoretical components and sent them to the participant's home, but the participant did not look at the report and threw it away. Finally, measurement issues may be the culprit in that the instrument assessing the mediator may not have been sensitive enough to detect a change, the timing of the assessment may have been too late to detect the impact, or some kind of bias (e.g., social desirability) may have influenced the measurement.

8. Did the outcome change? This is (obviously) a necessary condition for any conclusions to be drawn that the intervention is indeed effective. In some instances the outcome changes, but the mediators do not. This can indicate that the theory does not address the right mechanisms. There are also cases in which the mediators change but the outcome does not. Here a careful examination is required before the theory is dismissed, as the measurement issues previously described relative to mediators also apply to outcomes.

This "litmus test," presented in abbreviated form in figure 4.1, not only is informative when one is evaluating interventions, but also is a useful guide for planning a program or intervention. The more affirmatively this set of questions is answered, the more likely it is that the theory has been used to best effect. Being aware of these items when planning programs or interventions will ensure a more thorough approach, decrease the likelihood of missing important components, and provide a more informative evaluation.

Applying the Litmus Test to Interventions

On the basis of the litmus test just presented, we evaluated 20 theory-based physical activity interventions in children and adolescents (see table 4.1 for study descriptions). None of the studies that we identified met all requirements of the proposed litmus test, suggesting that most youth physical activity intervention

1) Is a theory identified?

2) Is the (entire) theory described?

3) Are all of the theory components translated into the intervention?

4) Are all of the intervention components implemented?

5) Are all of the theory components assessed?

6) Are the theory variables and the outcome congruent?

7) Did the mediators change during the intervention?

8) Did the outcome change?

Greater affirmation in set of responses ⟶ more appropriate use of theory

Figure 4.1 This abbreviated version of the litmus test can serve as a useful checklist for readers planning their own research or evaluating that of others.

research is not totally theory driven but may better be characterized as theory inspired. Most studies identified a specific theory as driving the intervention, with the exception of Fun 5 (Battista et al., 2005); however, we know that Fun 5 is based on a structural ecological model (Cohen, Scribner, & Farley, 2000). The bases of structural ecological models are to change behavior beyond the individual level and influence the social and physical environment in which the interventions are placed. Structural ecological models extend typical ecological models of health behavior change by specifying structures whereby population-level factors effect change in individual-level factors (Cohen et al., 2000). These interventions are usually conducted within existing structures, such as schools, communities, and organizations. Examples of other interventions that have applied such approaches include the Middle-School Physical Activity and Nutrition study (M-SPAN; Sallis et al., 2003) and the Sports, Play, and Active Recreation for Kids study (SPARK; Sallis et al., 1997).

With regard to items 2 through 5 of the litmus test, there was evidence that two studies met at least two of these requirements. These studies include the Baylor GEMS study and the Healthy Habits study (Baranowski et al., 2003a; Saelens et al., 2002). The Baylor GEMS study did not describe the entire theory (item 2) in detail or assess all theoretical components (item 5), but did translate and implement all components within the intervention (items 3 and 4). The Healthy Habits intervention (Saelens et al., 2002) involved several behavioral modification techniques including self-monitoring, goal setting, problem solving, stimulus control, self-reward, preplanning, and meeting requirements for items 2 and 3 of the litmus test. It also appears that all the intervention components were implemented within the intervention (item 4); however, Saelens and colleagues (2002) did not assess all components that were applied within the intervention (item 5).

Item 6 ("Are the theory variables and the outcome congruent?") was satisfied by most of the interventions reviewed, with the exception of those providing only follow-up data on behavioral measures (e.g., physical activity, obesity, sedentary behavior) (Battista et al., 2005; Fitzgibbon et al., 2005; Ford et al., 2002; Gortmaker et al., 1999b; Nigg et al., 2006; Robinson, 1999; Roemmich, Gurgol, & Epstein, 2004). Researchers less frequently report results related to the psychosocial variables. The only two interventions that successfully affected mediators (item 7) were Healthy Habits (Saelens et al., 2002) and LEAP (Dishman et al., 2004; Pate et al., 2003). Groups of studies were identified that successfully changed outcome measures (item 8) such as BMI or physical activity, including the CATCH studies (Kelder et al., 2003; Nader et al., 1999), Project FAB (Jamner et al., 2004), M-SPAN (Sallis et al., 2003), LEAP (Pate et al., 2005), Planet Health (Gortmaker et al, 1999b), Hip-Hop to Health Jr. (Fitzgibbon et al., 2005), Fun 5 (Battista et al., 2005; Nigg et al., 2006), Stanford GEMS (Robinson et al., 2003), Be Smart (Warren et al., 2003), Healthy Habits (Roemmich et al., 2004), and two other interventions (Ford et al., 2002; Robinson, 1999).

Overall, we notice that interventions designed to change activity or inactivity are producing significant results. However, a consistent trend in the interventions is that mediators are not changing. In addition, researchers are not adequately describing intervention components, evaluating entire theories, or measuring important theoretical components. More focus, attention, and research are needed in this area.

Improving Our Theoretical Understanding

A number of theory-based interventions focus on changing current levels of physical activity, but few if any address sedentary behavior. It has yet to be determined if the theories and models that are useful for physical activity promotion are also applicable to sedentary behavior. An important theoretical distinction between utilitarian (nonleisure) and disposable (leisure) inactive time needs to be made (Buckworth & Nigg, 2004). When attempting to decrease sedentary behavior, we should not be affecting homework time (which may include time in front of a computer or screen) but rather should decrease inactivity that is devoted to leisure pursuits such as computer games or watching TV for entertainment (which also includes time in front of a computer or screen). The effectiveness of these investigations can be maximized if we increase our efforts to develop a holistic approach to our work.

A Holistic Approach

Our field is developing by amassing information that can be integrated into what we already know and by identifying areas that have not received adequate attention (such as theories of reducing sedentary behaviors). It would be useful to study multiple theories to empirically integrate their salient components in an effort to create a more complete or holistic theory of health behavior change (Nigg et al., 2002). However, before theorists can begin to integrate multiple theories, we must first examine and assess existing theoretical models within the context of our interventions. Although these approaches can increase participant burden given the lengthiness of data collection, without complete approaches we risk the chance of not appropriately evaluating our theoretical models (Baranowski et al., 2003b). The success of this endeavor depends on the construction and use of valid measures of a range of mediators and outcomes (Traub, 1994).

Another effort to establish a more holistic understanding of youth physical activity is grounded within ecological conceptual frameworks. Individual psychology is just one element within ecological models and community interventions. This effort is reflected in the recognition of the importance of multilevel models and the impact of the environment in health promotion (e.g., McLeroy et al., 1988). Holistic approaches include the promotion of trails, pathways, and parks as means of increasing physical activity, decreasing inactivity, and reducing sedentary behavior in youth. Trails and pathways designed and created in local communities can be easily connected to local parks; open space areas; and school, library, and other public lands. The adoption of the Safe Routes to School programs encourages the connection of communities to schools. These programs promote safe and convenient exercise among children and adolescents. If we begin to implement these programs on a broader level, more children will be able to walk or ride their bikes to school, potentially decreasing inactivity and reducing sedentary behavior (Ege & Krag, 1999; Sallis, Bauman, & Pratt, 1998). Moreover, bicycles provide vital mobility for children; bicycles also represent fun, freedom, exercise, and fresh air—issues that matter to most children (Sallis et al., 1998).

The policy and environmental components from the social ecological approaches are understudied in the school environment (Sallis et al., 2003). However, they do appear to be effective approaches to changing the school environment and potentially transfer into the homes of many adolescents. Structural and social ecological approaches implemented at school encourage collaboration among various health professionals and individuals from other sectors of the population. Such sectors include principals, community agencies and organizations, legislators, and the mass media. However, implementing interventions based on ecological models (e.g., policy and environmental changes) in the school environment is not easy. These interventions often require a great deal of planning and collaboration with school officials, teachers, parents, and community leaders.

Finally, health behavior and physical activity theory has been discussed alongside health behavior and physical activity *change* theory without clear differentiation (Nigg & Jordan, 2005). There are clear distinctions between the two (e.g., Glanz & Rimer, 1997). Describing and understanding behavior are not the same as changing it. Behavior theories identify why a behavior exists. Behavior change theories explain why *and* how changes come about, and later guide the development of interventions. Of course, these two types of theories are symbiotic in nature, but we should not confuse their underlying ideologies.

Directions for Future Research

We propose two main research agendas:

- The appropriateness of theory for increasing physical activity and reducing inactivity should be further explored. Exploratory studies investigating why youth adopt physical activity and decrease inactivity are required immediately to allow us to judge whether the current theories address the correct constructs. Techniques from anthropology and sociology, along with qualitative methods, are recommended for the pursuit of this descriptive work. Depending on the outcomes of these efforts, it may be necessary to develop behavior-specific theories targeting physical activity or inactivity separately.

- The second agenda, related to the first, is theory testing using experimental designs to empirically provide understanding of the active components of increasing activity and decreasing inactivity. This may reveal that
    - existing theories and models are applicable;
    - revisions to existing theories and models are required; or
    - existing theories and models are not appropriate, and theories and models specific to youth activity, inactivity, or both need to be developed.

Jeffery (2004) argued that many theories fail to influence behavior because they focus on predictors of motivation and fail to address opportunities or capabilities to change. He then suggests returning to classic learning theories and emphasizing interactions between the person and her or his environment. Rothman (2004) suggested that theory often fails because current protocols that are in place to apply theoretical constructs within the framework of interventions are poorly developed, resulting in poor application of theory. This proposition was supported by Kremers and colleagues (2006). Our second proposal recommends expanding, refining, or rejecting existing theories based on intervention rather than observational studies (Brug, Oenema, & Ferreira, 2005; Rothman, 2004). Recommendations have also been made to address behavior change instead of motivation, and to focus more on hypothetical causal pathways rather than associations. Therefore, well-designed intervention studies of theoretically based interventions of behavior change are warranted.

The *International Journal of Behavioral Nutrition and Physical Activity* reported on several theoretical debates and related commentary (Baranowski, 2006; Brug, 2006; Resnicow & Vaughan, 2006). Perhaps a "chaotic view" of behavioral change should be considered. Resnicow and Vaughan propose that behavior change should be viewed as a chaotic system in which change is influenced by complex interactions and does not always necessarily follow a linear pattern. This perspective suggests that multiple interactions exist that may vary across individuals, and that random external and intrapsychic events can significantly affect the system. Gladwell (2000) agrees with Resnicow and Vaughan's proposed concept of a "tipping point," or a dramatic change in a person's behavior that is usually unexpected and arises quickly. They report that many decisions to change are not planned events but rather arise depending upon motivation. However, reactions from theorists have challenged the notion of chaos theory with a more ordered, linear approach to understanding behavior change (Brug, 2006).

In activity research, psychosocial, personality, and environmental variables are important determinants of behavior. If multiple factors are measured over time and the interactions among them are considered, mathematical algorithms of behavior can be created. Small changes in knowledge, efficacy, attitudes, social and physical environments, and other constructs may have a dramatic impact on a young person's motivation to be sedentary or active.

It is clear from research debate that there is little understanding of how to identify the "tipping point" of behavior change. However, as researchers we should do our best to understand and explain behavior change. Perhaps we should explore modeling techniques similar to those used in the molecular sciences (e.g., plant genetics). To succeed in this endeavor we also must improve our existing measurement methods. Baranowski (2006) proposes the use of Item Response Theory (Wilson, 2005) as a way to determine which variables are being poorly measured and, more specifically, in what demographic group(s). In addition, perhaps measurement models such as latent class analysis or latent transition analysis would be useful for examining the chaotic nature of health behavior; and growth mixture modeling could be used to capture the clustering of behaviors over time and to determine whether those clusters evolve over time (Muthen & Muthen, 2000).

Finally, we should assess the "big picture" of theory as applied to youth physical activity and sedentary behavior. In the fields of social and behavioral sciences, there is an overabundance of theories; and deciding which theory works and in what population is a significant challenge. Therefore, as researchers we should ask ourselves several questions:

- Are the current health behavior theories addressing behavior *change*?
- Are current health behavior theories too complicated?
- Should we consider revising, extending, integrating, or abandoning current theories?
- Are more complicated statistical designs needed to adequately test health behavior theories?
- Are there simple examples in nature that we can learn from?

Essentially, as Thomas Kuhn (1970) suggested, we are in the early stages of scientific development, or the preparadigm stage. We are in the preparadigm stage for research that relates to physical activity or inactivity because we are currently using theories from different fields to describe or interpret phenomena in our own field. Scientific knowledge develops slowly during these stages because there is often little agreement among scientists due to confusion, frustration, the defense of theory and research, and power struggles among factions within the discipline (Hardy, 1978). Nonetheless, despite the fact that we are in the early stages of scientific development, great strides have been made to identify determinants of physical activity in children and adolescents. We need much more high-quality research to better understand determinants and mechanisms of reducing inactivity and decreasing sedentary behaviors. If there is nothing as practical as a good theory, we must focus our efforts, intelligence, and creativity on bringing about the best theory.

APPLICATIONS FOR RESEARCHERS

The most general recommendation is that any physical activity and inactivity research addressing youth be grounded in theory. Care should be taken to use entire theories, not just individual constructs. Attention should also be paid to the appropriateness of the current theories and models in explaining physical activity and inactivity in youth. One area of inquiry that may need to be revisited is theory development and redevelopment in the physical activity or inactivity domains. New or hybrid theories unique to these specific behaviors may be necessary. We may also have to look at other disciplines (physics, biology, economics, etc.) for potentially salient frameworks. Both theory integration and theory comparison research are warranted at this time. Finally, one should evaluate any theory-based intervention research with the litmus test described in this chapter before concluding that the theory works or does not.

APPLICATIONS FOR PROFESSIONALS

Although the theoretical foundation for studying youth physical activity, inactivity, and (especially) sedentary behavior is not well developed, it is important that interventions be designed and implemented using the current theoretical understanding. The reason is that theory-based interventions tend to be more effective than non–theory-based interventions, and they provide guidance on how to intervene, evaluate, and explain why change occurs—all of which are essential for effective decision making. Based on our current understanding, these interventions should integrate the individual, social, and physical environments. Both physical activity levels and sedentary behaviors need to be targeted in youth because both are independently related to health; and current alarming chronic disease rates and quality of life issues appear to be related to not being physically active on a regular basis and engaging in sedentary leisure-time behaviors.

References

Ajzen, I. (1988). *Attitudes, personality and behavior.* Milton Keynes: Open University Press.

Ajzen, I. (1991). The theory of planned behavior. *Organizational Behavior and Human Decision Processes, 57*, 179-211.

Arbeit, M.L., Johnson, C.C., Mott, D.S., Harsha, D.W., Nicklas, T.A., Webber, L.S., et al. (1992). The Heart Smart cardiovascular school health promotion: Behavior correlates of risk factor change. *Preventive Medicine, 21*, 18-32.

Bandura, A. (1986). *Social foundations of thought and action: A social cognitive theory.* Englewood Cliffs, NJ: Prentice Hall.

Baranowski, T. (2006). Crisis and chaos in behavioral nutrition and physical activity. *International Journal of Behavioral Nutrition and Physical Activity, 3,* 27.

Baranowski, T., Anderson, C., & Carmack, C. (1998). Mediating variable framework in physical activity interventions: How are we doing? how might we do better? *American Journal of Preventive Medicine, 15,* 266-297.

Baranowski, T., Baranowski, J.C., Cullen, K.W., Thompson, D.I., Nichlas, T., Zakeri, I.F., et al. (2003a). The fun, food, and fitness project (FFFP): The Baylor GEMS pilot study. *Ethnicity and Disease, 13,* S1-30-S1-39.

Baranowski, T., Cullen, K.W., Nicklas, T., Thompson, D., & Baranowski, J. (2003b). Are current health behavioral change models helpful in guiding prevention of weight gain efforts? *Obesity Research, 11,* 23S-43S.

Baranowski, T., & Jago, R. (2005). Understanding the mechanisms of change in children's physical activity programs. *Exercise and Sport Sciences Reviews, 33,* 163-168.

Baranowski, T., & Stables, G. (2000). Process evaluations of the 5-a-day projects. *Health Education and Behavior, 27,* 157-166.

Battista, J., Nigg, C.R., Chang, J.A., Yamashita, M., & Chung, R. (2005). Elementary after school programs: An opportunity to promote physical activity for children. *California Journal of Health Promotion, 3,* 108-118.

Biddle, S.J.H., Hagger, M.S., Chatzisarantis, N.L.D., & Lippke, S. (2007). Theoretical frameworks in exercise psychology. In G. Tenenbaum & R.C. Eklund (Eds.), *Handbook of sport psychology* (3rd ed., pp. 537-559). Hoboken, NJ: Wiley.

Biddle, S.J.H., & Nigg, C.R. (2000). Theories of exercise behavior. *International Journal of Sport Psychology, 31,* 290-304.

Brug, J. (2006). Order is needed to promote linear or quantum changes in nutrition and physical activity behaviors: A reaction to "A chaotic view of behavior change" by Resnicow and Vaughan. *International Journal of Behavioral Nutrition and Physical Activity, 3,* 29.

Brug, J., Oenema, A., & Ferreira, I. (2005). Theory, evidence and intervention mapping to improve behavioral nutrition and physical activity interventions. *International Journal of Behavioral Nutrition and Physical Activity, 2,* 2.

Buckworth, J., & Nigg, C.R. (2004). Physical activity, exercise, and sedentary behavior in college students. *Journal of American College Health, 53,* 28-34.

Bush, P.J., Zuckerman, A.E., Taggart, V.S., Theiss, P.K., Peleg, E.O., & Smith, S.A. (1989a). Cardiovascular risk factor prevention in black school children: The "Know Your Body" evaluation project. *Health Education Quarterly, 16,* 215-227.

Bush, P.J., Zuckerman, A.E., Theiss, P.K., Taggart, V.S., Horowitz, C., Sheridan, M.J., et al. (1989b). Cardiovascular risk factor prevention in Black schoolchildren: Two-year results of the "Know Your Body" program. *American Journal of Epidemiology, 129,* 466-482.

Caballero, B., Clay, T., Davis, S.M., Ethelbah, B., Rock, B.H., Lohman, T., et al. (2003). Pathways: A school-based, randomized controlled trial for the prevention of obesity in American Indian schoolchildren. *American Journal of Clinical Nutrition, 78,* 1030-1038.

Centers for Disease Control and Prevention. (1997). Guidelines for school and community programs to promote lifelong physical activity among young people. *Morbidity and Mortality Weekly Report, 46*(RR-6), 1-36.

Centers for Disease Control and Prevention. (2001). Increasing physical activity: A report on recommendations of the Task Force on Community Preventive Services. *Morbidity and Mortality Weekly Report, 50*(RR18), 1-16.

Centers for Disease Control and Prevention. (2004). *Physical activity and good nutrition: Essential elements to prevent chronic diseases and obesity.* Retrieved March 1, 2005, from www.cdc.gov.

Chatzisarantis, N., & Biddle, S.J.H. (1998). Functional significance of psychological variables that are included in the Theory of Planned Behavior: A self-determination theory approach to the study of attitudes, subjective norms, perceptions of control, and intentions. *European Journal of Social Psychology, 28*, 303-322.

Cohen, D.A., Scribner, R.A., & Farley, T.A. (2000). A structural model of health behavior: A pragmatic approach to explain and influence health behaviors at the population level. *Preventive Medicine, 30*, 146-154.

Craig, S., Goldberg, J., & Dietz, W. (1996). Psychosocial correlates of physical activity among fifth and eighth graders. *Preventive Medicine, 25*, 506-513.

Dale, D., Corbin, C.B., & Dale, K.S. (2000). Restricting opportunities to be active during school time: Do children compensate by increasing physical activity levels after school? *Research Quarterly for Exercise and Sport, 71*, 240-248.

Deci, E.L., & Ryan, R.M. (1985). *Intrinsic motivation and self-determination in human behavior.* New York: Plenum Press.

DeMattia, L., Lemont, L., & Muerer, M. (2006). Do interventions to limit sedentary behaviours change behaviour and reduce childhood obesity? A critical review of literature. *Obesity Reviews, 7*, 111-136.

Dennison, B., Russo, T., Burdick, P., & Jenkins, P. (2004). An intervention to reduce television viewing by preschool children. *Archives of Pediatric and Adolescent Medicine, 158*, 170-176.

Dishman, R.K. (Ed.). (1994). *Advances in exercise adherence.* Champaign, IL: Human Kinetics.

Dishman, R.K., Motl, R.W., Saunders, R.P., Dowda, M., Felton, G., Ward, D.S., et al. (2002). Factorial invariance and latent mean structure of questionnaires measuring social-cognitive determinants of physical activity among black and white adolescent girls. *Preventive Medicine, 34*, 100-108.

Dishman, R.K., Motl, R.W., Saunders, R., Felton, G., Ward, D.S., Dowda, M., et al. (2004). Self-efficacy partially mediates the effect of a school-based physical activity intervention among adolescent girls. *Preventive Medicine, 38*, 628-636.

Dowling, C. (2000). Getting them out of the doll corner. In *The frailty myth: Redefining the physical potential of women and girls* (pp. 84-112). New York: Random House.

Ege, C., & Krag, T. (1999). *Cycling will improve environment and health.* Retrieved June 3, 2003, from www.thomaskrag.com/Ege-Krag_Velo-city_2005_paper.pdf.

Epstein, L.H., Myers, M.D., Raynor, H.A., & Saelens, B.E. (1998). Treatment of pediatric obesity. *Pediatrics, 101*, 554-570.

Faith, M.S., Berman, N., Heo, M., Peitrobelli, A., Gallagher, D., Epstein, L., et al. (2001). Effects of contingent television on physical activity and television viewing in obese children. *Pediatrics, 107,* 1043-1048.

Faucette, N., Sallis, J.F., McKenzie, T., Alcaraz, J., Kolody, B., & Nugent, P. (1995). Comparison of fourth grade students' out-of-school physical activity levels and choices by gender: Project SPARK. *Journal of Health Education, 26,* S82-S90.

Felton, G., Saunders, R., Ward, D., Dishman, R., Dowda, M., & Pate, R. (2005). Promoting physical activity in girls: A case study of one school's success. *Journal of School Health, 75,* 57-62.

Fitzgibbon, M.L., Stolley, M.R., Schiffer, L., Van Horn, L., KauferChristoffel, K., & Dyer, A. (2005). Two-year follow-up results for Hip-Hop to Health Jr.: A randomized controlled trial for overweight prevention in preschool minority children. *Journal of Pediatrics, 146,* 618-625.

Ford, B.S., McDonald, T.E., Owens, A.S., & Robinson, T.N. (2002). Primary care interventions to reduce television viewing in African American children. *American Journal of Preventive Medicine, 22,* 106-109.

Gladwell, M. (2000). *The tipping point.* Boston: Little, Brown.

Glanz, K., & Rimer, B. (1997). *Theory at a glance: A guide for health promotion practice.* National Cancer Institute (NIH Pub. No. 92-3316, updated 27 February 2003). Retrieved June 3, 2004, from http://cancer.gov/cancerinformation/theory-at-a-glance.

Gortmaker, S.L., Cheung, L.W., Peterson, K.E., Chomitz, G., Cradle, J.H., Dart, H., et al. (1999a). Impact of a school-based interdisciplinary intervention on diet and physical activity among urban primary school children: Eat Well and Keep Moving. *Archives of Pediatric and Adolescent Medicine, 153,* 975-983.

Gortmaker, S.L., Peterson, K., Wiecha, J., Sobol, A.M., Dixit, S., Fox, M.K., et al. (1999b). Reducing obesity via a school-based interdisciplinary intervention among youth. *Archives of Pediatric and Adolescent Medicine, 153,* 409-418.

Green, L.W., & Kreuter, M.W. (1991). *Health promotion planning: An educational and environmental approach.* Toronto: Mayfield.

Hardy, M. (1978). Evaluating nursing theory. In *Theory development: What, why, how?* New York: National League for Nursing.

Hausenblas, H., Carron, A.V., & Mack, D.E. (1997). Application of the Theories of Reasoned Action and Planned Behavior to exercise behavior: A meta-analysis. *Journal of Sport and Exercise Psychology, 19,* 36-51.

Jamner, M.S., Metz, D.S., Bassin, S., & Cooper, D.M. (2004). A controlled evaluation of a school-based intervention to promote physical activity among sedentary adolescent females: Project FAB. *Journal of Adolescent Health, 34,* 279-289.

Jeffery, R.W. (2004). How can health behavior theory be made more useful for intervention research? *International Journal of Behavioral Nutrition and Physical Activity, 1,*10.

Kelder, S.H., Mitchell, P., McKenzie, T.L., Derby, C., Strikmiller, P.K., Luepker, R., et al. (2003). Long-term implementation of CATCH physical education program. *Health Education and Behavior, 30,* 463-475.

Kelder, S.H., Perry, C.L., & Klepp, K.I. (1993). Community-wide youth exercise promotion: Long-term outcomes of the Minnesota Heart Health Program and the Class of 1989 Study. *Journal of School Health, 63*, 218-223.

Kelder, S.H., Perry, C.L., Peters, R., Lytle, L.L., & Klepp, K.I. (1995). Gender differences in the Class of 1989 Study: The school component of the Minnesota Heart Health Program. *Journal of Health Education, 26*, S36-S44.

Kerlinger, F.H. (1973). *Foundations of behavioral research* (2nd ed.). New York: Holt, Rinehart and Winston.

Kimiecik, J.C., & Horn, T.S. (1998). Parental beliefs and children's moderate to vigorous physical activity. *Research Quarterly for Exercise and Sport, 69*, 163-175.

King, A.C., Stokols, D., Talen, E., Brassington, G.S., & Killingsworth, R. (2002). Theoretical approaches to the promotion of physical activity: Forging a transdisciplinary paradigm. *American Journal of Preventive Medicine, 23*(2S), 15-25.

King, I. (1978). The "why" of theory development. In *Theory development: What, why, how?* New York: National League for Nursing.

Kraemer, H.C., Wilson, G.T., Fairburn, C.G., & Agras, W.S. (2002). Mediators and moderators of treatment effects in randomized clinical trials. *Archives of General Psychiatry, 59*, 877-883.

Kremers, S.P.J., de Bruijn, G.J., Visscher, T.L.S., van Mechelen, W., de Vries, N.K., & Brug, J. (2006). Environmental influences on energy balance-related behaviors: A dual-process view. *International Journal of Behavioral Nutrition and Physical Activity, 3*, 9.

Kuhn, T.S. (1970). *The structure of scientific revolutions* (2nd ed.). Chicago: University of Chicago Press.

McLeroy, K.R., Bibeau, D., Steckler, A., & Glanz, K. (1988). An ecological perspective on health promotion programs. *Health Education Quarterly, 15*, 351-377.

Muthen, B., & Muthen, L. (2000). Integrating person-centered and variable-centered analysis: Growth mixture modeling with latent trajectory classes. *Alcoholism: Clinical and Experimental Research, 24*, 882-891.

Myers, L., Strikmiller, P.K., Webber, L.S., & Berenson, G.S. (1996). Physical and sedentary activity in school children grades 5-8: The Bogalusa Heart Study. *Medicine and Science in Sports and Exercise, 28*, 852-859.

Nader, P.R., Stone, E.J., Lytle, L.A., Perry, C.L., Osganian, S.K., Kelder, S.K., et al. (1999). Three-year maintenance of improved diet and physical activity. *Archives of Pediatric and Adolescent Medicine, 153*, 695-704.

Neumark-Sztainer, D., Story, M., Hanna, P.J., & Rex, J. (2003). New Moves: A school-based obesity prevention program for adolescent girls. *Preventive Medicine, 37*, 41-51.

Nigg, C.R., Allegrante, J.P., & Ory, M. (2002). Theory-comparison and multiple-behavior research: Common themes advancing health behavior research. *Health Education Research, 17*, 670-679.

Nigg, C.R., & Jordan, P.J. (2005). It's a difference of opinion that makes a horserace. *Health Education Research, 20*, 291-293.

Nigg, C.R., Kerr, N.A., Hottenstein, C., Yamashita, M., Inada, M., Paxton, R., et al. (2006). First year dissemination results: Fun 5—a physical activity and nutrition program for elementary after school programs. *Annals of Behavioral Medicine, 31*, S92.

Pate, R.R., Saunders, R., Ward, D., Felton, G., Trost, S., & Dowda, M. (2003). Evaluation of a community-based intervention to promote physical activity in youth: Lessons from active winners. *American Journal of Health Promotion, 17*, 171-182.

Pate, R.R., Ward, D.S., Saunders, R.P., Felton, G., Dishman, R.K., & Dowda, M. (2005). Promotion of physical activity among high-school girls: A randomized controlled trial. *American Journal of Public Health, 95*, 1582-1587.

Prochaska, J.O., & DiClemente, C.C. (1983). Stages and processes of self-change in smoking: Towards an integrative model of change. *Journal of Consulting and Clinical Psychology, 51*, 390-395.

Prochaska, J.O., & Marcus, B.H. (1994). The transtheoretical model: Applications to exercise. In R.K. Dishman (Ed.), *Advances in exercise adherence* (pp. 161-180). Champaign, IL: Human Kinetics.

Prokhorov, A.V., Perry, C.L., Kelder, S.H., & Klepp, K.I. (1993). Lifestyle values of adolescents: Results from Minnesota heart health youth program. *Adolescence, 28*(111), 637-647.

Resnicow, K., & Vaughan, R. (2006). A chaotic view of behavior change: A quantum leap for health promotion. *International Journal of Behavioral Nutrition and Physical Activity, 3*, 25.

Resnicow, K., Yaroch, A.L., Davis, A., Wang, D.T., Carter, S., Slaughter, L., et al. (2000). GO GIRLS!: Results from a nutrition and physical activity program for low-income, overweight African American adolescent females. *Health Education and Behavior, 27*, 616-631.

Robinson, T.N. (1999). Reducing children's television viewing to prevent obesity: A randomized controlled trial. *Journal of the American Medical Association, 282*, 1561-1567.

Robinson, T.N., Killen, J.D., Kraemer, H.C., Wilson, D.M., Matheson, D.M., Haskell, W.L., et al. (2003). Dance and reducing television viewing to prevent weight gain in African-American girls: The Stanford GEMS pilot study. *Ethnicity and Dance, 13*, S1-65-S1-77.

Roemmich, J.N., Gurgol, C.M., & Epstein, L.H. (2004). Open-loop feedback increases physical activity of youth. *Medicine and Science in Sports and Exercise, 36*, 668-673.

Rothman, A.J. (2004). "Is there nothing more practical than a good theory?": Why innovations and advances in health behavior change will arise if interventions are used to test and refine theory. *International Journal of Behavioral Nutrition and Physical Activity, 1*, 11.

Saelens, B.E., Sallis, J.S., Wilfley, D.E., Patrick, K., Cella, J.A., & Buchta, R. (2002). Behavioral weight control for overweight adolescents initiated in primary care. *Obesity Research, 10*, 22-32.

Sallis, J.F., Bauman, A., & Pratt, M. (1998). Environmental and policy interventions to promote physical activity. *American Journal of Preventive Medicine, 15*, 379-397.

Sallis, J.F., McKenzie, T.L., Alcaraz, J.E., Kolody, B., Faucette, N., & Hovell, M.F. (1997). Effects of a two-year health-related physical education program on physical activity and fitness in elementary school students: SPARK. *American Journal of Public Health, 87*, 1328-1334.

Sallis, J.F., McKenzie, T.L., Conway, T.L., Elder, J.P., Prochaska, J.J., Brown, M., et al. (2003). Environmental interventions for eating and physical activity: A randomized controlled trial in middle schools. *American Journal of Preventive Medicine, 24*, 209-217.

Sallis, J.F., McKenzie, T.L., Kolody, B., Lewis, M., Marshall, S., & Rosengard, P. (1999). Effects of health-related physical education on academic achievement: Project SPARK. *Research Quarterly for Exercise and Sport, 70*, 127-134.

Sallis, J.F., Prochaska, J.J., & Taylor, W.C. (2000). A review of correlates of physical activity of children and adolescents. *Medicine and Science in Sports and Exercise, 32*, 963-975.

Schwarzer, R. (1999). Self-regulatory processes in the adoption and maintenance of health behaviors: The role of optimism, goals, and threats. *Journal of Health Psychology, 4*, 115-127.

Sharma, M. (2006). School-based interventions for childhood and adolescent obesity. *Obesity Reviews*, 7, 261-269.

Simon, C., Wagner, A., DiVita, C., Rauscher, E., Klein-Platat, C., Arveiler, D., et al. (2004). Intervention centered on adolescents' physical activity and sedentary behaviour (ICAPS): Concepts and 6-month results. *International Journal of Obesity, 28*, S96-S103.

SPARK. (2004). *SPARK coordinated health*. Retrieved March 19, 2005, from www.sparkpe.org/csh.jsp.

Stevens, J., Story, M., Ring, K., Murray, D.M., Cornell, C.E., Juhaeri, et al. (2003). The impact of the Pathways intervention on psychosocial variables related to diet and physical activity in American Indian schoolchildren. *Preventive Medicine, 37*(Suppl. 1), S70-S79.

Stokols, D. (1996). Translating social ecological theory into guidelines for community health promotion. *American Journal of Health Promotion, 10*, 282-298.

Stone, E.J., McKenzie, T.L., Welk, G.J., & Booth, M.L. (1998). Effects of physical activity interventions in youth: Review and synthesis. *American Journal of Preventive Medicine, 15*, 298-315.

Theunissen, N.C.M., & Tates, K. (2004). Models and theories in studies on educating and counseling children about physical health: A systematic review. *Patient Education and Counseling, 55*, 316-330.

Traub, R. (1994). *Reliability for the social sciences: Theory and applications*. Thousand Oaks, CA: Sage.

Trost, S.G., Pate, R.R., Dowda, M., Ward, D.S., Pelton, O., & Saunders, R. (2002). Psychosocial correlates of physical activity in White and African-American girls. *Journal of Adolescent Health, 31*, 228-233.

Trost, S.G., Pate, R.R., Ward, S.S., Saunders, R., & Riner, W. (1999). Correlates of objectively measured physical activity in preadolescent youth. *American Journal of Preventive Medicine, 17*, 120-126.

Trost, S.G., Saunders, R., & Ward, D.S. (2002). Determinants of physical activity in middle school children. *American Journal of Health Behavior, 26*, 95-102.

U.S. Department of Health and Human Services. (1996). *Physical activity and health: A report of the Surgeon General.* Atlanta: Centers for Disease Control and Prevention, National Center for Chronic Disease Prevention and Health Promotion.

Warren, J.W., Henry, J.K., Lighttowler, H.J., Bradshaw, S.M., & Perwaiz, S. (2003). Evaluation of a pilot school programme aimed at the prevention of obesity in children. *Health Promotion International*, *18*, 287-296.

Welk, G.J. (1999). The youth physical activity promotion model: A conceptual bridge between theory and practice. *Quest*, *51*, 5-23.

Wikipedia, The free encyclopedia. (2006). *Theory.* Retrieved August 8, 2006, from http://en.wikipedia.org/wiki/Theory.

Wilson, M. (2005). *Constructing measures: An item response modeling approach.* Mahwah, NJ: Erlbaum.

CHAPTER

5

"Couch Potatoes" and "Wind-Up Dolls"?
A Critical Assessment of the Ethics of Youth Physical Activity Research

Michael Gard, PhD

I have been reading and making use of research on physical activity and young people for about 10 years. My purpose has primarily been to treat this area of research as one strand of a larger research agenda aimed at synthesizing the state of our knowledge about food, physical activity, body weight, and health. My approach has been to look for consistencies and inconsistencies within and between distinct but related areas of research. For example, how is it that in one branch of health science we are considered to be in the middle of an "obesity crisis," set to wipe years off life expectancy (Olshansky et al., 2005), when in other research circles Western populations are described as healthier, in terms of both quantity and quality of life, than at any other point in their history (Australian Institute of Health and Welfare, 2004), or life expectancy is said to be heading toward 100 (Balzer, 2005)? Even though until recently Australia was ranked as the second most overweight country in the world, improvements in Australian mortality rates show no sign of slowing, and the five years prior to 2001 saw the greatest decline since 1923 (Australian Institute of Health and Welfare, 2004, p. xii). Who is correct here? Do we have a health crisis on our hands or not?

Science, of course, has been and remains a colossal force for both good and ill in the world. The question addressed in my work has been whether science—by which I mean predominantly quantitative research published in scholarly journals and books and presented at scientific conferences—has produced useful or true knowledge about food, physical activity, body weight, and health *in particular*. It hardly needs to be stated that deciding whether knowledge is true or useful is an innately hazardous business. However, it is an indispensable part of the work that scholars do. And it is especially important, for reasons I take to be self-evident, that this evaluative meta-research be done from time to time by people *outside* of the research teams doing the original research.

Were someone else writing this chapter about ethics and youth physical activity research, that person may have dealt with the rights and physical safety of young people in the research context and other important practical considerations. However, in this chapter I will link ethics with wider theoretical, epistemological, and political questions. In this context, I take "ethics" to mean the real-world effects of our intellectual work: What are the consequences of claiming certain things to be true? Which moral or ideological agendas are served when we choose one theoretical model over another or one set of research methods over others? Who wins? Who loses? I ask these questions to promote discussion and debate about our collective role in physical activity–related disciplines. It goes without saying that my voice is but one in the research community that we share. I neither expect nor hope that all readers will agree with the arguments I make here, but I offer them in the spirit of robust academic discourse.

Couch Potatoes?

There are doubtless many different reasons why people research youth physical activity, but, given the limited space I have here, I want to focus on what appears to be the dominant reason for doing this research—the emergence of the idea in the West that the current generation of young people are, for the most part, "couch potatoes" (Conway, 2003). In the last decade a symbiotic relationship has developed between obesity science and the popular media around the issue of the physical activity levels of young people. In the process, a consensus has developed: Young people are doing less physical activity than they used to—certainly not enough to foster good health—and something needs to be done about it. If these things were true, that would probably be a good reason to research youth physical activity. If they were untrue, we might wonder about the ethics of branding a generation of children "couch potatoes."

A Dearth of Solid Evidence

That a decline in the amount of physical activity young people do is self-evident is asserted repeatedly in the scholarly literature (e.g., Boreham & Riddoch, 2003; Dietz, 2002; Dionne & Tremblay, 2000; Hill & Melanson, 1999; Koplan, 2000; Savage & Scott, 1998; Seidell, 2000; Van Mechelen et al., 2000). What is striking about this claim is the regularity with which it is made without supporting references. If the authors making this claim attempt to justify it at all, they usually do so, as with Boreham and Riddoch (p. 22) here, by invoking some kind of nostalgic stereotype:

> Today's 50-60-year-old adult is likely to have walked or cycled to school, played for hours outside the house, was not distracted too much by television, and eventually walked to work—which was probably a manual job. So any baseline measurements of physical activity or fitness may be misleading, as we know there has been an inexorable decrease in activity levels in all sectors of society over the past 50 years.

Boreham and Riddoch's assertion that "all sectors of society" have become less physically active is particularly striking for its grand generalizing sweep. The only explanation that I can offer for this consistent and widespread departure from the normal academic practice of providing supporting references is that the research community as a whole sees this kind of assertion as incontestable. Where provided, and this is extremely rare, the references offered usually have little or no bearing on the claim being made. For example, Pi-Sunyer (2003) references Kimm and colleagues, 2002, and Prentice and Jebb, 1995, to substantiate the blanket claim that physical activity is declining in all sections of the population. Checking these references reveals that Kimm and colleagues present research on the physical activity levels of a cohort of adolescent girls whom the researchers tracked as they got older. This paper offers no secular trend data whatsoever. Prentice and Jebb merely speculate about what they see as circumstantial evidence—for example, television use and numbers of people employed in manual labor—for a decline in physical activity.

What are the implications of characterizing youth as "couch potatoes" and uncritically promoting physical activity in young people?

Elsewhere, Peters and colleagues (2002) assert that people of all ages are doing less physical activity but offer only Hill and Peters, 1998, by way of evidence—a paper in which this claim is simply repeated rather than substantiated. Rippe and Hess (1998) reference Rippe and colleagues, 1992, to support their claim that children's physical activity went down in the 20 years prior to 1998. However, this latter paper offers no firm empirical conclusions about secular trends in children's physical activity. In fact, as so often happens in this area of study, while the paper cited strikes a cautious tone, any hint of uncertainty is absent in the citing paper where the phenomenon in question, declining physical activity levels among young people, morphs into an established research finding.

In all of this work, whether a downward secular trend is claimed or not, the explicit assertion or clear implication is that Western societies and, most worrying of all, their young people, are in the grip of an "epidemic of inactivity" (Rippe & Hess, 1998; Seefeldt, Malina, & Clark, 2002). And although the significance of this is not always spelled out, the idea of an "epidemic of inactivity" seems to assume two things: first, that this situation is different from a pre-epidemic point somewhere in the past and, second, that this situation has serious health consequences.

Methodological Difficulties

Confident, though mostly unreferenced, pronouncements by researchers about declining physical activity levels are all the more remarkable given the general lack of solid data and the well-known methodological difficulties in this area of study. For example, in the U.K. context, Sleap and colleagues (2000) describe the existing data for children under 10 as "negligible" (see also Durnin, 1992, for an earlier and highly equivocal assessment of secular trend evidence; Booth et al., 2002, and Magarey, Daniels, & Boulton, 2001, for a similar assessment of Australian data; and Pratt, Macera, & Blanton, 1999, for the United States). However, while it is true that secular trend data related to youth physical activity are scarce, data do exist. For example, having argued that the perception of an increasingly inactive U.S. adult population is not supported by available evidence, Pratt and colleagues (p. S531) write:

> The public perception of an increasingly sedentary way of life among children is even more widespread than for adults. However, there is even less good information available on national trends in youth physical activity or fitness than for adults. No fitness data exist since the completion of NCYFS II in 1986. The YRBS physical activity questions have been standardized only since 1993 and provide information only on young people attending school in grades 9–12. There has been no significant change in reported vigorous physical activity between 1993 and 1997: 1993, 65.8%; 1995, 63.7%; and 1997, 63.8%.

In the United Kingdom, the Schools Health Education Unit (SHEU, 2004) has produced one of the more comprehensive attempts to measure physical activity secular trends in young people. As the SHEU report notes, the instruments used to collect large amounts of self-report data for children aged 10 to 15 years have evolved over a period of 16 years, meaning that comparisons are not always straightforward. According to some media reports, the research shows that the so-called couch potato generation "are actually fighting fit" (Halpin, 2004). My assessment is that the results suggest, perhaps not surprisingly, that some forms of physical activity have gone up, others have gone down, and others have stayed the same. For example, the children surveyed reported cycling less and dancing more, whereas swimming and gymnastics were unchanged. There were some upward trends in jogging and soccer and slightly downward trends in rugby. Perhaps more importantly, there was no overall change in the percentages of children who did none of the surveyed physical activity types; there were increases in the percentage exercising vigorously as well as increases in those describing themselves as "unfit." The latter finding might easily be due to increased publicity about a "couch potato generation," publicity that few young people are likely to have been lucky enough to avoid.

Population-level physical activity research is complex and time-consuming, and conclusions must inevitably be cautious. In my view the SHEU report lends support to the conclusion that there has been no precipitous or even small decline in young people's overall physical activity levels in the United Kingdom since the late 1980s. However, my point here is not so much to argue for a particular truth, but to highlight the way perfectly useful data are ignored when researchers talk about youth physical activity levels.

Given the scarcity of supporting evidence, some scientists point to other factors that they assume might have some bearing on youth physical activity levels. For example, reports of declining amounts of school class time for physical education, particularly in the United Kingdom and the United States, are held up as a possible cause for increasing childhood obesity (Hill & Melanson, 1999; Savage & Scott, 1998). Scientists who do this seem unaware of a quite longstanding finding in the literature that participation in official physical education lessons appears to make little or no difference to young people's physical activity levels (in particular, see Booth et al., 2002; Simons-Morton et al., 1993; Stratton, 1997; Warburton & Woods, 1996; Yelling, Penney, & Swaine, 2000).

In other publications (Gard, 2005; Gard & Wright, 2005) I have argued that obesity science seems guided by a reflex that may be nearly as old as civilization itself. That is, as societies change and become more (for want of a better term) "technological," the tendency to celebrate these advances seems always to have gone hand in hand with a more suspicious impulse—the sense that modernization makes us weak, soft, and a pale imitation of our robust and resourceful ancestors. Sometimes obesity scientists romanticize prehistoric hunter-gatherers while, as we saw earlier, others look to more recent times. For Boreham and Riddoch (2003), it is the 1950s. But, as often happens, empirical findings rarely come to the rescue of dewy-eyed stereotypes. For example, while the concern was with children aged 6 to 18 months and, therefore, technically not youth (a term that seems most commonly defined in the literature as those aged 2 to 18 years), Lawrence and colleagues (1991) found that a group of U.K. children were considerably more active than a group living in a Gambian village. Once again, few would suggest that this single paper presents us with an obvious conclusion. By contrast, Singh and colleagues (1989) have published data on the average caloric expenditure of Gambian women, which the authors claim is higher than that of adults living in more affluent societies. But Lawrence and colleagues' paper should at least remind us how much more complicated the world *might* be than our cultural stereotypes. In reviewing the literature concerned with the physical activity levels of hunter-gatherer and other non-Western societies, Panter-Brick (2002, 2003) argues that it is simply unclear whether these groups led or lead more physically active lives than modern Western children or adults; some studies' findings point one way and some the other.

It is possible to take a step back here. The idea that children are doing less physical activity than they used to is preceded conceptually by the idea that the amount of physical activity young people do has important health consequences and that we can be reasonably confident about the line that separates sufficient and insufficient physical activity for youth. If this were not so, it would be hard to know why the amount of physical activity that young people do has created so much scholarly and popular interest as opposed to, say, the time they spend going to church or art galleries.

Healthy Body, Healthy Mind

In my view, the link between physical activity and young people's health is an even more fundamental and ethically charged issue than questions about the alleged decline in youth physical activity. This is so because lurking behind all our discussions about youth physical activity and health is the idea that physically fit and active young people are more likely to be well behaved, well adjusted, responsible, and morally "good" members of society. For example, it is quite common in the literature for physical activity participation to be linked with emotional well-being as well as lower rates of teenage pregnancy, drug use, and academic failure (Federal Interagency Forum on Child and Family Statistics, 2005; Kimm et al., 2002; Morris et al., 2003; Savage & Scott, 1998; Steptoe & Butler, 1996). In my experience, there are few clearer examples of the tendency to confuse correlation with causation than in this area of research and, by extension, the tendency for researchers to allow their moral and ideological assumptions to infect their scholarship. In short, we might call this the "healthy body, healthy mind discourse," and one needs only to undertake a brief survey of the Western philosophical canon or the work of social historians to know that this is one of Western civilization's most cherished myths. We might also notice the way modern elite athletes are paraded in front of schoolchildren as "good role models" to appreciate the enduring power of physically active young bodies to signify high moral standards in our culture.

Some readers will object that this has nothing to do with the study of youth physical activity and medical health. This may be true in theory, but the case for the ability of childhood physical activity to increase health is not obviously stronger than the case for, say, the ability of childhood physical activity to reduce teenage pregnancy. To take just one example, the current clamor in a number of Western countries to use schools to promote childhood physical activity seems grossly out of proportion with physical activity's demonstrated benefits. Are we dealing with the sober implementation of conclusive research findings or something else here?

Boreham and Riddoch (2003) seem almost to concede this point when they write, "Although we feel instinctively that physical activity ought to be beneficial to the health of children, there is surprisingly little empirical evidence to support this notion" (p. 17). Earlier, they note, "It is important to emphasize at the outset that most of what can be written on this topic remains speculative. No study exists which has recorded adequate birth-to-death information relating physical activity to health" (p. 14). This point is made in more detail by Twisk (2001) in his review of the literature concerning the health benefits of childhood physical activity. Twisk's conclusion is that there are no empirical grounds upon which to base physical activity guidelines for children and that, even if there were, health guidelines are unlikely to motivate children to be more physically active. It is difficult to see how this conclusion could be disputed. As we have seen, there are limited data concerning the amount of physical activity that young people are actually doing (or did in the past), a situation which *by itself* makes it impossible to draw meaningful conclusions about the connection between childhood physical activity and short- or long-term health.

Perhaps an even more telling point is that the connection between physical activity per se and good health is far from clear beyond the general (and, I would argue, widely understood) point that some amount of regular moderate or vigorous activity is probably good for most of us. But, as is widely acknowledged in the literature, questions about how much and at what point(s) in the life cycle physical activity confers health benefits remain unanswered. Some have even provocatively suggested that physical activity delivers health benefits only for older people, meaning that physical activity done by younger people is not "money in the bank" (Coleman, 1998; Nicholl, Coleman, & Brazier, 1994). This is not a completely implausible point of view. As Twisk points out, there is no evidence to suggest that adults who have been sedentary all their life are likely to suffer worse health than sedentary adults who were physically active in their youth. And this is to say nothing of the widely acknowledged inconclusiveness of research findings connecting childhood overweight or obesity with inactivity (Fogelholm et al., 1999; Rippe & Hess, 1998) or with disease risk (Flegal, 1999; Micic, 2001; Sleap, Warburton, & Waring, 2000).

We live at a moment in history when the physical activity that young people do is being transformed into, at best, a kind of medicine or, at worst, a form of hard-labor punishment for their allegedly slothful ways. And despite what to me would seem the absence of anything approaching compelling scientific evidence, researchers and journals casually announce or suggest that children who do not meet minimum physical activity guidelines place themselves at significantly greater risk of life-threatening disease (Savage & Scott, 1998), an idea echoed and amplified by eager popular media.

It is surely worth remembering that young people are the most physically active section of Western populations and that many papers report high levels of physical activity among youth (e.g., Booth et al., 2002; Rehor & Cottam, 2000; Savage & Scott, 1998; Van Mechelen et al., 2000). In fact, some authors seem genuinely surprised when their findings turn out not to be so disastrous after all. This probably explains why the discussion sections of these papers often include speculation about the methodological limitations that might have produced such high numbers or reasons why we should resist celebrating the findings. To a research community that has already made up its mind that a couch potato generation is upon us, physically active young people will always be the exception to the rule.

Wind-Up Dolls?

Grouping all young people together under the label "couch potato generation" means that many, probably most, children are incorrectly categorized and unfairly stigmatized. It also makes the creation and implementation of carefully targeted interventions less likely. Although I am aware of this happening elsewhere, I have recently worked in the Canadian province of Ontario, where the provincial government has mandated that all elementary school children will do 20 min of vigorous physical activity during each school day. Already struggling with impossibly crowded curricula, heavy administrative loads, and mounting pressure to increase academic test results, Ontario elementary schools find themselves charged with solving yet another of society's problems.

This has happened despite consistent research findings suggesting that many children are already physically active (see earlier). It has also happened despite the well-established research finding that many elementary teachers lack the expertise, and often the enthusiasm, to implement high-quality physical education programs (e.g., Taggart, Medland, & Alexander, 1995; Williams, 2000). Of course, resources will be consumed implementing and reporting on this policy. And what of the experiences of children? What kinds of experiences will they have in a context in which eradicating "couch potatoes" is the name of the game? Which students will have fun? Which will not? It is already apparent that at least some people who research youth physical activity think that a "high-quality" physical education lesson is not one in which the students learn anything, but rather one in which heart rates remain elevated for as long as possible (e.g., Brown & Brown, 1996; McKenzie et al., 2000; Simons-Morton et al., 1994).

In a particularly egregious example of scientists encroaching on the professional autonomy of others, Thomas and colleagues (2004) have gone as far as to recommend that physical education teachers concentrate more on "exercise" and less on skill development in their classes. One wonders whether researchers who advocate for such policies (and I am not aware of any voices of dissent from the scientific research community) have consulted research on what children feel about being made to participate in physical activity that is either boring or not appropriate for their current fitness level (e.g., Carlson, 1995; Hopple & Graham, 1995). What kinds of programs do the researchers imagine springing up in busy schools following these government mandates? Given that it will be teachers' working lives that will be affected, not scientists' lives, what are we to make of the absence of evidence supporting the efficacy of such policies? Is it possible that more affluent schools with more facilities for physical activity, fewer concerns about student discipline and teacher burnout, and students who are likely to be the most active already are likely to be most able to implement extra mandatory physical activity?

In passing, it is difficult not to wonder how the medical and sport science communities would react if the tables were turned and school teachers began pronouncing upon the ways scientists did their jobs and the goals that scientists should strive for. If nothing else, the issue of youth physical activity is a reminder that different fields of knowledge are regarded by most people as having distinct positions in a hierarchy of value; we seem to have no trouble accepting the right of scientists telling teachers how to teach, but the idea of elementary teachers telling researchers how to do research is almost unthinkable.

Mechanistic Tendencies

Perhaps part of the reason why such crude public policies might find favor relates to the way the youth physical activity research community appears to conceptualize young people and physical activity. I would be committing the same error of overgeneralization that I have pointed to in others if I suggested that all members of this community work in the same way using the same conceptual tools. However, it is apparent that a great deal of research in this area proposes that we think of people, young and old, as machines operating in a mechanical world. This tendency is manifest in at least two ways.

The "Habits" Thesis

First, it now appears to be the dominant view in the obesity and physical activity literature that young people should be the focus of research and intervention. The logic here runs that, once gained, body weight is difficult to lose, and therefore it is better to encourage young people to be physically active in order to stop them from gaining excess weight as they mature. This focus on the young seems to be shared by many who research physical activity as opposed to obesity per se, though the rationale in this case is that a "habit" of physical activity needs to be established at an early age in order to increase the likelihood of it becoming a long-term feature of a person's "lifestyle." As Salbe and Ravussin (2000, p. 84) put it,

> It is well known that activity habits are established at a young age; studies on intervention strategies in adults show that those people who can most readily be persuaded to take up more activity are those who, as children, were physically active at school and during their leisure.

Is it well known that activity habits are established at a young age? Is it useful or true to call physical activity a "habit"? An unkind reader of Salbe and Ravussin's paper might ask why, if physical activity actually were a "habit" "established at a young age," physically active young people would ever need to be "persuaded to take up physical activity" later in life. The word "habit" suggests a kind of automatic (i.e., machine-like) behavior, something that is done without thinking. The idea of young people forming physical activity "habits" is widely believed within my own discipline of physical education, and it conjures images of youth as wind-up dolls, or as computers that we can "program" for a life of physical activity.

If supporting evidence for this mechanical view of young people and physical activity were to come from anywhere, it would be from the "tracking" literature, research that seeks to measure the "durability" of physically active behavior over the life course. On the whole, tracking research offers limited support for the "habits" hypothesis, a point that seems to be generally acknowledged (De Bourdeaudhuij, Sallis, & Vandelanotte, 2002; Malina, 1996). Varying levels have been reported, but physical activity tracking is usually described as "low" or "moderate." Why the level of tracking is not actually zero is unknown, but it is at least plausible that this is due to factors like economic capacity (wealthy children grow up to be wealthy adults and have enough money to participate in physical activity at any age; see Green, 2004, for a fuller exploration of this point) or that children who play sport establish personal links with other sporting people that, at least to some extent, endure over time. In my view, neither of these two plausible, although speculative, scenarios would support the "habits" thesis. At any rate, as we have already seen, researchers report that physical activity drops rapidly as young people get older (Malina, 1996; Telama & Yang, 2000).

In sum, it is extremely difficult to see how we might reconcile the idea that youth physical activity establishes lifestyle habits, on the one hand, with the reported "erosion" of physical activity as people age, on the other. This is a critical point. A more efficacious point of view might be to accept that at different phases in their lives people find physical activity more or less affordable, more or less enjoyable, or more or less possible. If this admittedly fuzzier conceptualization of physical activity were true, and if young people really are the most physically active group in Western societies, then the focus of research and interventions on young people might not seem so axiomatic. Perhaps we should leave children alone and focus on adults. This would be especially indicated if those arguing that the health benefits of physical activity do not "kick in" until middle age turned out to be correct.

At this point, at least one objection to my line of argument might legitimately be raised. That is, while it is true that some researchers have identified young people as a particular area of concern, surely it is possible and preferable to study all groups of the population, including children and adolescents. In other words, shouldn't we cast a wide net? By way of response I would make two points. First, as mentioned previously, there are serious questions about the empirical case for focusing on youth physical activity. For example, what *exactly* is the problem with youth physical activity that needs to be fixed? What *exactly* do researchers believe will be the benefits of this focus, and what evidence exists for these beliefs? What is the record of success up to now? Second, the mechanistic approach of many researchers in this field does not bode well for better understandings about youth physical activity in the future, a point to which I now turn.

The "Determinants" Thesis

Apart from the "habits" thesis, the dominance of mechanistic thinking with respect to physical activity is also exemplified in the so-called determinants literature. Here, a human's relationship with physical activity is conceptualized as consisting of a set of factors or "determinants." With respect to adult physical activity, a review of the literature by leaders in the field (Trost et al., 2002) has produced a list of no less than 75 possible determinants that have been studied. In passing, it is interesting to note that the title of the paper uses the word "correlates," not "determinants," while the authors go on to use the two words interchangeably throughout the paper. This is a curious, if not disturbing, example of the ubiquitous and erroneous jump from correlation to causation in this literature.

The theoretical antecedents of this approach to physical activity research seem fairly obvious. It draws its inspiration from an archaic form of functionalism, long since discarded in the social sciences, whereby human societies and the humans in them are seen as essentially mechanical. The "advantage" of this model is that the scientist can imagine her- or himself sifting through a finite selection of buttons and levers that, when pressed or pulled, will produce predictable changes in the society or person. The purpose of research, therefore, is to compile an exhaustive list of buttons and levers (determinants) and establish which ones will have the greatest positive effect. Use of this model is widespread in research on youth physical activity. In particular, authors of papers repeatedly write that findings in this area of research will help us to design youth physical activity interventions. But will they?

Let us take an example from this literature. Neumark-Sztainer and colleagues (2003) use a standard set of research protocols but add the feature of concern with the decline of adolescent girls' physical activity, a particular preoccupation in this literature. The authors' belief in the correlates/determinants model is stated clearly and early in the paper: "To prevent this decline in physical activity among adolescent girls, it is essential to identify factors correlated with physical activity that are amenable to change and can be addressed within interventions" (pp. 803-804).

Adolescent girls were surveyed three times over an eight-month period about the amount of physical activity they do, their reasons for doing or not doing physical activity, and other measures such as body image. The purpose of the study was to see which factors were most closely associated with *changes* in reported physical activity over the life of the study. The researchers adopted the standard practice of using survey instruments employed in previous studies. With respect to physical activity, survey instruments asked research participants to rate the importance of a range of "factors" identified by *other researchers in previous studies.* The study showed that "The 2 strongest and most consistent factors associated with change in physical activity were time constraints and support for physical activity from peers, parents, and teachers" (p. 803). The authors conclude that interventions aimed at adolescent girls might be enhanced by engaging the support of friends and family.

On one hand, we might see this as an unsurprising finding, though I would argue that there is no disgrace in research confirming the apparently self-evident. But, on the other hand, is even this conservative conclusion justified? Although the authors claim that they have identified factors associated with physical activity change, this is a debatable assertion. Indeed, the data might more correctly be described as the research participants' level of agreement or disagreement with the factors offered to them by the researchers. In other words, the adolescent girls may simply have chosen the factors they saw as most plausible in explaining physical activity behavior change generally. The study's results may simply be the product of two sets of speculation; the researchers' and the participants'. Rather than being concerned with what factors influence youth physical activity, the study seems more revealing about the way researchers' preconceptions influence the response of research participants. I suggest that this is an example of research methods not matching the research question.

Compounding the error, Neumark-Sztainer and colleagues (2003) proceed to quantify an entity they call "perceived support" into a unit "U." They claim, "An increase of 2.0 U in perceived support for physical activity (possible range, 3.0-12.0 U) would be expected to lead to an increase of 35 minutes of moderate to vigorous physical activity per week (95% confidence interval, 13-56 minutes)" (p. 803). Even if we concede that the researchers do not mean this section of the paper's abstract to be taken literally, it is difficult know where to begin with a conclusion of this kind. At no stage do the researchers define what *they* mean by "support." At no stage do they seek to find out what the research participants mean by "support" or if different participants mean the same thing. And despite the authors' conclusion (quoted earlier), there is no evidence in this study that "support," however defined, would have a positive *effect* on physical activity. The reason is that the study was designed merely to establish statistical correlations between a predetermined list of "factors" and reported changes in physical activity behavior. It is not a study that can say anything about the *causes* of behavior change over time.

Furthermore, at no stage do the researchers seem to entertain the possibility that young people do not know the reasons they do or do not participate in physical activity or are unable to explain their reasons. This is not a trivial point. The complexity of and motivations underlying human behavior have occupied the minds of philosophers, scientists, and social and cultural theorists for centuries. Researchers in the area of physical activity might do well to remember that no simple or comprehensive theory of human behavior has yet emerged. The conclusion I have reached about this style of research is that its conceptual framework and logical slides could be justified only if the world really were mechanical and the people in it machines.

Unspoken Assumptions

On the question of adolescent girls' physical activity, it is worth pointing out that it is not at all clear why this arouses the concern it does in the scholarly literature. Although there are variations from country to country, there is evidence that female adults are often as physically active as adult males (or more so), are less likely to be obese, live longer and healthier lives, and participate in physical activity that is (at least according to some researchers) more health

Pleasure in movement is evident on this girl's face. What motivates her to move? Is it evident to her what makes her enjoy moving?

enhancing (Australian Institute of Health and Welfare, 2004; Brown, 2001; Paxton, Sculthorpe, & Gibbons, 1994; Telama & Yang, 2000; Van Mechelen et al., 2000; Whitfield et al., 2002).

There is also clear evidence that youth physical activity researchers have often simply assumed that being "active" means being involved in the kinds of activities that young males stereotypically do (for full discussions of this point see Green, 2004, and Wright, 1997). A robust case could easily be made to the effect that the so-called girl problem in physical activity is simply an artifact of sexist research bias. Proponents of the correlates/determinants model might also consider a relatively recent Centers for Disease Control and Prevention (2001) report showing that boys watched more television than girls and were more likely to have played a team sport during the last year, carried a weapon, smoked cigarettes, been overweight, eaten greater than or equal to five servings of fruits and vegetables and drunk three glasses of milk per day, and reported doing "sufficient" vigorous physical activity. My reaction to a finding of this kind is to wonder what, if any, conclusion we can draw from such correlational statistical findings and whether, in fact, a statistical approach to young people's lives can actually create "problems" that do not exist. In short, if there is a problem with girls', as opposed to boys', physical activity, it is time that the precise nature of this problem was articulated.

Epistemological issues aside, it is also difficult to see how Neumark-Sztainer and colleagues' (2003) findings could actually be used in an intervention—should parents and friends be encouraged to say more positive things whenever the adolescent girls in their lives mention physical activity? Although I accept that this is not a conclusion many others will share, it appears that the search for the true and presumably most useful determinants of youth physical activity has failed. No "smoking gun" has emerged. As a result, some scholars have called for a greater focus on "inactivity" (e.g., Gordon-Larsen, McMurray, & Popkin, 2000).

This call to focus on inactivity seems premised once again on the highly mechanistic assumption that time spent being physically inactive necessarily displaces activity. Perhaps not surprisingly, televisions, computers, and video games have been preemptively and repeatedly demonized in the literature for causing childhood inactivity and obesity. Once again, this is done mostly without supporting references. And yet, as a number of researchers have found, televisions and computers have also not turned out to be the key to the puzzle that some originally thought they would be (Crawford, Jeffrey, & French, 1999; Gorely, Marshall, & Biddle, 2004; Marshall et al., 2004). Many children score highly on both sedentary behavior and physical activity scales while study after study has shown no correlation between television or computer use and physical activity or physical fitness in young people (see Gard & Wright, 2005, for a summary). In fact, in their review, Gorely and colleagues found that "inactivity appears more strongly related to sociodemographic factors than modifiable factors such as psychosocial or behavioral variables" (p. 159), a point that does not auger well for the correlates/determinants literature finding a cure for "couch potato-ism."

The quantitative determinants of physical activity approach is flawed, not only because of its conceptual limitations, but also because it is essentially authoritarian in nature. That is, it seeks to tell people how they should live. The determinants literature singles out specific kinds of people (such as girls or children from disadvantaged areas) who, it is explicitly argued, need to change their behavior and become more like other groups. For governments wanting to be seen to be doing something without thinking too hard about the problem or spending too much money, the scientific community's identification of statistically "at-risk" groups is a godsend because these individuals can now be blamed for their behavior. For example, faced with the common statistical finding that people with higher education levels are less likely to be obese and physically inactive, rather than advocating for wider access to education, some researchers in the field focus on the failings of individuals. Alternatives to this approach, of course, do exist.

Some Alternatives

A number of ethnographically minded researchers have tried to understand the place of physical activity in the lives of young people by asking them to talk (or sometimes use methods other than talk, such as photography) more generally about their lives. This has the advantage that researchers are not presuming to know the answers, or the list of all possible answers, before the research actually takes place. In some of this work, physical activity emerges as being connected to young people's sense of themselves—their identity, if you will (Green, 2004; Lee, 2003; Wright, Macdonald, & Groom, 2003). This means that young people, particularly but not only adolescents, are not really responding to "factors" in their life, but rather building an image of themselves that draws on the huge varieties of signs, symbols, labels, styles and, in short, identities that Western cultures make available. It is for this reason that the term "physical culture" and its intersections with "youth culture" have been used to explain and explore how and why young people use or avoid using physical activity in their lives (Green, 2004).

The idea of "physical culture" is not simply a different way of describing the same thing that "determinants" researchers look at. The crucial question is whether physical activity is best thought of as an aspect of our culture or as a form of behavior with a finite list of determinants, each with a fairly predictable and quantifiable influence on behavior. In my view, seeing physical activity as an aspect of culture has both a strength and a weakness. The strength is that this view more closely approximates the complicated and changing worlds in which young people live. It gives us greater scope to account for such "variables" as ethnic diversity or level of disability and the cultural biases and prejudices that these young people face, particularly (but not only) in settings like the physical education classroom and competitive sport.

It is interesting to note that the determinants literature very rarely addresses racial prejudice or straightforward sexism with regard to the influences on young people's activity choices. There is nothing essential about quantitative methodologies that prevents them from doing so, but the mechanical model used makes this much more difficult because we do not normally think of machines as having emotional lives lived in the context of cultural politics. The weakness of the term "physical culture" is that it does not offer us clear direction for intervening in the lives of young people. "Culture" is a big concept, and its material forms are partly fueled by enormous global forces (such as the media) and entities (such as corporations). But if "physical culture" discourages us from direct intervention in people's lives, perhaps this is not such a bad thing.

Research by Frank and colleagues (2003) in the United States suggests that the determinants of physical activity may not be such a mysterious puzzle requiring the generation of long lists of "factors." Their work on community design argues that in modern urban America, physical activity tends to be most difficult for the poor, the old, and the young, often for reasons of cost, safety, geographical distance, and lack of access to a motor vehicle. This work is valuable because it resists the temptation to blame individuals for either being too lazy to exercise or being ignorant of the benefits of physical activity.

If this work were to stimulate any kind of intervention, it would *not* be by insisting that parents limit children's television watching or that adolescent girls change the way they allocate their time, as is espoused repeatedly in the literature. Rather, it would stimulate us to advocate for creating livable communities that are not built, first and foremost, with economic efficiency in mind; for providing safe, cheap, and numerous spaces for recreation; for achieving a level of income parity with a view to ameliorating the worsening plight of the culturally, geographically, and economically disconnected "working poor" in Western countries; and for establishing well-funded schools, particularly in disadvantaged areas, so that opportunities for physical activity might be maximized.

To reduce the possibility of ineffective policy direction, researchers would then spend more time researching how young people live, listening to their stories about their lives, understanding a little more about how they order their priorities. With this knowledge we might have a better, and probably very different, idea about what our roles as researchers should be and less enthusiasm for "behavior change." There is a great deal of qualitative and quantitative research waiting to be done into the ways new cultural and economic conditions are shaping people's lifestyles and the nature of the communities that are taking form.

This does not mean that we should ignore the plight of chronically overweight young people; specific clinical work will always have its place. In fact, programs that target the specific day-to-day challenges faced by specific communities is exactly what I am arguing for. Youth physical activity research is

Physical activity opportunities tend to be limited for children of the poor in comparison to children of the upper middle class. This nicely uniformed group of preschool T-ball players in the United States is typical of opportunities available to advantaged young people from an early age.

currently hamstrung by a generalizing tendency, which means that crude anti-"couch potato" policies and programs are being rolled out at a rapid rate. My call in this chapter is for researchers to be much more methodologically precise and epistemologically rigorous when defining the nature of the problem they claim to be addressing, the population of people "at risk," and the rationale for whatever course of action is proposed. In short, I am saying that doing the highest possible quality research, free of sloppy generalizations and conceptual slipperiness, is part of doing ethical research.

One of the dangers at the moment is also that "be active" and "lose weight" messages will be interpreted by young people as condescending, overly serious, or simply old news. Unless they refer to some obviously fatal and dramatic health risk (such as HIV-AIDS or bird flu), there is not much chance that public health initiatives will convince young people en masse to adopt new lifestyles. There is research suggesting that behavior change may be more likely when young people feel in control of their decisions (e.g., Johnson & Birch, 1994), and this should at least temper our scholarly obsessions for seeing young people as machines, name-calling ("couch potatoes"), and generalized moral disapprobation ("epidemic of inactivity").

Ethics

Although all forms of research present us with ideological and moral choices, this is perhaps especially so when we study young people. The young often serve as the screen upon which the fears and anxieties of adults, including researchers, are projected. And though many would see the study of health and its various dimensions, including physical activity, as a generally objective scientific endeavor, it too is both enriched and compromised by inevitable moral biases. After all, those who study human health offer us answers about how we "should" live, and though some scientists may insist on the conceptual distinction between being "well" and being "good," being "healthy" and "active" will probably always be closely associated with moral rectitude in Western culture.

I do not doubt that youth physical activity researchers do their work with the best of intentions in mind. But I am also professionally and personally aware of a wide range of highly dubious interventions now being conducted by universities, schools, and community organizations in the name of promoting physical activity and, never too far away, weight loss. "Fat camps" in the public arena are gaining in popularity while fitness testing in schools is resurgent following a period in which many physical educators had agreed that fitness testing was a bad idea and a simple waste of school time. Very often physical activity programs resemble "crash courses" in physical activity, sometimes superfluously directed at youth who are already physically active and other times directed at youth who *are* inactive and potentially not equipped to handle program demands.

It is precisely because the popular media and academic community have traded in the language of a "generation of couch potatoes" that many physical activity professionals believe we need to "up" the pressure on young people to understand the benefits of physical activity and to stop avoiding it. Physical and health education curriculum documents in Australia, Canada, New Zealand, the United Kingdom, and the United States now insist that young people acknowledge the health benefits of physical activity, monitor their own physical activity habits, and set their own personal fitness goals. And all of this is being done in the mistaken belief that young people are like "wind-up dolls"—mindless machines that will simply behave in the way we program them to.

These things are done despite the debatable scientific foundations these imperatives rest on and, apparently, without any concern about the effects of presenting young people with weighty decisions about the long-term quality of their lives and, as they are constantly reminded, the possibility of an early death. Is all of this necessary, and are we sure that emphasizing "healthy decision making" is an efficacious or ethical course of action? Of course, focusing on "healthy decision making" has the benefit of placing the responsibility for health on the individual. While it is true that some dissenting voices can be heard in the literature, I am aware of very few who have been prepared to challenge the field's preoccupation with the behavior of individuals in public forums and, instead, to advocate for more macro-level policies that shape the physical conditions in which young people live. Indeed, it often appears that the research community is leading the "couch potato" chorus. Perhaps it has been going on for so long that scientists and other experts do not notice their tendency to tell us how we should live our daily lives. If nothing else, it is probably useful from time to time to draw attention to the "habits" of those who would lecture others about theirs.

APPLICATIONS FOR RESEARCHERS

Part of researchers' ethical responsibility is to offer society a considered and balanced view of the world rather than adding to the torrent of hyperbolic and inaccurate chatter that fuels contemporary public comment about youth physical activity. We need to remember that when we generalize about or categorize or judge youth we are talking about real people, many of whom live very different lives from our own, sometimes by choice, sometimes by necessity. Acting ethically also means thinking about the ways in which preconceived ideas shape our research questions, the methods we use, and the conclusions we reach. Greater understanding among youth physical activity researchers of the philosophical and epistemological assumptions *and* potential practical consequences of particular kinds of research would probably be a good thing. At the very least, researchers should provide evidence for their views on youth physical activity even, and perhaps especially, when they see these views as self-evident.

APPLICATIONS FOR PROFESSIONALS

It is reasonable to expect people who wish to intervene in young people's lives to articulate a clear and credible justification for their proposed intervention. This means being precise about the problem to be addressed and the rationale for specific kinds of interventions, and being sure that the proposed intervention will do as little harm to young people as possible. A lack of precision in youth physical activity programs probably accounts for our general lack of success in the past. But perhaps if we also demonized young people a little less and were more cautious about the claims we make for the benefits of physical activity, we might make physical activity sound a little less like bad-tasting medicine and more like a normal part of life. Youth physical activity is not a matter of life and death, and trying to convince young people that this is the case serves no purpose.

References

Australian Institute of Health and Welfare. (2004). *National report on health sector performance indicators 2003.* Canberra: AIHW.

Balzer, K. (2005). Menschen ziehen den Jahrhundertweg. *Pflege Zeitschrift, 58*(8), 471.

Booth, M.L., Okely, A.D., Chey, T., Bauman, A.E., & Macaskill, P. (2002). Epidemiology of physical activity participation among New South Wales school students. *Australian and New Zealand Journal of Public Health, 26*(4), 371-374.

Boreham, C., & Riddoch, C. (2003). Physical activity and health through the lifespan. In J. McKenna & C. Riddoch (Eds.), *Perspectives on health and exercise* (pp. 11-30). Basingstoke: Palgrave.

Brown, W. (2001). Couch potato: No quick fix. *Sport Health, 19*(1), 9-10.

Brown, W.J., & Brown, P.R. (1996). Children, physical activity and better health. *ACHPER Healthy Lifestyles Journal, 43*(1), 19-24.

Carlson, T.B. (1995). We hate gym: Student alienation from physical education. *Journal of Teaching in Physical Education, 14*(4), 467-477.

Centers for Disease Control and Prevention. (2001). *Youth risk behavior surveillance—surveillance summaries* (Vol. 51, No. SS-4). Atlanta: CDC.

Coleman, P. (1998). I'll take the health benefits of exercise without the risks please. *Lancet, 352*(9126), 492.

Conway, D. (2003). "Aussie sport is in crisis." *Geelong Advertiser,* 4 December, 1, 54.

Crawford, D.A., Jeffrey, R.W., & French, S.A. (1999). Television viewing, physical inactivity and obesity. *International Journal of Obesity, 23*(4), 437-440.

De Bourdeaudhuij, I., Sallis, J., & Vandelanotte, C. (2002). Tracking and explanation of physical activity in young adults over a 7-year period. *Research Quarterly for Exercise and Sport, 73*(4), 376-385.

Dietz, W.H. (2002). Foreword. In W. Burniat, T. Cole, I. Lissau, & E. Poskitt (Eds.), *Child and adolescent obesity: Cause and consequences, prevention and management* (pp. xv-xvii). Cambridge: Cambridge University Press.

Dionne, I., & Tremblay, A. (2000). Human energy and nutrient balance. In C. Bouchard (Ed.), *Physical activity and obesity* (pp. 151-179). Champaign, IL: Human Kinetics.

Durnin, J.V.G.A. (1992). Physical activity levels—past and present. In N.G. Norgan (Ed.), *Physical activity and health: 34th symposium volume of the Society for the Study of Human Biology* (pp. 20-27). Cambridge: Cambridge University Press.

Federal Interagency Forum on Child and Family Statistics. (2005). *America's children: Key national indicators of well-being.* Washington, DC: U.S. Government Printing Office.

Flegal, K.M. (1999). The obesity epidemic in children and adults: Current evidence and research issues. *Medicine and Science in Sports and Exercise, 31*(Suppl. 11), S509-S514.

Fogelholm, M., Nuutinen, O., Pasanen, M., Myöhänen, E., & Säätelä, T. (1999). Parent-child relationship of physical activity patterns and obesity. *International Journal of Obesity, 23*(12), 1262-1268.

Frank, L.D., Engelke, P.O., & Schmid, T.L. (2003). *Health and community design: The impact of the built environment on physical activity.* Washington, DC: Island Press.

Gard, M. (2005). A reply to Hancox: The problem with medical and scientific thinking about obesity. *Children's Issues: Journal of the Children's Issues Centre, 9*(1), 37-39.

Gard, M., & Wright, J. (2005). *The obesity epidemic: Science, morality and ideology*. London: Routledge.

Gordon-Larsen, P., McMurray, R.G., & Popkin, B.M. (2000). Determinants of adolescent physical activity and inactivity patterns. *Pediatrics, 105*(6), 301-306.

Gorely, T., Marshall, S.J., & Biddle, S.J.H. (2004). Couch kids: Correlates of television viewing among youth. *International Journal of Behavioral Medicine, 11*(3), 152-163.

Green, K. (2004). Physical education, lifelong participation and "the couch potato society." *Physical Education and Sport Pedagogy, 9*(1), 73-86.

Halpin, T. (2004). Couch potato children "are actually fighting fit." *The Times*, 14 December, 5.

Hill, J.O., & Melanson, E.L. (1999). Overview of the determinants of overweight and obesity: Current evidence and research issues. *Medicine and Science in Sports and Exercise, 31*(11), S515-S521.

Hill, J.O., & Peters, J.C. (1998). Environmental contributions to the obesity epidemic. *Science, 280*(5368), 1371-1374.

Hopple, C., & Graham, G. (1995). What children think, feel, and know about physical fitness testing. *Journal of Teaching in Physical Education, 14*(4), 408-417.

Johnson, S.L., & Birch, L.L. (1994). Parents' and children's adiposity and eating style. *Pediatrics, 94*(5), 653-661.

Kimm, S.Y.S., Glynn, N.W., Kriska, A.M., Barton, B.A., Kronsberg, S.S., Daniels, S.R., Crawford, P.B., Sabry, Z.I., & Liu, K. (2002). Decline in physical activity in black girls and white girls during adolescence. *New England Journal of Medicine, 347*(10), 709-715.

Koplan, J.P. (2000). The obesity epidemic: Trends and solutions. *Sports Medicine Bulletin, 35*(3), 8.

Lawrence, M., Lawrence, F., Durnin, J.V.G.A., & Whitehead, R.G. (1991). A comparison of physical activity in Gambian and UK children aged 6-18 months. *European Journal of Clinical Nutrition, 45*(5), 243-252.

Lee, J. (2003). The place and meaning of physical activity, physical education, and physical culture in the lives of young people living in rural Queensland. *Education in Rural Australia, 13*(2), 27-46.

Magarey, A.M., Daniels, L.A., & Boulton, T.J.C. (2001). Prevalence of overweight and obesity in Australian children and adolescents: Reassessment of 1985 and 1995 data against new standard international definitions. *Medical Journal of Australia, 174*(11), 561-564.

Malina, R.M. (1996). Tracking of physical activity and physical fitness across the lifespan. *Research Quarterly for Exercise and Sport, 67*(Suppl. 3), S48-S57.

Marshall, S.J., Biddle, S.J.H., Gorely, T., Cameron, N., & Murdey, I. (2004). Relationships between media use, body fatness and physical activity in children and youth: A meta-analysis. *International Journal of Obesity, 28*(10), 1238-1246.

McKenzie, T.L., Marshall, S.J., Sallis, J.F., & Conway, T.L. (2000). Student activity levels, lesson context, and teacher behavior during middle school physical education. *Research Quarterly for Exercise and Sport, 71*(3), 249-259.

Micic, D. (2001). Obesity in children and adolescents—a new epidemic? Consequences in adult life. *Journal of Pediatric Endocrinology and Metabolism, 14*(Suppl. 5), 1345-1352.

Morris, L., Sallybanks, J., Willis, K., & Makkai, T. (2003). *Sport, physical activity and antisocial behaviour in youth.* Canberra: Australian Institute of Criminology.

Neumark-Sztainer, D., Story, M., Hannan, P.J., Tharp, T., & Rex, J. (2003). Factors associated with changes in physical activity: A cohort study of inactive adolescent girls. *Archives of Pediatrics and Adolescent Medicine, 157*(8), 803-810.

Nicholl, J.P., Coleman, P., & Brazier, J.E. (1994). Health and healthcare costs and benefits of exercise. *Pharmacoeconomics, 5*(2), 109-122.

Olshansky, S.J., Passaro, D.J., Hershow, R.C., Layden, J., Carnes, B.A., Brody, J., Hayflick, L., Butler, R.N., Allison, D.B., & Ludwig, D.S. (2005). A potential decline in life expectancy in the United States in the 21st century. *New England Journal of Medicine, 352*(11), 1138-1145.

Panter-Brick, C. (2002). Sexual division of labor: Energetic and evolutionary scenarios. *American Journal of Human Biology, 14*(5), 627-640.

Panter-Brick, C. (2003). The anthropology of physical activity. In J. McKenna & C. Riddoch (Eds.), *Perspectives on health and exercise* (pp. 263-284). Basingstoke: Palgrave Macmillan.

Paxton, S.J., Sculthorpe, A., & Gibbons, K. (1994). Weight-loss strategies and beliefs in high and low socioeconomic areas of Melbourne. *Australian Journal of Public Health, 18*(4), 412-417.

Peters, J.C., Wyatt, H.R., Donahoo, W.T., & Hill, J.O. (2002). From instinct to intellect: The challenge of maintaining healthy weight in the modern world. *Obesity Reviews, 3*(2), 69-74.

Pi-Sunyer, X. (2003). A clinical view of the obesity problem. *Science, 299*(5608), 859-860.

Pratt, M., Macera, C.A., & Blanton, C. (1999). Levels of physical activity and inactivity in children and adults in the United States: Current evidence and research issues. *Medicine and Science in Sports and Exercise, 31*(Suppl. 11), S526-S533.

Prentice, A.M., & Jebb, S.A. (1995). Obesity in Britain: Gluttony or sloth? *British Medical Journal, 311*(7002), 437-439.

Rehor, P., & Cottam, B. (2000). Physical activity levels of northern Tasmanian high school students. *ACHPER Healthy Lifestyles Journal, 47*(1), 14-17.

Rippe, J.M., Blair, S.N., Freedson, P., Micheli, L., Morrow, J., Pate, R., Plowman, S., & Rowland, T. (1992). Childhood health and fitness in the United States: Current status and future challenges, part I. *Medicine, Exercise, Nutrition and Health, 1*, 97-104.

Rippe, J.M., & Hess, S. (1998). The role of physical activity in the prevention and management of obesity. *Journal of the American Dietetic Association, 98*(10 Suppl. 2), S31-S38.

Salbe, A.D., & Ravussin, E. (2000). The determinants of obesity. In C. Bouchard (Ed.), *Physical activity and obesity* (pp. 69-102). Champaign, IL: Human Kinetics.

Savage, M.P., & Scott, L.B. (1998). Physical activity and rural middle school adolescents. *Journal of Youth and Adolescence, 27*(2), 245-249.

Schools Health Education Unit. (2004). *Trends—young people and physical activity: Attitudes to and participation in exercise and sport 1987-2003.* Exeter: SHEU.

Seefeldt, V., Malina, R.M., & Clark, M.A. (2002). Factors affecting levels of physical activity in adults. *Sports Medicine, 32*(3), 143-168.

Seidell, J.C. (2000). The current epidemic of obesity. In C. Bouchard (Ed.), *Physical activity and obesity* (pp. 21-30). Champaign, IL: Human Kinetics.

Simons-Morton, B.G., Taylor, W.G., Snider, S.A., & Huang, I.W. (1993). The physical activity of fifth-grade students during physical education. *American Journal of Public Health, 83*(2), 262-264.

Simons-Morton, B.G., Taylor, W.G., Snider, S.A., Huang, I.W., & Fulton, J.E. (1994). Observed level of elementary and middle school children's physical activity during physical education classes. *Preventive Medicine, 23*(4), 437-441.

Singh, J., Prentice, A.M., Diaz, E., Coward, W.A., Ashford, J., Sawyer, M., & Whitehead, R.G. (1989). Energy expenditure of Gambian women during peak agricultural activity measured by the doubly-labelled water method. *British Journal of Nutrition, 62*, 315-329.

Sleap, M., Warburton, P., & Waring, M. (2000). Couch potato kids and lazy layabouts: The role of primary schools in relation to physical activity among children. In A. Williams (Ed.), *Primary school physical education: Research into practice* (pp. 31-50). London: Routledge/Falmer.

Steptoe, A., & Butler, N. (1996). Sports participation and emotional wellbeing in adolescents. *Lancet, 347*(9018), 1789-1792.

Stratton, G. (1997). Children's heart rates during British physical education lessons. *Journal of Teaching in Physical Education, 16*(3), 357-367.

Taggart, A., Medland, A., & Alexander, K. (1995). "Goodbye Superteacher!": Teaching sport education in the primary school. *ACHPER Healthy Lifestyles Journal, 42*(4), 16-18.

Telama, R., & Yang, X. (2000). Decline of physical activity from youth to young adulthood in Finland. *Medicine and Science in Sports and Exercise, 32*(9), 1617-1622.

Thomas, H., Ciliska, D., Wilson-Abra, J., Micucci, S., Dobbins, M., & Dwyer, J. (2004). *Effectiveness of physical activity enhancement and obesity prevention programs in children and youth.* Ottawa: Health Canada.

Trost, S.G., Owen, N., Bauman, A.E., Sallis, J.F., & Brown, W. (2002). Correlates of adults' participation in physical activity: Review and update. *Medicine and Science in Sports and Exercise, 34*(12), 1996-2001.

Twisk, J.W.R. (2001). Physical activity guidelines for children and adolescents: A critical review. *Sports Medicine, 31*(8), 617-627.

Van Mechelen, W., Twisk, J.W.R., Post, G.B., Snel, J., & Kemper, H.C.G. (2000). Habitual physical activity of young Dutch males and females: 15 years of follow-up in the Amsterdam Growth and Health Study. *Medicine and Science in Sports and Exercise, 32*(9), 1610-1616.

Warburton, P., & Woods, J. (1996). Observation of children's physical activity levels during primary school physical education lessons. *European Journal of Physical Education, 1*(1), 55-65.

Whitfield, K.E., Weidner, G., Clark, R., & Anderson, N.B. (2002). Sociodemographic diversity and behavioural medicine. *Journal of Consulting and Clinical Psychology, 70*(3), 463-481.

Williams, A. (2000). Research and the primary school teacher. In A. Williams (Ed.), *Primary school physical education: Research into practice* (pp. 1-15). London: Routledge/ Falmer.

Wright, J. (1997). Fundamental motor skills testing as problematic practice: A feminist analysis. *ACHPER Healthy Lifestyles Journal, 44*(4), 18-20.

Wright, J., Macdonald, D., & Groom, L. (2003). Physical activity and young people: Beyond participation. *Sport, Education and Society, 8*(1), 17-33.

Yelling, M., Penney, D., & Swaine, I.L. (2000). Physical activity in physical education. *European Journal of Physical Education, 5*(1), 45-66.

Developmental and Psychological Factors in Youth Physical Activity and Sedentary Behavior

Part I provided key foundation messages, including information on measurement, health outcomes, trends, and concepts. Part II builds on this by providing a comprehensive account of personal factors likely to be associated with physical activity and sedentary behavior in young people. Typically this has involved a psychological perspective; but in this book we have adopted a wider view by also including developmental issues related to maturation, ethnicity, environments, and genetics, thus offering an important biocultural view (chapter 6). This provides context for more focused individual psychological approaches to physical activity and sedentary behavior in young people.

Specifically, we consider fundamental issues associated with what young people might *think* about physical activity ("attitudes," chapter 7), how they might be *energized* into being active ("motivation," chapter 8), and how young people *view themselves* generally and physically as well as how this might be related to their physical activity ("the self," chapter 9). Finally, it is important to recognize that not all young people are the same and that some have movement difficulties (chapter 10). This is discussed in the context of disability alongside well-tested motivational theories, such as those adopting perspectives centered on competence.

CHAPTER

6

Biocultural Factors in Developing Physical Activity Levels

Robert M. Malina, PhD

Physical activity and sedentary behavior are topics of considerable interest in public health, medicine, and education. The public health and biomedical perspectives focus on physical activity in the context of health promotion and disease prevention, as well as physical inactivity as a major risk factor, among others, for several degenerative diseases. The educational perspective highlights activity in the context of physical education as a major component of the overall school experiences of youth.

Physical activity and sedentary behaviors occur in many contexts, and both are important avenues for learning, enjoyment, social interactions, and self-understanding. Currently, however, evidence and opinion suggest an imbalance in the direction of increased sedentary behaviors and reduced activity levels underlying the current epidemic of obesity and the emergence of components of the metabolic syndrome and risk factors for cardiovascular disease in youth (Malina, Bouchard, & Bar-Or, 2004; Strong et al., 2005).

Physical activity is a multidimensional behavior. It is viewed most often in terms of energy expenditure and the stresses and strains associated with weight bearing and ground reaction forces. Fitness (performance and health related) and skill (proficiency) are other important dimensions. Context is an important dimension of physical activity that is often overlooked. Context refers to types and settings and includes play, physical education, exercise, sport, work, and others. Contexts and the meanings attached to them vary among and within different cultural groups.

Physical inactivity also has several dimensions. Public health and medicine view inactivity in terms of insufficient energy expenditure, force generation, and health-related fitness. Physical activity levels are significantly influenced by culture, and many forms of sedentary behavior are highly valued by societies—school, study, reading, music, art, television viewing, video games, and the like. Motorized transport is a form of sedentary behavior that is also valued by major segments of society.

Physical activity and sedentary behavior represent a repertoire of behaviors performed in a societal context, and many forms of each have high valence in society. Physically active and sedentary behaviors span the spectrum of maximal to minimal energy expenditure. The present chapter considers the development of this spectrum of behaviors in the context of the demands of childhood and adolescence. It emphasizes the interdependence of biology and culture, broadly defined.

The business of growing up in which all children and adolescents are engaged involves three primary tasks:

- To **grow—**to increase in the size of the body as a whole and of its parts and systems
- To **mature—**to progress toward the biologically mature state, which is an operational concept because the mature state varies with the body system
- To **develop—**to learn the appropriate cognitive, social, affective, moral, motor, and other behaviors expected by society

Growth and maturation are biological processes, while development is a behavioral process often expressed by the phrase, "socialization specific to a culture." The three processes are distinct, though related and interacting, tasks that dominate the daily lives of youth for approximately the first two decades of life (Malina et al., 2004). Motor development, though included among developmental tasks, clearly involves interactions among the three processes: growth in size, proportions, and muscle mass; maturation of the neuromuscular system; and cultural demands and expectations associated with the early movements of children. Interactions among growth, maturation, and development thus vary during childhood and adolescence, among individuals, and within and between cultural groups. These interactions in turn influence active and sedentary behaviors, among others.

The terms *biosocial* and *biopsychosocial* are often used in discussions of active and sedentary behaviors. The term *biocultural* is more inclusive, as culture extends beyond psychological and social domains. Though definitions of culture vary, it can be viewed as a system of meanings and associated feelings in which individuals and groups are embedded. The system of meanings is an amalgam of symbols, values, and behaviors that characterize a population (Kroeber & Kluckhohn, 1952). Meanings influence beliefs, sanctions, definitions, and ultimately human behaviors, including physically active and sedentary behaviors in their many forms. Potential interrelationships among culture, the individual, the

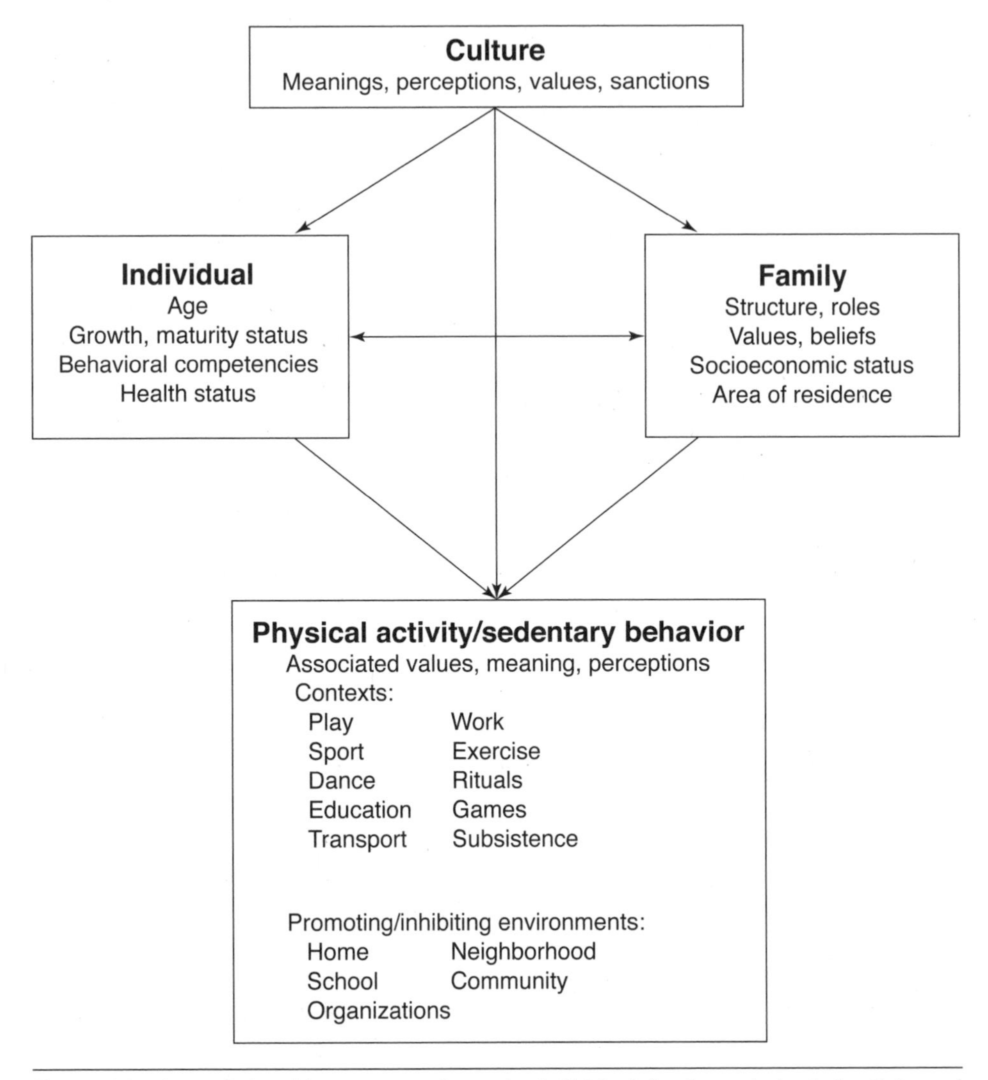

Figure 6.1 Interrelationships among culture, the individual, family, and physical activity and sedentary behavior.

family, and physical activity and sedentary behavior are schematically illustrated in figure 6.1. Children and adolescents grow, mature, and develop within this complex of interrelated domains.

Movement skills are the substrate of physical activity, "any bodily movement produced by skeletal muscles and resulting in energy expenditure" (Bouchard & Shephard, 1990, p. 6). The development of proficiency in movement skills (behaviors) is influenced by the child's own play behaviors and activity level (Butcher & Eaton, 1989) and by interactions of growth and maturation with these behaviors and the environments in which the individual is reared.

A Biocultural Perspective

Many factors influence physical activity levels and sedentary behaviors of children and adolescents. Discussions tend to focus on physical activity levels and by inference sedentary behavior. A number of correlates of activity, both positive and negative, have been identified for children and adolescents, arbitrarily categorized as ages 4 to 12 and 13 to 18 years, respectively (Sallis, Prochaska, & Taylor, 2000). The correlates are labeled as demographic and biological; psychological, cognitive, and emotional; behavioral attributes and skills; social and cultural; and physical environment factors. Except for sex of the individual (which is genetically determined) and the body mass index (BMI, kg/m^2), biological factors are not ordinarily included in large-scale surveys and observational studies of physical activity levels and sedentary behavior. Heredity, indicators of growth and maturation, nutritional status, and physical fitness are notably lacking. The BMI is often used as a proxy for fatness. However, because the BMI is about equally related to fat-free mass, fat mass, and percentage fat in children and adolescents it is more appropriately an index of heaviness (Malina et al., 2004). At its low and high extremes, the BMI most likely reflects, respectively, deficiency in muscle mass and excessive fatness.

The term *physical activity* may have different meanings to those in the biomedical and behavioral sciences and also to individuals who are subjects in many surveys. There is a need to consider the meaning of the constructs physical activity and sedentary behavior in youth and how they change according to age and sex during childhood and adolescence into adulthood. Table 6.1 summarizes responses of middle school youth from a rural community in the United States to a single question administered during language arts (English) classes: "What does the term 'Physical Activity' mean to you?" The sample included youth ages 11 to 14 years in grades 6 to 8 of American White, Black, and Hispanic ancestry. The question was open-ended, and youth could write as much or as little as they preferred. Answers were subsequently summarized inductively into several categories of descriptors used to characterize the construct physical activity. All children indicated at least one descriptive label (the "First" column in the table), including eight boys and four girls who noted that they did not know what physical activity meant! A number of children indicated two or three descriptors (the "Second" and "Third" columns in the table). There was no pattern in the distribution of descriptors by grade or ethnicity. The variety of descriptors used by youth to describe physical activity is considerable. The most commonly reported descriptor for physical activity in both sexes was sport(s). More girls than boys described the meaning of physical activity as "moving/doing something" and "staying/getting in shape." Though relatively few, more girls than boys also indicated a contrasting descriptor (i.e., not being lazy or inactive).

Correlates of activity are different in childhood and adolescence, and vary with ethnicity and socioeconomic status. Biological correlates are not often

Table 6.1 Categories of Descriptors Used by Youth Ages 11 to 14 Years in Response to the Question "What does the term 'Physical Activity' mean to you?"[1]

	BOYS						GIRLS					
	First (n = 92)[2]		Second (n = 65)		Third (n = 36)		First (n = 119)		Second (n = 92)		Third (n = 58)	
Category	f*	%	f*	%	f*	%	f*	%	f*	%	f*	%
Sports[3]	31	34	16	25	9	25	28	24	22	24	15	26
Moving/doing something[4]	12	13	6	9	2	6	21	18	6	7	4	7
Exercise	8	9	3	5	1	3	15	13	11	12	4	7
Do not know	8	9	—		—		4	3	—		—	
Fun	7	8	8	12	1	3	5	4	6	7	3	5
Staying/getting in shape[5]	6	7	9	14	7	19	21	18	12	13	5	9
Outdoor activity	6	7	1	2	1	3	6	5	1	1	—	
Physical education	6	7	1	2	—		4	3	5	5	2	3
Fitness activities[6]	5	5	11	17	4	11	7	6	17	18	8	14
Play	1	1	5	8	2	6	1	1	4	4	3	5
Not being lazy/inactive	1	1	—		1	3	1	1	2	2	4	7
Games	—		2	3	5	14	1	1	—		4	7
Chores/work	—		2	3	1	3	2	2	2	2	3	5
Others[7]	1	1	1	2	2	6	3	3	4	4	3	5

*Frequency

[1]R. Malina unpublished data. The question was presented to all youth in attendance in middle school language arts/English classes following standard informed consent procedures. Answers were collated into the indicated categories independently on two separate occasions.

[2]All children gave at least one answer to the question, while others gave two or more. The column labeled First refers to the single descriptor and the first of several descriptors. The column labeled Second refers to the second descriptor; the column labeled Third refers to the third descriptor where applicable.

[3]Specific sports were collapsed into the sport(s) category.

[4]Moving/doing something includes descriptors such as letting all of your energy fly, doing something with your body, and something that gets you moving.

[5]Staying in shape includes "working out."

[6]Fitness activities include running, jogging, swimming, bicycling, and walking.

[7]Others includes skateboarding in both sexes and dance and rodeo in girls.

considered, and "social and cultural" correlates focus primarily on parents and only incidentally on siblings and peers, which reflects a very narrow view of culture. There is a need to broaden our perceptions of potential correlates and to integrate the biological and the cultural for a more complete understanding of youth in the context of physical activity and sedentary behavior. In other words, children and adolescents can be considered in neither an exclusively biological nor an exclusively cultural manner; a biocultural perspective is warranted. In this perspective, biological and cultural variables interact so that biology can influence behavior (i.e., development) and behavior can influence biology. The subsequent discussion highlights the biocultural perspective in the context of several factors associated with physical activity levels and sedentary behavior.

Factors Related to Physical Activity Levels

A number of factors influence physical activity levels. Among the most significant of these are age and sex, maturity status, ethnicity, and genetics. These are characteristics of the individual. They are treated separately here, but in reality they interact with each other and with the cultural conditions and environments within which the individual is reared and lives.

Age and Sex

Total daily energy expenditure (TDEE), assessed by the doubly labeled water method, shows a decline with age beginning at about 3 years but accelerating by 6 to 7 years of age (Torun et al., 1996). About 50% to 60% of the decline is associated with the age-related decline in basal metabolic rate (Malina et al., 2004). Daily energy expenditure is, on average, higher in boys than in girls at all ages. The sex difference is small in early childhood and becomes greater with age (Torun et al., 1996), likely reflecting the sex difference in fat-free mass, specifically muscle mass (Malina et al., 2004).

Energy expended in physical activity is a significant and variable component of TDEE. Data on activity energy expenditure (AEE = TDEE – resting metabolic rate, adjusted for thermal effect of food) spanning early childhood through adolescence are becoming available. Absolute AEE increases with age and is greater in boys than in girls (DeLany et al., 2002; Grund et al., 2000; Hoos et al., 2003; Spadano et al., 2005). These trends reflect age-associated changes in body size and specifically fat-free mass and the sex difference in size and composition (Malina et al., 2004).

The doubly labeled water method and the associated measure of AEE provide global estimates and do not capture components of physical activity related to duration, frequency, and intensity that may vary with age and sex. Estimated AEE, for example, was not related to estimated physical activity derived from a maternal questionnaire and television time (sedentary behavior) in two samples of children 5.3 ± 0.9 and 6.3 ± 0.9 years of age (Goran et al., 1997), but was significantly, though moderately, related to movement counts (accelerometry) in youth of both sexes roughly 10 and 18 years of age (Ekelund et al., 2004).

Estimates of physical activity levels based on questionnaires, movement monitors or counters, and heart rate are more available than measures of TDEE and AEE. Evidence indicates greater levels of movement activity in males than in females beginning prenatally. The trend in effect sizes from a meta-analysis indicates an increase in sex differences with age: prenatal, 0.33 ± 0.15; infancy, 0.29 ± 0.11; preschool, 0.44 ± 0.06; and school age, 0.64 ± 0.06 (Eaton & Enns, 1986). Parental (preschool, primary grades)- and self (older grades)-reported levels of physical activity are reasonably stable during childhood and in many studies reach a peak during the transition into adolescence, about 10 to 12 years of age in girls and 11 to 14 years of age in boys, followed by a decline. There is, however, considerable variation among studies.

The magnitude of age- and sex-associated variation in physical activity levels varies with instrument and context, especially in adolescence. In a longitudinal study of Dutch adolescents, total activity (activities >4 METs, expressed as METs/week) declined, on average, from ages 13 through 16 years with little absolute change to 21 years, and males were more active than females (Van Mechelen et al., 2000). The trends, however, vary by context of activity. Males are more active than females in organized sport, but there are no changes with age across adolescence into young adulthood. Males are also more active than females in nonorganized sport, but there is a major decline in activity in males from ages 13 to 16 years, a smaller decline in females, and a continued decline in both sexes to 21 years. On the other hand, other activities (nonsport) show no consistent sex difference across adolescence into young adulthood, with stable levels from ages 13 to 16 years and an increase to young adulthood.

Similar variation with age and sex is evident among adolescents and young adults ages 12 to 21 years in the 1991-1992 U.S. National Health Interview Study (Caspersen, Pereira, & Curran, 2000). Percentages of males and females engaged in regular sustained light-to-moderate activity (five or more days/week and 30 min/occasion walking, bicycling) decline with age from 12 to 17 years and are then stable through 21 years. Percentages of boys engaged in regular vigorous activity (three or more days/week running, jogging, swimming) increase from ages 12 to 14 years and then decline through ages 19 to 21 years; corresponding percentages of girls decline linearly with age from 12 through 20 years. Percentages of boys and girls engaged in strengthening and stretching activities (three or more days/week) are reasonably stable from 12 to 16 years and then decline. On the other hand, percentages of physically inactive (inactivity was simply defined as "no participation in vigorous or moderate physical activity") boys and girls are rather stable from ages 12 to 15 years and then increase gradually to about 20% at 20 years of age. Unfortunately, specific sedentary behaviors were not identified. The trend in physical inactivity across adolescence into young adulthood is generally consistent with the suggestion that adolescence may be a risk factor for inactivity (Rowland, 1999).

Over ages 12 to 21 years, the prevalence of specific activities declines differentially by activity and sex. The prevalence of regular vigorous activity and stretching activities declines more in females (36% and 28%, respectively) than in males (29% and 19%, respectively), while the prevalence of regular sustained activity declines more in males (16%) than in females (10%). In contrast, the prevalence of strengthening activities declines to about the same extent in males (19%) and females (21%).

Males are more often involved than females in strengthening, regular vigorous, and regular sustained activities; the estimated magnitude of the sex differences are 18%, 11%, and 5%, respectively. On the other hand, the sex difference in percentages of physically inactive youth and youth engaged in stretching activities is negligible (Caspersen et al., 2000).

With the exception of BMI, studies of correlates of physical activity levels have not ordinarily considered potential associations between morphological characteristics of youth and participation in specific activities. This issue was addressed in a sample of Portuguese adolescents, 387 males and 410 females, 15.5 to 18.4 years of age (Coelho e Silva, Sobral, & Malina, 2003). Participation in physical activity over the previous year was estimated using a protocol modified after that of Aaron and colleagues (1993, 1995). In addition to an overall score, the protocol provided estimates for time in organized sport, recreational activities, and fitness-health activities. Morphological variables included indicators of body size and proportions and a sum of eight skinfolds. Results of canonical correlation analyses indicated differences by sex. Among males, the first canonical correlate indicated an association of smaller body size (height, weight, BMI) and a lower androgyny index (less masculine shoulder-hip proportions) with lower participation in physical activity, primarily organized sport and to a lesser extent fitness activities. Among girls, one significant canonical correlate highlighted an association of greater height, a lower BMI and sum of skinfolds, and a higher androgyny index with increased participation in organized sport. Otherwise, the influence of morphology on recreational and fitness-health activities of girls is negligible.

The results suggest an association between physical characteristics of adolescents and specific contexts of physical activity, with associations varying by context—organized sport, recreational activities, and fitness-health activities. The trends are generally consistent with our knowledge of the characteristics of young athletes. Among boys, small size was strongly associated with low participation in organized sport. With the exception of gymnasts, adolescent male athletes tend to have heights and weights that are equal to and above reference values for the general population and to have proportionally broader shoulders (Malina et al., 2004). On the other hand, larger size, leanness, and more masculine shoulder–hip relationships were associated with increased participation of adolescent females in organized sport, which is also consistent with observations on the size and physique characteristics of these athletes (Malina et al., 2004).

Maturity Status

Age for age, females are, on average, advanced in maturity status compared to males. The sex difference is apparent prenatally and, on average, reaches its largest magnitude in adolescence (Malina et al., 2004). Age- and sex-associated variation in physical activity levels and sedentary behavior may be influenced by differences in maturity status. Data on this issue are limited. Part of the problem relates to the failure to include an indicator of maturity status, especially one that is noninvasive, in physical activity surveys.

Percentage of predicted mature (adult) height was used as a maturity indicator in a study of the relationship among maturity status, sex, and activity level in children ages 5 to 9 years (Eaton & Yu, 1989). This maturity indicator utilizes age, height and weight of the child, and midparent height (average of paternal and maternal heights) to predict mature height. The current height of the child is then expressed as a percentage of her or his predicted mature height to provide an estimate of biological maturity status. The rationale is as follows: Two children of the same age can have the same height, but one is closer to mature height than the other. The individual who is closer to mature height is advanced in maturity status compared to the individual who is more removed from mature height (Malina et al., 2004). Among the children ages 5 to 9, sex of the child accounted for 25% of the variance in activity level, males being more active than females. Controlling for maturity, however, reduced the contribution of sex of the child to explained variance in activity level to 15%. By inference, maturity status tempers the effect of sex on activity level. Subsequent analyses indicated that a maturity effect on activity level was evident in boys but not girls, with advanced maturity of boys associated with lower activity levels (Eaton & Yu, 1989). The results thus suggest that variation in biological maturity status at these young ages may influence activity level and that the effect may differ in boys and girls.

Age- and sex-associated variation in physical activity levels and sedentary behavior may also be influenced by the differential timing of the adolescent growth spurt and sexual maturation in males and females. This issue was examined in the Saskatchewan Bone Mineral Accrual Study, a longitudinal study of Canadian adolescents (Thompson et al., 2003). Physical activity levels declined linearly with age from 10 through 18 years, males were more active than females, and the magnitude of the sex difference was constant. However, a different pattern of sex differences emerged when activity scores were expressed in terms of years before and after age at peak height velocity (PHV, age at maximum growth rate in height). Peak height velocity occurs, on average, two years earlier in girls than in boys (Malina et al., 2004). The magnitude of the sex difference in activity level declined from three years before PHV to the time at PHV, and there were no sex differences in physical activity at PHV and for five years after PHV. Using a biological marker in contrast to chronological age thus provides a different perspective of sex differences in physical activity levels during adolescence. This observation, of course, needs to be verified in other samples.

In addition to the sex difference in the timing of biological maturation (earlier in females), behavioral development associated with the transition into puberty and sexual maturation differs between the sexes and perhaps among cultural groups. It is reasonably well documented that self-worth, or self-esteem, declines in American girls during the transition into puberty, and the decline is especially marked in girls who enter puberty early compared to girls who enter puberty on time and later (Brooks-Gunn & Peterson, 1983; Simmons & Blyth, 1987). The key appears to be variation in maturational timing among girls. There is also ethnic variation. The developmental pattern for self-esteem differs for American Black girls compared to American White girls (Brown et al., 1998). Global self-worth was seen to decline in White girls but was stable in Black girls from ages 9 to 14 years. Adjusting for stage of sexual maturation (a proxy for maturational timing; Black girls mature in advance of White girls), BMI (Black girls tend to have a higher BMI), and household income (tends to be lower in Blacks) did not alter the trends. Further, as BMI increased, scores for perceived physical appearance and perceived social acceptance decreased, however, at distinct rates, with the trend weaker in Black than in White girls (Brown et al., 1998). Implications of these results for patterns of physical activity and sedentary behavior during adolescence need study. Interactions between differential timing of biological maturation, associated changes in perception of self-worth, and changes in active and sedentary behaviors during the transition into adolescence should be more closely examined. Allowing for individual and population differences, such studies should begin among girls as early as 9 years of age.

Ethnicity

Within the U.S. culture complex, ethnicity is most often classified on the basis of color, surname, geographic origin, or some combination of these: White (non-Hispanic White, European ancestry), Black (non-Hispanic Black, African ancestry), Hispanic (Mexican, Cuban, Puerto Rican, Central or South American ancestry), Native American (Amerindians), and Asian and Pacific Islanders. Classifications vary among countries; for example, the protocol of the Canadian Community Health Survey (2001) refers to the "cultural and racial backgrounds" of respondents and includes 12 categories with an additional one for other designations.

The term race is often used synonymously with ethnicity, but there are subtle differences between the terms. Ethnicity implies a group with cultural characteristics in common (culturally distinct), while race implies a group with biological characteristics in common (biologically distinct). Quite often, cultural and biological homogeneity overlap or coincide as in linguistic and religious groups who share a common ancestry or groups of color (see Damon, 1969).

Nevertheless, the terms ethnicity and race encompass genetic and cultural heterogeneity that reflect historical (political, immigration), geographic, and social factors among others. For example, in the United States, ethnicity/race is often coterminous with poverty or low socioeconomic status (or both) in minorities of color.

The literature relating ethnicity to physical activity levels among children and adolescents, though limited, is equivocal (Sallis et al., 2000). Nevertheless, some recent data may provide insights. Results from the 2003 Youth Risk Behavior Surveillance System (YRBSS) of high school youth in the United States indicate ethnic variation in physical activity and sedentary behaviors (Grunbaum et al., 2004). Variation in physical activities of different forms and intensities were noted. Percentages of American Black girls reporting participation in moderate and vigorous activities were lower than corresponding percentages of American White and Hispanic girls, while percentages of Black girls reporting no moderate or vigorous activity and watching ≥3 h per day of television were higher than those of White and Hispanic girls. Trends were similar in American Black boys, specifically for television viewing, but the differences were not as marked as among girls.

Of potential relevance to patterns of physical activity and sedentary behavior is ethnic variation in the prevalence of overweight and obesity among American adolescents. Both overweight (BMI > 85th and < 95th age- and sex-specific percentiles) and obesity (BMI ≥ 95th age- and sex-specific percentiles) are of greatest prevalence in American Black, followed by American Hispanic and then American White high school girls. Of interest, fewer Black girls describe themselves as overweight and report trying to lose weight compared to White and Hispanic girls. On the other hand, the prevalence of overweight and obesity is highest among Hispanic boys, followed by Black and then White boys. More Hispanic boys describe themselves as overweight and report trying to lose weight, followed by White and then Black boys (Grunbaum et al., 2004).

The trends in physically active and sedentary behaviors, overweight and obesity, perceptions of overweight, and attempts at weight loss among high school students suggest differences among ethnic groups in the United States. The variation is especially apparent in the perception of overweight among Black girls and pressures for slenderness among White girls. The Hispanic community in the United States is culturally more heterogeneous than American White and Black communities in the sense that individuals tend to retain contacts with their ancestral roots in Mexico and other Latin American and Caribbean countries. Hispanic families vary in degree of acculturation—"those phenomena which result when groups of individuals having different cultures come into continuous first-hand contact, with subsequent changes in the original cultural patterns of either or both groups" (Redfield, Linton, & Herskovits, 1936, p. 149). An associated concept is assimilation, the degree of integration or adaptation into the mainstream culture. Changes associated with acculturation occur at the individual and group levels. There is a need to map the course of acculturation in different groups to determine specific aspects of the process (and their correlates) that may influence physical activity levels and sedentary behavior.

Data on the association between acculturation and physical activity levels are limited. Overweight American Hispanic youth (age 11.2 ± 1.7 years), most likely Mexican American, did not differ in estimates of physical activity by degree of acculturation (foreign born, first generation, second-third generation), but peak $\dot{V}O_2$ was higher in youth from more acculturated families (Crespo et al., 2006). Degree of acculturation was based on birthplace of children and parents. The association between BMI and degree of acculturation (based on ethnic identification, language, food, television and music preferences, ethnicity of peers/friends, birthplace of parents and grandparents) varied between Mexican American girls and boys ages 6 to 12 years resident in Minnesota in the early 1990s (Eschway, 1994). Body mass index was, on average, lowest in girls from families of high acculturation compared to girls from families of low and medium acculturation. On the other hand, it increased in boys from families of low to medium to high acculturation. In contrast to BMI, two estimates of relative fat distribution (waist–hip ratio and trunk–extremity skinfold ratio) did not vary with acculturation status of the family (Eschway, 1994).

In a multiethnic, low-income, inner-city sample of Montreal youth 9 to 13 years of age, predictors of inactivity in both sexes included older age, Asian ancestry, low self-efficacy, and no participation on school teams or teams outside of school (O'Loughlin et al., 1999). Additional predictors of inactivity in boys included no paternal encouragement and infrequent use of video games; additional predictors in girls included no maternal encouragement and no maternal participation in sport.

Data dealing with physical activity levels and sedentary behavior in youth from different ethnic backgrounds are limited. It would be informative to address ethnic variation in perceptions and values attached to different contexts of physical activity, specifically organized sport, and to different forms of sedentary behavior, such as study time, education in the arts, and screen time.

Genetic Considerations

Physical activity levels are behavioral phenotypes. Results of several studies of twins and family members are compatible with the notion that genetic or cultural factors (or both) transmitted across generations may predispose an individual to be more or less active (Bouchard, Malina, & Perusse, 1997). A model commonly used in the analysis of twin and family data attempts to partition the observed variance into its genetic (i.e., additive genetic effects or heritability) and environmental components. The latter are often partitioned into common environmental effects (i.e., variation associated with a shared environment common to or equal for both members of a twin pair or members of a family) and unique environmental effects (i.e., variation associated with specific or different environments unique to each member of a twin pair or each member of the family unit). The common environment, for example, includes general living conditions, rearing conditions, household diet, and so on. Unique environments, for example, can include an active or inactive lifestyle for one member of a twin pair in contrast to the other, different friends or peers, and so on.

Several examples suffice to illustrate the potential role of genotype in physical activity levels. Variation in components of energy expenditure has a genetic basis (Bouchard et al., 1997). A recent study of twins 18 to 39 years of age used doubly labeled water to estimate energy expenditure and accelerometry to measure physical activity in the controlled environment of a respiration chamber and in free-living daily life (Joosen et al., 2005). Within the confined environment of the chamber, the majority of variation in activity energy expenditure was attributed to the common environment (68%). In contrast, most of the variation in activity energy expenditure under free-living conditions was accounted for by genetic factors (72%). Variation in daily physical activity (accelerometry) in the chamber was exclusively associated with common (41%) and unique (59%) environmental factors. In contrast, under free-living conditions, genetic factors accounted for the majority of the physical activity variance (78%). The results highlight the role of the restricted environment of the chamber in limiting physical activity, as well as the role of individual differences, including genotype, in physical activity as a component of daily living.

Participation in organized and nonorganized sport activities is commonly used as an indicator of physical activity levels in many European studies. The estimated heritability of sport participation scores is higher in male (77%) than female (35%) Dutch adolescent twins (Boomsma et al., 1989), while a more comprehensive analysis of sport participation scores in Flemish adolescent twins (15 years) indicates a larger role for additive genetic factors (i.e., independent effects of specific alleles) in sport participation scores of males (83% of the variance) than of females (44%) (Beunen & Thomis, 1999). Common environmental factors (e.g., familial pressures for similarity, peers, social dimensions of the activities) explain 54% of the variance in sport participation in females in contrast to none of the variance in males, while environmental factors unique to the individual explain 17% of the variance in males and only 2% in females (Beunen & Thomis, 1999). Thus, the results suggest a sex difference in sources of variation in sport participation scores of adolescent twins. Common environmental factors followed by genetic factors are apparently major determinants in adolescent girls. In contrast, genetic factors are the major source of variation in adolescent males with a small but significant contribution of environmental factors unique to individual males (e.g., motivation, skill).

Estimates of habitual physical activity and exercise participation (planned, structured moderate-to-vigorous activity) show significant familial similarity in the Quebec Family Study with some variation across generations (Perusse et al., 1989). Nontransmissible environmental factors account for most of the variance, 71% in habitual activity and 88% in exercise participation. Transmissible effects vary across generations, 29% for genetic transmission of habitual physical activity and 12% for cultural transmission of exercise participation (Perusse et al., 1989). The familial aggregation of habitual physical activity and exercise participation suggests an inherited propensity to be physically active or inactive.

There is a need for further study of factors associated with common environmental and cultural transmission of physical activity levels. Much emphasis focuses on the family environment (Malina, 1996). During childhood and less so during adolescence, a child's activity pattern is related, in part, to parental activity levels and attitudes. Although parents are generally viewed as providing role models for activity, evidence supporting the modeling hypothesis is not convincing (see chapter 11). Correlations between activity levels and estimated energy expenditure of parents and their offspring are generally low, less than 0.3 (Bouchard et al., 1997). The low correlations may reflect limitations of physical activity assessment protocols, but also highlight the need to consider other factors common to the family environment, such as peer and sibling modeling independent of parents. When activity levels were examined in more detail among children 4 to 7 years of age and their parents, children with active parents were more likely to be active (Moore et al., 1991). Active fathers or mothers were more likely to have active children compared to inactive fathers or mothers, with odds ratios of 3.5 and 2.0, respectively. But when both parents were active, their children were 5.8 times more likely to be active than children of two inactive parents. The process of translating parents' activity behaviors to activities in their children needs to be studied. Parents and children share genes in common, so a genetic effect is reasonable. The interaction between genotype and specific behaviors such as modeling and sharing activities, encouragement, emphasis on active behaviors, and others needs evaluation.

Clearly, much needs to be done to identify components of the environment that influence the predisposition to be physically active or inactive and to greater or lesser levels of energy expenditure. Evidence is clear that these predispositions are in part genetic, but the extent of the genetic contribution to habitual physical activity and the nature of the genes involved remain to be further investigated. Linkage studies utilize extended family pedigrees to identify where in the genome there *may* be loci on specific chromosomes harboring genes that influence a particular trait, while association studies look for specific genes (alleles) that may be associated with a particular trait (Bouchard et al., 1997). Detailed analyses of the human genome based on linkage and association studies have identified potential candidate genes for physical activity phenotypes. For example, analyses of the Quebec Family Study pedigree indicate a strong linkage between physical inactivity and loci on chromosome 2, as well as suggestive linkages for total daily activity and moderate-to-strenuous activity with loci on chromosome 13 and for inactivity and moderate-to-vigorous activity with loci on chromosome 7 (Wolfarth et al., 2005).

Activity Levels, Fitness, and Sport

Many studies have been done on the relationships among physical activity levels, physical fitness, and sport participation. Some conclusions may seem unexpected; many issues remain either unresolved or unexplored. The subsequent discussion considers what has been done in three broad areas and also suggests directions for what remains to be done; the three areas are relationships between activity and fitness, environments for physical activity, and sport participation. Other topics could certainly be added, but these highlight interactions between biological and cultural determinants and the need to consider both in discussions of physical activity in youth.

Physical Activity and Physical Fitness

It is generally assumed that individuals who are physically fit are more physically active than others and, conversely, that individuals who are physically active are more fit. The available data, however, do not support such generalizations. Activity level and fitness are related, but relationships are moderate at best and vary with indicators of activity and fitness. Growth and maturation, among others, are confounding factors. Indicators of fitness change with normal growth and maturation whether or not the child or adolescent is active. The current focus is more on health-related fitness than on performance-related fitness (Malina et al., 2004).

Correlational studies indicate generally low-to-moderate relationships between activity level and measures of health-related fitness in children (Pate, Dowda, & Ross, 1990; Sallis, McKenzie, & Alcaraz, 1993) and adolescents (Aaron et al., 1993; Katzmarzyk et al., 1998a). Subsequent multivariate analyses of data from children indicate that physical activity accounts for a relatively small percentage of variance in fitness variables, 3% to 21% (Pate et al., 1990; Sallis et al., 1993). Results of a corresponding analysis of youth ages 9 to 18 years indicate that TDEE, energy expenditure in moderate-to-vigorous physical activity, and time viewing television (sedentary behavior) account for 11% to 21% of the variance in measures of health-related fitness (Katzmarzyk et al., 1998b). Although relationships between indicators of physical activity, sedentary behavior, and fitness are significant, a large part of the variability in health-related fitness is not accounted for by physical activity or sedentary behavior.

It is possible that these relationships are masked, in part, by the normal range of variability in heterogeneous samples of children and adolescents. Relationships may be more apparent at the extremes of the physical fitness continuum. Are more fit youth in fact more active than unfit youth? This question was addressed in Taiwanese youth 12 to 14 years of age who were classified as fit (highest quartile) or unfit (lowest quartile) for four components of health-related physical fitness (Huang & Malina, 2002). Estimated daily energy expenditure (kcal/kg per day) in boys and girls classified as fit in the 1-mile run and sit and reach was greater than in those classified as unfit in these items. In contrast, those fit and unfit in sit-ups and sum of skinfolds did not differ in estimated energy expenditure. Estimated energy expenditure differences thus varied with the indicator of fitness used for the comparison. Moreover, the activity status of youth classified as fit and unfit in each test item overlapped considerably. The observations of Lange Andersen and colleagues (1984, p. 435) in a study relating habitual physical activity to maximal aerobic power in 14- to 18-year-old boys and girls are appropriate: ". . . among those with the poorest fitness, there are sedentary, moderately active and very active children. Similarly, there are sedentary, moderately active and very active children among those who are in excellent physical condition."

The same approach can be used in comparing the fitness of youth classified as physically active and inactive (less active is probably a more appropriate label). Data are more available for boys than for girls, but there is one reasonably consistent observation in cross-sectional and longitudinal comparisons. More active youth are more fit in cardiorespiratory endurance tasks, measured primarily in the form of endurance runs and peak VO_2 (Beunen et al., 1992; Blair et al., 1989; Huang & Malina, 2002; Mirwald & Bailey, 1986; Verschuur, 1987). Comparisons for other components of physical fitness are equivocal, suggesting that level of habitual physical activity is only one of several factors that influence physical fitness.

Relationships between measures of health-related (pulse recovery to a 1 min run, 60 s sit-ups, sit and reach) and performance-related (25 m dash, vertical jump, standing long jump, grip strength) physical fitness with participation in specific physical activities were considered in a sample of Portuguese adolescents (Coelho e Silva et al., 2003). Among boys, a positive association was observed between scores on all fitness items and participation in organized sport, and to a lesser extent participation in recreational and fitness-health activities. For girls, fitness scores were positively associated with scores for participation in the three modes of activity: organized sport, recreational activities, and fitness-health activities.

In contrast to studies addressing the association between habitual physical activity and measures of physical fitness, experimental evidence has established clearly that activity programs result in improved aerobic fitness and muscular strength and endurance. Protocols for the improvement of aerobic fitness ordinarily involve vigorous activities (about 80% of maximal heart rate), 30 to 45 min in duration, more than three days per week (Strong et al., 2005). Protocols for the improvement of muscular strength and endurance involve a variety of progressive resistance activities of 30 to 45 min duration, two or three days per week (Malina, 2006). For strength, emphasis is on high resistance and low repetition, while for endurance, emphasis is on high repetition and low resistance.

Television viewing has received considerable attention as a sedentary behavior that may influence obesity and physical activity and to a lesser extent physical fitness. However, the relationship between time spent viewing television and

Many sedentary behaviors, such as participating in a chess tournament, prove beneficial. The issue is not to eliminate these behaviors but to encourage a balance of physically active and sedentary behaviors that foster well-being.

activity (Robinson et al., 1993; Wolf et al., 1993) as well as indicators of health-related fitness, daily energy expenditure, and moderate to vigorous physical activity (Katzmarzyk et al., 1998b) is rather weak in children and adolescents. Moreover, there does not appear to be a pattern by age-group and sex. Television viewing is only one form of sedentary behavior; other sedentary behaviors such as playing video games, personal computer activities, use of DVDs, use of cell phones, homework, extracurricular classes (tutoring, art, music), and motorized transport to school and other organized activities need systematic evaluation. Of relevance, these sedentary behaviors, and perhaps others, are seemingly highly valued by society and are often encouraged both directly and indirectly so that time available for physical activities is reduced.

Environments for Physical Activity

Environments available for physical activity reflect the broader cultural environment of a population. Studies relating environmental factors to physical activity levels among children and adolescents, though limited, give mixed results (Sallis et al., 2000). The results may reflect overly generic definitions of "physical environment factors" such as access to facilities or programs; transportation; season; urban or rural milieu; safety and time outdoors for children and equipment or supplies; or opportunities and sport media for adolescents. Moreover, composite estimates of physical activity, rather than particular forms of activity, may lack the specificity needed for a more precise evaluation of relationships with the environmental factors.

More recently, emphasis has shifted to the integration of the built environment, land use and zoning, and community planning in the development of physical spaces and buildings that encourage physical activity and reduce sedentary behavior, specifically in urban centers (see Sallis et al., 2005). Discussions of the built environment largely focus on physically active transport (i.e., walking and cycling) and play spaces for children in an effort to increase overall physical activity and reduce sedentary behavior with the hope of reducing overweight and obesity. Of course, the individuals who live in and use these environments must also be included in the equation.

The relevance of environmental factors for opportunities to be physically active or sedentary is seemingly obvious. There is also a need to consider specific components of the environment such as the spatial (e.g., playgrounds, open spaces, swimming pools, gymnasiums), material (e.g., bicycles, roller or in-line skates, balls, rackets, ropes), and social (e.g., sport clubs, recreation centers, folk and theater groups, church organizations) components. Spatial, material, and social aspects of the environment may determine or modify contexts of physical activity among youth. Associations between participation in specific forms of physical activity (organized sport, recreational, fitness-health) and components of the environment (spatial, material, social) were considered in the sample of Portuguese adolescents described earlier (Coelho e Silva et al., 2003). Lower levels of participation in sport, recreational, and fitness activities were associated with lower scores on the spatial, material, and social components of the environment in both sexes. Limited material stimuli were associated with reduced levels of participation in organized sport among males and with reduced levels of participation in recreational physical activities in females.

Sport Participation

Participation in sport is perhaps the primary form of organized physical activity for youth of both sexes. Indeed, youth often identify physical activity with sport (table 6.1). Organized sport provides opportunity for physical activity on a regular basis and in a safe environment. Of relevance, therefore, are the activity levels of youth involved and not involved in sport. Surprisingly, such data for youth sport participants are limited. Among youth 12 to 14 years of age, boys and girls involved in organized youth sport expend, on average, more overall energy (TDEE, absolute and per unit body mass) and energy in moderate-to-vigorous activities (≥4.8 METs) than nonparticipants (Katzmarzyk & Malina, 1998). Of interest, youth sport participants also indicate less television viewing time. Though limited to a single community in mid-Michigan that was surveyed in January and February (there may be variation by season of the year), the results suggest a greater level of physical activity and less time in one form of sedentary behavior in organized youth sport participants compared to nonparticipants (Katzmarzyk & Malina, 1998).

Evidence from Finland indicates that membership (by inference, participation) in sport clubs tracks better than other indicators of physical activity across adolescence into young adulthood (Telama, Laakso, & Yang, 1994; Telama et al., 1997; see Malina, 2001). The higher interage correlations for participation in sport clubs suggest that more attention should be given to this context of physical activity among adolescents. Moreover, frequency of participation in sport at 14 years of age (Tammelin et al., 2003), membership in sport clubs at 16 years of age (Barnekow-Bergkvist et al., 2001), and sport club training and competition during adolescence (Telama et al., 2006) significantly predict physical activity in young adulthood in both sexes. How participation in sport during adolescence translates into a young adult lifestyle that includes a greater level of habitual physical activity needs study. An association between sport participation during adolescence and "psychological readiness" for physical activity in adulthood has been proposed (Engstrom, 1986, 1991).

Of interest, the preceding data are from Scandinavian countries. Sport clubs vary among countries, in accessibility and cost, and in degree of sport specialization and participant selectivity (Heinemann, 1999). In addition, many European countries have adopted a "sport for all" theme that contrasts with the thrust of youth and interscholastic sport programs in the United States, which become quite exclusive during adolescence. As a result of this exclusivity, sport offerings for youth with lesser skill or with less interest in elite competition are limited in many U.S. communities (Malina, 2001).

In short, sport participation occurs within a cultural context. Sports provide a medium in which developmental processes occur within a specific culture and also may facilitate the assimilation process for minority groups within the broader mainstream culture. Of relevance to the present discussion is the following question: How are children and adolescents socialized into a physically active lifestyle that will persist into adulthood? For example, as children move into school, there is greater independence from the family; and the influence of peers becomes more important in the context of sport and physical activity and perhaps interacts with parental influences (see chapter 12). With adolescence, parental influences may be reduced somewhat, while peer influences strengthen. Nevertheless, participation in sport and other organized activity programs often depends on a parent's willingness and ability to take the child to the venue and to provide equipment and other resources.

In the multiethnic, low-income sample of inner-city Montreal youth ages 9 to 13 years, significant predictors of participation in sports outside of school included non-Asian ancestry, moderate to high self-efficacy, and participation on school teams in both sexes (O'Loughlin et al., 1999). Additional predictors for boys included father's participation in sport and daily playing of video games; other predictors for girls included maternal encouragement of and participation in sport, family income, and educational level of the father.

Social behaviors are a major component of the developmental processes that occur in infancy, childhood, and adolescence. The development of social behaviors (i.e., socialization) involves three primary components: socializing agents or significant others, social environments or situations, and role learners. The process has received relatively more attention in the context of sport compared to other physical activities and compared to moral and affective behaviors associated with sport. Accordingly, the individual is socialized into a specific role, such as that of an active participant (player) or a sedentary participant (spectator) or both. The individual is also socialized through these roles into the learning of more general attitudes and values toward physically active or sedentary participation.

Sport activities provide a variety of experiences or interactions—social, psychological, physical—that may influence whether or not the individual persists in this form of activity or other activities. Studies of youth sport participants, however, focus more often on the role of sport experiences in the motivation to participate in and to discontinue participation in sport, and not as a medium for socializing youth into a physically active lifestyle.

Evidence from tracking studies of physical activity (Barnekow-Bergkvist et al., 2001; Tammelin et al., 2003; Telama et al., 1994, 1997, 2006) highlights the importance of sport participation during adolescence as a significant predictor of adult physical activity levels. Adolescence, of course, is a developmental period that involves complex and time-sensitive interactions among biological and behavioral characteristics. For example, interindividual variation in the timing and tempo of the growth spurt and sexual maturation enters the matrix of factors that influence social status and the socialization process during adolescence. Individual differences in biological growth and maturation and associated changes are a major component of the backdrop against which youth evaluate and interpret their social status among peers. Physical performance and success in sport constitute an important aspect of the evaluative process, particularly among boys. Biological maturation significantly influences physical skills among boys and perhaps the value attached to these skills by peers. Moreover, early maturation in boys is also associated with success in many youth sports (Malina, 2002), with resulting advantages in social status.

The situation for adolescent girls is different. Late-maturing girls more often perform better, experience success in sport, and persist in sport through adolescence (Malina, 2002). Further, differences in strength and performance among girls of contrasting maturity status are not as apparent as among boys during the transition into adolescence and do not persist into adolescence as in boys (Malina et al., 2004). This raises several questions that merit study. For example, how do individual differences in the timing of sexual maturation affect participation and persistence in sport and other forms of physical activity? Are late-maturing girls socialized into such active behaviors, or are early-maturing girls socialized away from them?

Changes associated with sexual maturity in females are often accompanied by a reduction in physical activity. This association can be explained, in part, by psychosocial factors associated with the transition into puberty and with sexual maturation, including a decline in self-esteem and changing interests in pursuits other than physical activities and sport. A related factor may be persistent public perception that physical activity, especially in the context of sport, is associated with "masculinity," though this perception certainly varies in substance and degree among cultures. The reduction in activity may also be related to biological changes associated with puberty and the growth spurt per se, and perhaps differential timing of these events of adolescence. Sexual maturation and the growth spurt are associated with an increase in absolute and relative fatness; changing body proportions (relative broadening of the pelvis, an increase in the angle between the femur and the tibia that may affect movement mechanics); and establishment of regular menstrual cycles, which may have associated discomfort.

Implications

Demands placed on children and adolescents are many and can be summarized in terms of physical growth, biological maturation, and behavioral development. These processes vary during childhood and adolescence, among individuals, and within and between cultural groups. The processes interact to influence physically active and sedentary behaviors.

There is a need to integrate measures of growth, maturation, and development in studies of physical activity and sedentary behavior to better understand their specific contributions and interactions. The assessment of biological maturation is problematic, as the more commonly used indicators are invasive. Several noninvasive approaches to assessing maturity status are available and need to be further refined. Chronological age is not an indicator of maturity status.

Contexts of physical activity and sedentary behavior vary during childhood and adolescence. It is imperative that specific active and sedentary behaviors be addressed to provide further insights into developmental processes and associated factors that influence these behaviors. Research into physical activity levels needs to move beyond global estimates of habitual activity or energy expenditure and inactivity to the study of more specific behaviors.

Physical activity levels and sedentary behaviors have important implications for health, education, and overall well-being. A biocultural perspective provides an ideal framework for identifying the specific contributions of biology and culture to active and sedentary behaviors and associated outcomes, as well as potential influences of biology and culture upon each other.

Applications for Researchers

Growth, maturation, and development are the primary tasks of children and adolescents. These processes and their interactions influence physically active and sedentary behaviors. Measures of growth, maturation, and development should be incorporated into studies of youth activity levels in the context of a biocultural perspective that recognizes the interactions of biological and behavioral processes.

Cultural influences on physical activity levels need further study. Worldwide trends in emigration and immigration highlight the need for awareness of cultural variation and potential impact on lifestyle, including active and sedentary behaviors. There is a need for more systematic study of sport, especially during adolescence, when many sports become exclusive or discriminatory and focus on the talented in contrast to the general population of youth.

Advances in research provide new and often better ways of conceptualizing questions related to the study of activity levels. These advances are especially apparent in genetic research that incorporates active and inactive phenotypes.

Applications for Professionals

Youth are commonly viewed through the lens of public health, biomedical, and educational communities with an eye toward their present and future health and success in the adult world. Youth are, however, children and adolescents with the specific needs of children and adolescents. Physical activity levels must be in relation to the context and demands of the universal tasks of childhood and adolescence—the processes of physical growth, biological maturation, and behavioral development. The biological and behavioral processes interact, and these interactions may influence physical activity levels and sedentary behaviors. Culture-specific demands are superimposed upon these processes and interactions. Physically active and sedentary behaviors take a variety of forms, and many are highly valued by society. There is more to these behaviors than energy expenditure or lack thereof; they are behaviors with important developmental, educational, and health outcomes. Thus, as professionals design ways to promote physical activity, goals should not be limited, for example, to the achievement by each child of a certain number of minutes per week of activity resulting in a target heart rate, or a specific number of steps per day. Activity promotion must also consider other equally important outcomes such as enjoyment of the activities and effects on perceived competence in the context of individual differences in movement abilities and the developmental appropriateness of the activities.

References

Aaron, D.J., Kriska, A.M., Dearwater, S.R., Anderson, R.L., Olsen, T.L., Cauley, J.A., et al. (1993). The epidemiology of leisure physical activity in an adolescent population. *Medicine and Science in Sports and Exercise, 25*, 847-853.

Aaron, D.J., Kriska, A.M., Dearwater, S.R., Cauley, J.A., Metz, K.F., & Laporte, R.E. (1995). Reproducibility and validity of an epidemiologic questionnaire to assess past year physical activity in adolescents. *American Journal of Epidemiology, 142*, 191-201.

Barnekow-Bergkvist, M., Hedberg, G., Janlert, U., & Jansson, E. (2001). Adolescent determinants of cardiovascular risk factors in adult men and women. *Scandinavian Journal of Public Health, 29*, 208-217.

Beunen, G.P., Malina, R.M., Renson, R., Simons, J., Ostyn, M., & Lefevre, J. (1992). Physical activity and growth, maturation and performance: A longitudinal study. *Medicine and Science in Sports and Exercise, 24*, 576-585.

Beunen, G., & Thomis, M. (1999). Genetic determinants of sport participation and daily physical activity. *International Journal of Obesity, 23*(Suppl. 3), S55-S63.

Blair, S.N., Clark, D.G., Cureton, K.J., & Powell, K.E. (1989). Exercise and fitness in childhood: Implications for a lifetime of health. In C.V. Gisolfi & D.R. Lamb (Eds.), *Perspectives in exercise science and sports medicine*. Vol. II. *Youth, exercise, and sport* (pp. 401-430). Indianapolis: Benchmark Press.

Boomsma, D.I., van den Bree, M.B.M., Orlebeke, J.F., & Molenaar, P.C.M. (1989). Resemblances of parents and twins in sports participation and heart rate. *Behavior Genetics, 19*, 123-141.

Bouchard, C., Malina, R.M., & Perusse, L. (1997). *Genetics of fitness and physical performance*. Champaign, IL: Human Kinetics.

Bouchard, C., & Shephard, R.J. (1990). Physical activity, fitness and health: The model and key concepts. In C. Bouchard, R.J. Shephard, & T. Stephens (Eds.), *Physical activity, fitness, and health: International proceedings and consensus statement* (pp. 77-88). Champaign, IL: Human Kinetics.

Brooks-Gunn, J., & Peterson, A. (1983). *Girls at puberty: Biological and psychosocial perspectives*. New York: Plenum Press.

Brown, K.M., McMahon, R.P., Biro, F.M., Crawford, P., Schreiber, G.B., Similo, S.L., et al. (1998). Changes in self-esteem in Black and White girls between the ages of 9 and 14 years: The NHLBI Growth and Health Study. *Journal of Adolescent Health, 23*, 7-19.

Butcher, J.E., & Eaton, W.O. (1989). Gross and fine motor proficiency in preschoolers: Relationships with free play behavior and activity level. *Journal of Human Movement Studies, 16*, 27-36.

Canadian Community Health Survey. (2001). *Questionnaire for cycle 1.1*. Retrieved March 1, 2006, from www.statcan.ca.

Caspersen, C.J., Pereira, M.A., & Curran, K.M. (2000). Changes in physical activity patterns in the United States, by sex and cross-sectional age. *Medicine and Science in Sports and Exercise, 32*, 1601-1609.

Coelho e Silva, M.J., Sobral, F., & Malina, R.M. (2003). *Determinância sociogeográfica da prática desportiva na adolescência*. Coimbra: Universidade de Coimbra, Centro de

Estudos do Desporto Infanto-Juvenil, Faculdade de Ciências do Desporto e Educação Física.

Crespo, N.C., Ball, G.D.C., Shaibi, G.Q., Cruz, M.L., Weigensberg, M.J., & Goran, M.I. (2006). Acculturation is associated with higher $\dot{V}O_{2max}$ in overweight Hispanic children. *Pediatric Exercise Science, 17*, 89-100.

Damon, A. (1969). Race, ethnic group, and disease. *Social Biology, 16*, 69-80.

DeLany, J.P., Bray, G.A., Harsha, D.W., & Volaufova, J. (2002). Energy expenditure in preadolescent African American and White boys and girls: The Baton Rouge Children's Study. *American Journal of Clinical Nutrition, 75*, 705-713.

Eaton, W.O., & Enns, L.R. (1986). Sex differences in human motor activity level. *Psychological Bulletin, 100*, 19-28.

Eaton, W.O., & Yu, A.P. (1989). Are sex differences in child motor activity level a function of sex differences in maturational status? *Child Development, 60*, 1005-1011.

Ekelund, U., Yngve, A., Brage, S., Westerterp, K., & Sjöström, M. (2004). Body movement and physical activity energy expenditure in children and adolescents: How to adjust for differences in body size and age. *American Journal of Clinical Nutrition, 79*, 851-856.

Engstrom, L-M. (1986). The process of socialization into keep-fit activities. *Journal of Sports Sciences, 8*, 89-97.

Engstrom, L-M. (1991). Exercise adherence in sport for all from youth to adulthood. In P. Oja & R. Telama (Eds.), *Sport for all* (pp. 473-483). Amsterdam: Elsevier Science.

Eschway, M.J. (1994). *The relationship of acculturation to the growth status of Mexican American children 6 through 12 years residing in Minnesota.* Doctoral dissertation, University of Texas at Austin.

Goran, M.I., Hunter, G., Nagy, T.R., & Johnson, R. (1997). Physical activity related energy expenditure and fat mass in young children. *International Journal of Obesity, 21*, 171-178.

Grunbaum, J.A., Kann, L., Kinchen, S., Ross, J., Hawkins, J., Lowry, R., et al. (2004). Youth risk behavior surveillance—United States, 2003. *Morbidity and Mortality Weekly Report, 53*(SS-2), 1-96.

Grund, A., Vollbrecht, H., Frandsen, W., Krause, H., Siewers, M., Riechert, H., et al. (2000). No effect of gender on different components of daily energy expenditure in free living prepubertal children. *International Journal of Obesity, 24*, 299-305.

Heinemann, K. (Ed.). (1999). *Sport clubs in various European countries.* Schorndorf, Germany: Karl Hoffman (Series Club of Cologne, Vol. 1).

Hoos, M.B., Gerver, W.J.M., Kester, A.D., & Westerterp, K.R. (2003). Physical activity levels in children and adolescents. *International Journal of Obesity, 27*, 605-609.

Huang, Y.C., & Malina, R.M. (2002). Physical activity and health-related fitness in Taiwanese adolescents. *Journal of Physiological Anthropology, 21*, 11-19.

Joosen, A.M.C.P., Gielen, M., Vlietinck, R., & Westerterp, K.R. (2005). Genetic analysis of physical activity in twins. *American Journal of Clinical Nutrition, 82*, 1253-1259.

Katzmarzyk, P.T., & Malina, R.M. (1998). Contribution of organized sports participation to estimated daily energy expenditure in youth. *Pediatric Exercise Science, 10*, 378-386.

Katzmarzyk, P.T., Malina, R.M., Song, T.M.K., & Bouchard, C. (1998a). Physical activity and health-related fitness in youth: A multivariate analysis. *Medicine and Science in Sports and Exercise, 30*, 709-714.

Katzmarzyk, P.T., Malina, R.M., Song, T.M.K., & Bouchard, C. (1998b). Television viewing, physical activity, and health-related fitness of youth in the Quebec Family Study. *Journal of Adolescent Health, 23*, 318-325.

Kroeber, A.L., & Kluckhohn, C., with Unterreiner, W. (1952). Culture: A critical review of concepts and definitions. *Papers of the Peabody Museum of American Archaeology and Ethnology, 47/1*. Cambridge, MA: Peabody Museum, Harvard University.

Lange Andersen, K., Ilmarinen, J., Rutenfranz, J., Ottman, W., Berndt, I., Kylian, H., et al. (1984). Leisure time sport activities and maximal aerobic power during late adolescence. *European Journal of Applied Physiology, 52*, 431-436.

Malina, R.M. (1996). Familial factors in physical activity and performance of children and youth. *Journal of Human Ecology*, Special Issue No. 4, 131-143.

Malina, R.M. (2001). Tracking of physical activity across the lifespan. *President's Council on Physical Fitness and Sports Research Digest*, Series 3, No. 14, 1-8.

Malina, R.M. (2002). The young athlete: Biological growth and maturation in a biocultural context. In F.L. Smoll & R.E. Smith (Eds.), *Children and youth in sport: A biopsychosocial perspective* (2nd ed., pp. 261-292). Dubuque, IA: Kendall/Hunt.

Malina, R.M. (2006). Weight training in youth—growth, maturation, and safety: An evidence-based review. *Clinical Journal of Sports Medicine, 16*, 478-487.

Malina, R.M., Bouchard, C., & Bar-Or, O. (2004). *Growth, maturation, and physical activity* (2nd ed.). Champaign, IL: Human Kinetics.

Mirwald, R.L., & Bailey, D.A. (1986). *Maximal aerobic power*. London, ON: Sport Dynamics.

Moore, L.L., Lombardi, E., White, M.J., Campbell, J.L., Oliveira, S.A., & Ellison, R.C. (1991). Influence of parents' physical activity levels on activity levels of young children. *Journal of Pediatrics, 118*, 215-219.

O'Loughlin, J., Paradis, G., Kishchuk, N., Barnett, T., & Renaud, L. (1999). Prevalence and correlates of physical activity behaviors among elementary schoolchildren in multiethnic, low income, inner-city neighborhoods in Montreal, Canada. *Annals of Epidemiology, 9*, 397-407.

Pate, R.R., Dowda, M., & Ross, J.G. (1990). Associations between physical activity and physical fitness in American children. *American Journal of Diseases of Children, 144*, 1123-1129.

Perusse, L., Tremblay, A., Leblanc, C., & Bouchard, C. (1989). Genetic and environmental influences on level of habitual physical activity and exercise participation. *American Journal of Epidemiology, 129*, 1012-1022.

Redfield, R., Linton, R., & Herskovits, M.J. (1936). Memorandum for the study of acculturation. *American Anthropologist, 38*, 149-152.

Robinson, T.N., Hammer, L.D., Killen, J.D., Kraemer, H.C., Wilson, D.M., Hayward, C., et al. (1993). Does television viewing increase obesity and reduce physical activity? Cross-sectional and longitudinal analyses among adolescent girls. *Pediatrics, 91*, 273-280.

Rowland, T.W. (1999). Adolescence: A "risk factor" for physical inactivity. *President's Council on Physical Fitness and Sport*, Series 3, No. 6, 1-8.

Sallis, J.F., McKenzie, T.L., & Alcaraz, J.E. (1993). Habitual physical activity and health-related physical fitness in fourth grade children. *American Journal of Diseases of Children, 147*, 890-896.

Sallis, J.F., Moudon, A.V., Linton, L.S., & Powell, K.E. (Eds.). (2005). Active living research. *American Journal of Preventive Medicine, 28*(Suppl. 2), 93-219.

Sallis, J.F., Prochaska, J.J., & Taylor, W.C. (2000). A review of correlates of physical activity of children and adolescents. *Medicine and Science in Sports and Exercise, 32*, 963-975.

Simmons, R.G., & Blyth, D.A. (1987). *Moving into adolescence: The impact of pubertal change and social context.* New York: Aldine de Gruyter.

Spadano, J.L., Bandini, L.G., Must, A., Dallal, G.E., & Dietz, W.H. (2005). Longitudinal changes in energy expenditure in girls from late childhood through mid-adolescence. *American Journal of Clinical Nutrition, 81*, 1102-1109.

Strong, W.B., Malina, R.M., Blimkie, C.J.R., Daniels, S.R., Dishman, R.K., Gutin, B., et al. (2005). Evidence based physical activity for school-age youth. *Journal of Pediatrics, 146*, 732-737.

Tammelin, T., Nayha, S., Hills, A.P., & Jarvelin, M-R. (2003). Adolescent participation in sports and adult physical activity. *American Journal of Preventive Medicine, 24*, 22-28.

Telama, R., Laakso, L., & Yang, X. (1994). Physical activity and participation in sports of young people in Finland. *Scandinavian Journal of Medicine and Science in Sports, 4*, 65-74.

Telama, R., Laakso, L., Yang, X., & Vikari, J. (1997). Physical activity in childhood and adolescence as predictor of physical activity in young adulthood. *American Journal of Preventive Medicine, 13*, 317-323.

Telama, R., Yang, X., Hirvensalo, M., & Raitakari, O. (2006). Participation in organized youth sport as a predictor of adult physical activity: A 21-year longitudinal study. *Pediatric Exercise Science, 17*, 76-88.

Thompson, A.M., Baxter-Jones, A.D.G., Mirwald, R.L., & Bailey, D.A. (2003). Comparison of physical activity in male and female children: Does maturation matter? *Medicine and Science in Sports and Exercise, 35*, 1684-1690.

Torun, B., Davies, P.S.W., Livingstone, M.B.E., Paolisso, M., Sackett, R., & Spurr, G.B. (1996). Energy requirements and dietary energy recommendations for children and adolescents 1 to 18 years old. *European Journal of Clinical Nutrition, 50*, S37-S81.

Van Mechelen, W., Twisk, J.W.R., Post, G.B., & Kemper, H.C.G. (2000). Physical activity of young people: The Amsterdam Longitudinal Growth and Health Study. *Medicine and Science in Sports and Exercise, 32*, 1610-1616.

Verschuur, R. (1987). *Daily physical activity: Longitudinal changes during the teenage period.* Haarlem, The Netherlands: Uitgeverrij de Vrieseborch.

Wolf, A.M., Gortmaker, S.L., Cheung, L., Gray, H.M., Herzog, D.B., & Colditz, G.A. (1993). Activity, inactivity, and obesity: Racial, ethnic, and age differences among school-girls. *American Journal of Public Health, 83*, 1625-1627.

Wolfarth, B., Bray, M.S., Hagberg, J.M., Perusse, L., Rauramaa, R., Rivera, M.A., et al. (2005). The human gene map for performance and health-related fitness phenotypes: The 2004 update. *Medicine and Science in Sports and Exercise, 37*, 881-903.

CHAPTER

7

Youth Attitudes

Martin S. Hagger, PhD ▪ Nikos L.D. Chatzisarantis, PhD

Attitude is considered one of the most important concepts in social psychology and has a high-profile status in many applied social psychological contexts in which motivation and intentional behavior are focal, such as physical activity (Hagger, Chatzisarantis, & Biddle, 2002b). The importance of attitudes can be largely attributed to the common assumption by researchers and laypeople alike that attitudes play an important role in determining behavior (Wicker, 1969), even though there is considerable evidence to suggest that such a simple, deterministic relationship is erroneous. However, there has been a resurgence of interest in the field of attitude in the past two decades with the development of social cognitive theories that incorporate the attitude construct and attempt to explain the *processes* that underlie the attitude–behavior relationship in applied contexts (Hagger & Chatzisarantis, 2005). The development of these models has contributed to understanding of the determinants and, equally important, the processes that result in people's participation in active pastimes. Furthermore, these models have been applied to the understanding of the processes by which attitudes lead to physical activity behavior in young people.

The aim of this chapter is to outline attitudinal and social cognitive approaches to explain the processes that lead to physical activity intentions and behavior in young people. We begin with a definition of attitudes and proceed to a discussion of the importance of attitudes applied to physical activity behavior. We discuss the importance of social cognitive constructs as sources of social information that affect decision making and the ways in which attitudes are a focal construct in social cognition. Specifically, we focus on how attitude-based social cognitive approaches, like the theories of reasoned action and planned behavior, have utility in identifying not only the psychological constructs influential in explaining youth physical activity behavior, but also the processes behind their influence, and particularly the role of intentions and the intention–behavior relationship. This will form the basis of the argument that social cognitive models can underpin interventions to change young people's physical activity intentions and behavior. In addition, we outline the importance of social cognitive approaches in accounting for unique variance in youth physical activity behavior, as well as the stability of physical activity behavior over time and the role of past physical activity behavior in predicting current behavior. This will be contrasted with sedentary behaviors and physical inactivity in young people.

Attitudes in Social Science Research

Attitude is commonly defined as an individual's favorable or unfavorable evaluation of an attitude object or target behavior (Ajzen, 1991; Eagly & Chaiken, 1993). Attitudes are often viewed as multidimensional, with affective (emotional), instrumental (cognitive), and functional (behavioral) aspects. Recent social cognitive models have generally incorporated both affective and instrumental aspects (or dimensions) of attitudes, while the behavioral aspect is usually associated with motivational constructs such as intention. In physical activity, young people are likely to report enjoying certain sports or active pastimes that likely reflect their affective attitudes. In addition, they might also have certain beliefs that these pastimes will have some instrumental benefits to them, such as improving their skills or making new friends, and these are cognitive attitudes. Young people might also form action plans to engage in these pastimes in the future, illustrating a behavioral aspect of their attitudes.

Measuring Attitudes

Central to attitude research is valid and reliable measurement. Attitudes are generally measured "directly" through multi-item questionnaires that ask for respondents' opinions of an attitude object or behavior using semantic differential scales and bipolar adjectives (Ajzen, 2001). For example, measures of attitudes have often reflected the instrumental (e.g., *harmful-beneficial*) and affective (e.g., *pleasant-unpleasant*) aspects of the target behavior (Ajzen & Driver, 1991). Attitudes have also been "indirectly" measured using expectancy-value models, focusing on specific beliefs regarding the attitude object or behavior rather than global evaluations. Such measures are often used in research employing social cognitive theories (e.g., theories of reasoned action and planned behavior) to validate direct measures by correlating the composite expectancy-value items with the direct measures.

Indirect, expectancy-value models of attitudes have been used to measure youth attitudes toward physical activity. For example, Hagger, Chatzisarantis, and Biddle (2001) conducted an open-ended elicitation study with adolescents to identify the salient advantages and disadvantages of physical activity participation. Young people were asked to list benefits and disadvantages of doing physical activities. Responses included enjoyment and having fun, improving skills, increasing fitness, making friends, getting hot and sweaty, and getting an injury. These were then used to develop measures of expectancies (e.g., "I think doing physical activity will improve my fitness.") and values (e.g., "Getting fit is very important.") for use with larger samples of young people to predict physical activity behavior.

Attitude Theories

The theories of reasoned action (Ajzen & Fishbein, 1980) and planned behavior (Ajzen, 1985) are focal social cognitive perspectives on social behavior with attitudes as a central construct. These theories are important because they represent a culmination of research on attitudes and attitude structure; they incorporate all aspects of the tripartite model of attitudes described earlier and specify exactly how these are formed and the processes by which these attitudinal constructs determine behavior.

The Theory of Reasoned Action

The main aim of the theory is to explain intentional behaviors, and it is assumed that the theoretical components are universal across all behaviors of this type. Physical activity behavior is considered intentional and not subject to automatic or habitual processes. Importantly, the perspectives focus on a specific behavior or set of behaviors (e.g., engaging in a sporting activity like tennis or soccer or visiting the gym to perform aerobics, run on the treadmill, or perform weight training) rather than a broad behavioral category like inactive pastimes. They postulate that intentions are the most proximal predictor of physical activity. Intention reflects the degree of planning and effort that individuals expect to invest in engaging in future participation in physical activity.

In the theory of reasoned action, intention is proposed to be a function of a set of personal and normative expectations regarding the performance of physical activity, termed attitudes and subjective norms, respectively. Attitudes represent an overall positive or negative evaluation of physical activity and have a dominant role in forming intentions. Subjective norms represent perceptions of the influence that significant others have on the execution of physical activity. In addition, it is proposed that intentions mediate the effects of attitudes and subjective norms on behavior such that intentions are the process by which attitudes and subjective norms are converted into physical activity behavior. The major predictions of the theory of reasoned action are given in figure 7.1 (paths bolded and labeled "TRA").

Meta-analyses of studies adopting the theory of reasoned action in physical activity settings support the major premises of the theory (e.g., Hagger et al., 2002b; Hausenblas, Carron, & Mack, 1997; Symons-Downs & Hausenblas, 2005). To summarize the meta-analytic findings across studies, intentions have a strong relationship with physical activity behavior, and attitudes have a strong relationship with intentions, with a lesser role for subjective norms (Hagger et al., 2002b). Furthermore, belief-based studies among young people have indicated that beliefs (for example, that getting fit and exercising will help them make friends and be with peer groups) affect attitudes, corroborating Ajzen's (1985)

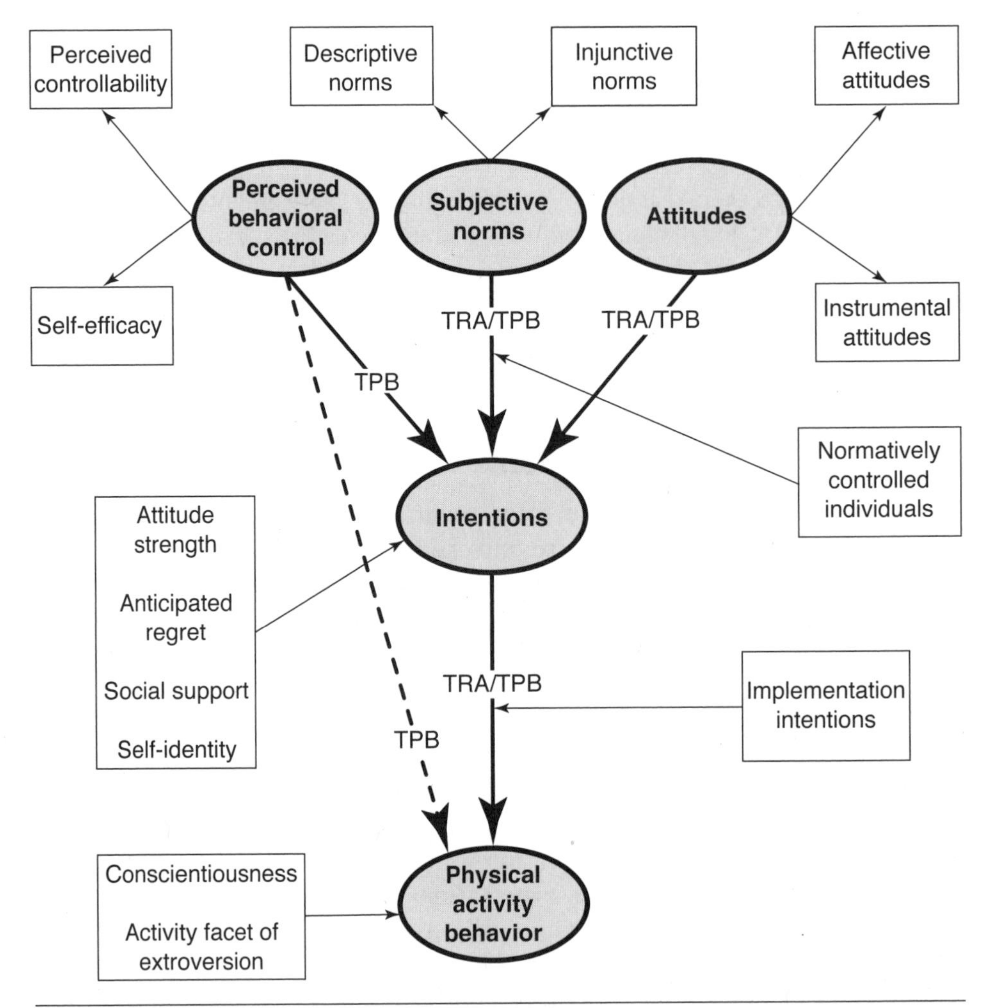

Figure 7.1 Extended version of the theory of planned behavior. Note that the original theory of reasoned action (TRA) and theory of planned behavior (TPB) appear in bold with relationships labeled.

contention that attitudes are underpinned by a set of belief-based evaluations of physical activity and its expected outcomes (Hagger et al., 2001).

The Theory of Planned Behavior

While the theory of reasoned action has worked well to explain various behaviors, its effectiveness is limited in that not all behaviors are under the volitional control of the individual. In order to resolve this limitation, and to increase the universality of the theory, Ajzen (1985) proposed a modified version known as the theory of planned behavior. The major premises of this theory were identical to those of the theory of reasoned action. However, the theory of planned behavior included perceived behavioral control as an additional construct affecting intentions and behavior when an individual has limited personal control over behavior.

Perceived behavioral control represents an individual's subjective judgments regarding her or his capacity and ability to engage in physical activity in the future, and incorporates judgments such as self-efficacy and perceived barriers to doing physical activity (Ajzen & Driver, 1991). Perceived behavioral control is important because it reflects the personal and environmental factors, real or perceived, that affect performance of a behavior such as physical activity (Ajzen, 1985). Importantly, perceived behavioral control was hypothesized to have a dual effect within the theory of planned behavior:

- A direct effect on intentions, in addition to the effects of attitude and subjective norm; relative to physical activity, it is proposed that this effect reflects how personal resources and capacities influence intentions to engage in physical activity.
- A direct effect on behavior, reflecting the influences of actual constraints or barriers to participation in physical activity. Under such circumstances, perceived behavioral control serves as a proxy measure of *actual* behavioral control (Ajzen, 1991).

The proposed relationships among the theory of planned behavior constructs are shown in figure 7.1 (paths bolded and labeled "TPB"). The direct effect of perceived behavioral control is represented by the broken arrow in the diagram because it reflects an influence that is dependent on the degree of actual control an individual has over the behavior.

Validity of the TRA and the TPB

A number of studies support the predictions of the theory of planned behavior and suggest that the modifications (from the theory of reasoned action) result in a larger proportion of explained variance in physical activity (Dzewaltowski, Noble, & Shaw, 1990; Hagger et al., 2002b). Indeed, studies have shown that the relative contributions of the attitude and perceived behavioral control constructs to the explanation of physical activity are roughly equal (Hagger, Chatzisarantis, & Biddle, 2002a; Hagger & Chatzisarantis, 2005). As with the theory of reasoned action, subjective norm is a substantially weaker predictor of physical activity than attitude and perceived behavioral control.

Physical activity studies examining the pattern of associations in the theory of planned behavior with younger samples (e.g., Hagger, Chatzisarantis, & Biddle, 2001; Hagger et al., 2007; Martin et al., 2005; Motl et al., 2002; Pate et al., 1997; Rhodes, MacDonald, & McKay, 2006; Saunders et al., 2004; Trost et al., 1996, 2002), mainly adolescents, have shown that attitudes and perceived behavioral control tend to exhibit the strongest associations with intentions and that the intention–behavior relationship tends to be significant and consistent in size (Hagger et al., 2002b). As with adult samples, the relationship of subjective norms with intentions to engage in physical activity is substantially smaller and is often nonsignificant. However, Hagger and colleagues (2002b) found that age moderated the intention–behavior relationship across these studies, with adolescents having a significantly lower averaged corrected correlation ($r_c = .48$, $p < .01$) compared with older samples ($r_c = .57$, $p < .01$). However, the size of the effect was small, and relations among the other theory constructs were not significantly variant across adolescent and older samples. Recent research has also indicated that young people from different cultures do not differ in the size and pattern of effects in the theory of planned behavior in exercise (Hagger et al., 2007). These findings for age and culture corroborate Ajzen's (1985) premise that the pattern of effects in the theory of planned behavior tends to be universal across people and demographic groups, and support individual studies of the theory of planned behavior in physical activity contexts with children (e.g., Hagger, Cale, & Ashford, 1997; Theodorakis et al., 1991) and adolescents (e.g., Hagger, Chatzisarantis, & Biddle, 2001; Hagger et al., 2001).

In terms of the belief systems that underpin the theory of planned behavior constructs, research has also shown little variation in the modal beliefs underlying attitudes in studies of younger and older people. Adults as well as adolescents have cited fitness, enjoyment, and making friends as the most frequent advantages for engaging in physical activity (Bozionelos & Bennett, 1999; Godin & Shephard, 1984; Hagger, Chatzisarantis, & Biddle, 2001), though older adults also tend to cite outcomes such as losing weight and improving physique as advantages that are not cited by younger people to the same degree. This may be different in special populations such as those with high levels of juvenile obesity.

It is important to note that attitudinal models like the theory of planned behavior focus exclusively on intentions to adopt and maintain a specified target behavior, which is why measures tend to make explicit reference to the context, time, and target of the behavior. There is no research adopting these models on physical *in*activity because their focus is entirely on behavior uptake and maintenance. The theory of planned behavior could very well be adopted to predict inactive pastimes, like watching TV and video game playing, but that would not be very informative of the constructs that contribute to the uptake of and adherence to physical activity behavior, which is the behavior most relevant to producing desirable health outcomes in young people.

That is not to say that beliefs about inactive pursuits play no role in attitudinal models. Such beliefs may very well be implicated in the belief systems that underlie some of the predictors of intention. For example, an adolescent may believe that doing physical activity may be boring and less desirable than playing computer games. Such beliefs may then have a detrimental effect on the adolescent's intentions to engage in physical activity in the future.

Unsurprisingly, beliefs about inactive pastimes are not among the most frequently cited beliefs of physically active children. Therefore, beliefs about inactive pastimes are neither salient nor accessible to researchers studying active young people and are not considered in investigations of the formation of active young people's intentions to be physically active in the future. Furthermore, it is unlikely that people engage in inactive pastimes with the intention of being inactive; rather they are likely to engage in such pursuits because of the interesting or enjoyable features and expected valued outcomes of the target behavior.

Modifications of the Theory of Planned Behavior

Despite its success, the theory of planned behavior does not account for all of the variance in physical activity behavior. For example, given the moderate corrected meta-analytic correlation (r = .48) between intentions and behavior among studies adopting the theory of planned behavior with younger samples, and assuming that none of the other theory of planned behavior constructs directly affect behavior, the amount of unexplained variance in behavior is a rather substantial 77%. Furthermore, the research has suggested that there are other "external variables," such as personality and belief-based constructs, with independent effects on intentions and behavior that are not mediated by attitudes, subjective norms, perceived behavioral control, and intention (Bagozzi & Kimmel, 1995; Conner & Abraham, 2001; Conner & Armitage, 1998; Rhodes, Courneya, & Jones, 2002). Ajzen (1991) states that the inclusion of additional constructs within the framework of the theory of planned behavior is a legitimate practice when a meaningful theoretical justification can be provided and the additional constructs substantially contribute to the prediction of physical activity intentions or behavior (or both). However, many of these additional variables have not been tested in younger samples. While research comparing theory of planned behavior relationships across adult and youth samples suggests that findings can generally be extrapolated from older to younger samples, until empirical evidence is available, the potential of age to moderate the effects of additional variables cannot be ruled out.

The following sections focus on modifications of the theory of planned behavior in physical activity contexts that are based on the three core antecedents of intention in the theory: attitudes, subjective norms, and perceived behavioral control. As demonstrated in the previous section, these constructs are underpinned by sets of beliefs, and research has suggested that these beliefs may determine different aspects of the global construct. As such, multidimensional models of attitudes, subjective norms, and perceived behavioral control have been proposed. These modifications are presented with specific reference to research in the physical activity domain and are illustrated in figure 7.1. Also, table 7.1 provides a series of sample items that should serve as a reference guide for how to measure the constructs included in the modified versions of the theory of planned behavior.

Table 7.1 Sample Items Used to Measure Constructs Included in Modified Versions of the Theory of Planned Behavior

Global construct	Subfacet	Sample questionnaire item
Attitude strength		How certain are you of your feelings regarding doing physical activity in the next 2 weeks? *(extremely certain–extremely uncertain)*[1]
Attitude	Affective attitude	For me, doing vigorous physical activities for 20 minutes at a time in the next 2 weeks is . . . *(pleasant–unpleasant)*[2]
	Instrumental attitude	For me, doing vigorous physical activities for 20 minutes at a time in the next 2 weeks is . . . *(useful–useless, beneficial–not beneficial)*[2]
Anticipated regret		I would regret it if I did not do vigorous physical activities for 20 minutes at a time in the next 2 weeks. *(strongly agree–strongly disagree)*[3]
Subjective norm	Injunctive norm	Most people who are important to me would *want* me to do vigorous physical activities for 20 minutes per day in the next 2 weeks. *(strongly agree–strongly disagree)*[2]
	Descriptive norm	Most people I know do vigorous physical activities on a regular basis. *(extremely true–extremely false)*[2]
Social support		How much support do you receive for participating in regular physical activity from the people closest to you? *(very much–none at all)*[4]
Perceived behavioral control	Perceived controllability	Overall, how much control do you have over doing vigorous physical activities for 20 minutes in the next 2 weeks? *(high control–low control)*[2]
	Self-efficacy	I believe I have the ability to participate in physical activities for 20 minutes at a time in the next 2 weeks. *(strongly agree–strongly disagree)*[2]
Conscientiousness		I often push myself very hard when trying to achieve a goal. *(strongly agree–strongly disagree)*[5]
Activity facet of extraversion		Please describe yourself as accurately as possible as you are typically or generally compared to other people you know of the same age and gender (am always busy, am always on the go, do a lot in my spare time, manage many things at the same time, react quickly). *(extremely accurate–extremely inaccurate for each adjective)*[6]
Self-identity		I think of myself as someone who participates in physical activities regularly. *(strongly agree–strongly disagree)*[7]

[1]Adapted from Bassili 1996; [2]Hagger and Chatzisarantis 2005; [3]Abraham and Sheeran 2004; [4]Courneya, Plotnikoff, Hotz, and Birkett 2001; [5]Ashton et al. 2004; [6]Saucier and Ostendorf 1999; [7]Hagger and Chatzisarantis 2006.

Attitude Based

Given the prominent role that attitudes play in the theory of planned behavior, much research has focused on the properties of attitudes and how they affect the attitude–intention relationship. These include the strength of the attitude construct and the distinction between affective and instrumental (cognitive) attitudes. Attitude strength has been an important addition to studies of attitudes. Such studies indicate that the predictive efficacy of the attitude construct is limited by the strength of the attitude in an individual's memory (Doll & Ajzen, 1992). Studies in physical activity have included an attitude strength construct in the theory of planned behavior. For example, Theodorakis (1994) found that attitude strength explained additional variance in intentions above and beyond the traditional attitude component. This is illustrated in figure 7.1 with a direct effect of attitude strength on intentions. Further, an example of items to measure the attitude strength construct within the context of the theory of planned behavior is provided in table 7.1.

In addition to attitude strength, research has examined the importance of the emotional (affective) and instrumental (cognitive) components of attitude within the theory of planned behavior. Affective attitudes reflect emotive beliefs regarding participating in a target behavior, for example the belief that physical activity will provide enjoyable experiences. This could be measured by scales with endpoints such as "happy" versus "sad" or "pleasant" versus "unpleasant." Cognitive attitudes reflect beliefs regarding the instrumentality or usefulness of the behavior to produce positive outcomes. An example of an instrumental belief is the belief that physical activity results in increasing fitness levels, which could be measured using scales with endpoints such as "useful" versus "useless" or "beneficial" versus "not beneficial" (see table 7.1).

Trafimow and Sheeran (1998) argue in favor of a multidimensional conceptualization of attitudes on grounds that the unidimensional model neglects the unique effects of the instrumental and emotional aspects of attitude on behavior (Eagly & Chaiken, 1993). Indeed, studies have shown that this distinction can be made on the basis of the belief systems that underpin attitudes (Trafimow & Sheeran, 1998) as well as in terms of the direct or global measures of attitudes (Hagger & Chatzisarantis, 2005). Studies have shown that affective attitudes exert unique effects on intentions in a physical activity context (Lowe, Eves, & Carroll, 2002). Interestingly, Lowe and colleagues showed that the affective component exerts a direct effect on exercise after behavior intentions and past behavior were controlled for. This suggests that the affective component of attitudes facilitates spontaneous or unplanned behavior (Fazio, 1990), and reinforces the necessity of testing both the affective and instrumental components of attitudes within the theory of planned behavior.

Although the conceptual and empirical distinction between affective and instrumental attitudes appears sound, there has been a recent resurgence of interest in the conceptualization of the attitude construct within the theory of planned behavior, particularly in its constituent components. As argued by

Hagger and Chatzisarantis (2005), a principal problem is that affective and instrumental attitudes are not always consistent. For example, someone might enjoy eating a tasty ice cream and hence have a positive affective attitude toward eating such a treat, but may also believe that eating an ice cream is unhealthy and may therefore not have a positive instrumental attitude toward ice cream consumption. However, the affective and instrumental components have been shown to be highly consistent in the context of physical activity, with substantial correlations between them (Hagger & Chatzisarantis, 2005). These components should therefore be typically measured as indicators of a global or "second-order" attitude component, rather than distinct constructs (Hagger & Chatzisarantis, 2005). In contrast, when affective and instrumental components are inconsistent, such as in the ice cream example, they should be considered distinct and be permitted to exert independent effects on intentions. The basic argument then seems to be that while a model that acknowledges independence between affective and instrumental attitudes cannot be rejected on statistical or theoretical grounds, the global or second-order model can be favored in the physical activity domain because it gives the most parsimonious account of the attitude influences on intention (see figure 7.1).

While researchers have made the distinction between the affective and instrumental components of attitude at the global and beliefs levels, some theorists suggest that affective attitudes do not encompass all of the emotional or affective influences on the intention component. It seems that affective attitudes may not incorporate beliefs concerning future anticipated emotional reactions, particularly regret—aspects of emotion that have not been adequately represented in the theory of planned behavior (Conner & Armitage, 1998). The concept of anticipated regret is featured in regret theory (Loomes & Sugden, 1982), which proposes that people's decisions, in certain situations, can be based on feelings and emotions that they expect to experience from rejecting alternative behavioral courses, rather than on a rational analysis of the costs and benefits associated with behavioral engagement. Therefore, people familiar with postbehavioral feelings of regret and guilt are likely to reject alternatives that will make them feel regretful because they are motivated to avoid such feelings.

Several studies have supported the inclusion of anticipated regret in the theory of planned behavior (e.g., Parker, Manstead, & Stradling, 1995), including studies of college-age participants (Bozionelos & Bennett, 1999). However, no studies of adolescents or children have included this construct. In the physical activity domain, Abraham and Sheeran (2003) showed that participants were most likely to exercise if they both intended to exercise and reported high levels of anticipated regret if they failed to exercise. In addition, Abraham and Sheeran (2004) demonstrated that manipulations of anticipated regret, prompting people to focus on negative emotions that follow decisions not to exercise, can strengthen exercise intentions (see figure 7.1). A sample item for tapping anticipated regret appears in table 7.1. Although these effects require investigation in youth samples, extrapolating these findings to young people might mean getting young people to rehearse their feelings of regret if they choose not to exercise. However, as attitudes tend to have stronger effects on intentions than anticipated regret, such recommendations should be added to attitude change interventions, not take their place.

Social Based

Research adopting the theory of planned behavior in physical activity contexts has frequently cited a lesser role for subjective norms in the prediction of intentions compared with attitude and perceived behavioral control (e.g., Armitage & Conner, 2001; Conner & Armitage, 1998; Hagger, Chatzisarantis, & Biddle, 2001). This is surprising because it has often been thought that social influences would be very pervasive in the decisions of young people to engage in physical activities. The reason is that young people are likely to attach high value to social influences given the identity crises that they encounter in adolescence, the rapid changes they undergo as they seek their social identity, and the importance of their self-esteem within their peer groups (Heaven, 2001). Empirically, attitudes and perceived behavioral control contribute substantially to the explanation of physical activity intentions, but subjective norms have not (e.g., Hagger, Chatzisarantis, & Biddle, 2001; Hagger et al., 2002b). A possible explanation is that only a minority of individuals form intentions on the basis of subjective norms (so-called normatively controlled individuals) (Trafimow & Finlay, 1996), whereas most form intentions on the basis of attitudes and perceived behavioral control (Sheeran et al., 2002). Studies using within- and between-participants designs have shown that subjective norms are particularly important for a small number of people in the physical activity domain (Finlay, Trafimow, & Villarreal, 2002). Therefore, there are individual differences in the degree to which people form intentions on the basis of social or attitudinal influences, and this affects their future plans, or intentions, to engage in physical activity.

Another reason why studies have not *always* observed significant prediction of intentions by subjective norms may be that the subjective norms construct insufficiently captures all aspects of social influence on physical activity (Hagger et al., 2002a). Several studies distinguish subjective or *injunctive norms*, the perceived pressure from significant others to engage in the target behavior, and *descriptive norms*, the extent to which significant others engage in the target behavior (Rivis & Sheeran, 2003). Research has shown injunctive and descriptive norms to be conceptually and empirically distinct constructs (Hagger & Chatzisarantis, 2005). Studies examining the independent effects of descriptive norms in the theory of planned behavior have shown a positive and additive influence of the construct on intentions independent of the traditional subjective norms construct (e.g., Rivis & Sheeran, 2003). However, research has also indicated that these normative-based perceptions can be subsumed by a global subjective norms factor. In such a case, unique influences of descriptive norms on intentions are accounted for by the global factor (Hagger & Chatzisarantis, 2005).

Examples of items tapping the injunctive and descriptive norms constructs are given in table 7.1. The significant association between descriptive norms and intentions has been supported in the physical activity domain (Baker, Little, & Brownell, 2003). To illustrate, young people are likely to make decisions to participate in physical activity not only because their parents or friends want them too, but also because they perceive that important others are regular exercisers. In effect, these individuals provide an example or "role model" for these young people. It is therefore important to acknowledge the distinction between these different types of norms when one is examining the social influences on physical activity intentions and behavior within the theory of planned behavior (see figure 7.1).

Some researchers have argued that making the distinction between injunctive norms and descriptive norms may not sufficiently capture all potential social influences on physical activity intentions. Courneya and colleagues (2000) argued that the extent to which significant others are perceived to *assist* performance of behavior can also exert unique effects on intentions (Courneya et al., 2000; Rhodes, Jones, & Courneya, 2002). Social support (see figure 7.1) provides additional impetus toward forming intentions, beyond beliefs that the behavior is beneficial (attitudes), that significant others want the person to engage in it (subjective norms), and that the actor is capable of doing so (perceived behavioral control) (Courneya et al., 2000). Several studies have shown that social support and subjective norms achieve discriminant validity (Rhodes, Jones, & Courneya, 2002), and that social support predicts intentions over and above the core theory of planned behavior constructs (Courneya et al., 2000). An example of a social support item is given in table 7.1. Despite the unique contribution that social support appears to make to the prediction of intentions, it is important to note that the size of these effects is small compared to that exerted by attitudes and perceived behavioral control (Courneya et al., 2000).

Control Based

A large body of research makes the distinction between two subcomponents of the perceived behavioral control construct. Generally, researchers differentiate between *perceived controllability*, defined as the extent to which an individual has access to the means to exert control over the target behavior (Ajzen, 2002), and *self-efficacy*, defined as an individual's estimate of ability and personal capacity to engage in the behavior (Terry & O'Leary, 1995). Perceived controllability is frequently characterized by the *subjective control* an individual has over her or his participation in physical activity, while self-efficacy is often composed of statements pertaining to the perceived *capacities* and *abilities* of the actor to participate in the physical activity behavior.

Table 7.1 provides sample items of measures adopted in studies distinguishing between perceived controllability and self-efficacy. Studies have made the explicit distinction between perceived controllability and self-efficacy (see figure 7.1) and have provided conceptual and empirical evidence in support of the distinction in the physical activity domain (Hagger & Chatzisarantis, 2005; Terry & O'Leary, 1995). In addition, strong support for the distinction between perceived control (control perceptions related to voluntary control, similar to self-efficacy) and perceived difficulty (similar to perceived controllability) has been obtained experimentally on the basis of the belief systems that underlie these constructs (Trafimow et al., 2002). Notwithstanding the increasing body of evidence in support of this distinction, there is little conclusive evidence to support a clear, unambiguous pattern of predictions for perceived behavioral control and self-efficacy within the theory of planned behavior (Ajzen, 2002). Moreover, such distinctions may not be salient to youth, particularly young children, who are less cognitively mature and as a result are likely to have less clear and distinct beliefs compared with adolescents and adults. Therefore, making these distinctions may be less meaningful in younger samples, and further evidence is required to confirm the need to do so.

Individual Differences in the Theory of Planned Behavior

In addition to research that has focused on the properties of the core variables in the theory of planned behavior, other research has emphasized the contribution of individual differences or traitlike constructs within the theory. The inclusion of such constructs is justified on the basis that the belief systems underpinning the core predictors of intention do not sufficiently tap dispositional influences on behavioral engagement. Dispositional constructs are likely to be personality constructs or psychological variables that share similar characteristics with personality. Such constraints are relatively enduring and stable rather than changeable and unstable and affecting behavior across a variety of contexts. Such dispositional variables are likely to influence intentions in a manner that is beyond the conscious or planned decision-making processes mapped by the original theory. To the extent that these dispositional constructs have a unique effect on intention or behavior, and are not mediated by the core theory variables of attitude, subjective norm, and perceived behavioral control, the researcher has evidence to support the inclusion of such a construct within the theory. For example, a number of dispositional constructs have been found to have a unique effect on intentions or behavior or both, including personality (Courneya, Bobick, & Schinke, 1999) and self-identity (Sparks & Shepherd, 1992).

Personality

There has been a resurgence of interest in the effect of personality in decision-making models such as the theory of planned behavior. This is the case because researchers are interested in whether such stable, dispositional, traitlike variables are mediated by the theory of planned behavior as hypothesized by Ajzen (1991). Confirmation of such a mediation hypothesis would imply that people make future decisions to engage in physical activity because traitlike factors are involved in the formation of antecedents of intention like attitudes and perceived behavioral control (Courneya et al., 1999). In some cases, dispositional constructs like locus of control were found to affect intentions and behavior only via the mediation of the theory constructs (Armitage, Norman, & Conner, 2002; Hagger & Armitage, 2004). However, recent research on the "big five" personality constructs has indicated that some personality aspects uniquely and directly predict physical activity behavior (see figure 7.1). Prominent among these factors are the personality traits conscientiousness (Conner & Abraham, 2001) and "activity," a specific subfacet of the extroversion personality factor (Rhodes, Courneya, & Jones, 2002). These researchers explain that such influences reflect more spontaneous and less deliberative routes to behavioral engagement, so that the deliberative planning involved in the theory of planned behavior is not relevant to such behavioral enactment.

Examples of items measuring these personality constructs are given in table 7.1. It must, however, be stressed that these routes were in addition to, not a substitute for, the deliberative influences outlined by intentions and their antecedents in the theory of planned behavior. This supports the dual-route models of behavioral engagement that have recently been put forward (e.g., Fazio, 1990; Hagger, Chatzisarantis, & Harris, 2006; Strack & Deutsch, 2004). Such models imply that people make decisions on the basis of deliberative processes, such as attitudes and perceived behavioral control, but also may decide to engage in a behavior without the involvement of planning. Such spontaneous behavioral engagement may occur when the appropriate conditions and opportunities unexpectedly arise, as when a passerby joins in with her or his friends playing an informal soccer game in the park.

Self-Identity

Another socially defined construct that has been included in the theory of planned behavior is self-identity. Self-identity or "role identity" is a person's identification of her- or himself as the type of person who typically engages in the target behavior. It is strongly linked to the social categorization and social comparison processes involved in social identity theory (Hagger & Chatzisarantis, 2006). For example, a young person may consider herself to be a "sporty" person who typically participates in lots of sporting activities—a role that is likely to strongly influence her intentions to participate in physical activities in the future. Self-identity has been found to be an independent predictor of intentions (see figure 7.1) and conceptually and empirically distinct from subjective norms (e.g., Armitage & Conner, 1999; Sparks & Guthrie, 1998).

A sample questionnaire item used to measure self-identity in the context of the theory of planned behavior is given in table 7.1. The inclusion of self-identity has been based on hypotheses developed from social identity theory with respect to the socially defined roles that people consider when engaging in a behavior (Charng, Piliavin, & Callero, 1988; Terry, Hogg, & White, 1999).

Recent research has suggested that some people may be more oriented toward their self-identity when making decisions to act than toward the traditional components of the theory of planned behavior. Such people can be termed "self-identity oriented." A self-identity orientation can be considered a trait-like individual difference variable that is likely to affect intention formation across numerous behavioral contexts (Hagger & Chatzisarantis, 2006). Using within-participants analyses, we found that self-identity–oriented people have a stronger influence of self-identity on their intentions than people who are more oriented toward the other predictors of intention in the theory of planned behavior. Therefore, self-identity orientation is an individual difference construct, and this is manifested in the strong effects of self-identity on intentions in many different behavioral contexts including physical activity (Hagger & Chatzisarantis, 2006).

Recommended Interventions Derived From the TPB

The findings of research using the theory of planned behavior have informed interventions to modify physical activity behavior in young people. Given the strong effects of attitudes and perceived behavioral control on physical activity (Hagger et al., 2002b), it can be suggested that exercise interventions target these constructs. It would be less advisable to target subjective norm because its influence on physical activity intentions is weak in the physical activity domain on the basis of meta-analytic evidence from studies in youth physical activity. In addition, it is not advisable to devise an intervention focusing on dispositional constructs, such as the activity facet of extroversion or conscientiousness, because these are very stable, traitlike constructs that are difficult to change. However, interventions that take into account anticipated emotions and self-identity may be successful because these have been shown to have unique and significant effects on intentions independent of attitudes and subjective norms.

Core Theory of Planned Behavior Constructs

The theories of reasoned action and planned behavior propose that attitudes and perceived behavioral control can be changed through modification of the beliefs that influence these constructs. According to Ajzen (1991), changes in beliefs are most likely to produce demonstrable changes in attitudes and perceptions of control when the modal salient beliefs are targeted. Modal beliefs are the most frequently cited beliefs regarding the behavior as elicited from the target population. Modal salient beliefs can be identified through the use of open-ended questionnaire techniques that require individuals to recall and list beliefs about the target behavior, as illustrated in studies that have elicited beliefs about youth physical activity using the theory of planned behavior (e.g., Hagger, Chatzisarantis, & Biddle, 2001).

Once salient beliefs are identified, pamphlets, face-to-face discussion, observational modeling, or any other effective method could be used to send persuasive communications (Ajzen & Fishbein, 1980). Persuasive communications are appeals that involve arguments endorsing the positive aspects and outcomes of the target behavior while at the same time downplaying the negative aspects. For example, a persuasive appeal that aims to change the attitudes of adolescents toward exercise might take the form of the following text that highlights the advantages and downplays disadvantages of exercise:

> Participating in regular exercise has many benefits. You might learn how to play a new game or sport as well as improve your general level of fitness and well-being at the same time. Exercise can be great fun as well. It does not necessarily cause injuries or make you feel uncomfortably hot and sweaty if you exercise at an intensity you feel comfortable with.

This persuasive appeal should be effective in changing young people's attitudes because it targets the accessible behavioral beliefs of young people identified in formative research. For example, a recent intervention using this method with young people presented participants with a persuasive message targeting modal salient behavioral beliefs with the aim of changing intentions by influencing a change in attitudes (Chatzisarantis & Hagger, 2005). Results of this experiment are given in the path model in figure 7.2, which shows the standardized regression coefficients of the effects of the intervention on the theory of planned behavior constructs and physical activity behavior. Specifically, the persuasive communication that targeted salient behavioral beliefs resulted in more posi-

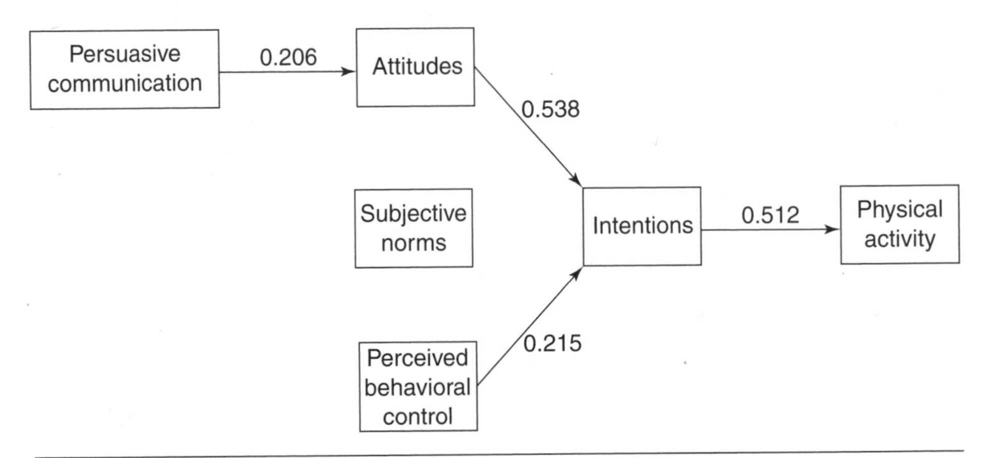

Figure 7.2 Path model of a persuasive communication intervention designed to influence attitudes. Only statistically significant path coefficients are shown.

Reproduced and adapted, by permission from N.L.D. Chatzisarantis and M.S. Hagger, 2005, "Effects of a brief intervention based on the Theory of Planned Behavior on leisure-time physical activity participation," *Journal of Sport and Exercise Physiology* 27(4): 480.

tive attitudes and, in turn, stronger intentions than when the communications targeted nonsalient behavioral beliefs.

A major limitation of interventions based on the theory of planned behavior is that they can be directed only at individuals who do not intend to perform a behavior or have low levels of intentions. Interventions based on these theories aim to produce positive intentions among nonintenders or strengthen the intentions of those who have low levels of intention. They achieve this by changing the beliefs that underlie the attitudes and perceived behavioral control construct from the theory. Such an approach is not useful for people who possess strong intentions to engage in physical activity behavior. This is so because the theory of planned behavior is an intentional theory that can facilitate only the *formation*, rather than enactment, of intentions. In contrast, theories that focus on how intentions might be translated into behavior (the "implemental phase") after an intention has been formed are most effective in facilitating the enactment of behavioral intentions and can be successfully employed alongside interventions based on the theory of planned behavior.

Recommendations Based on Implementation Intentions

Despite the modifications to the existing theory of planned behavior constructs and the addition of individual difference constructs as outlined earlier, the intention–behavior relationship is not perfectly modeled. The theory does not fully explain the processes by which intentions are translated into action. A recent line of research suggests that this is the case because people often forget to carry out their intentions (Gollwitzer, 1999; Orbell, 2000; Orbell, Hodgkins, & Sheeran, 1997; Sheeran & Orbell, 1999). Alternatively, individuals' execution of their intentions may be interrupted because other competing goals gain priority over the original intended behavior (Verplanken & Faes, 1999). Social cognitive theories, like the theories of reasoned action and planned behavior, do not address these difficulties associated with enactment of intentions, and as a result may not fully explain the intention–behavior relationship.

One approach that has been put forward to resolve the inadequacies of the theory of planned behavior in explaining the intention–behavior relationship is that of *implementation intentions.* Implementation intentions are self-regulatory strategies that involve the formation of concrete plans specifying when, how, and where performance of behavior will take place. Indeed, recent research has indicated that planning is implicated in the conversion of intentions into behavior in a physical activity context (Norman & Conner, 2005). Experimental paradigms using implementation intention strategies require research participants to specify explicitly *when* and *where* they will engage in an intended behavior to achieve their behavioral goals (Orbell, 2000). According to Gollwitzer (1999), implementation intentions help people move from a motivational phase to a volitional phase, ensuring that intentions are converted into action.

Research indicates that forming implementation intentions decreases the probability that people will forget to execute their goal-directed intentions at the point of initiation (Orbell, 2000; Orbell et al., 1997; Sheeran & Orbell, 1999). The reason is that planning when and where to initiate a future action strengthens the mental association between representations of situations and representations of actions. This influence is represented in figure 7.1 not as a direct effect, but as an effect that changes or alters the effect of intentions on behavior. For example, a person may have a strong intention to engage in physical activity in the next few weeks but not have thought of when or where. If an intervention required that person to write down precisely when and where physical activity would take place (e.g., "at five o'clock, after work, at the gym on the way home"), when such cues arose the person would be more likely to enact the intention. Such an intervention is likely to be useful to young people, as it enables them to form concrete ideas on the location and timing of physical activity. Research also shows that increased accessibility of situational representations in memory increases the probability that action opportunities will get noticed, and that action initiation will occur given that the mere perception of action opportunities can automatically trigger a behavioral response (Orbell et al., 1997; Sheeran & Orbell, 1999). Importantly, implementation intentions increase behavioral engagement through these postdecisional, automatic processes, and not by concomitant increases in motivation or intention (Orbell et al., 1997).

Recent research has evaluated the effectiveness of interventions that combine motivational techniques with volitional techniques, such as implementation intentions, to influence the performance of behavior (Koestner et al., 2002; Milne, Orbell, & Sheeran, 2002; Prestwich, Lawton, & Conner, 2003; Sheeran & Silverman, 2003). The rationale behind this combined approach is that motivational strategies focus on increasing intention levels but do not facilitate the enactment of intentions, while volitional strategies, such as implementation intentions, increase the probability that these strong intentions will be converted into action. For example, Prestwich and colleagues demonstrated that an intervention adopting a rational decision-making strategy (decisional balance sheet, a strategy that enhances behavioral beliefs and attitudes toward physical activity) and implementation intentions was more effective in promoting physical activity than either of the strategies alone.

These results support the existence of two distinct phases of motivation: a motivational (predecisional) phase, during which people decide whether or not to perform a behavior, and a volitional (postdecisional, or implemental) phase, during which people plan when and where they will convert their intentions into behavior (Gollwitzer, 1999; Norman & Conner, 2005). As a consequence, interventions that combine motivational and volitional techniques are likely to be most effective in promoting physical activity. As of mid-2008, such combined approaches had not been used in youth contexts. Future researchers should endeavor to replicate these findings in a youth context to evaluate whether such additive effects are found in young people. To speculate, such strategies are likely to be effective because they help young people structure their lives in a way that enables them to act on their motives.

Given the success of implementation intentions in augmenting the theory of planned behavior and converting intentions into behavior, we recommend that interventions focus not only on enhancing beliefs but also on the implemental phase. This will result in the greatest changes in behavior. Therefore, interventions to promote physical activity would do well to include some persuasive communications focusing on behavioral and control beliefs to increase attitudes and perceived behavioral control toward physical activity among young people. In addition, they should provide an implementation intention that facilitates the conversion of intentions into behavior.

For example, social agents involved in promoting physical activity among young people would do well to provide young people with opportunities to participate in physical activities that address the outcome beliefs typically cited by children, such as to increase skills and fitness and make friends. In addition, they might also draw up a "contract" with each child by suggesting how to plan when and where the child will participate in these activities. This need not be done in an authoritarian manner but can be done in a fun, informal manner.

Theoretically, the provision of these interventions should be staggered such that interventions focusing on enhancing attitudes and intentions are presented first, followed by an implementation intention to facilitate the efficient conversion of an intention into action. This dual-phase intervention represents the best practice in accordance with research based on modified versions of the theory of planned behavior and implementation intentions. However, these interventions can be executed in a single session without staggering the "phases" temporally, as synergistic efforts including attitude-based interventions to increase intentions and implementation intentions have been effective (e.g., Prestwich et al., 2003).

Conclusion

This chapter has focused on attitude-based theories of physical activity behavior, how they have been developed and modified, and their effectiveness in explaining physical activity in young people. Also considered are the provision of other variables and approaches to maximize the prediction of intentions and subsequent physical activity, and interventions to promote physical activity behavior in young people. Research suggests that attitude-based models, like the theory of planned behavior, explain significant variance in physical activity intentions and behavior, with attitudes and perceived behavioral control playing a major role (Hagger et al., 2002b). Further, the attitude, subjective norm, and perceived behavioral control constructs are multidimensional and can be split into affective and instrumental attitudes, injunctive and descriptive norms, and perceived controllability and self-efficacy, respectively (Hagger & Chatzisarantis, 2005).

In addition, research augmenting these theories indicates that a number of other constructs can explain additional variance in physical activity intention and behavior, namely attitude strength, anticipated affect, conscientiousness, the activity facet of extroversion, social support, and self-identity (Conner & Armitage, 1998). Implemental theories indicate that the conversion of intentions into behavior can be augmented by an implementation intention, a simple intervention requiring individuals to state when and where they will perform physical activity (Gollwitzer, 1999). Research suggests that interventions to promote attitudes and intentions toward engaging in physical activity behavior should focus on the belief systems that underline the core influential constructs within the theory of planned behavior (Chatzisarantis & Hagger, 2005). But interventions should also consider the inclusion of implementation intentions to optimize the effectiveness of physical activity behavior change (Prestwich et al., 2003).

APPLICATIONS FOR RESEARCHERS

Researchers applying attitudinal approaches to youth physical activity are encouraged to use models such as the theory of planned behavior that focus on processes rather than mere description, and to incorporate recent modifications to comprehensively capture the precise processes underpinning youths' physical activity intentions. For example, researchers should distinguish between the subcomponents of attitudes, subjective norms, and perceived behavioral control to illustrate which constructs are most influential. Research is needed that incorporates other constructs explaining intentions and behavior (e.g., attitude strength, anticipated regret, social support, personality, self-identity) and adopts intervention strategies that both promote intentions and enhance their implementation. Finally, much attitudinal research addresses youth 14 years or older (Hagger et al., 2002b). Researchers would do well to examine the validity of attitude constructs and theories in younger children, carefully considering the challenges associated with self-report measures of attitudes in this population. Considering the real barriers and constraints placed on young children's physical activity by caregivers would be essential to this work.

APPLICATIONS FOR PROFESSIONALS

Practitioners are encouraged to target the major psychological influences on physical activity intentions, attitudes, and perceived behavioral control. Interventions should provide opportunities for young people to engage in activities that are congruent with their typical beliefs about doing physical activities, such as enjoyment, getting fit, improving skills, and being with friends. It is also important that practitioners address other influences on physical activity such as social support. For example, practitioners may wish to find a way of reaching or including parents in the decision-making process when prescribing physical activity outside of structured sessions (e.g., physical education class). Practitioners also should engage young people in decision making by getting them to state when and where they will perform physical activities in the future. This implementation intention approach will help convert intentions into behavior more effectively. For example, youth sport leaders might get participants to state when and where they will attend certain practices.

References

Abraham, C., & Sheeran, P. (2003). Acting on intentions: The role of anticipated regret. *British Journal of Social Psychology, 42*, 495-511.

Abraham, C., & Sheeran, P. (2004). Deciding to exercise: The role of anticipated regret. *British Journal of Health Psychology, 9*, 269-278.

Ajzen, I. (1985). From intentions to actions: A theory of planned behavior. In J. Kuhl & J. Beckmann (Eds.), *Action-control: From cognition to behavior* (pp. 11-39). Heidelberg: Springer.

Ajzen, I. (1991). The theory of planned behavior. *Organizational Behavior and Human Decision Processes, 50*, 179-211.

Ajzen, I. (2001). Nature and operation of attitudes. *Annual Review of Psychology, 52*, 27-58.

Ajzen, I. (2002). Perceived behavioral control, self-efficacy, locus of control, and the theory of planned behavior. *Journal of Applied Social Psychology, 32*, 1-20.

Ajzen, I., & Driver, B.E. (1991). Prediction of leisure participation from behavioral, normative, and control beliefs: An application of the theory of planned behavior. *Leisure Sciences, 13*, 185-204.

Ajzen, I., & Fishbein, M. (1980). *Understanding attitudes and predicting social behavior.* Englewood Cliffs, NJ: Prentice Hall.

Armitage, C.J., & Conner, M. (1999). The theory of planned behavior: Assessment of predictive validity and "perceived control." *British Journal of Social Psychology, 38*, 35-54.

Armitage, C.J., & Conner, M. (2001). Efficacy of the theory of planned behaviour: A meta-analytic review. *British Journal of Social Psychology, 40*, 471-499.

Armitage, C.J., Norman, P., & Conner, M. (2002). Can the theory of planned behaviour mediate the effects of age, gender and multidimensional locus of control? *British Journal of Health Psychology*, 7, 299-316.

Ashton, M.C., Lee, K., Perugini, M., Szarota, P., de Vries, R.E., Di Blas, L., et al. (2004). A six-factor structure of personality-descriptive adjectives: Solutions from psycholexical studies in seven languages. *Journal of Personality and Social Psychology, 86*, 356-366.

Bagozzi, R.P., & Kimmel, S.K. (1995). A comparison of leading theories for the prediction of goal directed behaviours. *British Journal of Social Psychology, 34*, 437-461.

Baker, C.W., Little, T.D., & Brownell, K.D. (2003). Predicting adolescent eating and activity behaviors: The role of social norms and personal agency. *Health Psychology, 22*, 189-198.

Bassili, J.N. (1996). Meta-judgmental versus operative indices of psychological properties: The case of measures of attitude strength. *Journal of Personality and Social Psychology, 71*, 637-653.

Bozionelos, G., & Bennett, P. (1999). The theory of planned behaviour as predictor of exercise: The moderating influence of beliefs and personality variables. *Journal of Health Psychology, 4*, 517-529.

Charng, H.-W., Piliavin, J.A., & Callero, P.L. (1988). Role identity and reasoned action in the prediction of repeated behavior. *Social Psychology Quarterly, 51*, 303-317.

Chatzisarantis, N.L.D., & Hagger, M.S. (2005). Effects of a brief intervention based on the theory of planned behavior on leisure time physical activity participation. *Journal of Sport and Exercise Psychology*, 27, 470-487.

Conner, M., & Abraham, C. (2001). Conscientiousness and the theory of planned behavior: Toward a more complete model of the antecedents of intentions and behavior. *Personality and Social Psychology Bulletin*, 27, 1547-1561.

Conner, M., & Armitage, C.J. (1998). Extending the theory of planned behavior: A review and avenues for further research. *Journal of Applied Social Psychology, 28*, 1429-1464.

Courneya, K.S., Bobick, T.M., & Schinke, R.J. (1999). Does the theory of planned behavior mediate the relationship between personality and exercise behavior? *Basic and Applied Social Psychology, 21*, 317-324.

Courneya, K.S., Plotnikoff, R.C., Hotz, S.B., & Birkett, N.J. (2000). Social support and the theory of planned behavior in the exercise domain. *American Journal of Health Behavior, 24*, 300-308.

Courneya, K.S., Plotnikoff, R.C., Hotz, S.B., & Birkett, N.J. (2001). Predicting exercise stage transitions over two consecutive 6-month periods: A test of the theory of planned behaviour in a population-based sample. *British Journal of Health Psychology, 6*, 135-150.

Doll, J., & Ajzen, I. (1992). Accessibility and stability of predictors in the theory of planned behavior. *Journal of Personality and Social Psychology, 63*, 754-765.

Dzewaltowski, D.A., Noble, J.M., & Shaw, J.M. (1990). Physical activity participation: Social cognitive theory vs. the theories of reasoned action and planned behavior. *Journal of Sport and Exercise Psychology, 12*, 388-405.

Eagly, A.H., & Chaiken, S. (1993). *The psychology of attitudes.* San Diego: Harcourt, Brace and Jovanovich.

Fazio, R.H. (1990). Multiple processes by which attitudes guide behavior: The MODE model as an integrative framework. In M.P. Zanna (Ed.), *Advances in experimental social psychology* (Vol. 23, pp. 75-109). San Diego: Academic Press.

Finlay, K.A., Trafimow, D., & Villarreal, A. (2002). Predicting exercise and health behavioral intentions: Attitudes, subjective norms, and other behavioral determinants. *Journal of Applied Social Psychology, 32*, 342-358.

Godin, G., & Shephard, R.J. (1984). Normative beliefs of school children concerning regular exercise. *Journal of School Health, 54*(11), 443-445.

Goldberg, L.R., Johnson, J.A., Eber, H.W., Hogan, R., Ashton, M.C., Cloninger, C.R., & Gough, H.C. (2006). The International Personality Item Pool and the future of public-domain personality measures. *Journal of Research in Personality, 40*, 84-96.

Gollwitzer, P.M. (1999). Implementation intentions: Strong effects of simple plans. *American Psychologist, 54*, 493-503.

Hagger, M.S., & Armitage, C. (2004). The influence of perceived loci of control and causality in the theory of planned behavior in a leisure-time exercise context. *Journal of Applied Biobehavioral Research, 9*, 45-64.

Hagger, M.S., Cale, L.A., & Ashford, B.A. (1997). Children's physical activity levels and attitudes towards physical activity. *European Physical Education Review, 3*, 144-164.

Hagger, M.S., & Chatzisarantis, N.L.D. (2005). First- and higher-order models of attitudes, normative influence, and perceived behavioural control in the theory of planned behaviour. *British Journal of Social Psychology, 44*, 513-535.

Hagger, M.S., & Chatzisarantis, N.L.D. (2006). Self-identity and the theory of planned behaviour: Between-and within-participants analyses. *British Journal of Social Psychology, 45*, 731-757.

Hagger, M.S., Chatzisarantis, N.L.D., Barkoukis, V., Wang, C.K.J., Hein, V., Pihu, M., et al. (2007). Cross-cultural generalizability of the theory of planned behavior among young people in a physical activity context. *Journal of Sport and Exercise Psychology, 29*, 2-20.

Hagger, M.S., Chatzisarantis, N., & Biddle, S.J.H. (2001). The influence of self-efficacy and past behaviour on the physical activity intentions of young people. *Journal of Sports Sciences, 19*, 711-725.

Hagger, M.S., Chatzisarantis, N., & Biddle, S.J.H. (2002a). The influence of autonomous and controlling motives on physical activity intentions within the theory of planned behaviour. *British Journal of Health Psychology, 7*, 283-297.

Hagger, M.S., Chatzisarantis, N., & Biddle, S.J.H. (2002b). A meta-analytic review of the theories of reasoned action and planned behavior in physical activity: Predictive validity and the contribution of additional variables. *Journal of Sport and Exercise Psychology, 24*, 3-32.

Hagger, M.S., Chatzisarantis, N., Biddle, S.J.H., & Orbell, S. (2001). Antecedents of children's physical activity intentions and behaviour: Predictive validity and longitudinal effects. *Psychology and Health, 16*, 391-407.

Hagger, M.S., Chatzisarantis, N.L.D., & Harris, J. (2006). From psychological need satisfaction to intentional behavior: Testing a motivational sequence in two behavioral contexts. *Personality and Social Psychology Bulletin, 32*, 131-138.

Hausenblas, H.A., Carron, A.V., & Mack, D.E. (1997). Application of the theories of reasoned action and planned behavior to exercise behavior: A meta analysis. *Journal of Sport and Exercise Psychology, 19*, 36-41.

Heaven, P.C.L. (2001). *The social psychology of adolescence.* New York: Palgrave Macmillan.

Koestner, R., Lekes, N., Powers, T.A., & Chicoine, E. (2002). Attaining personal goals: Self-concordance plus implementation intentions equals success. *Journal of Personality and Social Psychology, 83*, 231-244.

Loomes, G., & Sugden, R. (1982). Regret theory: An alternative theory of rational choice under uncertainty. *Economic Journal, 92*, 805-824.

Lowe, R., Eves, F., & Carroll, D. (2002). The influence of affective and instrumental beliefs on exercise intentions and behavior: A longitudinal analysis. *Journal of Applied Social Psychology, 32*, 1241-1252.

Martin, J.J., Hodges Kulinna, P., McCaughtry, N., Cothran, D., Dake, J., & Fahoome, G. (2005). The theory of planned behavior: Predicting physical activity and cardiorespiratory fitness in African American children. *Journal of Sport and Exercise Psychology, 27*, 456-469.

Milne, S.E., Orbell, S., & Sheeran, P. (2002). Combining motivational and volitional interventions to promote exercise participation: Protection motivation theory and implementation intentions. *British Journal of Health Psychology*, 7, 163-184.

Motl, R.W., Dishman, R.K., Saunders, R.P., Dowda, M., Felton, G., Ward, D.S., et al. (2002). Examining social-cognitive determinants of intention and physical activity among Black and White adolescent girls using structural equation modeling. *Health Psychology, 21*, 459-467.

Norman, P., & Conner, M. (2005). The theory of planned behavior and exercise: Evidence for the mediating and moderating roles of planning on intention-behavior relationships. *Journal of Sport and Exercise Psychology*, 27, 488-504.

Orbell, S. (2000). Motivational and volitional components in action initiation: A field study of the role of implementation intentions. *Journal of Applied Social Psychology, 30*, 780-797.

Orbell, S., Hodgkins, S., & Sheeran, P. (1997). Implementation intentions and the theory of planned behavior. *Personality and Social Psychology Bulletin, 23*, 945-954.

Parker, D., Manstead, A.S.R., & Stradling, S.G. (1995). Extending the theory of planned behaviour: The role of personal norm. *British Journal of Social Psychology, 34*, 127-137.

Pate, R.R., Trost, S.G., Felton, G., Ward, D.S., Dowda, M., & Saunders, R. (1997). Correlates of physical activity behavior in rural youth. *Research Quarterly for Exercise and Sport, 68*, 241-248.

Prestwich, A., Lawton, R., & Conner, M. (2003). The use of implementation intentions and the decision balance sheet in promoting exercise behaviour. *Psychology and Health, 10*, 707-721.

Rhodes, R.E., Courneya, K.S., & Jones, L.W. (2002). Personality, the theory of planned behavior, and exercise: A unique role for extroversion's activity facet. *Journal of Applied Social Psychology, 32*, 1721-1736.

Rhodes, R.E., Jones, L.W., & Courneya, K.S. (2002). Extending the theory of planned behavior in the exercise domain: A comparison of social support and subjective norm. *Research Quarterly for Exercise and Sport, 73*, 193-199.

Rhodes, R.E., MacDonald, H.M., & McKay, H.A. (2006). Predicting physical activity intention and behaviour among children in a longitudinal sample. *Social Science and Medicine, 62*, 3146-3156.

Rivis, A., & Sheeran, P. (2003). Descriptive norms as an additional predictor in the theory of planned behaviour: A meta-analysis. *Current Psychology, 22*, 218-233.

Saunders, R.P., Motl, R.W., Dowda, M., Dishman, R.K., & Pate, R.R. (2004). Comparison of social variables for understanding physical activity in adolescent girls. *American Journal of Health Behavior, 28*, 426-436.

Sheeran, P., & Orbell, S. (1999). Implementation intentions and repeated behaviour: Augmenting the predictive validity of the theory of planned behaviour. *European Journal of Social Psychology, 29*, 349-369.

Sheeran, P., & Silverman, M. (2003). Evaluation of three interventions to promote workplace health and safety: Evidence for the utility of implementation intentions. *Social Science and Medicine, 56*, 2153-2163.

Sheeran, P., Trafimow, D., Finlay, K.A., & Norman, P. (2002). Evidence that the type of person affects the strength of the perceived behavioural control-intention relationship. *British Journal of Social Psychology, 41*, 253-270.

Sparks, P., & Guthrie, C.A. (1998). Self-identity and the theory of planned behaviour: A useful addition or an unhelpful artifice. *Journal of Applied Social Psychology, 27*, 1393-1410.

Sparks, P., & Shepherd, R. (1992). Self-identity and the theory of planned behavior: Assessing the role of identification with "green consumerism." *Social Psychology Quarterly, 55*, 388-399.

Strack, F., & Deutsch, R. (2004). Reflective and impulsive determinants of social behavior. *Personality and Social Psychology Review, 8*, 220-247.

Symons-Downs, D., & Hausenblas, H.A. (2005). The theories of reasoned action and planned behavior applied to exercise: A meta-analytic update. *Journal of Physical Activity and Health, 2*, 76-97.

Terry, D.J., Hogg, M.A., & White, K.M. (1999). The theory of planned behaviour: Self-identity, social identity and group norms. *British Journal of Social Psychology, 38*, 225-244.

Terry, D.J., & O'Leary, J.E. (1995). The theory of planned behaviour: The effects of perceived behavioural control and self-efficacy. *British Journal of Social Psychology, 34*, 199-220.

Theodorakis, Y. (1994). Planned behavior, attitude strength, role identity, and the prediction of exercise behavior. *Sport Psychologist, 8*, 149-165.

Theodorakis, Y., Doganis, G., Bagiatis, K., & Gouthas, M. (1991). Preliminary study of the ability of Reasoned Action Model in predicting exercise behavior in young children. *Perceptual and Motor Skills, 72*, 51-58.

Trafimow, D., & Finlay, K.A. (1996). The importance of subjective norms for a minority of people: Between-subjects and within-subjects effects. *Personality and Social Psychology Bulletin, 22*, 820-828.

Trafimow, D., & Sheeran, P. (1998). Some tests of the distinction between cognitive and affective beliefs. *Journal of Experimental Social Psychology, 34*, 378-397.

Trafimow, D., Sheeran, P., Conner, M., & Finlay, K.A. (2002). Evidence that perceived behavioral control is a multidimensional construct: Perceived control and perceived difficulty. *British Journal of Social Psychology, 41*, 101-121.

Trost, S.G., Pate, R.R., Dowda, M., Saunders, R., Ward, D.S., & Felton, G. (1996). Gender differences in physical activity and determinants of physical activity in rural 5th grade children. *Journal of School Health, 66*, 145-150.

Trost, S.G., Pate, R.R., Dowda, M., Ward, D.S., Felton, G., & Saunders, R. (2002). Psychosocial correlates of physical activity in White and African-American girls. *Journal of Adolescent Health, 31*, 226-233.

Verplanken, B., & Faes, S. (1999). Good intentions, bad habits, and effects of forming implementation intentions on healthy eating. *European Journal of Social Psychology, 29*, 591-604.

Wicker, A.W. (1969). Attitudes versus actions: The relationship of verbal and overt behavioral responses to attitude objects. *Journal of Social Issues, 25*, 41-78.

CHAPTER

8

Motivational Characteristics

Stuart J.H. Biddle, PhD ▪ Darren C. Treasure, PhD ▪ C.K. John Wang, PhD

When it comes to the topic of motivation of young people in physical activity and sedentary behaviors, it is tempting to identify the lack of activity as a simple issue and therefore to seek a simple answer. For example, it is common to "blame" inadequate quality or quantity of school physical education or TV sets in children's bedrooms for causing inactivity or negative health outcomes such as obesity. However, this area is highly complex, and simplistic approaches are doomed to fail in substantially predicting behavior.

In this chapter, therefore, we seek to clarify approaches that have been used to understand physically active and sedentary behaviors in young people—approaches capturing the so-called *why* of behavior. Specifically, we address different motivational frameworks centered on competence and autonomy. Other psychological approaches that emphasize attitudes (chapter 7) or self-perceptions more broadly (chapter 9) are dealt with elsewhere in this book, as are more generic issues concerning correlates, interventions, and theory (chapter 4).

Descriptive Approaches

The research on young people's physical activity participation often includes competitive sport and focuses less on other aspects of exercise and physical activity. This is not entirely surprising because children are less likely to participate in fitness pursuits currently favored by adults, at least not until mid- to late adolescence. It is important, however, that we understand more fully the reasons children give for participation or nonparticipation in various physical activities, such as recreational play or the use of nonmotorized forms of transport.

Interviews with young people and their parents in England (Mulvihill, Rivers, & Aggleton, 2000) have shown that children aged 5 to 11 years are often physically active and are enthusiastic about activity. They are motivated by enjoyment and social elements of participation, whereas for those aged 11 to 15 years, enjoyment remained important but was enhanced when an element of choice was evident. Motives for weight control start to emerge in girls at this age as well.

Data from over 3000 children in Northern Ireland (Van Wersch, 1997; Van Wersch, Trew, & Turner, 1992) have shown that "interest in physical education" remains relatively constant for boys from age 11 to 19 years, whereas during the same period interest declines sharply for girls. Data from the English Sports Council's survey of over 4000 respondents, 6 to 16 years old (Mason, 1995), showed that motives are diverse, ranging from general enjoyment to fitness and friendships.

Reviews have concluded that children are motivated to engage in physical activity for diverse reasons, including fun and enjoyment, learning and improving skills, being with friends, having success and winning, and physical fitness and health (Biddle, 1995). The latter factor might also include weight control and body appearance for older youth. However, more research is needed to increase understanding of the differences in motives across activities, levels of participation, and developmental stages, though so far the research shows some similarity in motives across settings and groups.

As with motives for participation, there appear to be numerous barriers to physical activity in children and adolescents. Time has been reported as a barrier for 5- to 11-year-olds (Mulvihill et al., 2000), perhaps reflecting less discretionary time allowed by parents (Sturm, 2005). Environmental barriers, such as road traffic and fear for safety, may be interrelated with such barriers (Davis & Jones, 1996, 1997; Gomez et al., 2004). These barriers to physical activity may also contribute to more time inside, predicting greater amounts of sedentary behavior. Somewhat surprisingly, Sallis, Prochaska, and Taylor (2000) found that only 33% of studies on potential barriers for adolescents showed actual associations with reduced levels of physical activity. All three studies they located for children, however, did show the expected relationship. Our own review of correlates of physical activity for adolescent girls revealed that time barriers were associated with less physical activity, as were barriers related to schoolwork and the perceived effort required to be active (Biddle et al., 2005).

Such descriptive approaches are a good place to start, but not to finish! More explanatory approaches, sometimes using theories from related disciplines, are required. A theory is "a set of interrelated constructs (concepts), definitions, and propositions that present a systematic view of phenomena by specifying relations among variables, with the purpose of explaining and predicting phenomena" (Kerlinger, 1973, p. 9). Physical activity behavior researchers have adopted theories and models from general, social, educational, and health psychology and have tested and applied them in the context of physical activity. The rest of this chapter addresses such approaches, with a focus on perceptions of competence and autonomy.

Framework for Theoretical Perspectives

To assist in the organization of approaches and theories, we have placed key theoretical frameworks into a classification system that allows types of theories to be clustered (Biddle et al., 2007; Biddle & Nigg, 2000). For example, some theories are based more on constructs of competence while others are based more on attitudes. The framework is shown in figure 8.1. Such a system is a heuristic, and there is overlap between categories. For the purposes of the present chapter, we cover only competence- and control-based approaches.

Competence-Based Approaches

Motivation is strongly associated with competence. Rarely, if ever, do we freely choose what might be considered nonessential behaviors that we feel particularly incompetent about. For this reason, it is important to understand the role of competence in physical activity motivation. Several approaches have been proposed.

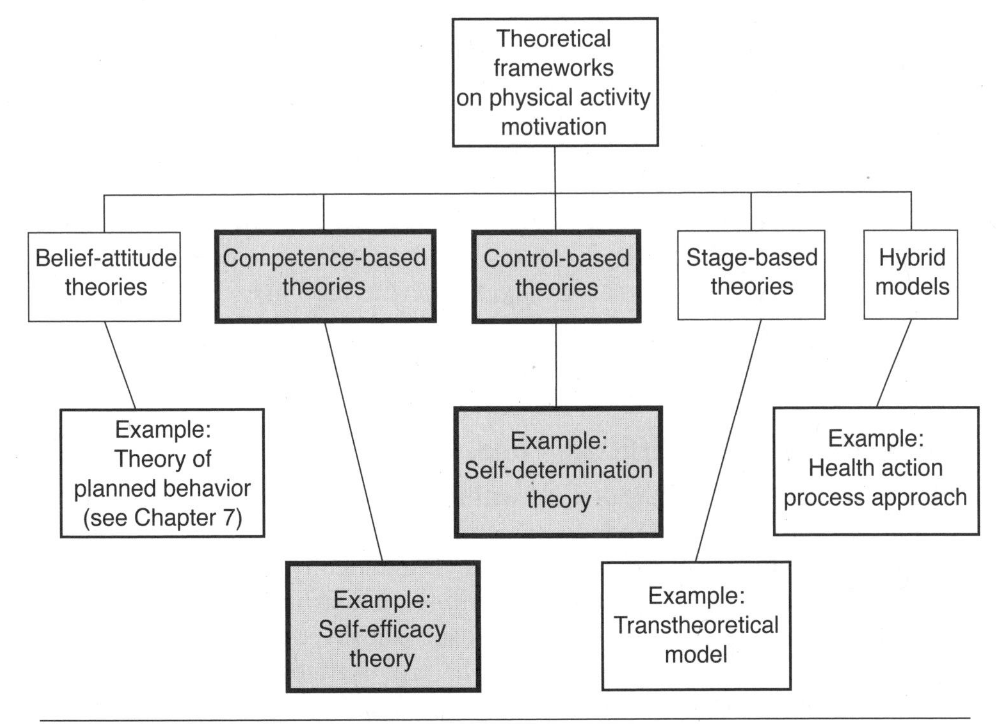

Figure 8.1 A framework for classifying theories of physical activity. Current chapter focus is shaded.

Reprinted, by permission, from S. Biddle et al., 2007, "Theoretical frameworks in exercise psychology." In *Handbook of sport psychology,* 3rd ed., edited by G. Tenenbaum and R.C. Eklund (Hoboken, NJ: John Wiley and Sons), 538; adapted, by permission, from S.J.H. Biddle and C.R. Nigg, "Theories of exercise behavior," *International Journal of Sport Psychology* 31 290-304.

Competence I: Competence Motivation Theory

Harter's competence motivation theory (Harter, 1978, 1981) has made a significant impact on research in motivation among youth in the physical activity and sport domains and deserves a brief discussion. Harter argued that the "effectance motivation" model of White (1959) was too global and proposed that competence should be considered as multidimensional with distinct domains such as scholastic, social, and athletic competence. She suggested that individuals are motivated to demonstrate competence and avoid incompetence in achievement domains, which is essential for the development of their self-worth. As children get older, their competence becomes more differentiated, reflected in an ever-increasing number of competence domains that they use to judge themselves. Also, the sources of information used to judge competence vary in number and salience with age.

Harter (1978, 1981) considered the social environment to be important in affecting feelings of competence. Feedback from the social environment regarding one's actions directly influences a person's perceptions of competence either positively or negatively. This, in turn, may increase or decrease the likelihood of the behavior being repeated. Moreover, Harter linked the perception of control to self-perceptions of competence. In her model of mastery motivation in children, Harter (1981) suggests that if an individual perceives competence and internal control, the resultant affect will be pleasure and low anxiety.

Harter's (1978, 1981) perspective has influenced research in the field and directly underpins more contemporary studies of physical activity (Weiss & Williams, 2004), including those on self-esteem (Fox & Corbin, 1989). Indeed, many social cognitive theories have identified confidence and competence as important constructs at the anecdotal and empirical level in exercise and health research. Some of these theories include self-efficacy theory, achievement goal theory, and self-theories of ability. These theories will be discussed in subsequent subsections.

Competence II: Self-Efficacy Theory

Self-efficacy theory (SET) is defined within Bandura's social cognitive theory (Bandura, 1986) as "people's judgments of their capabilities to organize and execute courses of action required to attain designated types of performances" (p. 391). It is a situation-specific self-confidence that indicates the strength and level at which a person believes he or she can successfully perform a skill or task. One's self-efficacy toward a specific task is affected by the contextual variables surrounding the task, such as the physical activity environment, and one's perceptions of ability to cope with these factors. For example, self-efficacy concerning walking to school may be affected by local traffic density, distance, and personal fitness or energy.

According to SET, behavioral choice, performance, and persistence in a specific activity are influenced by two types of expectancies of the behavior:

outcome expectations and efficacy expectations. Efficacy expectations concern whether people believe that they can execute the behavior, such as taking part in exercise on a regular basis. Outcome expectations concern whether people believe their actions will produce certain outcomes, such as fitness gain (see figure 8.2).

There are several sources from which self-efficacy information can be derived. The three most relevant to physical activity are performance accomplishments, vicarious experiences (modeling), and verbal and social persuasion. Performance accomplishments are the most influential source of efficacy information because they are based on the outcomes of one's past experiences. Vicarious sources of efficacy information depend on observers' self-comparison with people they observe and the outcomes attained by those they observe. Persuasive information includes verbal persuasion, self-talk, and imagery. Although most studies are conducted with adults, the application of SET across younger age-groups has received support (Nigg & Courneya, 1998).

Research has shown that exercise self-efficacy (a) can be increased through intervention; (b) will predict participation, particularly in the early stages of an exercise program; (c) declines after a period of inactivity; and (d) is associated with positive exercise emotion (Hagger, Chatzisarantis, & Biddle, 2001; McAuley & Blissmer, 2000; Trost et al., 1999). Self-efficacy emerges as one of the most consistent predictors of physical activity behaviors and is key to promoting activity in young people. Using the known sources of self-efficacy, it is recommended that those working with young people seek to enhance self-efficacy through positive experiences, such as verbal encouragement and opportunities for self-improvement; give young people opportunities to observe success of others who are similar to themselves; and provide a nonthreatening atmosphere. Whether self-efficacy is an important construct in sedentary behaviors remains to be tested. Usually self-efficacy is required for undertaking challenging tasks, and this may not apply to sedentary pursuits. Perhaps one exception might be computer games, where competence could play a motivating or reinforcing role for some young people.

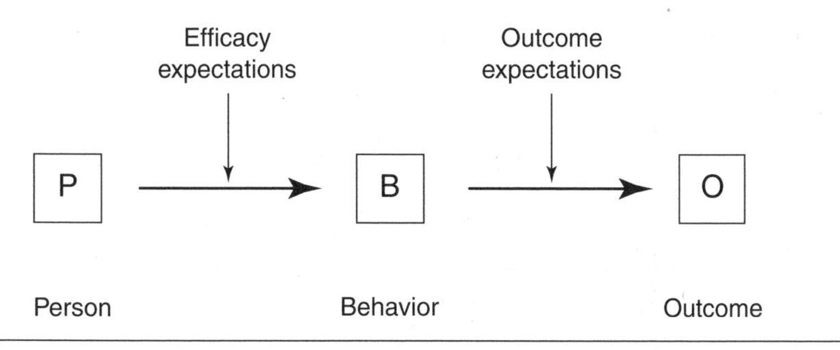

Figure 8.2 The distinction between efficacy and outcome expectations.

Competence III: Achievement Goals

Perhaps the most popular contemporary approach to examining the motivation of young people in physical activity settings such as sport and physical education is achievement goal theory (Nicholls, 1989). This is a social cognitive approach to understanding human behavior and assumes that individuals act in an intentional, goal-directed, and rational way. Achievement goals are postulated to govern achievement-related beliefs and guide subsequent decision making and behavior. It is argued that in order to understand the motivation of young people in physical activity settings, one must account for the function and meaning of the behavior. It is this meaning that reflects the purpose of achievement striving and determines the integrated pattern of beliefs that form the basis of approach and avoidance strategies, differing engagement levels, and other social cognitive and behavioral responses.

An achievement goal approach to understanding behavior of young people proposes that differences in the way an individual judges her or his competence and perceives success are important factors in determining the intensity and direction of behavior in achievement activities (Nicholls, 1989). This approach contends that the goal of action in achievement settings is the demonstration of ability and that an individual may adopt two goal states of involvement, namely *task* and *ego*. When a task state of involvement is adopted, ability is demonstrated when learning and mastery at the task are achieved and high effort is exerted. Therefore, in the case of young people and physical activity, their assessment of their ability in, say, sport or physical education is self-referenced and success is perceived when mastery is demonstrated. However, when an ego state of involvement is adopted, ability is demonstrated when people exceed the performance of others, as in winning in sport.

Whether a person is task or ego involved is dependent on the dispositional goal orientation of the individual as well as the perception of achievement cues within the environment. Literature in physical education and sport on dispositional goal orientations has shown task orientation to be linked to an adaptive (positive) pattern of motivational responses, and much of this literature is on young people. In addition, ego orientation may be linked to positive outcomes if perceived competence is high (Nicholls, 1989). Our systematic review (Biddle et al., 2003b) showed associations of varying magnitude between a task orientation and

- beliefs that effort produces success (positive association: +);
- motives of skill development and team membership (+);
- beliefs that the purpose of sport and physical education is for mastery, fitness, and self-esteem (+);
- perceptions of competence (+);
- positive affect (+);
- negative affect (negative association: –);

- parental task orientation (+); and
- various measures or markers of behavior (+).

Associations of varying magnitude were found between an ego orientation and

- beliefs that ability produces success (+),
- motives of status or recognition and competition (+),
- beliefs that the purpose of sport and physical education is for social status and being a good citizen (+),
- perceptions of competence (+), and
- parental ego orientation (+).

A fundamental part of achievement goal theory is the central role played by perceptions of the context (Nicholls, 1989). Consistent with other motivation research that has emphasized the situational determinants of behavior (Deci & Ryan, 1985), research from an achievement goal perspective has examined how the structure of the environment can make it more or less likely that achievement behaviors, thoughts, and feelings associated with a particular achievement goal are adopted. The premise of this line of research is that the nature of an individual's experience influences the degree to which task and ego criteria are perceived as salient within the context, and the resultant perception, therefore, of a mastery (task) or performance (ego) motivational atmosphere or climate (Ames, 1992). This is then assumed to affect the achievement responses (behavioral, cognitive, and affective) through the individual's perception of the behaviors necessary to achieve success. In other words, we are influenced in our responses by previous experiences. If parents ask only about the outcome of their child's sport participation, the child will soon learn that the outcome is more important than participation or other "process" factors. Such children are then more likely to value an ego orientation.

The literature in physical education and sport suggests that the creation of a mastery motivational climate is likely to be important in optimizing positive (e.g., persistence, intrinsic motivation) and attenuating negative (e.g., self-handicapping) responses (Parish & Treasure, 2003; Standage, Duda, & Ntoumanis, 2003). The evidence, therefore, supports the position that perceptions of a mastery motivational climate are associated with more positive motivational and affective responses than perceptions of a performance climate (Ntoumanis & Biddle, 1999).

The research described so far focuses almost exclusively on the antecedents and consequences of two orthogonal (uncorrelated) achievement goals, namely task and ego. Some researchers, however, have recently begun to examine whether individuals focus on striving to avoid demonstrating incompetence as much as or more than they are striving to demonstrate competence. The proposal is that it is possible to differentiate goals based on their valence, or the degree to which the focal outcome is pleasant or unpleasant. Initial research has examined a tripartite model of achievement goals comprising what have been defined as mastery, performance approach, and performance avoidance goals (Elliot & Harackiewicz, 1996). This research has shown some promise in highlighting the role that approach and avoidance tendencies may play in determining achievement behaviors for young people in physical activity. It is unclear at the present time, however, whether the addition of approach and avoidance to performance goals explains more variance than the interaction of ego goals and perceived ability (high/low) that is grounded in Nicholls' achievement goal theory.

A recent suggestion is that achievement goals should consider both the definition of competence and the valence of the striving (Elliot, 1999; Elliot & Conroy, 2005). This suggestion leads to two definitions of competence (i.e., mastery vs. performance) and two valences of striving (i.e., approaching competence vs. avoiding incompetence) and yields a 2 × 2 model of achievement goals, comprising what Elliott and his colleagues have defined as mastery approach, mastery avoidance, performance approach, and performance avoidance goals (see figure 8.3).

Our own research (Wang, Biddle, & Elliot, 2007) on secondary school students in Singapore showed that the structure of a 2 × 2 achievement goal questionnaire, modified for the physical education context, was supported with

	Approach valence	Avoidance valence
Mastery goals	Mastery-approach goal: I want to learn as much as possible from PE class.	Mastery-avoidance goal: I worry that I may not learn all that I possibly could in PE class.
Performance goals	Performance-approach goal: It is important for me to do better than other students in PE class.	Performance-avoidance goal: I just want to avoid doing poorly in PE class.

Figure 8.3 The 2 × 2 achievement goal framework, with sample items from the 2 × 2 Achievement Goals in Physical Education Questionnaire.

Adapted from Wang, Biddle, and Elliot 2007.

the hypothesized four-factor structure consisting of mastery approach, performance approach, mastery avoidance, and performance avoidance goals. In the second study of our research, we used cluster analysis to identify intraindividual achievement goal profiles and to examine their links to various psychological characteristics and outcomes.

Four achievement goal clusters were identified in the data, and it was the cluster consisting of high scores on all four achievement goals that was linked to the most positive set of characteristics and outcomes. Young people in this cluster compared to others were more likely to be male athletes, and this is consistent with research investigating task and ego goals (Hodge & Petlichkoff, 2000; Wang & Biddle, 2001). The Singaporean youth in our "high" cluster clearly sought out both mastery and performance goals and adopted both approach and avoidance strategies. The group had the highest social relatedness and perceived competence scores, and the lowest amotivation score. They also reported the most effort in, the least boredom with, the most participation in, and the most enjoyment of physical education activities. Conversely, the cluster consisting of low scores on all four achievement goals ("low achievement goals" cluster) was linked to the least positive set of characteristics and outcomes. Participants here evidenced the lowest autonomy, relatedness, and perceived competence and the highest amotivation, and also reported the least effort in, the most boredom with, the least participation in, and the least enjoyment of physical education activities. However, more work is needed to show how the four goals differentiate young people according to motivational measures such as participation. For example, in our study in Singapore we were unable to show clear differences in the interaction between approach-avoidance orientations and mastery and performance goals. Rather, as in the two clusters described, many young people adopted either all or none of the goals. This needs clarifying.

Competence IV: Self-Theories of Ability

Carol Dweck and her colleagues (Dweck, 1999; Dweck, Chiu, & Hong, 1995) propose that individuals' implicit theories (beliefs) regarding intelligence provide a pivotal role in interpreting and understanding achievement behavior in the classroom. Two different perceptions of intelligence underpin the adoption of achievement goals. These two beliefs center on the way people view the malleability, or changeability, of attributes:

- Entity belief views an attribute as fixed and relatively stable. Dweck (1999) suggests that an entity theory of intelligence should lead to the adoption of performance goals (ego involvement) because individuals focus on the notion of fixed intelligence. When faced with setbacks, they tend to set performance goals and exhibit a maladaptive (negative) motivational pattern, characterized by negative cognitions, negative affect, and a sharp decline in performance.

- Incremental theory tends to view an attribute as more changeable and open to development. With its idea of malleable intelligence, an incremental belief should promote learning goals (task involvement) because individuals with that understanding will usually strive to increase their ability or "intelligence." They tend to demonstrate more adaptive (positive) patterns, characterized by positive thoughts, positive affect, and effective problem-solving strategies (Dweck, 1999).

In replicating and extending Dweck and Leggett's (1988) research, Sarrazin and colleagues (1996) developed the Conceptions of the Nature of Athletic Ability Questionnaire (CNAAQ) to assess the nature of sport ability. The 21-item scale assessed six subdomains:

1. *Learning* (sport ability is the product of learning)
2. *Incremental/improvement* (sport ability can change)
3. *Specific* (sport ability is specific to certain sports or groups of sports)
4. *General* (sport ability generalizes across many sports)
5. *Stable* (sport ability is stable over time)
6. *Gift* (sport ability is a gift, i.e., "God-given")

Some support was found for the relationship between beliefs concerning the nature of athletic ability and the adoption of different goals in sport for children aged 11 to 12 years. Those choosing a "learning" (task) goal were more likely to endorse incremental beliefs about sport ability than those adopting performance (ego) goals.

We (Biddle et al., 2003a) examined the psychometric properties of the CNAAQ with over 3400 English children and youth aged 11 to 19 years. The original CNAAQ was reduced to 12 items with two higher-order factors (incremental and entity) and four first-order factors (learning, improvement, gift, and stable) and labeled the CNAAQ-2. Overall, the results showed strong psychometric support for the multidimensional hierarchical structure. In other studies we provided support for the invariance of the CNAAQ-2 measurement model across gender and age (Biddle et al., 2003a) and culture (Wang et al., 2005).

Further studies using the CNAAQ-2 in physical activity settings have shown that high incremental beliefs are associated with a positive motivational profile (Biddle et al., 2003a; Wang & Biddle, 2001; Wang et al., 2002). Entity or fixed beliefs, however, resulted in less positive motivational profiles. Potential causal links were confirmed by recent experimental evidence (Spray et al., 2006).

The findings associated with these self-theories of ability may translate to practical strategies to increase physical activity participation among youth. Exercise leaders and educators should cultivate incremental beliefs among young participants, particularly in the skill-learning stages. They could do this by providing vicarious experiences or modeling and verbal persuasion. Through the reinforcement that ability is dynamic and can be improved through effort and practice, participants with lower perceived competence will be more satisfied with their involvement and experience less negative emotions in physical activity settings. Those with higher competence may strive to learn and do even better in the future. Having high incremental beliefs will also help participants to cope more effectively with failure and setbacks, which are unavoidable in the sport and physical activity domains. Spray and colleagues (2006) showed that verbal persuasion could be successful in making a particular belief system salient.

Self-Determination Theory

Perceptions of self-determination and autonomy—feeling that you have choice—are central to feelings of high intrinsic motivation and well-being (Ryan & Deci, 2000b, 2001). Self-determination theory (SDT) is a control-based approach to understanding "why we do what we do" (Deci & Flaste, 1995). It enables us to understand human behavior undertaken for more or less self-determined reasons (Deci & Ryan, 1985; Ryan & Deci, 2000c). The theory examines why an individual acts (i.e., the level to which one's motivation is more or less self-determined), how various types of motivation lead to different outcomes, and the social conditions that support or undermine various aspects of human functioning and well-being via the satisfaction of three basic, innate, and universal psychological needs (Ryan & Deci, 2000a):

- **Autonomy:** the need to endorse and be the origin of one's behavior
- **Competence:** the need to interact effectively within the environment
- **Relatedness:** the need to feel connected with, cared for by, and close to others and one's community

It is proposed that the degree to which the three psychological needs are satisfied explains variations in the quality of motivation, well-being, and various aspects of human functioning (Deci & Ryan, 2000).

Motivation

Understanding the differing intrinsic and extrinsic reasons why people act has been a central theme of SDT research. According to SDT, different types of motivation fall along a continuum of self-determination (see figure 8.4). At the self-determined pole of the continuum is intrinsic motivation. This underlies behaviors that are performed for the activity's sake with no external contingency (i.e., done for the interest and pleasure the activity provides). In the middle of the continuum reside various forms of extrinsic motivation that differ in their degree of relative autonomy. Ranging from high to low autonomy, the regulations usually studied with young people are identified regulation, introjected regulation, and external regulation. Identified regulation occurs when one freely chooses to carry out an activity that is not necessarily considered enjoyable, but is thought of as important to and consistent with one's sense of self or identity. Introjected regulation refers to the incomplete internalization of a regulation

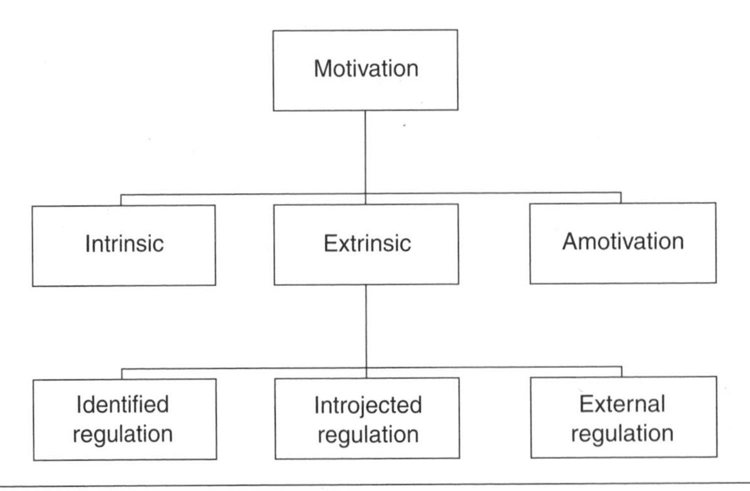

Figure 8.4 Motivational regulations from self-determination theory.

that was previously solely external. The behavior is performed to avoid feelings of guilt or for ego enhancement. External regulation occurs when an individual engages in a behavior solely to receive a reward or to avoid punishment. Amotivation is a lack of motivation, in which no contingency between actions and outcomes is perceived and there is no perceived purpose in engaging in the activity (Deci & Ryan, 1985). Vallerand and Fortier (1998) suggest that the study of amotivation "may prove helpful in predicting lack of persistence in sport and physical activity" (p. 85).

The Motivation–Outcome Connection

The type of motivation an individual possesses influences the selection of activities, attitude toward an activity, effort and persistence, and affective responses. Past work has often shown that intrinsic motivation and identified regulation predict positive behavioral, cognitive, and affective outcomes (Deci & Ryan, 2002). Self-determination theory would predict that autonomously motivated youngsters—those possessing intrinsic motivation and identification—would demonstrate a strong interest in the activity, choose to continue the activity when given a choice, and exhibit a high degree of effort. Empirical work has supported this by showing that self-determined forms of motivation positively correspond to a number of desirable responses in physical education, youth sport, and physical activity. These correlates include positive affect (Ntoumanis, 2005; Standage, Duda, & Ntoumanis, 2005), effort (Ntoumanis, 2001), interest (Goudas, Biddle, & Fox, 1994), intention to be physically active in leisure time (Standage et al., 2003), and objectively assessed physical activity (Vierling, Standage, & Treasure, 2007). In contrast, nonautonomous forms of motivation (i.e., amotivation, external regulation, or both) have been shown to be positively related to less desirable outcomes, such as boredom, and negatively related to attitudes toward physical activity and intention to be physically active (Standage et al., 2003; Vierling et al., in press).

Social Conditions and the Three Basic Needs

Recent research in the context of youth sport and physical activity has examined various aspects of the environment that undermine or facilitate the satisfaction of the three basic needs listed and defined earlier: autonomy, competence, and relatedness (Reinboth, Duda, & Ntoumanis, 2004). This is an important area of study because it is proposed that basic needs mediate the influence of social conditions on motivation and functioning. In a series of studies, Standage and colleagues have examined the impact of an autonomy-supportive environment on basic needs (Standage et al., 2003, 2005). A consistent finding in this research has been the positive association between autonomy support and the basic needs for competence and relatedness, as well as autonomy, of course.

Although little research has addressed the basic need satisfaction aspect of SDT in the context of youth physical activity, a growing body of literature has emerged on the influence of social conditions on different types of motivation. Specifically, researchers have examined the assumption that an autonomy-supportive environment (one that supports choice, initiation, and understanding), as opposed to a controlling environment (one defined as authoritarian, pressuring, and dictating), facilitates self-determined motivation, healthy development, and optimal psychological functioning. A consistent pattern of findings has emerged that reveals a positive association between an autonomy-supportive environment and self-determined types of motivation (Hagger et al., 2003, 2005; Prusak et al., 2004; Reinboth et al., 2004; Vierling et al., 2007). To date, most of the research has focused on the effects of autonomy-supportive versus controlling environments created by social agents such as teachers, coaches, and parents. This approach may overlook social factors that may contribute to the satisfaction of the basic needs of competence and relatedness. To this end, Standage and colleagues (2005) coined the term *need-supporting context* and have examined the influence of competence support and relatedness support aspects of the environment on physical education students' psychological need satisfaction. The findings demonstrate that social support elements have a strong positive influence on psychological need satisfaction.

Suggestions for Action

Although there has been extensive research activity concerning the constructs covered in this chapter, several issues require further thought.

- **Combine approaches.** Researchers have often studied single theories or perspectives when investigating the motivation of youth in physical activity. We propose that all frameworks outlined here are informative and that a case can be made for looking at their combined effect. For example, we studied a large group of youth from across the United Kingdom and found that motivation could be described with reference to a combination of achievement goals, ability beliefs, and self-determination (Wang & Biddle, 2001). Table 8.1 shows the characteristics of five clusters that were identified. At least three main conclusions are evident from table 8.1:

- First, some characteristics can coexist, such as task with ego goal orientations and incremental with entity beliefs. For some clusters, goal orientations were positively correlated, and for others these were orthogonal, suggesting that interventions may need to account for both in some groups. The same was true for ability beliefs.
- Second, gender seemed to moderate the clusters. "Highly motivated" youth were predominantly boys, whereas "poorly motivated" and "amotivated" young people were largely girls.

Table 8.1 Motivational Characteristics of 12- to 15-Year-Old British Youth Surveyed by Wang and Biddle (2001)

Name of cluster	Psychological characteristics defining the cluster	Other characteristics of the cluster
Self-determined	High task goal High incremental belief Low entity belief High RAI Low amotivation	33% of the sample Equal gender split
Highly motivated	High task goal High ego goal High incremental belief Low entity belief High perceived competence High RAI Low amotivation	11% of the sample 67% males High physical activity High physical self-worth
Poorly motivated	Low task goal Low ego goal Low incremental belief Low perceived competence	17% of the sample 66% females
Moderately motivated externals	A "flat profile" with a tendency to show slightly high ego orientation	26% of the sample Equal gender split
Amotivated	Low task goal Low incremental belief High entity belief Low perceived competence Low RAI High amotivation	14% of the sample 66% females Slightly older than other clusters Low physical activity Low physical self-worth

N = 2510; RAI = relative autonomy index; higher RAI scores correspond with more self-determined (intrinsic) motivation.

Based on Wang and Biddle 2001.

- Finally, only 31% of the sample (the poorly motivated and amotivated clusters) were what one might describe as motivationally "at risk," suggesting that many young people are motivated toward physical activity. If their participation is less than we would like, as data sometimes suggest, then maybe we also need to look at social and environmental influences on participation or, more likely, the interaction between such factors and individual psychological variables and frameworks described in this chapter.

▪ **Investigate the motivation–physical activity connection.** It seems obvious that "motivation" is needed in order for young people to participate in physical activity, particularly when effort is required and there are competing attractions in their free time. In addition, we know a great deal about human motivation, as evidenced by the material covered in this chapter. However, we remain largely unsuccessful in achieving population-wide behavior change in physical activity for both adults and young people. This may be due to a powerful environment that inhibits physical activity and reinforces sedentary living (Swinburn & Egger, 2004), or it may reflect relatively weak effects for individual motivational constructs. Time will tell, but we need to do better in pursuing studies that combine different theoretical approaches and to assess motivation against a good measure of behavior. This measure will not always be a so-called "objective" measure of physical activity such as those derived from a movement sensor or pedometer. Such measures can be highly valuable in quantifying more precisely how much activity young people are doing, but rarely tell us *what* they are doing. With respect to the role of motivation, we need to know what young people are choosing to do. Equally, we need studies that measure specific forms of activity, such as active travel or structured exercise, and then look at motivational influences in a more behavior-specific way than we have in the past. Moreover, we need studies investigating a variety of sedentary behaviors, not just TV viewing (Biddle et al., 2004). It is far from clear whether the motivational constructs discussed in this chapter, such as achievement goals, are relevant at all for low-effort sedentary pursuits because such behaviors do not involve elements of "achievement" or "ability." We need to know more about this and to look for alternative frameworks.

▪ **Investigate the inactivity–sedentary behavior distinction.** A key issue that requires attention is the distinction between physical activity and sedentary behavior and therefore the role that motivation may play in physical inactivity (see chapter 1). It is often thought that sedentary behaviors displace more physically active pursuits—the so-called displacement hypothesis (Mutz, Roberts, & Vuuren, 1993); yet the degree to which sedentary behaviors prohibit an active lifestyle is unclear (Marshall et al., 2002). To better understand sedentary behaviors it is important to investigate different sedentary pursuits alongside more active ones. Evidence to date has focused on the incidence and prevalence of sedentary behavior when defined as a lack of physical activity or when using highly selective sedentary pursuits, such as TV viewing. Defining a sedentary lifestyle as an absence of physical activity fails to identify what young people are actually doing while they are sedentary, thus precluding an understanding of *why* they are sedentary. Equally, by studying sedentary behaviors in isolation, we may misrepresent what a sedentary or active lifestyle actually entails. Other sedentary behavior issues requiring further study are the following:

- Very little is known about motivation toward sedentary living, as opposed to barriers to physical activity. However, we do know that more activity is likely in favorable environments, such as opportunities to be outside and not living on a busy road, though these factors may be associated with only certain types of activity, such as active commuting or play. Yet, to what extent such factors influence motivation directly, or operate relatively subconsciously, remains unknown.
- A highly prevalent sedentary behavior for children and adolescents is TV viewing. In a recent review of ours (Gorely, Marshall, & Biddle, 2004), we found that variables consistently associated with TV or video viewing were ethnicity (non-White +), parent income (–), parent education (–), body weight (+), between-meal snacking (+), number of parents in the house (–), parents' TV viewing habits (+), weekend (+), and having a TV in the bedroom (+). With so few modifiable correlates identified, further research should aim to determine correlates that can be changed if interventions are to be successfully tailored to reduce this one aspect of inactivity among youth. The role of motivation, therefore, remains unclear. Indeed, with regard to TV viewing alone, there are some anomalies—for example, as levels of obesity rise across the teenage years, physical activity and TV viewing both decline. Moreover, girls have activity levels lower than boys yet also watch less TV. This leads us to two main conclusions: First, the study of single sedentary behaviors may be misleading, and second, the area of sedentary behaviors is clearly complex. In addition, the area lacks unifying theory. Motivational perspectives that might require investigation in this domain include those of SDT, social influence, flow, addiction, and habit.

Conclusion

In conclusion, the theoretical frameworks covered in this chapter have been found to be highly relevant and useful in helping us better understand the motivation of young people in physical activity contexts. However, we need to know more about their relevance in explaining the appeal of sedentary behaviors, as well as whether other motivational or nonmotivational frameworks are also required. Finally, the theories covered here have been successful in explaining cross-sectional associations between psychological constructs, but less time has been devoted to the study of experimental or longitudinal "causal" linkages between indices of motivation and good measures of physical activity behavior.

APPLICATIONS FOR RESEARCHERS

Motivation has been a popular area of investigation for researchers interested in youth physical activity. More research is needed using motivational perspectives in the study of sedentary behaviors. Contemporary theoretical approaches of motivation center on social cognitive frameworks, but these require further refining through investigation of the interactions between overlapping constructs such as competence, confidence, self-determination, goals, and beliefs. Recent research that has identified clusters of young people with similar motivational profiles needs to be continued, with an extension into experimental manipulation of motivational constructs and investigation of effects on specific types of physical activity and sedentary behaviors. Researchers might also want to study the interaction between positive and negative motivation operating within facilitative and inhibiting environments, whether this be for active or sedentary pursuits.

APPLICATIONS FOR PROFESSIONALS

Despite sometimes complex theorizing and terminology, the key motivational constructs described in this chapter are essentially practical. Understanding of these constructs allows us to address several issues. First, self-determination theory enables an understanding of how young people can feel that "**I** did that!" (rather than "I did that"). This distinction centers on the strong feeling of personal autonomy and control in the first phrase. Second, self-efficacy theory and incremental ability beliefs help young people adopt the attitude taken by the "little engine that could"—"I think I can, I think I can, I know I can!" Third, achievement goals enable understanding of how to take part in an activity for enjoyment and personal success, such that we want to come back for more. Creating an activity atmosphere that values self-improvement, personal competence, autonomy and choice, and social interaction will go a long way toward keeping the children who initially love activity involved throughout their adolescence and into adulthood.

References

Ames, C. (1992). Achievement goals, motivational climate, and motivational processes. In G.C. Roberts (Ed.), *Motivation in sport and exercise* (pp. 161-176). Champaign, IL: Human Kinetics.

Bandura, A. (1986). *Social foundations of thought and action: A social cognitive theory.* Englewood Cliffs, NJ: Prentice Hall.

Biddle, S.J.H. (1995). Exercise motivation across the lifespan. In S.J.H. Biddle (Ed.), *European perspectives on exercise and sport psychology* (pp. 5-25). Champaign, IL: Human Kinetics.

Biddle, S.J.H., Gorely, T., Marshall, S.J., Murdey, I., & Cameron, N. (2004). Physical activity and sedentary behaviours in youth: Issues and controversies. *Journal of the Royal Society for the Promotion of Health, 124*(1), 29-33.

Biddle, S.J.H., Hagger, M.S., Chatzisarantis, N.L.D., & Lippke, S. (2007). Theoretical frameworks in exercise psychology. In G. Tenenbaum & R.C. Eklund (Eds.), *Handbook of sport psychology* (3rd ed., pp. 537-559). Hoboken, NJ: Wiley.

Biddle, S.J.H., & Nigg, C.R. (2000). Theories of exercise behavior. *International Journal of Sport Psychology, 31*, 290-304.

Biddle, S.J.H., Wang, C.K.J., Chatzisarantis, N.L.D., & Spray, C.M. (2003a). Motivation for physical activity in young people: Entity and incremental beliefs about athletic ability. *Journal of Sports Sciences, 21*, 973-989.

Biddle, S.J.H., Wang, C.K.J., Kavussanu, M., & Spray, C.M. (2003b). Correlates of achievement goal orientations in physical activity: A systematic review of research. *European Journal of Sport Science, 3*(5), www.humankinetics.com/ejss.

Biddle, S.J.H., Whitehead, S.H., O'Donovan, T.M., & Nevill, M.E. (2005). Correlates of participation in physical activity for adolescent girls: A systematic review of recent literature. *Journal of Physical Activity and Health, 2*, 423-434.

Davis, A., & Jones, L. (1996). Children in the urban environment: An issue for the new public health agenda. *Health and Place, 2*, 107-113.

Davis, A., & Jones, L. (1997). Whose neighbourhood? Whose quality of life? Developing a new agenda for children's health in urban settings. *Health Education Journal, 56*, 350-363.

Deci, E.L., & Flaste, R. (1995). *Why we do what we do: Understanding self-motivation.* New York: Penguin.

Deci, E.L., & Ryan, R.M. (1985). *Intrinsic motivation and self-determination in human behavior.* New York: Plenum Press.

Deci, E.L., & Ryan, R.M. (2000). The "what" and "why" of goal pursuits: Human needs and the self-determination of behavior. *Psychological Inquiry, 11*, 227-268.

Deci, E.L., & Ryan, R.M. (Eds.). (2002). *Handbook of self-determination research.* Rochester, NY: University of Rochester Press.

Dweck, C. (1999). *Self-theories: Their role in motivation, personality, and development.* Philadelphia: Taylor & Francis.

Dweck, C., Chiu, C.Y., & Hong, Y.Y. (1995). Implicit theories and their role in judgments and reactions: A world from two perspectives. *Psychological Inquiry, 6*, 267-285.

Dweck, C., & Leggett, E. (1988). A social-cognitive approach to motivation and personality. *Psychological Review, 95*, 256-273.

Elliot, A.J. (1999). Approach and avoidance motivation and achievement goals. *Educational Psychologist, 34*, 169-189.

Elliot, A.J., & Conroy, D.E. (2005). Beyond the dichotomous model of achievement goals in sport and exercise psychology. *Sport and Exercise Psychology Review, 1*, 17-25.

Elliot, A.J., & Harackiewicz, J.M. (1996). Approach and avoidance achievement goals and intrinsic motivation: A mediational analysis. *Journal of Personality and Social Psychology, 70*, 461-475.

Fox, K.R., & Corbin, C.B. (1989). The Physical Self Perception Profile: Development and preliminary validation. *Journal of Sport and Exercise Psychology, 11*, 408-430.

Gomez, J.E., Johnson, B.A., Selva, M., & Sallis, J.F. (2004). Violent crime and outdoor physical activity among inner-city youth. *Preventive Medicine, 39*, 876-881.

Gorely, T., Marshall, S.J., & Biddle, S.J.H. (2004). Couch kids: Correlates of television viewing among youth. *International Journal of Behavioural Medicine, 11*, 152-163.

Goudas, M., Biddle, S., & Fox, K. (1994). Perceived locus of causality, goal orientations, and perceived competence in school physical education classes. *British Journal of Educational Psychology, 64*, 453-463.

Hagger, M.S., Biddle, S.J.H., Chow, E.W., Stambulova, N., & Kavussanu, M. (2003). Physical self-perceptions in adolescence: Generalizability of a hierarchical multidimensional model across three cultures. *Journal of Cross-Cultural Psychology, 34*, 611-628.

Hagger, M.S., Chatzisarantis, N.L.D., Barkoukis, V., Wang, C.K.J., & Baranowski, J. (2005). Perceived autonomy support in physical education and leisure-time physical activity: A cross-cultural evaluation of the trans-contextual model. *Journal of Educational Psychology, 97*, 376-390.

Hagger, M.S., Chatzisarantis, N., & Biddle, S.J.H. (2001). The influence of self-efficacy and past behaviour on the physical activity intentions of young people. *Journal of Sports Sciences, 19*, 711-725.

Harter, S. (1978). Effectance motivation reconsidered: Toward a developmental model. *Human Development, 21*, 34-64.

Harter, S. (1981). The development of competence motivation in the mastery of cognitive and physical skills: Is there a place for joy? In G.C. Roberts & D.M. Landers (Eds.), *Psychology of motor behavior and sport—1980* (pp. 3-29). Champaign, IL: Human Kinetics.

Hodge, K., & Petlichkoff, L. (2000). Goal profiles in sport motivation: A cluster analysis. *Journal of Sport and Exercise Psychology, 22*, 256-272.

Kerlinger, F. (1973). *Foundations of behavioral research.* New York: Holt, Rinehart & Winston.

Marshall, S.J., Biddle, S.J.H., Sallis, J.F., McKenzie, T.L., & Conway, T.L. (2002). Clustering of sedentary behaviors and physical activity among youth: A cross-national study. *Pediatric Exercise Science, 14*, 401-417.

Mason, V. (1995). *Young people and sport in England, 1994.* London: Sports Council.

McAuley, E., & Blissmer, B. (2000). Self-efficacy determinants and consequences of physical activity. *Exercise and Sport Sciences Reviews, 28*, 85-88.

Mulvihill, C., Rivers, K., & Aggleton, P. (2000). *Physical activity "at our time": Qualitative research among young people aged 5 to 15 years and parents.* London: Health Education Authority.

Mutz, D.C., Roberts, D.F., & Vuuren, D.P. (1993). Reconsidering the displacement hypothesis: Television's influence on children's time use. *Communication Research, 20*, 51-75.

Nicholls, J.G. (1989). *The competitive ethos and democratic education.* Cambridge, MA: Harvard University Press.

Nigg, C., & Courneya, K. (1998). Transtheoretical model: Examining adolescent exercise behavior. *Journal of Adolescent Health, 22*, 214-224.

Ntoumanis, N. (2001). A self-determination approach to the understanding of motivation in physical education. *British Journal of Educational Psychology, 71*, 225-242.

Ntoumanis, N. (2005). A prospective study of participation in optional school physical education using a self-determination theory framework. *Journal of Educational Psychology, 97*, 444-453.

Ntoumanis, N., & Biddle, S. (1999). A review of motivational climate in physical activity. *Journal of Sports Sciences, 17*, 643-665.

Parish, L.E., & Treasure, D.C. (2003). Physical activity and situational motivation in physical education: Influence of the motivational climate and perceived ability. *Research Quarterly for Exercise and Sport, 74*, 173-182.

Prusak, K.A., Treasure, D.C., Darst, P.W., & Pangrazi, R.P. (2004). The effects of choice on the motivation of adolescent females in physical education. *Journal of Teaching Physical Education, 23*, 19-29.

Reinboth, M., Duda, J.L., & Ntoumanis, N. (2004). Dimensions of coaching behavior, need satisfaction, and the psychological and physical welfare of young athletes. *Motivation and Emotion, 28*, 297-313.

Ryan, R.M., & Deci, E.L. (2000a). The darker and brighter sides of human exercise: Basic psychological needs as a unifying concept. *Psychological Inquiry, 11*, 319-338.

Ryan, R.M., & Deci, E.L. (2000b). Intrinsic and extrinsic motivations: Classic definitions and new directions. *Contemporary Educational Psychology, 25*, 54-67.

Ryan, R.M., & Deci, E.L. (2000c). Self-determination theory and the facilitation of intrinsic motivation, social development, and well-being. *American Psychologist, 55*, 68-78.

Ryan, R.M., & Deci, E.L. (2001). On happiness and human potentials: A review of research on hedonic and eudaimonic well-being. *Annual Review of Psychology, 52*, 141-166.

Sallis, J.F., Prochaska, J.J., & Taylor, W.C. (2000). A review of correlates of physical activity of children and adolescents. *Medicine and Science in Sports and Exercise, 32*, 963-975.

Sarrazin, P., Biddle, S., Famose, J.P., Cury, F., Fox, K., & Durand, M. (1996). Goal orientations and conceptions of the nature of sport ability in children: A social cognitive approach. *British Journal of Social Psychology, 35*, 399-414.

Spray, C.M., Wang, C.K.J., Biddle, S.J.H., Chatzisarantis, N.L.D., & Warburton, V.E. (2006). An experimental test of self-theories of ability in youth sport. *Psychology of Sport and Exercise*, 7(3), 255-267.

Standage, M., Duda, J.L., & Ntoumanis, N. (2003). A model of contextual motivation in physical education: Using constructs from self-determination and achievement goal theories to predict physical activity intentions. *Journal of Educational Psychology, 95*, 97-110.

Standage, M., Duda, J.L., & Ntoumanis, N. (2005). A test of self-determination theory in school physical education. *British Journal of Educational Psychology, 75*, 411-433.

Sturm, R. (2005). Childhood obesity: What we can learn from existing data on societal trends, part 1. *Preventing Chronic Disease* [Serial online], 2(1), January, www.cdc.gov/pcd/issues/2005/jan/04_0038.htm.

Swinburn, B., & Egger, G. (2004). The runaway weight gain train: Too many accelerators, not enough brakes. *British Medical Journal, 329*(7468), 736-739.

Trost, S.G., Pate, R.R., Ward, D.S., Saunders, R., & Riner, W. (1999). Determinants of physical activity in active and low-active, sixth grade African-American youth. *Journal of School Health, 69*(1), 29-34.

Vallerand, R.J., & Fortier, M.S. (1998). Measures of intrinsic and extrinsic motivation in sport and physical activity: A review and critique. In J.L. Duda (Ed.), *Advances in sport and exercise psychology measurement* (pp. 81-101). Morgantown, WV: Fitness Information Technology.

Van Wersch, A. (1997). Individual differences and intrinsic motivations for sport participation. In J. Kremer, K. Trew, & S. Ogle (Eds.), *Young people's involvement in sport* (pp. 57-77). London: Routledge.

Van Wersch, A., Trew, K., & Turner, I. (1992). Post-primary school pupils interest in physical education: Age and gender differences. *British Journal of Educational Psychology, 62*, 56-72.

Vierling, K., Standage, M., & Treasure, D.C. (2007). Predicting attitudes and objective physical activity in an "at-risk" minority youth sample: A test of self-determination theory. *Psychology of Sport and Exercise, 8*, 795-817.

Wang, C.K.J., & Biddle, S.J.H. (2001). Young people's motivational profiles in physical activity: A cluster analysis. *Journal of Sport and Exercise Psychology, 23*, 1-22.

Wang, C.K.J., Biddle, S.J.H., & Elliot, A.J. (2007). The 2 × 2 achievement goal framework in a physical education context. *Psychology of Sport and Exercise, 8*, 147-168.

Wang, C.K.J., Chatzisarantis, N.L.D., Spray, C.M., & Biddle, S.J.H. (2002). Achievement goal profiles in school physical education: Differences in self-determination, sport ability beliefs, and physical activity. *British Journal of Educational Psychology, 72*, 433-445.

Wang, C.K.J., Liu, W.C., Biddle, S.J.H., & Spray, C.M. (2005). Cross-cultural validation of the Conceptions of the Nature of Athletic Ability Questionnaire Version 2. *Personality and Individual Differences, 38*, 1245-1256.

Weiss, M.R., & Williams, L. (2004). The *why* of youth sport involvement: A developmental perspective on motivational processes. In *Developmental sport and exercise psychology: A lifespan perspective* (pp. 223-268). Morgantown, WV: Fitness Information Technology.

White, R.W. (1959). Motivation reconsidered: The concept of competence. *Psychological Review, 66*, 297-333.

CHAPTER

9

The Role of the Self

Peter R.E. Crocker, PhD ▪ Kent C. Kowalski, PhD ▪ Valerie Hadd, MA

For many years, researchers and practitioners have believed that participation in physical activity can enhance self-esteem in children and adolescents (see Gruber, 1986; Horn, 2004; Sonstroem, 1997). Furthermore, cross-sectional research, combined with theoretical propositions, suggests that positive judgments of the global and physical self influence the selection and maintenance of physical activity behavior (see Harter, 1999; Horn, 2004; Weiss & Williams, 2004). Therefore, many researchers and practitioners are interested in developing effective interventions that will enhance global and physical self-perceptions in youth (Horn, 2004). There are, however, major challenges in developing such interventions. These include incomplete understanding of the concept of the self and its relationship to physical activity; the direction of causality of effects between the self and physical activity; and gender and cultural influences on the construction of the physical self and its link to physical activity.

The purpose of this chapter is to review theoretical and empirical literature on the self and physical activity in youth. The chapter begins with a discussion of general conceptions of the self, including an identification of the notion of self-reflexivity as a key part of the self. Important terms are defined, and we include a short discussion of the role of culture and cognitive development in the development of self. We then address how the global self can be described as multidimensional, with the physical self being a key component, and discuss issues related to the measurement of the physical self, the direction of causal flow, and whether the perceived importance of particular domains of the self affects global self-esteem and motivated behavior. Next we cover research on physical activity and self-esteem processes. In particular, we focus on the predictive relationships between specific subdomains of the physical self and physical activity, exploring gender and cultural differences. In the final section we provide longitudinal evidence linking changes in the physical self and changes in physical activity.

General Conceptions of the Self

In reviewing the general self literature, Leary and Tangney (2003) identified over 60 self-related constructs, processes, and phenomena, ranging from ego identity to self-esteem. However, despite the lack of clarity around the concept of the self, they boldly concluded that the self has potential as an organizing construct in the behavioral and social sciences. In this section we discuss general conceptions of the self, addressing issues related to key characteristics and the role of culture and social cognitive development.

Self-Reflexivity

Any conversation around self and physical activity or inactivity needs to begin with a discussion of general conceptions of the self, which might not be as straightforward as appears at first glance. Many different terms in physical activity research include the word "self," including self-esteem, self-worth, self-concept, self-evaluation, self-schema, self-perception, and self-efficacy. These terms vary not only with regard to specificity of individuals' beliefs about themselves and their relationships to others, but also with regard to stability over time. Because several of these terms appear in this chapter, it would seem useful to provide the following definitions.

- **Self-concept** refers to a relatively stable description of personal attributes such as abilities (Harter, 1999). These descriptive judgments typically occur in discrete domains like physical competence (e.g., I have good athletic skills).
- **Self-esteem** or **self-worth** is also thought to be relatively stable, with a greater focus on one's global sense of worth (e.g., I am a good person).
- **Self-efficacy,** on the other hand, reflects beliefs in one's capabilities to organize and execute the courses of action required to produce given attainments (Bandura, 1997).

Although self-efficacy is influenced by perceptions of personal attributes, it does not in itself reflect descriptive or evaluative components of the self. Thus this chapter focuses on the relationship between physical activity and components of the self reflected in terms like self-esteem and self-concept.

Leary and Tangney (2003) identified the self as that which provides our ability for reflexive thinking, our ability to think about and attend to our self. Self-reflexivity is important not only because of its role in the way we interpret the world around us, but also because of its role in influencing behaviors we use to adapt to the world. Leary and Tangney's review of the self literature identified three general ways in which the term "self" has been used that capture people's ability for reflective thinking. First, the self has been described as the psychological process that is responsible for self-awareness and self-knowledge. Second, the self has been described as one's perceptions, thoughts, and feelings about oneself. Third, the self has been considered as a self-control process and as a regulator of one's behavior.

The concept of self has both attentional and self-regulation properties—properties that have implications for understanding some aspects of physical activity. Youth have the ability to develop symbolic mental representations about themselves (Leary, 2001). This process leads not only to self-awareness, but also to awareness of one's relationship to others. Thinking about oneself and how one is being evaluated by significant others can produce self-conscious emotions (e.g., anxiety, guilt, shame, pride) and motivated states that can lead one to engage in or avoid behaviors such as physical activity. For example, Sue may evaluate her physical self as "overweight" and unfit. As she reflects on this self-concept, she is also aware that being thin and fit is desirable to significant male peers and also the desired social norm of her female peers. This awareness of herself, and also of herself in relation to others, produces anxiety and shame. To cope with these emotional states, Sue may decide to engage in a number of adaptive and maladaptive behaviors including exercise, appearance management, dieting, and drug use (see Sabiston et al., 2007).

The self develops and is shaped through one's relationships with the social world via self-reflexivity, social interactions, and shared languages and meanings that provide a variety of perspectives (Stets & Burke, 2003). Self-perceptions related to specific attributes are influenced by three primary sources of information: self-comparison, social comparison, and evaluative feedback from significant others (see Harter, 1999, 2003; Horn, 2004). Self-comparison involves comparing actions with internal standards of behavior. Social comparison involves comparing behavior to that of relevant peers. Evaluative feedback from significant others includes positive and negative feedback from peers, parents, and teachers and other significant adults. These three sources provide important information about competence in specific areas that allow aspects of the self to be formed over time. Information from significant others is especially critical in identifying what self attributes (such as body shape, athletic skills, or intellectual ability) are valued in society.

Culture

Cultural context is an important aspect of the social environment in which the self develops (Cross & Gore, 2003; Heine, 2001). Most importantly, as described by Cross and Gore, beliefs about the nature of a person can vary widely across cultures. These differences can be reflected in a number of ways. For example, there may be differences in perceived boundaries between the self and others (e.g., whether one is interdependent and interconnected with others or independent and separate), in the time span and experiences that are considered self-related (e.g., whether past life, early-infancy memories, or dreams are used as sources of information about the self), in the societal ideals that members of the culture value and strive to achieve (e.g., personal or collective goals), and in the perceived stability of the self and identity (e.g., whether the self is stable or dynamic across time and situations). For example, traditional Japanese culture emphasizes self-improvement whereas much of North American culture emphasizes self-enhancement (e.g., thinking you are great) (Hamamura & Heine, in press; Heine, 2001). Another example might be the value attached to specific types of physical activities. Many cultures value the demonstration of physicality in males but not in females (Malszecki & Cavar, 2005). The evaluative feedback received from significant others will reflect the cultural value attached to activities that allow youth to demonstrate physical prowess. Negative evaluative feedback to young women, along with the absence of female role models, might promote the adoption of more sedentary activities that are valued within a given culture.

For individuals who have multiple cultural identities (which is especially true in many major cities in North America and Europe), there might be contradicting values and beliefs across various cultural contexts (Fleming et al., 2006). For example, in the study by Fleming and colleagues, three young women who identified themselves as mixed-race spoke about cultural differences between their Aboriginal culture and urban White culture with respect to appropriate clothes to wear, body size, and eating habits. Although not discussed in the research, these types of conflicting cultural expectations could emerge as salient to young women in physical activities, such as track and field or swimming; here, not only is the attire revealing but also the body is on display. Unfortunately, much of the literature on the self and the physical self has not considered cultural factors.

Cognitive Development

Finally, we would be remiss not also to acknowledge important developmental considerations in the formation of the self. Harter (2003) identified the key processes from early childhood to late adolescence that influence the development of self and identity and emphasized that the self is both a cognitive and a social construction. Although a review of this literature is beyond the scope of the chapter (see Horn, 2004, and Harter, 2003, for greater detail), some of the most important processes include the emergence of cognitive abilities that allow for the creation of self-evaluations in different contexts, the ability to use different sources of information to develop the self, and the ability to construct higher-order generalizations about the self.

Developmental changes in attention and memory systems, as well as critical thinking and problem-solving skills, influence the accuracy of children's and adolescents' internal representations of self and their processing of information (Gallagher et al., 1996; Harter, 2003; Keating, 1990; Santrock, 1998). Elementary school–aged children use more concrete operational thinking (Santrock & Yussen, 1992). Because of this less mature cognitive structure, younger children tend to rely heavily on evaluative feedback from significant others to form competence perceptions (Horn, 2004). Harter (2003) notes that the ability to evaluate one's worth as a person does not emerge until middle childhood, but younger children are able to make evaluative self-judgments in particular domains such as physical ability and social competence. Between the ages of 11 and 15 years, adolescents begin to engage in abstract and logical thinking (Santrock, 1998). This maturation of complex cognitive skills will influence the accuracy of information utilized in forming self-related beliefs. Adolescents are more likely to consider a situation from a variety of perspectives and to consider the credibility of sources of competence information. With increased cognitive maturation, adolescents are able to use multiple sources of competence information, including self-comparison, social comparison, and external sources of evaluation (Horn, 2004). This higher level of cognitive functioning enables the formation of a more complex and increasingly differentiated multidimensional sense of self (Harter, 1999, 2003).

The Physical Self

The self is often described as both multidimensional and hierarchical. A multidimensional view of the self considers the self not only at a broad, global level, but also at the domain level (e.g., physical, social, and academic) because the domain level is more strongly related to motivation for specific types of behavior (see figure 9.1). A great deal of evidence over the previous 30 years supports the concept of the multidimensional self (Hattie & Fletcher, 2005). Most relevant to this chapter is the clear identification of the physical domain as important to the self (e.g., Fox, 1997; Harter, 2003; Hattie, 1992; Sonstroem & Morgan, 1989). As Fox (1997) stated,

> Although there have been many different attempts to express and document this multidimensionality, virtually all have made some reference to the importance of the *physical self*. For many theorists, the physical self has become the *major* component of our self-expression and interaction with the world, and it is seen to hold a key to our understanding of the total self. (p. v)

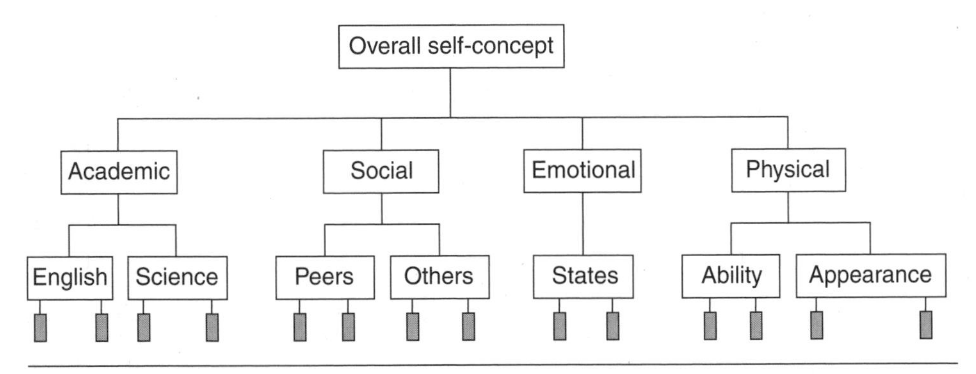

Figure 9.1 A multidimensional model of the self.

Reprinted, by permission, from R.J. Shavelson, J.J. Hubner, and G.C. Stanton, 1976, "Self-concept: Validation of construct interpretations," *Review of Educational Research* 46(3): 407-771.

Dimensions and Assessment

In a hierarchical model, various domains of self are nested under a more global self (Harter, 1999; Shavelson, Hubner, & Stanton, 1976). Each domain can also be further differentiated into distinct but interrelated subdomains, with individual experiences at the base of the hierarchy (see Hattie & Marsh, 1996).

Consistent with the proposed multidimensional and hierarchical self, self-perceptions can be used to increasingly differentiate levels of the physical self. The levels are created based on the degree of specificity of their content, with lower levels becoming more specific to a task (Fox, 1998). For example, global physical self-esteem might represent a first level of the physical self, with a specific subdomain such as perceptions of sport skills competence as a subdomain (see figure 9.2). As one proceeds to greater differentiation and specificity, the major components of the subdomains are broken down into facets (e.g., competence in a particular sport like soccer) and subfacets (e.g., shooting ability in soccer).

Instruments often used to assess the physical self, such as the Physical Self-Description Questionnaire (PSDQ; Marsh et al., 1994) and the Physical Self-Perception Profile (PSPP; Fox & Corbin, 1989), were developed based on the notion of a multidimensional, hierarchical self. However, despite the recognition

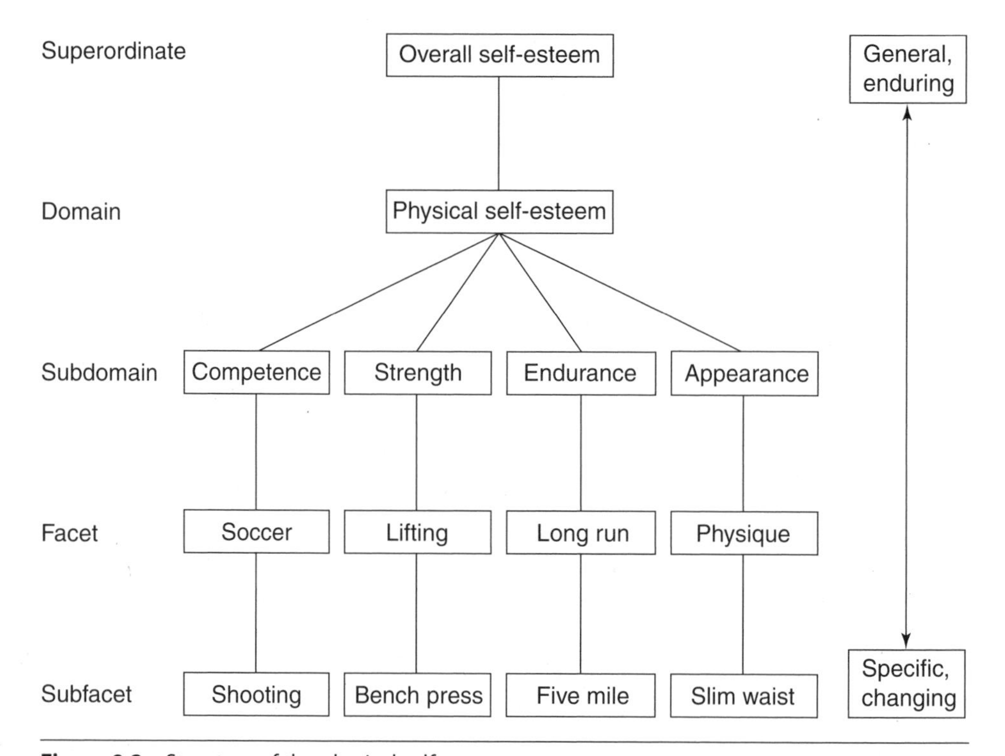

Figure 9.2 Structure of the physical self.

Reprinted, by permission, from J. Buckworth and R.K. Dishman, 2002, *Exercise Psychology* (Champaign, IL: Human Kinetics), 159.

of the importance of the physical self, what the physical self specifically consists of differs across models. For example, in Marsh and colleagues' PSDQ, physical self-concept is represented by 10 unique physical self dimensions (in addition to including a global self-esteem dimension): strength, body fat, activity, endurance/fitness, sports competence, coordination, health, appearance, flexibility, and general physical self-concept. Alternatively, Fox and Corbin's model underlying the PSPP specifies four subdomains of physical self-perceptions (sports competence, attractive body, physical strength, and physical conditioning) nested under the more global domain of physical self-worth. Children and youth versions of the PSPP also assess these subdomains (Eklund, Whitehead, & Welk, 1997; Welk & Eklund, 2005). Harter has suggested that, in older children and adolescents, physical self-perceptions can be compartmentalized into two primary domains: physical/athletic and body appearance (Harter, 1988). The PSDQ and PSPP are the two most widely used instruments in sport and exercise research on the physical self (Marsh et al., 2006).

Direction of Causal Flow

Although the multidimensionality of self-concept has been supported in the literature, the influence of various levels of the hierarchical structure on other levels is much less clear (Hattie & Fletcher, 2005; Kowalski et al., 2003). More specifically, there is no consistent theoretical or empirical answer to the question of the direction of causal flow in the hierarchy (Harter, 1999). A bottom-up hierarchical model hypothesizes that situation-specific experiences influence specific subdomain perceptions of competence, which then influence domain-specific self-concept and in turn global self-worth (Shavelson & Bolus, 1982; Sonstroem & Morgan, 1989). Thus, changes in a specific facet (e.g., distance running capability) influence a physical subdomain (e.g., physical conditioning), resulting in changes to the global physical self that then works to change global self-esteem. In this sense, physical activity interventions might be justified in that the changes in physical attributes (i.e., skills, conditioning, and strength) will ultimately enhance global self-esteem. Alternatively, a top-down model hypothesizes that the flow of change is from the top to lower-order domains and then to facets (Brown, 1993). Thus interventions might be directed toward more global aspects of self, in the hope that enhanced self-esteem will enhance lower-level self-perceptions and subsequently specific types of motivated behavior such as physical activity. For example, individuals with high self-esteem might have more confidence to engage in physical activity behavior even though they might have limited experience and perceived competence in that domain.

Recent research has shed some light on the hierarchical structure of self-concept and the issue of direction of causal flow. When the physical self is looked at longitudinally (i.e., through multiple assessments over time), there appears to be little support for either a top-down or a bottom-up causal flow model. Marsh and Yeung (1998) tested the top-down, bottom-up, and horizontal effects of self-concept in the physical domain using the PSDQ. They found little support for either global self-concept factors influencing specific self-concept factors (top-down effects) or specific self-concept factors influencing global self-concept factors (bottom-up effects) over a one-year period for their adolescent sample. Using the PSPP, Kowalski and colleagues (2003) also examined the direction of causal flow between global and specific dimensions of self-concept, obtaining similar results with a sample of adolescent females over a one-year period. There was little support for either top-down or bottom-up effects. Alternatively, both studies supported a horizontal effects model in which each self-concept factor at time 2 was primarily a function of its time 1 status. This work questions interventions targeting one level of the self with a goal toward influencing another level of self. However, since these studies did not involve interventions, it is necessary to be cautious in making generalizations. Further, little systematic work has evaluated interventions that should affect just the physical subdomains. This is critical, since there is evidence that it is the physical self subdomain level that is most strongly related to physical activity behavior in youth (see section on physical self and physical activity for details).

Another question that consistently emerges in the self literature is whether perceived importance moderates the association of domains with more global aspects of self and behavior. Fox (1997), among others, has discussed how the relationship between global and specific domains of the self might be influenced by the importance an individual attaches to specific domains. That is, being competent in a highly valued domain might have a greater impact on self-esteem than being competent in a less valued domain. Despite empirical evidence raising questions about the use of importance ratings in self-concept models (e.g., Marsh, 1994; Marsh & Sonstroem, 1995), some researchers contend—and provide supporting evidence—that one must consider the perceived importance of specific domains when trying to understand more global self constructs (e.g., Harter, 2003).

To shed light on this issue, Hattie and Fletcher (2005) used a variety of assessment methods to study the value of considering domain importance in understanding global self-worth. They looked at four self-concept domains (family, friends, academic, and physical) in relation to various self-worth scales with a sample of adolescents. Their results suggest that it may not be possible to understand self-esteem as based on a self-concept that comprises a weighting of different conceptions of self. Thus, the issue of the importance of any particular domain in predicting external behavior such as physical activity is unresolved.

The Self and Physical Activity

For many years, physical and health education professionals have argued that sport, physical education, and dance involvement can contribute to young people's psychological development, especially the enhancement of self-esteem (see Corbin, 2002; Fox, 1990, 1997; Smith, Smoll, & Curtis, 1979; Whitehead & Corbin, 1997). The following sections review the empirical evidence about the relationship between physical activity involvement and aspects of the global and physical self. What this evidence reveals is that the relationship is complex, possibly moderated by gender and culture, and is still not well understood.

Global Self-Esteem

Three meta-analyses have examined physical activity and global self-esteem in children and adolescents (Calfas & Taylor, 1994; Ekeland, Heian, & Hagen, 2005; Gruber, 1986). Gruber's article was highly influential as it provided strong empirical evidence that structured play or physical education (or both) caused increases in children's self-esteem. He examined the effect sizes of 27 "controlled" experimental studies and reported an average positive effect of .41 (slightly less than one-half standard deviation). The effect of physical activity on self-esteem was stronger for children with physical or mental disabilities and for participants involved in fitness-oriented programs. There was no evidence for gender differences. Calfas and Taylor (1994) examined 20 articles in an attempt to determine the effects of physical activity on specific psychological variables in adolescents. The studies included in the meta-analysis used various research designs including randomized controlled, prospective observational, and cross-sectional observational. Calfas and Taylor found a weak positive effect (.12) of physical activity on self-esteem.

Recently Ekeland and colleagues (2005) conducted a rigorous systematic review of randomized and quasi-randomized controlled studies on the effects of physical activity programs on global self-esteem in children and adolescents. They found essentially that the interventions produced a positive short-term effect (about a 10% improvement). However, they noted that very few studies (13) could be included in the review because of methodological or statistical problems. The authors argued that there was a need for more rigorous research on the effectiveness of physical activity programs in improving self-esteem.

The findings from Gruber (1986), Calfas and Taylor (1994), and Ekeland and colleagues (2005) seem to support the role of physical activity in the development of children's and adolescents' self-esteem. Nevertheless, there are many reasons to be cautious about espousing a causal link between physical activity and global self-esteem. First, Sonstroem (1984) identified numerous research design problems (e.g., lack of proper control groups, unidimensional measures of self-esteem) with this research originating in the 1960s and 1970s. Second, since 1983, few well-controlled experimental studies have presented statistically significant findings indicating that physical activity enhances global self-esteem. Indeed, Ekeland and colleagues (2005) argued that the vast majority of research studies are of poor quality. Third, recent research utilizing correlational designs has shown only weak or nonsignificant relationships between physical activity and global self-esteem (Crocker, Eklund, & Kowalski, 2000; Crocker et al., 2003; Dunton, Jamner, & Cooper, 2003; Tremblay, Inman, & Willms, 2000).

It is hardly surprising that it might be difficult to establish a causal link between global self-esteem and physical activity in youth populations. First, global self-esteem is influenced by multiple, potentially important domains in addition to the physical domain (i.e., social, academic, behavioral). Second, physical attractiveness is the physical domain that is typically most strongly associated with global self-esteem (Fox, 1997; Harter, 1999). Although physical activity can have some impact on body attractiveness, the two generally are not strongly related (we address this issue later). Furthermore, specific domains and subdomains of the self are more closely related to types of behavior or cognition that are similar. For example, conditioning self-perceptions typically are more closely related to distance running activity than to weightlifting. Therefore, rather than examining global self-esteem, it might be more fruitful to examine relationships between physical activity and the physical self and its specific subdomains.

The Physical Self-Concept

Sport and exercise psychologists have recognized that general physical self-perceptions (i.e., physical self-worth, athletic/physical perceived competence) predict motivated behavior in physical activity domains (for reviews see Horn, 2004; Weiss & Williams, 2004). Research studies typically demonstrate weak to moderate correlations between global physical self-perceptions and physical activity levels in children and adolescents (Burkhalter & Wendt, 2001; Craft, Pfeiffer, & Pivarnik, 2003; Crocker et al., 2000, 2003; Raudsepp, 2002; Whitehead, 1995). The empirical evidence is clear, however, that physical activity is related to specific subdomains of the physical self.

Physical Self Subdomains

Research with children and adolescents has shown that specific physical self-perceptions are related to field indicators of components of physical fitness and physical activity. Whitehead (1995) demonstrated that physical self-perceptions of body, sport competence, physical conditioning, and general physical self-worth were all related to a number of field indicators of anaerobic, aerobic, and muscular strength in 7th- and 8th-grade U.S. students. Biddle and colleagues (1993) found that a field measure of fitness (shuttle run) was positively related to perceptions of sport competence, physical conditioning, and physical self-worth in British schoolchildren. Recent work by Welk and Eklund (2005) in young U.S. children (ages 8-12 years) showed inconsistent relationships between specific subdomains and field indicators. For example, attractive body perceptions were moderately associated with body fat percentage, and conditioning perceptions were moderately associated with $\dot{V}O_2max$ in both boys and girls. However, strength perceptions were not associated with field indicators of strength, whereas sport competence was associated with VO_2 in boys but not girls. The inconsistent findings across studies are of some concern; however, both laboratory and field indicators of physical fitness often demonstrate inconsistent relationships with measures of habitual physical activity in youth (Armstrong & Welsman, 1997; Armstrong et al., 1991).

Research has consistently demonstrated positive simple relations and predictions between the physical self and physical activity, with differential strength of relationships reported for various physical self-perceptions (Crocker et al., 2003, 2006; Hagger, Ashford, & Stambulova, 1998; Marsh, 1998; Sonstroem, Harlow, & Josephs, 1994). Many researchers have attempted to determine which subdomains are the most influential or effective in predicting physical activity. In children and adolescents, perceptions of physical conditioning, and sometimes sport competence, are typically the dominant predictors of physical activity levels (Biddle et al., 1993; Crocker et al., 2000, 2006; Dunton et al., 2003; Hagger et al., 1998; Sabiston & Crocker, 2005; Welk & Eklund, 2005).

Predictive equations of physical activity, however, must be viewed with caution for two reasons. Firstly, most subdomains of the physical self are moderately intercorrelated. Depending on the type of statistical analysis, specific physical subdomains might be included or excluded in the final predictive equation. Secondly, research findings will also be biased by (a) the type of physical activity assessed by particular physical activity measures and (b) the type of activities typically engaged in by children and youth. The most popular activities (e.g., basketball, American football, soccer, rugby, volleyball, gymnastics, swimming, skiing, and snowboarding) require physical conditioning, sport skills, and strength for success. Therefore, it is not surprising that most physical self domains are significantly correlated with physical activity levels (Crocker et al., 2000; Welk & Eklund, 2005).

Body Attractiveness

It might be informative to consider the subdomain of body appearance or attractiveness separately from the other physical self subdomains. Body appearance can be viewed under the general construct of body image, a multidimensional construct that includes perceptions of body appearance, thoughts and beliefs regarding body shape and appearance, and behaviors related to appearance (Bane & McAuley, 1998; Cash & Pruzinsky, 2002). Adolescent body image concerns are influenced by perceived physical appearance evaluations and subsequent affective reactions to these evaluations (Cash & Fleming, 2002). In the sport and exercise psychology field, the affective reaction is often studied under the construct of social physique anxiety, "a subtype of social anxiety that occurs as a result of the prospect or presence of interpersonal evaluation involving one's physique" (Hart, Leary, & Rejeski, 1989, p. 96). Research has shown that perceptions of social physique anxiety are strongly associated with body attractiveness or appearance perceptions in adolescents, especially in females (Crocker et al., 2003, 2000; Smith, 2004). Research also indicates that perceptions of physical attractiveness are a primary predictor of both global physical self-worth and global self-esteem across the life span (see Fox, 1998; Harter, 1999).

The experience of low physical attractiveness and related high body anxiety might act as a trigger to engage in exercise behavior. Adolescents could use physical activity as a strategy to lose weight, gain muscularity, and manage anxiety. For example, Sabiston and colleagues (2007) found that some adolescent females engaged in physical activity to deal with social physique anxiety and other negative emotions (e.g., guilt, shame) associated with body appearance concerns. Other researchers have established that dissatisfaction with body appearance is associated with physical activity–related thoughts and behaviors. For example, McCabe and Ricciardelli (2003) found that boys were more likely than girls to think about and engage in strategies to increase the size of their muscles. Unexpectedly, they also found that girls were interested in increasing their muscular tone along with losing weight and adopted multiple strategies to achieve these goals.

What is surprising is that perceptions of physical attractiveness and social physique anxiety appear to be, at best, weak correlates of physical activity in children and adolescents (Craft et al., 2003; Crocker et al., 2000, 2003; Duncan et al., 2004; Sabiston & Crocker, 2005; Welk & Eklund, 2005). The strength of the relationship does not seem to vary by age or gender, though the number of studies is limited. We must recognize that there are positive and negative cognitive and affective factors associated with using physical activity to modify or even maintain physical appearance. Some youth might avoid physical activities because of the anxiety associated with negative physical appearance. Other youth, however, might engage in activity to either maintain positive appearance or change appearance. It is critical to realize that using physical activity to change the body requires much time and effort. The weak relationship between physical activity levels and perceived physical attractiveness might be attributable to the fact that many strategies are available to either change attractiveness or manage the anxiety associated with it. Recently, Kowalski and colleagues have reported that adolescent boys and girls could use numerous cognitive and behavioral coping strategies to manage social physique anxiety (Kowalski et al., 2006; Sabiston et al., 2007). These strategies included avoidance, dieting, clothing management, makeup, and physical activity. These studies failed, however, to explain why adolescents choose one set of strategies over another. Overall, there is poor understanding of the link between physical activity and perceptions of body appearance or social physique anxiety.

Gender Differences

Gender has been identified as a key factor in the examination of physical self-perceptions and physical activity. Research consistently shows that boys report higher physical self-perceptions than girls (Biddle & Armstrong, 1992; Marsh, 1994, 1998; Morgan et al., 2003; Raudsepp, 2002; Whitehead & Corbin, 1997). Furthermore, researchers consistently report higher physical activity levels among boys than among girls in North American and European samples (Biddle & Armstrong, 1992; Kowalski et al., 1997; Ross & Pate, 1987; Sallis, 1994). Some have suggested that gender stereotypes and the domination of traditional boys' games in school curricula and community sport organizations create the social conditions for physical activity to become more valued, and thus more prominent, in the development of self-esteem for boys (Crain, 1996; Hattie, 1992; Marsh, 1989).

Despite gender differences in physical activity and specific physical self-perceptions, there is no consistent evidence of gender differences in the relationship between the physical self and physical activity. Some studies have indicated that the relationships between physical self-perceptions and physical activity are stronger in males than in females (Biddle & Armstrong, 1992; Hagger et al., 1998; Raustorp et al., 2005), whereas others have yielded no evidence for gender differences (Crocker et al., 2000; Welk & Eklund, 2005). The discrepant findings are difficult to explain, as all of these studies employed representative samples of larger social-cultural populations.

At this point, there appears to be little stability in gender differences with regard to specific relationships among physical self-perception subdomains and physical activity in children and adolescents. One primary weakness in the studies is the measurement of physical activity. Most physical activity measures assess general physical activity levels rather than the specific types of activity engaged in by a young person. As noted in chapter 1 of this volume, the measurement of physical activity is very challenging. Clearly, evaluating participation in specific types of physical activity might be more sensitive to gender differences—differences often influenced by social-cultural expectations. Many authors have noted that cultural beliefs and values can differentially reinforce (or inhibit) the preferred physical activities and sedentary activities of boys and girls (Malszecki & Cavar, 2005).

Cultural Differences

Culture can have a major impact on how individuals develop their global and physical self (Cross & Gore, 2003), and thus we would expect differences in the ways individuals in various cultures interpret the meaning of their body and the importance of physical activity. We have very little empirical knowledge about how such cultural factors influence the complex relationship between body perceptions and physical activity. Initial work by Hagger and colleagues (1998) investigated cross-cultural and gender differences in the extent to which physical self-perceptions predicted moderate physical activity. The study involved comparing Russian children, ages 13 to 14 years, with British children of similar age. The measures included the Godin Leisure Time Exercise Questionnaire (Godin & Shephard, 1985) and the children's version of the PSPP. All measures were translated into Russian. With regard to moderate physical activity, British and Russian boys were more active than girls. Gender differences in physical self-perceptions were observed within each nationality. Russian boys reported higher perceptions of sport competence, conditioning, strength, and physical self-worth compared to Russian girls, whereas British boys reported only higher body appearance perceptions compared to British girls. Perceptions of conditioning, followed by sport competence, best distinguished active and inactive children among both Russian boys and girls. Among the British children, strength and sport competence were the best discriminators for girls, whereas sport competence and conditioning were the primary discriminators for boys. Nevertheless, strength of the predictions was modest.

The study by Hagger and colleagues (1998) was a good first step toward understanding cultural differences in youth physical self-perceptions and physical activity. However, a limitation evident in this work is the use of nationality as a proxy for culture. Similar problems occur when sport researchers use ethnic or racial grouping as a measure of culture. Such approaches assume that culture has a monolithic effect throughout national or ethnic groups. As discussed earlier, cultural issues are complex and have a profound impact on the construction of the self and the adoption and reinforcement of specific behaviors by individuals (Cross & Gore, 2003). Researchers might want to assess such cultural values as filial piety, familism, collectivism, fatalism, and machismo (Unger et al., 2002) to consider how physical self-perceptions develop and how they might be related to physical activity. For example, Ramanathan and Crocker (2006) found that adolescent females from the Indian Diaspora believed that family values around filial piety (obedience to and support of parents and elders) and familism (obligation and connectedness to one's immediate family) influenced not only encouragement for engaging in particular physical activities (e.g., sport vs. religious dance) but also the value attached to physical activity as part of life. Differences in how parents and significant others modeled and encouraged physical activity could not be understood simply from looking at ethnicity. Understanding the processes involved in the ways in which culture affects the self and the adoption of specific types of physical activity will require the development of a sound conceptual framework.

How the culture of young people influences their physical activity is not well understood.

Longitudinal Analysis

A basic goal in research on the physical self and physical activity should be to determine the direction of causality. Theorists hold that self-concepts in specific domains influence behavioral choice and persistence in those domains (e.g., Harter, 1999). Therefore, strong physical self-concept should lead youth to engage in physical activity. This notion is consistent with the cross-sectional research reviewed thus far. However, it is plausible that the relationship between the physical self and physical activity could be bidirectional (Fox, 1997; Sonstroem, 1997). For example, if youth have high perceptions of sport skills and conditioning, this might lead them to participate in various physical activities, thus increasing their physical skills. Actual enhancement of various physical competencies could help maintain or even enhance physical self-perceptions, increasing the probability of participating in future physical activity. The same reciprocal effect might also be responsible for youth's disengaging from physical activity or approaching sedentary pursuits. Low perceptions of physical competence might lead to activity avoidance and thus decrease the opportunity to develop both positive physical self-perceptions and actual physical competencies. Coupled with high perceptions of competence in other domains that involve sedentary behaviors, this could result not only in an absence of physical activity, but also in the reinforcement of nonactive behaviors.

Researchers can examine changes, especially reciprocal effects, in physical activity and the self over time using longitudinal or repeated measures research designs. An interesting study by Sonstroem, Harlow, and Salisbury (1993) dealt with bidirectional effects between swimming performance and perceived physical competence over a one-year period in U.S. high school athletes. Sonstroem and colleagues evaluated the hierarchical exercise and self-esteem model (Sonstroem & Morgan, 1989) over three time periods (November, January, and March) by assessing global self-esteem, physical estimation/competence (perceptions of a well-conditioned body, stamina, and strength), perception of swimming skills, and actual performance. They found some evidence that perceived competence had an impact on swimming performance. There was, however, no evidence that better performance improved perceived competence. Furthermore, time 2 and 3 variable scores were largely related to previous scores for each specific variable. For example, perceptions of physical competence at time 3 were primarily predicted by perceptions of physical competence at time 2. Overall, this study suggested that physical self variables and performance do not change much over a one-year period and that there is little evidence for bidirectional effects. However, Sonstroem (1997) noted that one of the limitations of the study was that the participants were experienced competitive swimmers, which possibly resulted in physical self-perceptions and performance that were highly stable over time. Examining elite youth athletes might not be the best way to examine change in either the physical self or physical activity levels.

We know very little about how the physical self changes over time. In addition, there are significant gaps in our understanding of the nature of relationships between the physical self and physical activity over time. Crocker and colleagues completed a series of studies on the relationship between the physical self and health behaviors in Canadian adolescent females over a three-year period (Crocker et al., 2001, 2003, 2006; Kowalski et al., 2003). The program of research included assessment of smoking behavior, dietary restraint, social physical anxiety, and body mass index; however, here we focus solely on findings on the self and physical activity behavior.

Data from the three-year longitudinal study (Crocker et al., 2006) allowed analysis of cross-sectional effects, mean group changes over time for variables, and stability (autoregressive) effects for variables, as well as bidirectional effects of specific physical self variables and physical activity. The teenagers were assessed once per year starting in grade 9 (ages 13-14 years), with assessments approximately one year apart. Both expected and unexpected results were obtained. With regard to the cross-sectional results, for all three years the four subdomains of the physical self (as measured by the PSPP) were moderately correlated with physical self-worth (PSW) and global self-esteem (GSE). Body attractiveness and conditioning perceptions were the primary correlates with PSW and GSE. When the correlates of physical activity were examined, both conditioning and sport competence perceptions were prominent. Overall, the patterns of the cross-sectional results are relatively consistent with findings from previous cross-sectional research.

With regard to the longitudinal findings, there was a significant decrease in the group mean of physical activity levels during adolescence. At the same time, there was little change in the means of global self-esteem and the physical self variables. This raises a flag about the potential relationship between the physical self and physical activity. However, mean group change is a limited measure, as it does not reflect changes in individual scores over time. Stability scores produce a more sensitive measure of individual change. Physical activity stability was moderately low, suggesting there was much individual variability in activity over time. However, the physical self measures demonstrated moderately high stability over time; that is, the year 2 and 3 scores were related in large part to the previous year's scores.

To examine the bidirectional effects between the physical self and physical activity, Crocker and colleagues (2006) used autoregressive path analysis techniques. Given the potential complexity of models using multiple physical self subdomains, Crocker and colleagues focused on the physical conditioning perceptions and physical activity. The data indicated that both physical activity and conditioning scores were moderately associated with previous-year scores. However, the results provided evidence of a bidirectional causal relationship between conditioning perceptions and physical activity. Conditioning self-perceptions had a stronger influence on physical activity than physical activity had on conditioning self-perceptions, but both relationships were statistically significant. The autoregression analysis, however, indicated that conditioning self-perceptions play a role in the adoption or maintenance of future physical activity (or both). The data from Crocker and colleagues (2006) and Sonstroem and colleagues (1994) also highlight that cross-sectional analyses tend to substantially overestimate the relationship between the physical self and physical activity variables.

Conclusion

How children and adolescents view their self, especially their physical self, is related to the initiation and maintenance of physical activity. It should be apparent to the reader, however, that the attentional and motivational properties associated with the self are complex. The self can be viewed as multidimensional, with increasing differentiation from childhood to late adolescence. Cognitive maturation and social-cultural context play major roles in the dynamic construction of the self. Although there is some evidence that global self-esteem is related to physical activity, the physical self appears to be more critical in determining motivated physical activity behavior. Correlational research, both cross-sectional and longitudinal, demonstrates that specific physical self subdomains are moderate correlates of physical activity. There are, however, several significant gaps in our understanding. First, the causal link between physical self-perceptions and physical activity is not clear. Second, culture appears to be vital in the development and meaning of the self. Yet there is a dearth of knowledge about culture, self, and physical activity. Third, body appearance is the physical self component most strongly related to global self-esteem, but is a surprisingly weak predictor of physical activity behavior. This raises serious challenges for both researchers and practitioners.

APPLICATIONS FOR RESEARCHERS

Researchers need to think more critically about the self and its relationship to physical activity and sedentary behavior in children and adolescents. Rather than relying on passive observational cross-sectional designs, researchers need to utilize longitudinal and experimental designs to test specific hypotheses about the self and physical activity and inactivity. How do perceptions of the physical self act in concert with social-cultural and pedagogical variables to influence the choice of physical (and sedentary) activities, and is there a reciprocal relationship between the physical self and physical activity and inactivity? Answering such questions will require researchers to integrate the physical self into motivational models (Biddle, 1997) and gain a better appreciation for how culture affects not only the meaning of the self, but also the value and function of physical activity. Lastly, physical activity researchers need to integrate recent theoretical and empirical developments addressing the diverse functions and conceptions of the self (see Tangney & Leary, 2003).

APPLICATIONS FOR PROFESSIONALS

There appears to be a reciprocal effect between the physical self and physical activity. Children and adolescents use multiple information sources to form self-perceptions, such as self-comparison, social comparison, and evaluative feedback from significant others, with young children heavily influenced by evaluative feedback. Physical self components can be enhanced through involving youth in challenging physical activity paired with positive feedback and approval from significant others (see Horn, 2004). Although the causal connection is not clear, practitioners must be aware that perceptions of physical attractiveness are strongly linked to anxiety and global self-esteem. Because it is difficult to change body appearance, coaches and teachers must be careful when providing evaluative feedback about the body. In an increasingly multicultural world, practitioners are encouraged to take an individualized approach, seeking to understand how each youngster views her- or himself and how physical activity fits into her or his life.

References

Armstrong, N., & Welsman, J.R. (1997). *Young people and physical activity.* Oxford: Oxford University Press.

Armstrong, N., Williams, J., Balding, J., Gentle, P., & Kirby, B. (1991). Cardiovascular fitness, physical activity patterns, and selected coronary risk factor variables in 11- to 16-year olds. *Pediatric Exercise Science, 3*, 219-228.

Bandura, A. (1997). *Self-efficacy: The exercise of control.* New York: Freeman.

Bane, S., & McAuley, E. (1998). Body image and exercise. In J.L. Duda (Ed.), *Advances in sport and exercise psychology measurement* (pp. 311-322). Morgantown, WV: Fitness Information Technology.

Biddle, S.J.H. (1997). Cognitive theories of motivation and the physical self. In K.R. Fox (Ed.), *The physical self: From motivation to well-being* (pp. 59-82). Champaign, IL: Human Kinetics.

Biddle, S.J.H., & Armstrong, N. (1992). Children's physical activity: An exploratory study of psychological correlates. *Social Science and Medicine, 34*, 325-331.

Biddle, S.J.H., Page, A., Ashford, B., Jennings, D., & Fox, K.R. (1993). Assessment of children's physical self-perceptions. *International Journal of Adolescence and Youth, 4*, 93-109.

Brown, J.D. (1993). Self-esteem and self-evaluation: Feeling is believing. In J. Suls (Ed.), *Psychological perspectives on the self* (Vol. 4, pp. 27-58). Hillsdale, NJ: Erlbaum.

Burkhalter, N.A., & Wendt, J.C. (2001). Prediction of selected fitness indicators by gender, age, alienation, and perceived competence. *Journal of Teaching in Physical Education, 21*, 3-15.

Calfas, K.J., & Taylor, W.C. (1994). Effects of physical activity on psychological variables in adolescents. *Pediatric Exercise Science, 6*, 406-423.

Cash, T.F., & Fleming, E.C. (2002). The impact of body image experiences: Development of the body image quality of life inventory. *International Journal of Eating Disorders, 31*, 455-460.

Cash, T.F., & Pruzinsky, T. (2002). *Body image: A handbook of theory, research, and clinical practice.* New York: Guilford Press.

Corbin, C.B. (2002). Physical activity for everyone: What every physical educator should know about promoting lifelong physical activity. *Journal of Teaching Physical Education, 21*, 128-144.

Craft, L.L., Pfeiffer, K.A., & Pivarnik, J.M. (2003). Predictors of physical competence in adolescent girls. *Journal of Youth and Adolescence, 32*, 431-438.

Crain, R.M. (1996). The influence of age, race and gender on child and adolescent multi-dimensional self-concept. In B.A. Bracken (Ed.), *Handbook of self-concept: Developmental, social, and clinical considerations* (pp. 395-420). New York: Wiley.

Crocker, P.R.E., Eklund, R.C., & Kowalski, K.C. (2000). Children's physical activity and physical self-perceptions. *Journal of Sports Sciences, 18*, 383-394.

Crocker, P.R.E., Kowalski, N., Kowalski, K.C., Chad, K., Humbert, L., & Forrester, S. (2001). Smoking behavior and dietary restraint in young adolescent women: The role of physical self-perceptions. *Canadian Journal of Public Health, 92*, 428-432.

Crocker, P.R.E., Sabiston, C.M., Forrester, S., Kowalski, N., Kowalski, K.C., & McDonough, M.H. (2003). Predicting change in physical activity, dietary restraint, and physique anxiety in adolescent girls: Examining covariance in physical self-perceptions. *Canadian Journal of Public Health, 94*, 332-337.

Crocker, P.R.E., Sabiston, C.M., Kowalski, K.C., McDonough, M.H., & Kowalski, N. (2006). Longitudinal assessment of the relationship between physical self-concept and health-related behavior and emotion in adolescent girls. *Journal of Applied Sport Psychology, 18*, 185-200.

Crocker, P.R.E., Synder, J., Kowalski, K.C., & Hoar, S. (2000). Don't let me be fat or physically incompetent! The relationship between physical self-concept and social physique anxiety in Canadian high performance female adolescent athletes. *Avante, 6*, 1-8.

Cross, S.E., & Gore, J.S. (2003). Cultural models of the self. In M.R. Leary & J.P. Tangney (Eds.), *Handbook of self and identity* (pp. 536-564). New York: Guilford Press.

Duncan, M.J., Al, M.Y., Nevill, A., & Jones, M.V. (2004). Body image and physical activity in British secondary school children. *European Physical Education Review, 10*, 243-260.

Dunton, G.F., Jamner, M.S., & Cooper, D.M. (2003). Physical self-concept in adolescent girls: Behavioral and physiological correlates. *Research Quarterly for Exercise and Sport, 74*, 360-365.

Ekeland, E., Heian, F., & Hagen, K.B. (2005). Can exercise improve self esteem in children and young people? A systematic review of randomised controlled trials. *British Journal of Sports Medicine, 39*, 792-798.

Eklund, R.C., Whitehead, J.R., & Welk, G.J. (1997). Validity of the Children and Youth Physical Self-Perception Profile: A confirmatory factor analysis. *Research Quarterly for Exercise and Sport, 68*, 249-256.

Fleming, T., Kowalski, K.C., Humbert, M.L., Fagan, K.R., Cannon, M.J., & Girolami, T.M. (2006). Body-related emotional experiences of young aboriginal women. *Qualitative Health Research, 16*, 517-537.

Fox, K.R. (1990). *The Physical Self-Perception Profile manual.* DeKalb, IL: Northern Illinois University Office for Health Promotion.

Fox, K.R. (Ed.). (1997). *The physical self: From motivation to well-being.* Champaign, IL: Human Kinetics.

Fox, K.R. (1998). Advances in the measurement of the physical self. In J.L. Duda (Ed.), *Advances in sport and exercise psychology measurement* (pp. 295-310). Morgantown, WV: Fitness Information Technology.

Fox, K.R., & Corbin, C.B. (1989). The Physical Self-perception Profile: Development and preliminary validation. *Journal of Sport and Exercise Psychology, 11*, 408-430.

Gallagher, J.D., French, K.E., Thomas, K.T., & Thomas, J.R. (1996). Expertise in youth sport: Relations between knowledge and skill. In F.L. Smoll & R.E. Smith (Eds.), *Children and youth in sport: A biopsychosocial perspective* (pp. 338-358). Madison, WI: Brown & Benchmark.

Godin, G., & Shephard, R.J. (1985). A simple method to assess exercise behavior in the community. *Canadian Journal of Applied Sport Science, 10*, 141-146.

Gruber, J.J. (1986). Physical activity and self-esteem development in children: A meta analysis. In G. Stull & H. Eckert (Eds.), *Effects of physical activity on children* (pp. 30-48). Champaign, IL: Human Kinetics.

Hagger, M., Ashford, B., & Stambulova, N. (1998). Russian and British children's physical self-perceptions and physical activity participation. *Pediatric Exercise Science, 10*, 137-152.

Hamamura, T., & Heine, S.J. (in press). Self-enhancement, self-improvement, and face among Japanese. In E.C. Chang (Ed.), *Self-criticism and self-enhancement: Theory, research, and clinical implications.* Washington, DC: American Psychological Association.

Hart, E.A., Leary, M.R., & Rejeski, W.J. (1989). The measurement of social physique anxiety. *Journal of Sport and Exercise Psychology, 11*, 94-104.

Harter, S. (1988). *Manual of the Self-Perception Profile for Adolescents.* Denver: University of Denver.

Harter, S. (1999). *The construction of the self: A developmental perspective.* New York: Guilford Press.

Harter, S. (2003). The development of self-representations during childhood and adolescence. In M.R. Leary & J.P. Tangney (Eds.), *Handbook of self and identity* (pp. 610-642). New York: Guilford Press.

Hattie, J. (1992). *Self-concept.* Hillsdale, NJ: Erlbaum.

Hattie, J., & Fletcher, R. (2005). Self-esteem = success/pretensions: Assessing pretensions/importance in self-esteem. In H.W. Marsh, R.G. Craven, & D.M. McInerney (Eds.), *International advances in self research:* Vol. 2. *New frontiers for self research* (pp. 123-152). Greenwich, CT: Information Age.

Hattie, J., & Marsh, H.W. (1996). Future directions in self-concept research. In B.A. Bracken (Ed.), *Handbook of self-concept: Developmental, social, and clinical considerations* (pp. 421-462). New York: Wiley.

Heine, S.J. (2001). Self as a product of culture: An examination of East Asian and North American selves. *Journal of Personality, 69*, 881-906.

Horn, T.S. (2004). Developmental perspectives on self-perceptions in children and adolescents. In M.R. Weiss (Ed.), *Developmental sport and exercise psychology: A lifespan perspective* (pp. 101-144). Morgantown, WV: Fitness Information Technology.

Keating, D.P. (1990). Adolescent thinking. In S.S. Feldman & G. Elliott (Eds.), *At the threshold: The developing adolescent* (pp. 54-89). Cambridge, MA: Harvard University Press.

Kowalski, K.C., Crocker, P.R.E., & Kowalski, N. (1997). Convergent validity of the physical activity questionnaire for adolescents. *Pediatric Exercise Science, 9*, 342-352.

Kowalski, K.C., Crocker, P.R.E., Kowalski, N.P., Chad, K.E., & Humbert, M.L. (2003). Examining the physical self in adolescent girls over time: Further evidence against the hierarchical model. *Journal of Sport and Exercise Psychology, 25*, 5-18.

Kowalski, K.C., Mack, D.E., Crocker, P.R.E., Niefer, C.B., & Fleming, T.L. (2006). Coping with social physique anxiety in adolescence. *Journal of Adolescent Health, 39* [Online], 275, e9-275. e16.

Leary, M.R. (2001). The self we know and the self we show: Self-esteem, self-presentation, and the maintenance of interpersonal relationships. In G.J.O. Fletcher & M. Clark (Eds.), *Blackwell handbook of social psychology: Interpersonal processes* (pp. 457-477). Malden, MA: Blackwell.

Leary, M.R., & Tangney, J.P. (2003). The self as an organizing construct in the behavioral and social sciences. In M.R. Leary & J.P. Tangney (Eds.), *Handbook of self and identity* (pp. 3-14). New York: Guilford Press.

Malszecki, G., & Cavar, T. (2005). Men, masculinities, war, and sport. In N. Mandell (Ed.), *Feminist issues: Race, class, and sexuality* (4th ed., pp. 160-187). Toronto: Pearson Education Canada.

Marsh, H.W. (1989). Age and sex effects in multiple dimensions of self-concept: Preadolescence to early-childhood. *Journal of Educational Psychology, 81*, 417-430.

Marsh, H.W. (1994). The importance of being important: Theoretical models of relations between specific and global components of physical self-concept. *Journal of Sport and Exercise Psychology, 16*, 306-325.

Marsh, H.W. (1998). Age and gender effects in physical self-concepts for adolescent elite athletes and nonathletes: A multicohort-multioccasion design. *Journal of Sport and Exercise Psychology, 20*, 237-259.

Marsh, H.W., Bar-Eli, M., Zach, S., & Richards, G.E. (2006). Construction validation of Hebrew versions of three physical self-concept measures: An extended multitrait-multimethod analysis. *Journal of Sport and Exercise Psychology, 28*, 310-343.

Marsh, H.W., Richards, G.E., Johnson, S., Roche, L., & Tremayne, P. (1994). Physical Self-Description Questionnaire: Psychometric properties and a multitrait-multimethod analysis of relations to existing instruments. *Journal of Sport and Exercise Psychology, 16*, 270-305.

Marsh, H.W., & Sonstroem, R.J. (1995). Importance ratings and specific components of physical self-concept: Relevance to predicting global components of self-concept and exercise. *Journal of Sport and Exercise Psychology, 17*, 84-104.

Marsh, H.W., & Yeung, A.S. (1998). Top-down, bottom-up, and horizontal models: The direction of causality in multidimensional, hierarchical self-concept models. *Journal of Personality and Social Psychology, 75*, 509-527.

McCabe, M.P., & Ricciardelli, L.A. (2003). Body image and strategies to lose weight and increase muscle among boys and girls. *Health Psychology, 22*, 39-46.

Morgan, C.F., McKenzie, T.L., Sallis, J.F., Broyles, S.L., Zive, M.M., & Nader, P.R. (2003). Personal, social, and environmental correlates of physical activity in a bi-ethnic sample of adolescents. *Pediatric Exercise Science, 15*, 288-301.

Ramanathan, S., & Crocker, P.R.E. (2006, November). *The influence of cultural values on physical activity among female adolescents from the Indian Diaspora.* Paper presented at the meeting of the Canadian Society for Psychomotor Learning and Sport Psychology, Halifax, NS.

Raudsepp, L. (2002). Children's and adolescents' physical self-perceptions as related to moderate to vigorous physical activity and physical fitness. *Pediatric Exercise Science, 14*, 97-106.

Raustorp, A., Stahle, A., Gudasic, H., Kinnunen, A., & Mattsson, E. (2005). Physical activity and self-perception in school children assessed with the Children and Youth–Physical Self-Perception Profile. *Scandinavian Journal of Medicine and Science in Sports, 15*, 126-134.

Ross, J.G., & Pate, R.R. (1987). The national children and youth fitness study II: A summary of findings. *Journal of Physical Education, Recreation and Dance, 58*, 51-56.

Sabiston, C.M., & Crocker, P.R.E. (2005). Examining correlates of physical activity in adolescents: An expectancy-value perspective. *Journal of Sport and Exercise Psychology, 27*, S131.

Sabiston, C.M., Sedgwick, W., Crocker, P.R.E., Kowalski, K.C., & Mack, D. (2007). Social physique anxiety in adolescents: An examination of influences, coping strategies and health behaviours. *Journal of Adolescent Research, 22*, 78-101.

Sallis, J.F. (1994). Determinants of physical activity behavior in children. In R.R. Pate & R.C. Holn (Eds.), *Health and fitness through physical education* (pp. 31-43). Champaign, IL: Human Kinetics.

Santrock, J.W. (1998). *Adolescence* (5th ed.). Boston: McGraw-Hill.

Santrock, J.W., & Yussen, S.R. (1992). *Child development: An introduction* (5th ed.). Dubuque, IA: Brown.

Shavelson, R.J., & Bolus, R. (1982). Self-concept: The interplay of theory and methods. *Journal of Educational Psychology, 74*, 3-17.

Shavelson, R.J., Hubner, J.J., & Stanton, G.C. (1976). Self-concept: Validation of construct interpretations. *Review of Educational Research, 46*, 407-441.

Smith, A.L. (2004). Measurement of social physique anxiety in early adolescence. *Medicine and Science in Sports and Exercise, 36*, 475-483.

Smith, R.E., Smoll, F.L., & Curtis, B. (1979). Coach effectiveness training: A cognitive behavioral approach to enhancing relationship skills in youth sport coaches. *Journal of Sport Psychology*, 1, 59-75.

Sonstroem, R.J. (1984). Exercise and self-esteem. *Exercise and Sport Sciences Reviews, 12*, 123-155.

Sonstroem, R.J. (1997). The physical self-system: A mediator of exercise and self-esteem. In K.R. Fox (Ed.), *The physical self: From motivation to well-being* (pp. 3-26). Champaign, IL: Human Kinetics.

Sonstroem, R.J., Harlow, L., & Josephs, L. (1994). Exercise and self-esteem: Validity of model expansion and exercise associations. *Journal of Sport and Exercise Psychology, 16*, 29-42.

Sonstroem, R.J., Harlow, L., & Salisbury, K.S. (1993). Path analysis of a self-esteem model across a competitive swim season. *Research Quarterly for Exercise and Sport, 64*, 335-342.

Sonstroem, R.J., & Morgan, W.P. (1989). Exercise and self-esteem: Rationale and model. *Medicine and Science in Sports and Exercise, 21*, 329-337.

Stets, J.E., & Burke, P.J. (2003). A sociological approach to self and identity. In M.R. Leary & J.P. Tangney (Eds.), *Handbook of self and identity* (pp. 128-152). New York: Guilford Press.

Tangney, J.P., & Leary, M.R. (2003). The next generation of self-research. In M.R. Leary & J.P. Tangney (Eds.), *Handbook of self and identity* (pp. 667-674). New York: Guilford Press.

Tremblay, M.S., Inman, J.W., & Willms, J.D. (2000). The relationship between physical activity, self-esteem, and academic achievement in 12-year-old children. *Pediatric Exercise Science, 12*, 312-323.

Unger, J.B., Ritt-Olson, A., Teran, L., Huang, T., Hoffman, B.R., & Palmer, P. (2002). Cultural values and substance use in a multiethnic sample of California adolescents. *Addiction Research and Theory, 10*, 257-279.

Weiss, M.R., & Williams, L. (2004). The why of youth sport involvement: A developmental perspective on motivation processes. In M.R. Weiss (Ed.), *Developmental sport and exercise psychology: A lifespan perspective* (pp. 223-268). Morgantown, WV: Fitness Information Technology.

Welk, G.J., & Eklund, B. (2005). Validation of the Children and Youth Physical Self Perceptions Profile for young children. *Psychology of Sport and Exercise, 6*, 51-65.

Whitehead, J.R. (1995). A study of children's physical self-perceptions using an adapted Physical Self-Perception Profile questionnaire. *Pediatric Exercise Science, 7*, 132-151.

Whitehead, J.R., & Corbin, C.B. (1997). Self-esteem in children and youth: The role of sport and physical education. In K.R. Fox (Ed.), *The physical self: From motivation to well-being* (pp. 175-204). Champaign, IL: Human Kinetics.

CHAPTER

10

Youth With Movement Difficulties

Janice Causgrove Dunn, PhD ▪ Donna L. Goodwin, PhD

The health risks associated with a sedentary lifestyle have led to growing concerns among health professionals about endemic inactivity, obesity, and poor fitness levels among youth in industrialized nations. These concerns are even more relevant when considered in the context of youth with disabilities (Longmuir & Bar-Or, 1994, 2000). Several studies have revealed even lower activity levels and lower fitness levels among children and adolescents with a range of disabilities in comparison to peers without disabilities (e.g., Longmuir & Bar-Or, 1994; Steele et al., 1996; van den Berg-Emons et al., 2003). Research also shows that youth with disabilities are less likely to engage in organized sport and recreational free play activities (Cairney et al., 2005) and are more likely to engage in sedentary leisure activities such as watching television (Steele et al., 1996).

Admittedly, the functional ability to participate in traditional vigorous physical activities can be reduced by disabilities that limit mobility. This may partly explain why some youth with physical disabilities lead more sedentary lives than peers with sensory impairments (e.g., visual impairments, hearing impairments) or chronic medical conditions (e.g., kidney disease, cystic fibrosis) (Longmuir & Bar-Or, 1994, 2000). However, even youth with relatively mild movement difficulties tend to be less active (Bouffard et al., 1996; Cantell, Smyth, & Ahonen, 1994; Smyth & Anderson, 2000) and less fit (Hands & Larkin, 2002) than more movement-competent peers, despite having sufficient capabilities for physical activity.

Greater understanding of the factors that afford and constrain the involvement of youth with disabilities in physical activities is clearly needed. To this end, the current chapter examines psychosocial aspects of physical activity and inactivity in youth with movement disorders in an effort to increase appreciation of the impact of movement difficulties, in particular, on physical activity participation. Youth with movement disorders are those whose ability to participate in preferred movement skills is negatively affected by disabilities or health conditions that can range from relatively mild unexplained clumsiness (e.g., developmental coordination disorder [DCD]) to more severe neurological and musculoskeletal conditions that limit mobility (e.g., spina bifida, cerebral palsy). Table 10.1 provides a description of movement difficulties that may be associated with groups of participants who have disabilities as reflected in the research reviewed in this chapter.

Table 10.1 Movement Difficulties That May Be Associated With Research Participant Groups

Disability type	Definition	Movement difficulty
Cerebral palsy	A group of chronic, nonprogressive disorders of movement or posture (or both) caused by faulty development or damage to motor areas in the brain.[1]	Muscular tightness, muscular weakness, or associated movements Muscular imbalance Decreased range of motion Inefficient, poorly coordinated movements
Developmental coordination disorder	Impairment in the development of motor coordination, in the absence of a general medical condition (e.g., cerebral palsy, visual impairment), intellectual impairment, or pervasive developmental disorder (e.g., autism). It is diagnosed only if motor difficulties interfere with performance of academic skills or activities of daily living.[2]	Poor planning or coordination of movements or both Difficulty learning complex movements Awkward or clumsy movements
Muscular dystrophy	A hereditary disease characterized by progressive, diffuse muscular weakness and atrophy of muscle fibers (characterized by degeneration of muscle cells and replacement by fat and fibrous tissue).[1]	Progressive muscular weakness Muscle imbalance Awkward or clumsy movements
Spina bifida	A congenital birth defect of the spinal column that occurs within the first 6 weeks of pregnancy, caused by failure of the neural arch of one or more vertebrae to develop properly and enclose the spinal cord.[1]	Loss of sensation Muscle weakness or paralysis or both Muscle imbalance
Spinal cord injury	Acquired paralysis of the trunk following trauma to the spinal cord, and resulting in impaired function and sensation. The degree of paralysis depends on the location of injury to the spinal cord and the extent of the destruction of the neural fibers.[1]	Loss of sensation Muscle weakness or muscle paralysis or both Muscle imbalance

[1]Sherrill 2004. [2]American Psychiatric Association 2000.

Notwithstanding the converging evidence of an overall trend toward physical inactivity in youth with disabilities, it is important to recognize that some youth with disabilities are as active as, or more active than, their peers without disabilities. A few studies have shown similar physical activity engagement levels among youth with and without disabilities. For instance, Rosser Sandt and Frey (2005) found no difference between children with autistic spectrum disorders and classmates without disabilities in the amount of moderate-to-vigorous physical activity (MVPA) during recess, during physical education, and after school. Interestingly, despite the fact that none of the children with autistic spectrum disorders participated in organized sport or inclusive community programs, the majority engaged in at least 60 min of MVPA per day over the five monitored days. It should also be emphasized that much of the research comparing physical activity rates of youth with disabilities and youth without disabilities is based upon aggregate data, and the engagement levels of some physically active youth with disabilities are lost in the average score that is calculated for their group (i.e., the mean score for youth with disabilities).

Importance of Physical Activity for This Population

Youth with disabilities accrue many of the same fitness and health benefits from regular involvement in physical activity that youth without disabilities do, including increased cardiovascular endurance, strength, and flexibility; improved self-concept; and enhanced psychological well-being (Durstine et al., 2000). However, several researchers have argued that an active lifestyle is even more important for those who have disabilities (e.g., Cooper et al., 1999; Steele et al., 1996; van der Ploeg et al., 2004). The rationale for this argument is based primarily on the beneficial effects of physical activity on the functional abilities of persons with disabilities, combined with a corresponding reduction of risk for secondary conditions. Secondary conditions are physical, medical, cognitive, emotional, or psychosocial consequences that a person with a primary disabling condition (e.g., spina bifida) is predisposed to experience (Simeonsson, Sturtz McMillen, & Huntington, 2002). While some secondary conditions are not preventable, many can be prevented or managed with appropriate interventions such as regular involvement in physical activity. Secondary conditions found in youth with spina bifida include obesity, scoliosis, pressure sores, body image dissatisfaction, limited school roles, and social isolation (Simmeonsson et al., 2002). In general, secondary conditions decrease the level of functioning experienced by people with disabilities and may restrict participation in preferred activities. For many of the secondary conditions associated with disability in youth (e.g., obesity, diabetes, osteoporosis, urinary tract infections, respiratory problems, cardiovascular disease, restricted social contacts, diminished self-concept, and depression), the degree of risk is compounded by the interactive effects of disability and physical inactivity (Cooper et al., 1999; van der Ploeg et al., 2004).

Physical activity is also important to consider when one is attempting to understand the self-perception and self-definition challenges faced by young people with disabilities. Involvement in physical activity can function as a "normalizing experience" that facilitates identity development, enhances perceptions of competence and self-worth, and expands social networks (Taub & Greer, 2000). Physical activity often places youth with disabilities in contact with others who have disabilities and with peers who do not have disabilities in a social context that is germane to childhood, providing common life experiences in which they can share. It also affords an opportunity for youth with disabilities to be active agents in their own identity construction through the affirmation of physical competence and the bodily expressions of fitness, health, and vitality (Goodwin, Thurmeier, & Gustafson, 2004).

Theoretically Based Predictions and Research Findings

Physical inactivity is a common problem among youth with movement difficulties. To change this, we need to understand factors that facilitate and constrain participation, and then apply that knowledge to interventions that promote activity. What are the important psychosocial factors that are influencing the physical activity engagement of youth with movement difficulties? Researchers in adapted physical activity and related disciplines have used social cognitive models to guide their attempts to provide answers to this question. Consistent with the World Health Organization's (2001) view of disability as a problem with functioning that limits activities and results from a complex interaction between health condition, personal factors, and environmental factors, social cognitive theories assume that behavior is a function of both personal and environmental factors.

Competence Motivation Theory

Competence motivation theory (Harter, 1978, 1981) has been one of the most widely used theories in the area of physical activity and disability. Because it is discussed in some detail in chapter 8 (p. 196), here we simply note the prediction of the theory that positively reinforced successful mastery attempts lead to feelings of competence and intrinsic pleasure associated with increased motivation to engage in future mastery attempts. Failed mastery attempts, on the other hand, are expected to result in a different pattern of perceptions and motivation.

Thus, youth who experience movement difficulties are generally expected to develop low perceptions of athletic competence as a consequence of both recurring failure experiences and criticism or the absence of positive feedback from parents, peers, and others. The negative effects of low competence perceptions go hand in hand with and are compounded by low control perceptions over outcomes in the athletic domain. Negative affective reactions in the form of worry or anxiety are also expected, leading to an increased risk of avoidance of or withdrawal from mastery attempts in physical activity situations.

The theoretical predictions embedded in Harter's competence motivation theory (1978, 1981) have been a particular focus of researchers interested in children and adolescents with mild movement difficulties (DCD, clumsiness, awkwardness). Studies have primarily concerned levels and correlates of perceptions of athletic competence, though several studies have also included assessments of other specific competence domains and global self-worth. Youth with DCD have generally been found to report lower physical or athletic competence (e.g., Cantell et al., 1994; Piek, Baynam, & Barrett, 2006; Skinner & Piek, 2001) and lower levels of peer acceptance (e.g., Rose, Larkin, & Berger, 1997; Skinner & Piek, 2001) than their peers without disabilities. This is not surprising given that peer acceptance is frequently linked to physical or athletic competence (Horn, 2004). If a child with DCD is unable to excel at a valued activity such as sport, opportunities to gain peer acceptance may be minimized, further deterring that child from participating. Playground observations and semistructured interviews have revealed that youth with even mild movement difficulties often hold marginal positions in their peer groups and are subjected to ridicule, teasing, and even bullying within group physical activity settings (Mandich, Polatajko, & Rodger, 2003; Portman, 1995).

Negative affective reactions to social problems resulting from movement difficulties were described by Fitzpatrick and Watkinson (2003), who retrospectively interviewed adults about their experiences as physically awkward youth. The adults recalled ridicule and teasing that resulted from public displays of movement incompetence, leading to feelings of shame, embarrassment, humiliation, and emotional hurt. Numerous studies have also shown that children and adolescents with DCD report greater anxiety than peers without disabilities when faced with movement situations (Rose et al., 1997; Skinner & Piek, 2001). It is little wonder that children with movement difficulties withdraw from physical activity settings that have the potential to inflict these negative psychosocial outcomes.

Self-reports (e.g., Cairney et al., 2005; Cantell et al., 1994), parent interviews (e.g., Mandich et al., 2003), and retrospective reports from adults (e.g., Fitzpatrick & Watkinson, 2003) confirm that children with mild movement disorders such as DCD do not often participate in games with peers, organized sport, or recreational free play activities. Direct observations of children with DCD during school recess reveal that they typically engage in less vigorous play or social play than peers with at least average movement competence, and spend more time on their own and more time watching others play (Bouffard et al., 1996; Smyth & Anderson, 2000). Unfortunately, the nature of these observations prevents discussion of the degree to which these physical activity and social interaction levels are voluntary (i.e., the child with movement difficulties has chosen to withdraw) and the degree to which the child is actively excluded by others. Nevertheless, several qualitative studies of the experiences of young people with movement difficulties such as DCD have revealed that both withdrawal and exclusion contribute to the lower activity levels of these youth (Evans & Roberts, 1987; Fitzpatrick & Watkinson, 2003; Mandich et al., 2003).

Compulsory physical education would seem to be a particularly threatening environment for children with movement difficulties. Observations have shown that children with movement difficulties in grades 1 to 6, on average, spend less time than other classmates actively engaged in class, less time successfully engaged in assigned motor activities, and more time engaged in activities not related to lesson goals (e.g., not doing any motor task, misbehaving, and engaging in tasks other than those assigned; Causgrove Dunn & Dunn, 2006; Thompson et al., 1994).

Causgrove Dunn and Dunn (2006) observed specific avoidance behaviors of children with movement difficulties in physical education classes. These behaviors included taking extended water or bathroom breaks during the activity (and returning only when the teacher noticed the absence), feigning injury, assuming nonparticipant roles (e.g., umpire in a softball game), and hiding or pretending to play in a game (i.e., pseudo-playing). Similar behaviors were retrospectively reported by the adult participants in Fitzpatrick and Watkinson's (2003) study. The participants explained that the behaviors were purposeful attempts to minimize the risk of exposing their physical awkwardness to peers and teachers. Although the behaviors might be classified as functional avoidance coping strategies that minimize threat and protect against negative emotional experiences, these behaviors lead to further deficits in skill development. Continued avoidance of physical activity by children with movement difficulties results in a practice deficit (Bouffard et al., 1996) that produces, in turn, a developmental skill-learning gap between youth with and without movement difficulties (Wall, 2004). Therefore, as time goes on, children who lack movement competence become increasingly less likely to engage in physical activity.

Mediational Model of Global Self-Worth

Harter's (1987) more recent work on the construct of self-worth has led to the development of her mediational model of global self-worth. Global self-worth is individuals' overall evaluation of their worth or value as persons. Harter suggested that perceptions of competence in domains of importance and perceptions of support from significant others are independent determinants of global self-worth, which, in turn, influences individuals' affective states (i.e., feelings of cheerfulness and happiness or sadness and depression) and motivated behavior. Continued work on the model has led to a number of modifications, including the identification of two distinct clusters of competence domains, as well as positive relationships between the competence domains and different sources of social support (see Harter, 1999). The first cluster of competence includes physical appearance, peer acceptance or likeability, and athletic competence, which are more important for peer support than for parental support. In contrast, the second cluster, composed of scholastic competence and behavioral conduct, is more important for parental support than for peer support.

Given the reported low perceptions of athletic competence and peer support, combined with recent findings by Piek and colleagues (2006) of lower perceptions of scholastic competence, physical appearance, and behavioral conduct, reports of lower self-worth in children and adolescents with mild movement difficulties in comparison to other peers are not surprising (Cantell et al., 1994; Piek et al., 2000; Skinner & Piek, 2001). Piek and colleagues (2006) examined the relationships among perceptions of athletic competence, scholastic competence, and self-worth in children and adolescents with and without DCD. Perceived athletic competence was positively related to self-worth for boys with and without DCD. In contrast, perceived athletic competence was positively related to self-worth for girls with DCD, while perceived scholastic competence was positively related to self-worth for girls with and without DCD. Piek and colleagues suggested that this pattern of relationships among the variables puts girls with DCD at greater risk of developing low self-worth than boys; girls' self-worth may be compromised by gross motor problems (linked to low perceptions of athletic competence) as well as fine motor problems (linked to low perceptions of scholastic competence), whereas boys' self-worth appears to be compromised only by gross motor problems.

Figure 10.1 provides a summary of the research findings related to Harter's models of competence motivation (1978, 1981) and global self-worth (1987, 1999) that pertain to youth with movement difficulties. The social, psychological, and emotional problems associated with movement difficulties lead to withdrawal from physical activity and a skill-learning gap that only exacerbates the movement problems experienced by these youth, triggering a downward spiral of physical inactivity. While this figure depicts the perceptions and experiences of children with movement difficulties in general, as suggested in the opening of this chapter there is evidence that *some* children with DCD and other movement difficulties have positive affective, cognitive, and behavioral experiences in physical activity settings. For example, children with DCD in studies by Piek and colleagues (2000, 2006) reported, despite lower perceptions of athletic competence, perceptions of peer acceptance that were comparable to those of children without DCD. In addition, Smyth and Anderson's (2001) individual-level reexamination of aggregated playground participation data from an earlier paper (i.e., Smyth & Anderson, 2000) revealed a subgroup of children with DCD who actively participated in team games on the playground and were rarely socially isolated. It seems that the movement difficulties experienced by individuals with DCD do not necessarily lead to the scenario depicted in figure 10.1.

What enables some children with movement difficulties such as DCD to report positive self-perceptions, high levels of peer acceptance, and high physical activity engagement levels despite their movement difficulties? According to Harter (1999), unexpectedly favorable self-perceptions in children with disabilities may be attributed to socially desirable responses, unconscious denial,

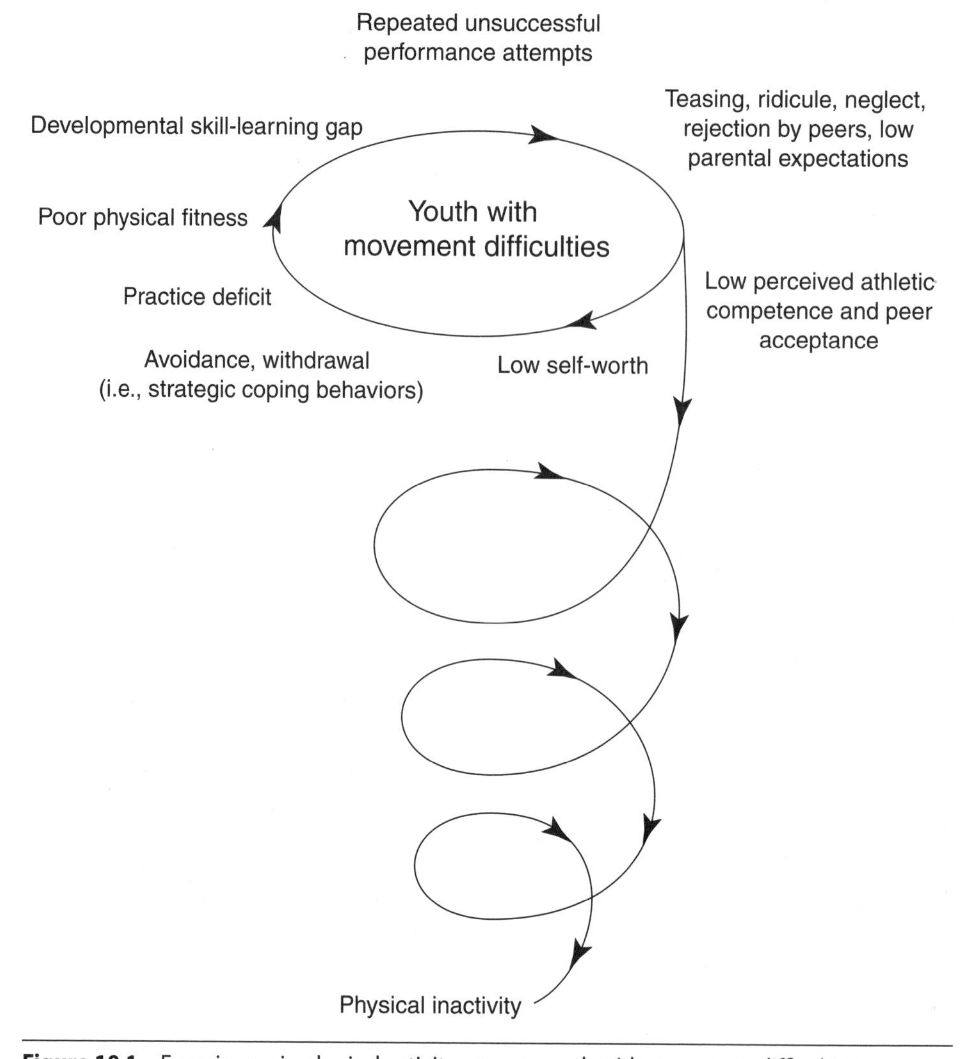

Figure 10.1 Experiences in physical activity among youth with movement difficulties.

confusion between the real and ideal self, comparisons to a similar social reference group, or "healthy adjustment to self-standards" (p. 139). The latter possibility has been explored in studies guided by achievement goal theory (Nicholls, 1989).

Achievement Goal Theory

Like Harter's (1978, 1981) competence motivation theory, achievement goal theory (Nicholls, 1989) assumes that individuals are driven to demonstrate competence, and that perceived competence is a critical determinant of motivated behavior. However, the latter theory assumes that individual differences in evaluating competence and defining success lead to differences in motivation.

Accordingly, this theory emphasizes goals that underlie behavior. Again, chapter 8 offers an extensive discussion of this theory (pp. 198-201).

This theory's distinction between *task goals* and *ego goals* is an important one for youth with movement difficulties. Given that task goals are associated with subjective self-referenced competence evaluations, even individuals who recognize that they are poorly skilled relative to others can maintain positive perceptions of competence and motivated behaviors when they approach physical challenges with task goals in mind. Because ego goals, on the other hand, revolve around social comparison and normative standards, youth with movement difficulties who approach physical activity with ego goals in mind are expected to develop low perceptions of competence and exhibit behaviors intended to prevent a public display of inadequate motor skills (e.g., withdrawal).

The degree to which individuals approach achievement settings with task and ego goals in mind is influenced by dispositional preferences for task and ego goals (goal orientations), perceptions of what goals are important in a given context (motivational climate), and the ability to use social comparison information to evaluate task difficulty and competence (level of cognitive development). Research has shown that most children develop this capability by about 12 years of age; prior to this they tend to rely on subjective, self-referenced information (Nicholls, 1989).

Motivational climates are structured through instructions, feedback, rewards, and explicit expectations that emphasize either task-involving goals (called a *mastery motivational climate*) or ego-involving goals (called a *performance motivational climate;* Ames, 1992; Nicholls, 1989). For example, situations devoid of normatively evaluative cues in which the emphasis is on performing a specific task, skill development, problem solving, or learning will likely promote the perception of a mastery climate. In contrast, situations that emphasize evaluation (e.g., tests of highly valued skills, interpersonal competition, social comparison, and the use of normative feedback) and factors that increase one's public self-awareness (e.g., the presence of an audience or video camera) tend to promote the perception of a performance climate (Nicholls, 1989). The theory suggests that children with movement difficulties would likely benefit the most from exposure to mastery motivational climates.

Causgrove Dunn (2000) examined the relationships between goal orientations, perceptions of the motivational climate, and perceptions of competence in physical education among 4th- to 6th-grade children with movement difficulties. In accordance with theoretical predictions (Nicholls, 1989), task orientation was positively related to perceptions of a mastery motivational climate, which in turn was positively associated with perceptions of competence in physical education. In contrast, ego orientation was positively related to perceptions of a performance motivational climate, which was negatively related to perceived competence in physical education. Findings of a related study (Causgrove Dunn & Dunn, 2006), involving the same children with movement difficulties as in Causgrove Dunn's (2000) original study, showed that participants' perceived competence in a physical education class were positively related to the amount of class time they spent engaged in adaptive behaviors (i.e., assigned tasks) and negatively related to the amount of class time spent engaged in maladaptive behaviors (i.e., off-task behaviors).

An additional finding from this latter study was that although perceptions of the motivational climate did not account for a significant amount of the variance in children's adaptive or maladaptive behaviors, there was a significant interaction effect between perceptions of the performance climate and perceptions of competence. Specifically, for children with movement difficulties who had low perceptions of competence (i.e., 1 SD below the mean), higher perceptions of a performance climate in physical education were associated with less adaptive and more maladaptive behavior during physical education. For children with high perceptions of competence (i.e., 1 SD above the mean), the perception of a performance climate did not have a significant impact on behavior. Although there is only limited evidence to date, studies grounded in achievement goal theory suggest that an emphasis on mastery rather than interpersonal comparisons or interpersonal competition (at least in integrated settings) may lead to positive self-perceptions and functional participation behaviors for young people with movement difficulties.

Self-Determination Theory

Self-determination theory—discussed extensively in chapter 8 (pp. 203-205)—has received very little attention from researchers examining physical activity patterns of youth with disabilities. However, on the basis of the findings of a growing number of investigations in the adapted physical activity area (e.g., Goodwin, 2001; Goodwin & Watkinson, 2000; Hutzler et al., 2002), we believe that self-determination theory (SDT) has the potential to be a very useful theoretical framework to explain the motivational factors underlying many of the inactive or sedentary behaviors of children and adolescents with disabilities.

Although the studies cited in the previous paragraph were not conducted within the specific context of an SDT framework, their findings are in accordance with many of the theoretical tenets of SDT. Specifically, these studies have shown that meaningful and positive experiences in physical activity for individuals with movement difficulties are characterized by perceptions of competence, the provision of appropriate support to enable and enhance participation, and a sense of belonging. For example, Goodwin and Watkinson (2000) conducted focus group interviews with nine youth who had physical disabilities (spina bifida, cerebral palsy, or amputation). Participants were asked about their experiences in integrated physical education. A number of themes were identified in the participants' responses, with experiences differing greatly on "bad days" and "good days." "Bad days" were experienced when participants felt socially isolated (i.e., were ridiculed, ignored, or perceived as objects of curiosity), their competence was questioned, or opportunities to participate were limited either through lack of support by the teacher, lack of engagement by classmates, or equipment or facility constraints. In contrast, "good days" were characterized by perceptions of skillful participation and belonging and by the provision of support needed in order to share in the benefits of physical activity. For these youth, it seemed that "bad days" and "good days" in physical education were associated with conditions that either thwarted or satisfied the basic psychological needs identified in SDT.

From the perspective of SDT, choice is a key determinant of intrinsic motivation. An individual's level of intrinsic motivation is likely to be enhanced if the individual perceives that the activity is freely chosen rather than externally imposed. This was apparent in a study of adherence to a six-week home-based strength training program for children and adolescents with cerebral palsy (Taylor et al., 2004). Interviews with participants and their parents at the conclusion of the program revealed that the provision of choice to engage in the program or not was the most important personal factor affecting adherence.

While it is clear that the provision of choice is important, the impact of an individual's perceived competence upon perceptions of choice or autonomy cannot be ignored, especially for children with movement difficulties. Consider a situation in which a teacher of an inclusive physical education program attempts to increase students' perceptions of autonomy by setting up different activity options around the gymnasium and then allowing each child to select from them. If an individual with a movement difficulty believes that he or she cannot do an activity successfully, or cannot do it in the same way as his or her peers, is that activity truly perceived as an option? This question clearly presents a challenge for physical education professionals who may wish to employ the tenets of SDT when attempting to create inclusive environments that cater to skills and abilities of all students.

Environmental Issues

The theoretical approaches discussed to this point focus primarily on personal factors (e.g., perceived competence, global self-worth, goal orientations, and perceptions of autonomy) that facilitate or constrain physical activity participation. However, there are other important supporting and limiting factors in the environment that influence physical activity participation of youth with disabilities. Among the most obvious of these are the physical characteristics of the environment, such as whether there is a ramp or elevator to enable access to a community facility for youth who use wheelchairs for locomotion. Less obvious perhaps is the impact of the psychosocial environment, which is the focus of our discussion in this section. For example, the motivational climate (discussed previously in relation to achievement goal theory) affects youths' achievement goals, perceptions of competence, and therefore their motivated behaviors. The attitudes of others toward youth with disabilities can influence opportunities for participation, expectations for involvement, and the degree of support provided (e.g., instruction, equipment, parental overprotection). Finally, as the ideology of inclusion is more widely implemented, youth with disabilities may find that they have little opportunity to interact with others with disabilities in school or community physical activity settings, which will influence their understanding of their physical potential in movement contexts.

Attitudes Toward Persons With Disabilities

Attitudinal factors have been among the most influential barriers to participation in physical activity faced by persons with disabilities (Rimmer et al., 2004). Attitudes are a product of beliefs about persons with disabilities, which subsequently influence behaviors toward them. Historically, the societal view of disability emphasizes loss of function, increased dependency, sickness, depression, vulnerability, weakness, and helplessness (Fine & Asch, 1988). Such a view has lowered expectations for persons with disabilities and engendered little motivation in (a) educators and health professionals to promote or reinforce self-reliance, independence, or physical well-being, and (b) researchers to study topics related to health promotion of people with disabilities (Hughes & Paterson, 1997; Stevens et al., 1996). Negative attitudes toward people with disabilities frequently result in negative, paternalistic, or apathetic behaviors toward them (Kennedy, Austin, & Smith, 1987), which in turn lower the self-perceptions and limit the physical activity opportunities of such individuals.

Societal beliefs about disability have certainly restricted opportunities for youth with disabilities to be physically active (Rimmer et al., 2004; Steele et al., 1996). For example, there is a dearth of accessible fitness and recreation programs and facilities, particularly in rural areas, for people with disabilities (Cooper et al., 1999; Rimmer et al., 2004). Fitness and recreation professionals and facility owners may view accessibility as unnecessary, believing instead that people with disabilities are unable to pursue or uninterested in pursuing activity programs (Rimmer et al., 2004). Not surprisingly, the majority of engagement in extracurricular physical activity by youth with disabilities occurs at home with family members (Rosser Sandt & Frey, 2005; Steele et al., 1996). This is in contrast to the engagement in community-based activities and organized sport more typical of peers without disabilities. Family members often facilitate participation by providing encouragement and by creating opportunities for youth with disabilities to engage in activities (Goodwin et al., 2004; Taylor et al., 2004). On the other hand, overprotection by parents (or caregivers) and other family members has also been identified as a prominent reason why young people with disabilities often do not participate in physical activity or play (Longmuir & Bar-Or, 1994; Rimmer et al., 2004; Steele et al., 1996).

In the context of physical education, children identified as low achievers tend to receive more criticism, less help, and less encouragement or praise from teachers in physical education classes than high achievers (Martinek & Karper,

1986; Portman, 1995). These factors undoubtedly function to limit skill development and participation motivation, particularly given the evidence that children recognize this differential treatment from teachers and interpret its meaning accordingly (Portman, 1995). According to findings by Goodwin and Watkinson, children with disabilities felt that teachers and peers engaged in behaviors that restricted opportunities for children with disabilities in physical education. Nevertheless, it should be recognized that teachers and peers on occasion also engaged in behaviors that supported or facilitated involvement for children with disabilities (Goodwin & Watkinson, 2000; Taub & Greer, 2000).

A key to understanding the physical activity context for youth with disabilities is to view their participation from the vantage of their perspective and the goals that may be set out by the instructor. For example, help can be interpreted as supportive if it is instrumental and assists in completion of the activity (e.g., retrieval of a piece of equipment that has rolled away so play can continue) or threatening if it is interpreted as having been offered because of perceptions of incompetence (e.g., a peer helping to cover a goal area when help is not wanted) (Goodwin, 2001).

Societal assumptions of dependency, sickness, and lack of productivity are slowly changing to encompass views of persons with disabilities as empowered, self-determined, and self-actualized. Recent models of disability emphasize the interaction of impairment, physical environment, and social contexts, thereby shifting responsibility for inactivity away from persons with disabilities and toward the impact of physical or psychosocial environmental barriers.

Ironically, participation in sport and other physical activities appears to be an effective means by which youth with physical disabilities can positively affect the perceptions and beliefs of others (Goodwin et al., 2004; Goodwin & Watkinson, 2000; Taub & Greer, 2000). People who are physically active tend to be perceived as healthy, vibrant, and able; and youth with disabilities have perceived or experienced these benefits when they participated in physical activity settings in the presence of others (Goodwin et al., 2004). Thus, it appears that societal attitudes toward people with disabilities, and the barriers they present, may be best overcome by increasing opportunities for people with disabilities to participate and achieve in a variety of physical activity settings.

Inclusive Versus Specialized Settings

Youth with disabilities participate in physical activity with youth who do not have disabilities (i.e., integrated or inclusive settings) and in settings with exclusively other individuals with disabilities (i.e., segregated or specialized settings). Considerable research attention has been devoted to comparing the attitudes, experiences, and performances of instructors and youth in segregated or specialized versus integrated or inclusive settings. Collectively, researchers in this area have attempted to determine if the conditions (e.g., teacher attitude, student ability, peer acceptance, programming) in one setting are more beneficial for individuals with disabilities than conditions of the other setting. Results have led to cautious claims about the benefits of inclusion. Unfortunately, by focusing on (a) how one "type" of learner affects other learners, (b) what constitutes a quality physical activity program (e.g., curriculum, instructor preparation), and (c) the attitudes of those teaching the programs, researchers have historically devoted less attention to examining how to best *support* children and adolescents with disabilities in physical activity programs. Nonetheless, a trend toward studying *how to* program effectively for students with disabilities in mainstream physical activity programs has been more evident in recent years.

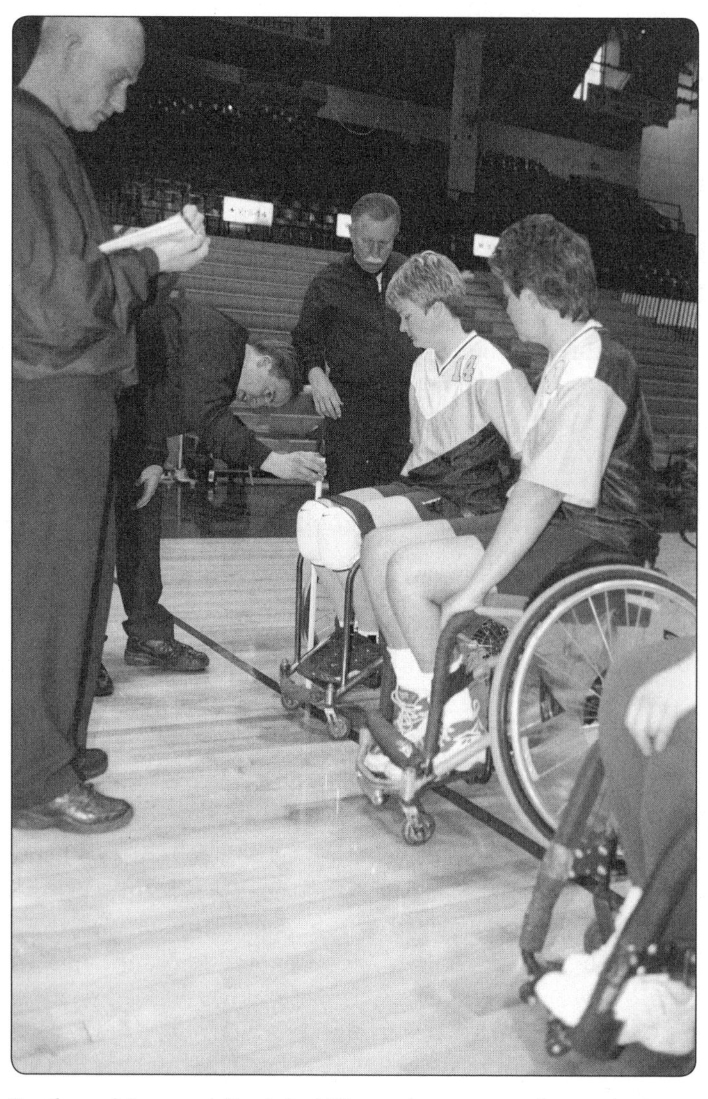

Youth receiving specialized disability equipment sport instruction.

Although youth with disabilities are increasingly being integrated into mainstream physical education and other physical activity settings at school, discussions of inclusion in physical education have not been without controversy. Block (1999), a strong advocate of inclusion for all students in physical education, gradually came to question the success of inclusive physical education programs. He suggested that *inclusion zealots* who did not accept anything less than full inclusion of all students with disabilities may have overlooked the needs of the student in their efforts to promote the philosophy of inclusion. Upon reflection, Block suggested that the assumptions held about regular physical education programs (e.g., that individualized instruction occurs, that teachers hold positive attitudes about inclusion of students with disabilities, that professional support and training are available to teachers) may be inaccurate, thereby making the implementation of inclusive physical education far more difficult than originally believed.

Although inclusion in physical activity settings can lead to numerous benefits for children with disabilities (e.g., increased physical activity levels, improved performance, expanded social networks, improved perceptions of peer acceptance and belonging, increased perceived competence), research suggests that placement in the environment does not guarantee these outcomes in itself (Goodwin & Watkinson, 2000). Indeed, the successful and safe inclusion of youth with different ability levels (and disabilities) in mainstream physical education, community sport, and recreation programs often requires adaptations and modifications to content, instructional strategies, techniques, game rules, and the environment or individual accommodations (e.g., use of wheelchairs) (Goodwin & Watkinson, 2000; Kalyvas & Reid, 2003; Taub & Greer, 2000).

Appropriate adaptations to sports and games can facilitate active and successful participation for children with disabilities *and* those without disabilities. For instance, Kalyvas and Reid (2003) found that children aged 7 to 12 years with and without disabilities were more active, were less inactive, and had more success executing skills during an adapted version of Newcomb (a lead-up game for volleyball) compared to a nonadapted game. Participants with disabilities and the younger children without disabilities enjoyed the adapted game more, though most also enjoyed the nonadapted game. In contrast, the older students (10- to 12-year-olds) without disabilities, especially the boys, found the adapted game less enjoyable, less challenging, and less interesting than the nonadapted game. Clearly, the motivational and affective experiences of all participants must be considered when activity settings are modified. The challenge facing practitioners is to find a way of structuring the physical activity environment such that positive motivational and affective opportunities are afforded to all participants. The use of adaptations should be considered a dynamic process, and adaptations may therefore need to be adjusted over time (e.g., to alter the challenge for higher- or lower-skilled participants).

As support for inclusion has grown, specialized programs such as the Special Olympics have been increasingly criticized for providing segregated sport

Participation in inclusive physical education class may require modifications or adaptations. Dance and creative movement activities offer many opportunities to accommodate varying movement ability levels and foster a cooperative activity environment.

training and competitions, and there are fewer opportunities to engage in segregated physical activity. At the same time, research on the experiences of youth with disabilities in specialized or segregated programs from the perspective of the participants has revealed important information about self-perceptions, social isolation, and sense of community (Taub & Greer, 2000). For example, Groff and Kleiber (2001) found that children with disabilities who were able to participate in physical activity with other peers with disabilities expressed a sense of freedom at being able to feel that they could "be themselves" and to experience a sense of community.

Similarly, Goodwin and Staples (2005) investigated the meaning of segregated or specialized summer camp experiences among nine youth with disabilities aged 14 to 19 years. Thematic analysis of data provided by semistructured interviews, document review, and field notes revealed that camp experiences provided the participants with a reprieve from perceptions of disability isolation that were often experienced in their home communities. The campers also experienced an increased sense of self-reliance and independence and believed that they developed new understandings of their physical potential. Through their recreational time and physical activity experiences at the camp, participants also changed their own views of what persons with disabilities were capable of achieving. The external influence of time spent with others who had disabilities appeared to positively influence the internal agency of participants and their own self-definition. They moved beyond physical activity limits previously achieved, in part because of expectations set by their peers with disabilities. After observing the successes of other youth with disabilities more significant than their own, the youth undertook greater challenges than they had previously been willing to accept.

Overall, it appears that segregated *and* integrated physical activity settings have value in the promotion of physically active lifestyles among youth with disabilities. In fact, the positive or negative impact of the setting on self-perceptions, affect, and physical activity behaviors of youth with disabilities seems to be primarily related to factors within the settings (e.g., attitudes, provision of support). That is, it is less the type of setting than the way in which the setting is constituted that dictates physical and psychosocial outcomes for youth with disabilities.

Intervention Research

Health promotion for persons with disabilities has only recently gained attention (Hogan, McLellan, & Bauman, 2000; Steele et al., 1996). Therefore, physical activity interventions for youth with disabilities, as well as studies of the effectiveness of these interventions, are relatively rare. Most of the research incorporating physical activity programs or interventions has been conducted for the purpose of identifying the health benefits of physical activity engagement for youth with disabilities. For example, Darrah and colleagues (1999) examined the effects on adolescents with cerebral palsy of a 10-week community fitness program that included aerobics, strength training, and stretching and found significant improvements in muscle strength and perceptions of physical appearance.

Many intervention strategies aimed at increasing activity levels among people with movement difficulties have involved attempts to increase movement competence. Underlying these interventions is the assumption that improved motor performance will lead to improved self-perceptions (e.g., perceived athletic competence) that in turn will increase participation in physical activities (Harter, 1978, 1999). Although there is growing evidence that task-specific approaches can improve performance on the specifically trained skills among samples of children with DCD, few studies have shown that the interventions result in positive changes in perceptions of athletic competence or global self-worth (Causgrove Dunn et al., 2005). Moreover, though performance in specific skills improves, DCD is not "cured" in that the general coordination difficulties are not eliminated.

Based on current hierarchical models of the self system (e.g., Harter, 1999), it may be unreasonable to expect that improvements in the performance of a few discrete motor skills is sufficient to change the perceptions of athletic competence of a young person with DCD. Rather, it would seem more likely that improved performance of a particular skill or task will alter (a) the individual's task- or skill-specific perceptions of competence and (b) her or his willingness to engage in that specific task in the future. In other words, learning to ride a bicycle may not significantly alter the overall perceptions of athletic competence of a child with DCD, but there is some evidence that such learning enhances the intrinsic desire to go riding, thereby increasing activity levels (Iversen et al., 2005; Mandich et al., 2003). Moreover, the acquisition and subsequent enjoyment of an age-appropriate and culturally valued specific skill like bicycling positively affect socialization opportunities with peers (Iversen et al., 2005), which may enhance perceptions of self-worth. Finally, successful mastery of a valued motor skill may positively influence a child's willingness to attempt mastery of other physical skills or activities (Iversen et al., 2005; Mandich et al., 2003).

While most physical activity interventions for children with movement difficulties have targeted motor performance, interventions aimed at the sociomotivational climate of the learning or performance environment can also help to combat withdrawal from physical activity settings or inactivity among children with movement difficulties. The assumption here is that changes in the way physical activity situations are structured will influence the way individuals define success and ability (e.g., in a task-involved manner rather than an ego-involved manner). A situation constructed to emphasize task goals will improve individuals' perceptions of competence and, in turn, increase their physical activity levels.

For example, Valentini and Rudisill (2004) reported the results of two intervention studies that examined the effects of a mastery climate on motor skill development and perceived competence of kindergarten-aged children with developmentally delayed motor skills. In both studies, greater improvements in locomotor and object-control skills were observed over the course of the 12-week program in the mastery climate intervention group than in the low-autonomy climate comparison group. At postintervention periods in both studies, the mastery climate group also outperformed the comparison group on the locomotor skills test. Indeed, a six-month follow-up assessment in the second study revealed that the mastery group maintained their skill development in both skill areas whereas the low-autonomy group's performances decreased. The perceptions of physical competence of the mastery climate group increased from pre- to postintervention in the first study while those of the comparison group did not. Also, in the second study the mastery climate group reported significant improvements in perceptions of physical competence following the intervention. These results support the efficacy of employing mastery climate interventions for the purpose of improving the perceptions of athletic or physical competence among children with movement difficulties.

Recommendations for Future Research

Although there is general agreement that regular physical activity involvement is important for reasons of health and well-being, our knowledge about why the majority of people with disabilities do not engage in physically active lifestyles is still quite limited (Rimmer et al., 2004). This chapter covered a number of theoretically important psychosocial factors associated with the physical activity involvement of youth with movement difficulties in the context of the available empirical evidence. Nevertheless, studies that expand our understanding of these factors and research that identifies additional influential factors are still needed.

Both qualitative (e.g., Goodwin & Watkinson, 2000) and quantitative (e.g., Causgrove Dunn & Dunn, 2006) studies point to the importance of perceived competence to physical activity behaviors of youth with disabilities. Although there is some evidence that perceptions of athletic competence of children with movement difficulties are amenable to motor skill interventions (Valentini & Rudisill, 2004), research is required to determine whether performance gains made in specific skills will alter overall motivation and physical activity patterns.

Longitudinal studies should be undertaken to carefully examine age-related changes in self-perceptions, sources of competence information, and conceptions of ability demonstrated by youth with movement disabilities. Harter (1999) herself has cautioned researchers about assuming that current frameworks of self-perceptions and assessment procedures developed with normative samples of typically developing Western youth are appropriate for use with different populations (such as those with movement difficulties). Moreover, longitudinal examination of age-related changes in the *relationships* between self-perceptions and physical activity involvement of youth with disabilities may identify if there are any key points of change that trigger one trajectory (i.e., physical activity withdrawal) or another (i.e., physical activity participation).

There is evidence that the physical activity levels of youth with disabilities vary by types of disabilities (Longmuir & Bar-Or, 2000). Consequently, researchers should not assume that the psychosocial factors affecting physical activity involvement are equivalent across disabilities. Certain factors (e.g., perceptions of belonging, support from parents) may be more relevant or important for youth with some types of disabilities than for others, and so different interventions to increase physical activity involvement may be needed. Individual variance can be dramatic, even among people with the same disability type, due to the interaction of the person and her or his disability with the environment and the nature of the physical activity. Other individual difference variables (e.g., gender, severity of disability, age at which a disability was acquired, past athletic experience) may also affect physical activity of youth with disabilities.

For example, Cairney and colleagues (2005) suggested that the social context surrounding girls with DCD may be different from that for boys with DCD so that different strategies may be required.

Ultimately, effective intervention strategies to promote physical activity among youth with movement difficulties (and other types of disabilities) are needed. Interventions designed to improve individuals' functional motor skills, especially those skills needed for participation in common everyday activities of childhood and adolescence (e.g., ball skills, bike riding), should continue to be developed and tested. The nature and severity of the motor disorder will dictate the substance of the intervention. For example, individuals with mild motor disorders may receive intense and specific instruction on traditional sport and game skills, whereas youth with more severe disabilities (e.g., those who require a wheelchair for mobility) may choose to receive specialized instruction in advanced wheelchair skills to facilitate participation in wheelchair basketball. Interventions also aimed at optimizing the psychosocial environment look promising (Valentini & Rudisill, 2004). Generally speaking, it seems that interventions such as these should not only increase perceptions of competence, but also facilitate increased perceptions of belonging and autonomy. Curricular adaptations, instructional modifications, and accommodations may also be necessary to maximize the motivational effects of interventions. There is clearly much to consider in future efforts to enhance the physical activity patterns of youth with disabilities.

APPLICATIONS FOR RESEARCHERS

Despite mounting evidence that many youth with movement difficulties have sedentary lifestyles and are at increased risk of negative health outcomes, including secondary disabilities, research on factors contributing to low physical activity levels among these youth remains rare. To date, theoretically driven studies have indicated the usefulness of several current motivation theories as frameworks for studying inactivity among these youth. Nevertheless, this work is just beginning and should expand to include new theoretical perspectives and variables. Developing effective interventions to increase physical activity motivation and behavior among these youth requires greater understanding of the complex network of individual *and* environmental factors surrounding participation. Greater emphasis on the influence of learning and performance environments on the physical activity perceptions and participation of these young people, with use of ecological models that consider both individual and contextual factors, is warranted. This would enable further investigation of such effects as the relative benefits of integrated and segregated settings on participation levels.

APPLICATIONS FOR PROFESSIONALS

Instructional climates emphasizing motor performance standards and competition can remove choices for youth with disabilities, thereby compromising perceptions of competence, autonomy, and belonging, along with motivation to participate. Avoidance behaviors such as repeated requests for water or bathroom breaks, misbehavior, or feigned injury may signal that the setting is causing participants to question their competence. Feelings of inadequacy reinforced by exclusion or negative peer feedback can interfere with a sense of belonging, social connectedness, and feelings of self-worth.

Traditional standards of motor performance should be broadened to incorporate individual differences among youth with and without movement disorders. When feedback focuses on improvement and skill mastery, even in competitive activities, participants' attention is directed toward their own performances and learning. This may foster positive self-standards, peer acceptance, and identity development for youth with movement difficulties. Moreover, providing choice (e.g., activity or equipment options) offers opportunities for all participants to engage in meaningful and challenging activities.

References

American Psychiatric Association. (2000). *DSM-IV-TR Diagnostic and statistical manual of mental disorders* (4th ed., text revision). Retrieved from www.psychiatryonline.com/content.aspx?aID=7390#7390.

Ames, C. (1992). Achievement goals, motivational climate, and motivational processes. In G.C. Roberts (Ed.), *Motivation in sport and exercise* (pp. 161-176). Champaign, IL: Human Kinetics.

Block, M. (1999). Did we jump on the wrong bandwagon? Part I: Problems with inclusion in physical education. *Palaestra, 15*(3), 30-36.

Bouffard, M., Watkinson, E.J., Thompson, L.P., Causgrove Dunn, J.L., & Romanow, S.K.E. (1996). A test of the activity deficit hypothesis with children with movement difficulties. *Adapted Physical Activity Quarterly, 13*, 61-73.

Cairney, J., Hay, J., Faught, B., Mandigo, J., & Flouris, A. (2005). Developmental coordination disorder, self-efficacy toward physical activity and play: Does gender matter? *Adapted Physical Activity Quarterly, 22*, 67-82.

Cantell, M.H., Smyth, M.M., & Ahonen, T.P. (1994). Clumsiness in adolescence: Educational, motor, and social outcomes of motor delay detected at 5 years. *Adapted Physical Activity Quarterly, 13*, 115-129.

Causgrove Dunn, J. (2000). Goal orientations, perceptions of the motivational climate, and perceived competence of children with movement difficulties. *Adapted Physical Activity Quarterly, 17*, 1-19.

Causgrove Dunn, J., & Dunn, J.G.H. (2006). Psychosocial determinants of physical education behavior in children with movement difficulties. *Adapted Physical Activity Quarterly, 23*, 293-309.

Causgrove Dunn, J., Magill-Evans, J., Klein, S., & Cavaliere, N. (2005). *A systematic review of motor skill interventions for children with DCD.* Paper presented at the 6th International Conference on Children with DCD, Trieste, Italy.

Cooper, R.A., Quatrano, L.A., Axelson, P.W., Harlan, W., Stineman, M., Franklin, B., et al. (1999). Research on physical activity and health among people with disabilities: A consensus statement. *Journal of Rehabilitation Research and Development, 36*, 142-154.

Darrah, J., Wessel, J., Nearingburg, P., & O'Connor, M. (1999). Evaluation of a community fitness program for adolescents with cerebral palsy. *Pediatric Physical Therapy, 11*, 18-23.

Durstine, J.L., Painter, P., Franklin, B.A., Morgan, D., Pitetti, K.H., & Roberts, S.O. (2000). Physical activity for the chronically ill and disabled. *Sports Medicine, 30*, 207-209.

Evans, J., & Roberts, G.C. (1987). Physical competence and the development of children's peer relations. *Quest, 39*, 23-35.

Fine, M., & Asch, A. (1988). Disability beyond stigma: Social interaction, discrimination, and activism. *Journal of Social Issues, 44*, 3-21.

Fitzpatrick, D.A., & Watkinson, E.J. (2003). The lived experience of physical awkwardness: Adults' retrospective views. *Adapted Physical Activity Quarterly, 20*, 279-297.

Goodwin, D.L. (2001). The meaning of help in PE: Perceptions of students with physical disabilities. *Adapted Physical Activity Quarterly, 18*, 289-303.

Goodwin, D.L., & Staples, K. (2005). The meaning of summer camp experiences to youth with disabilities. *Adapted Physical Activity Quarterly, 22*, 160-178.

Goodwin, D.L., Thurmeier, R., & Gustafson, P. (2004). Reactions to metaphors of disability: The mediating effects of physical activity. *Adapted Physical Activity Quarterly, 21*, 379-398.

Goodwin, D.L., & Watkinson, E.J. (2000). Inclusive physical education from the perspective of students with physical disabilities. *Adapted Physical Activity Quarterly, 17*, 144-160.

Groff, D., & Kleiber, D. (2001). Exploring the identity formation of youth involved in an adapted sports program. *Therapeutic Recreation Journal, 35*, 318-332.

Hands, B., & Larkin, D. (2002). Physical fitness and developmental coordination disorder. In S.A. Cermak & D. Larkin (Eds.), *Developmental coordination disorder* (pp. 172-184). Albany, NY: Delmar.

Harter, S. (1978). Effectance motivation reconsidered: Toward a developmental model. *Human Development, 21*, 34-64.

Harter, S. (1981). A model of intrinsic mastery motivation in children: Individual differences and developmental change. In W.A. Collins (Ed.), *Minnesota symposium on child psychology* (Vol. 14, pp. 215-255). Hillsdale, NJ: Erlbaum.

Harter, S. (1987). The determinants and mediational role of global self-worth in children. In N. Eisenberg (Ed.), *Contemporary topics in developmental psychology* (pp. 219-242). New York: Wiley.

Harter, S. (1999). *The construction of the self: A developmental perspective.* New York: Guilford Press.

Hogan, A., McLellan, L., & Bauman, A. (2000). Health promotion needs of young people with disabilities: A population study. *Disability and Rehabilitation, 22,* 352-357.

Horn, T.S. (2004). Developmental perspectives on self-perceptions in children and adolescents. In M.R. Weiss (Ed.), *Developmental sport and exercise psychology* (pp. 101-144). Morgantown, WV: Fitness Information Technology.

Hughes, B., & Paterson, K. (1997). The social model of disability and the disappearing body: Towards a sociology of impairment. *Disability and Society, 12,* 325-340.

Hutzler, Y., Fleiss, O., Chacham, A., & Vanden Auweele, Y. (2002). Perspectives of children with physical disabilities on inclusion and empowerment: Supporting and limiting factors. *Adapted Physical Activity Quarterly, 19,* 300-317.

Iversen, S., Ellertsen, B., Tytlandsvik, A., & Nødland, M. (2005). Intervention for 6-year-old children with motor coordination difficulties: Parental perspectives at follow-up in middle childhood. *Advances in Physiotherapy,* 7, 67-76.

Kalyvas, V., & Reid, G. (2003). Sport adaptation, participation, and enjoyment of students with and without physical disabilities. *Adapted Physical Activity Quarterly, 20,* 182-199.

Kennedy, D.W., Austin, D.R., & Smith, R.W. (1987). *Special recreation: Opportunities for persons with disabilities.* New York: Sanders College Publishing.

Longmuir, P.E., & Bar-Or, O. (1994). Physical activity of children and adolescents with a disability: Methodology and effects of age and gender. *Pediatric Exercise Science, 6,* 168-177.

Longmuir, P.E., & Bar-Or, O. (2000). Factors influencing the physical activity levels of youths with physical and sensory disabilities. *Adapted Physical Activity Quarterly, 17,* 40-53.

Mandich, A.D., Polatajko, H.J., & Rodger, S. (2003). Rites of passage: Understanding participation of children with developmental coordination disorder. *Human Movement Science, 22,* 583-595.

Martinek, T., & Karper, W. (1986). Motor ability and instructional contexts: Effects on teacher expectation and dyadic interactions in elementary physical education classes. *Journal of Classroom Interaction, 21,* 16-25.

Nicholls, J.G. (1989). *The competitive ethos and democratic education.* Cambridge, MA: Harvard University Press.

Piek, J.P., Baynam, G.B., & Barrett, N.C. (2006). The relationship between fine and gross motor ability, self-perceptions and self-worth in children and adolescents. *Human Movement Science, 25,* 65-75.

Piek, J.P., Dworkan, M., Barrett, N.C., & Coleman, R. (2000). Determinants of self-worth in children with and without Developmental Coordination Disorder. *International Journal of Disability Development and Education, 47,* 259-271.

Portman, P.A. (1995). Coping behaviors of low-skilled students in physical education: Avoid, announce, act out, and accept. *Physical Educator, 52,* 29-39.

Rimmer, J.H., Riley, B., Wang, E., Rauworth, A., & Jurkowski, J. (2004). Physical activity participation among persons with disabilities: Barriers and facilitators. *American Journal of Preventive Medicine, 26,* 419-425.

Rose, B., Larkin, D., & Berger, B.G. (1997). Coordination and gender influences on the perceived competence of children. *Adapted Physical Activity Quarterly, 14*, 210-221.

Rosser Sandt, D.D., & Frey, G.C. (2005). Comparison of physical activity levels between children with and without autistic spectrum disorders. *Adapted Physical Activity Quarterly, 22*, 146-159.

Sherrill, C. (2004). *Adapted physical activity, recreation, and sport. Crossdisciplinary and lifespan* (6th ed.). New York: McGraw-Hill.

Simeonsson, R.J., Sturtz McMillen, J., & Huntington, G.S. (2002). Secondary conditions in children with disabilities: Spina bifida as a case example. *Mental Retardation and Developmental Disabilities Research Reviews, 8*, 198-205.

Skinner, R.A., & Piek, J.P. (2001). Psychosocial implications of poor motor coordination in children and adolescents. *Human Movement Science, 20*, 73-94.

Smyth, M.M., & Anderson, H.I. (2000). Coping with clumsiness in the school playground: Social and physical play in children with coordination impairments. *British Journal of Developmental Psychology, 18*, 389-413.

Smyth, M.M., & Anderson, H.I. (2001). Football participation in the primary school playground: The role of coordination impairments. *British Journal of Developmental Psychology, 19*, 369-379.

Steele, C.A., Kalnins, I.V., Jutai, J.W., Stevens, S.E., Bortolussi, J.A., & Biggar, W.D. (1996). Lifestyle health behaviors of 11- to 16-year-old youth with physical disabilities. *Health Education Research, 11*, 173-186.

Stevens, S.E., Steele, C.A., Jutai, J.W., Kalnins, I.V., Bortolussi, J.A., & Biggar, D. (1996). The psychosocial health of 11 to 16-year-old youth with physical disabilities. *Journal of Adolescent Health, 19*, 157-164.

Taub, D.E., & Greer, K.R. (2000). Physical activity as a normalizing experience for school-age children with physical disability. *Journal of Sport and Social Issues, 24*, 395-414.

Taylor, N.F., Dodd, K.J., McBurney, H., & Graham, K. (2004). Factors influencing adherence to a home-based strength-training programme for young people with cerebral palsy. *Physiotherapy, 90*, 57-63.

Thompson, L.P., Bouffard, M., Watkinson, E.J., & Causgrove Dunn, J. (1994). Teaching children with movement difficulties: Highlighting the need for individualised instruction in regular physical education. *Physical Education Review, 17*, 152-159.

Valentini, N., & Rudisill, M. (2004). Motivational climate, motor-skill development, and perceived competence: Two studies of developmentally delayed kindergarten children. *Journal of Teaching in Physical Education, 23*, 216-234.

van den Berg-Emons, H.J.G., Bussmann, J.B.J., Meyerink, H.J., Roebroeck, M.E., & Stam, H.J. (2003). Body fat, fitness and level of everyday physical activity in adolescents and young adults with meningomyelocele. *Journal of Rehabilitation Medicine, 35*, 271-275.

van der Ploeg, H.P., van der Beek, A.J., van der Woude, L.H.V., & van Mechelen, W. (2004). Physical activity for people with a disability. *Sports Medicine, 34*, 639-649.

Wall, A.E. (2004). The developmental skill-learning gap hypothesis: Implications for children with movement difficulties. *Adapted Physical Activity Quarterly*, *21*, 197-218.

World Health Organization. (2001). ICF. *The international classification of functioning, disability and health (introduction).* Retrieved May 24, 2006, from www3.who.int/icf/icftemplate.cfm?myurl=introduction.html&mytitle=introduction.

PART III

Social and Contextual Factors in Youth Physical Activity and Sedentary Behavior

Part I having provided key foundation messages, and part II having offered a comprehensive account of personal factors likely to be associated with physical activity and sedentary behavior in young people, this section of the book takes us beyond the individual into the wider social and contextual aspects of physically active and sedentary living in young people. To start, two key social environments for youth activity are addressed in chapters on the family (chapter 11) and peers (chapter 12). These chapters review the conceptual and empirical bases for understanding the potential influence of these environments on young people's physical activity and sedentary behavior.

Moving from these social environments, coverage of the important contexts of school (chapter 13), out-of-school sport programs (chapter 14), and community physical activity (chapter 15) is provided before chapter 16 addresses the wider physical environment. These chapters highlight the importance of the diversity of contexts and environments that until recently were neglected in the study of youth physical activity and sedentary behavior. Finally, part III concludes with novel chapters for this field that address economic (chapter 17) and cultural (chapter 18) perspectives. Economists not only describe how physical activity levels are determined in the context of all other consumption decisions but also consider time allocation decisions. These could be made by both parents and their children in the context of youth physical activity. Finally, culturally sensitive research is highlighted as the context of ethnicity, including immigrant communities, is addressed. This is important for a more complete understanding of youth physical activity and sedentary behavior.

CHAPTER

11

The Family

Brian E. Saelens, PhD ▪ Jacqueline Kerr, PhD

Other things may change us, but we start and end with family
—Anthony Brandt

Children's physical activity levels and sedentary behavior are influenced by many factors, ranging from internal factors including the biologic and cognitive (e.g., attitudes, beliefs) to more macro-level factors outside the individual, such as policy (e.g., school physical education policy). Ecological models (Sallis & Owen, 1997), inherently multilevel, highlight the various levels and many potential factors within each level of influence, with family influences a ubiquitous part of these models. Family influences are often depicted in ecological models as among the most proximal influences on children's behavior or health status (Davison & Birch, 2001). Only characteristics such as a child's biologic factors or a child's own cognitions, attitudes, and affect about activity (themselves likely impacted by the family) are considered more proximal factors influencing physical activity and sedentary behavior. Other factors, including peers, the nonhome environment, and more macro-level factors, are considered more distal influences than family on the physical activity of young people. The magnitudes of influences on youth physical activity and inactivity are likely affected by developmental stage, but the prominent placement of family factors in theoretical models highlights their importance (Duncan et al., 2004).

The present chapter explores the role that family factors play in young people's activity, both physical activity and sedentary behavior. For the purposes of this chapter, a child's family is defined as any living immediate biologic relative (e.g., parent or caregiver, siblings) or other individual with whom a child lives. However, the majority of research focuses upon parental factors rather than siblings or the role that other nonparental family members have in affecting children's physical activity or sedentary behavior. Family factors are broadly considered; they include behaviors, attitudes, and other characteristics stemming from family sources either individually (e.g., mother's physical activity as an influence on child's activity) or collectively (e.g., home environment as an influence on child's activity). We examine youth activity ranging from overall physical and sedentary activity to context-specific activity, including active commuting and organized sport.

We also describe correlational and intervention research that constitutes a sizable empirical literature on various familial factors and their relations to children's physical activity. In most early studies the aim was to find familial determinants of children's total or *overall* physical activity, but more recent ecological models encourage examination of context-specific behaviors and their influences (Giles-Corti et al., 2005; Ommundsen et al., 2006). These models attempt to parse children's physical activity into various contexts or types, for example leisure-time nonorganized physical activity, active commuting to school, organized physical activity (e.g., sport), and sedentary behavior. We first consider familial influences on children's overall physical activity and then examine familial influences among these specific types or contexts for child activity. Finally, some likely moderators of family influence are discussed, as well as potential future research directions and implications based on current knowledge.

Types of Familial Influence

Most familial influences on youth activity can be categorized as either passive or active. Passive influences include parental modeling of physical and sedentary activities, demographic factors (e.g., family's socioeconomic status), and parental attitudes and beliefs about physical activity and sedentary behavior. More active influences include parental verbal encouragement for children to be active or logistic support (e.g., provision of transportation to physical activity opportunities) and other types of support. These various potential familial influences on young people's activity are themselves likely related to each other and therefore could exert both direct and indirect effects on youth activity (e.g., Davison, Cutting, & Birch, 2003; Trost et al., 2003).

As a starting point, we have examined reviews of the research literature on correlates of young people's activity. Sallis, Prochaska, and Taylor (2000) reviewed all determinants of youth's physical activity and familial factors featured in this discussion. Gustafson and Rhodes (2006) reviewed the literature from 1985 to 2003 on parental correlates of physical activity in children and adolescents; a more recent review by Ferreira and colleagues (2006) cataloged environmental factors related to youth physical activity up to the publication year 2004, including some family or household factors. In the more recent attempts to understand the importance of different activity domains, in particular active commuting to school (Tudor-Locke, Ainsworth, & Popkin, 2001), some parental factors are considered. We rely on individual studies to examine context-specific youth activity and family factors. Research has also turned its focus to sedentary behaviors, particularly TV watching. In this domain, parental factors have been considered and summarized in recent reviews examined here (e.g., Gorely, Marshall, & Biddle, 2004). In addition to relying on reviews, we provide details from recent (mostly from 2005 and 2006) research regarding family factors related to child activity.

Correlational Studies of Familial Influences

Family influences differ for overall versus specific types of physical activity and in fact may differ across specific types of physical activity. For instance, parental verbal encouragement may have a different level of impact on a child's organized sport participation than on a child's active commuting. We explore family factors associated with children's overall physical activity, organized physical activity or sport, active commuting and physical activity in the neighborhood, and sedentary behavior.

Overall Physical Activity

The following sections deal with various types of family influences on children's overall physical activity. We explore how direct parental modeling, parental support, and other family factors are related to children's overall physical activity.

Parental Modeling

The family factor that has received the majority of attention has been the influence of parents' own activity behaviors on youth activity. Based on a social cognitive model (Bandura, 1986), it has been hypothesized that more active parents will have more active children; conversely, parents engaging in more sedentary behavior will have children who do likewise. Sallis and colleagues (2000) describe findings from 29 studies among 4- to 12-year-old children and findings from 27 studies for 13- to 18-year-old adolescents that addressed this issue. In children, they reported that only 38% of the findings showed a positive and significant relationship between parent physical activity and child physical activity. In adolescents, findings were similarly weak, with only 33% demonstrating a positive and significant relationship between parent and child activity levels (Sallis et al., 2000). The Gustafson and Rhodes (2006) review considered 34 family correlate studies, 24 of which looked at parent behavior and modeling. Similarly, these authors report that only six of the studies showed a positive correlation between parent physical activity and child physical activity. In the review by Sallis and colleagues (2000), more study findings (5 of 10; 50%) showed a significant positive relationship between parent and child physical activity if the parent participated in the activity along with the child, but such participation is a less passive role than modeling. As with most of the associations observed between family factors and children's physical activity and sedentary behavior, effect sizes were mostly low to moderate.

Most studies used questionnaires rather than objective measures to assess physical activity, though the two studies cited in the Gustafson and Rhodes (2006) review that used accelerometers to measure children's and adults' physical activity (both studies included only children <9 years old) showed positive associations between parent and child physical activity. For instance, Moore and colleagues (1991) found that a child was 5.8 times more likely to be categorized as active than as inactive based on accelerometer measurement if both parents were similarly categorized as active according to accelerometer measurement of their activity. Two other, more recent, studies showed positive relationships between parent and child activity. Interestingly, both looked at adult leisure-time activity separately from other types of adult activity (e.g., work). In a population-based study of 3000 French 12-year-old children, with parents and children completing surveys, Wagner and colleagues (2004) found that parent sport participation was related to the child's activity levels outside of school when both parents (compared to neither parent) practiced sport. In another large nationally representative sample, this time in the United States, parents' participation along with their child was strongly related to youths' (ages 9 to 13 years) level of participation in free-time physical activity (Heitzler et al., 2006). In this sample, parental beliefs about the importance of physical activity participation were also positively related to youths' organized and free-time activity.

Overall, though, more recent studies have indicated that the relationship between parent activity and child activity may be weaker than previously expected (Kohl & Hobbs, 1998), particularly when the question was whether parent modeling of physical activity extended into the child's adult activity behavior. One study showed that changes in parents' physical activity over time were not related to changes in their children's physical activity from the ages of 13 to 21 years (Anderssen, Wold, & Torsheim, 2006). Another indicated that parental influences during childhood were not related to their children's physical activity later in life as adults (Trudeau, Laurencelle, & Shephard, 2004). A cross-sectional study in younger, preschool children also showed no relationship between parent and child physical activity (Trost et al., 2003). From a more biological perspective, one study indicated that parent physical activity and fitness levels were not related to child fitness level over time, but that paternal daily energy expenditure was related to boys' fitness levels (Campbell et al., 2001).

The picture regarding whether parent activity directly affects children's activity is complicated by the numerous pathways through which such influence may occur. In their review, Ferreira and colleagues (2006) conclude that such associations from parent activity to child activity are inconsistent and generally weak across gender and child age. However, there are numerous explanations for this lack or weakness of association. Considering overall activity of parents and children may not provide the best estimate of direct parental influence, as such influence may be specific to the type of activity and the context in which parents are engaging in activity. For example, if parents have an occupation that is active, this might curtail leisure-time physical activity even though their overall physical activity would be high. Their children's physical activity might be negatively affected (Osler et al., 2001), with the child likely being more responsive to parental leisure-time than work-related activity levels. Some studies examining parental activity by type support this idea (e.g., Wagner et al., 2004).

Parent Support

In addition to modeling a more active lifestyle or directly engaging in physical activity with their children, parents can provide many potential types of support for children's activity or sedentary behavior. Such support can range from simple verbal encouragement to be more active, to logistic support (e.g., providing transportation, paying fees) for physical activity participation, to the establishment of a home environment that is more or less conducive to physical activity or sedentary behaviors (see chapter 16).

Many studies have examined a general support factor; one example is a recent study that showed general parental support of physical activity to be positively related to physical activity among 8- to 10-year-old African American girls (Adkins et al., 2004). Overall, though, Sallis and colleagues in their review (2000) found that parental encouragement/persuasion (usually defined as verbal encouragement rather than encouragement such as transporting to activities) was related to physical activity outcomes in children in only 31% of studies dealing with this issue. Likewise, only 25% of studies exploring transportation to activity locations and payment of fees by parents (i.e., logistic support) showed a significant relationship with child physical activity. In adolescents, however, 67% to 75% of the studies reviewed by Sallis and colleagues indicated that greater parent support was related to greater overall physical activity. As reported in the Gustafson and Rhodes (2006) review, 19 studies investigated parental support and all but one showed a significant positive correlation with youth physical activity. In contrast to the findings of Sallis and colleagues (2000), this more recent review concludes that the association is stronger in younger than in older children. Biddle, Gorely, and Stensel (2004) reviewed the correlates of adolescent girls' physical activity and found that parental support was consistently related to physical activity. These contradictory review conclusions may be due to differences in sample gender.

Indeed, recent studies document associations between family support and youth physical activity, but highlight gender specificity. For example, in a 2004 study, family support (without reference to any specific family member) was positively related to moderate and vigorous physical activity among adolescent girls, but more strongly related to girls' team sport involvement (Saunders et al., 2004). Another investigation revealed that transportation to physical activity locations was more positively related to girls' than to boys' physical activity and participation in sport, though boys were transported significantly more often (Hoefer et al., 2001). In an Estonian study, father modeling of and logistic support for physical activity were positively related to adolescent physical activity (Raudsepp, 2006).

Other Family Factors

In addition to modeling and support, other factors that are filtered through a familial "lens" may influence young people's overall physical activity; that is, family members and particularly parents shape attitudes, beliefs, and rules about children's physical activity based in part on other nonfamilial factors. For example, safety, particularly safety from crime or injury in one's neighborhood, has become a more common empirical focus in child physical activity research. In a recent survey study that we conducted in Boston, Cincinnati, and San Diego with 96 parents of young children (<11 years old) and 191 parents of adolescents (11- to 18-year-olds), we found a high prevalence of parent concerns about their children being harmed or abducted by strangers as they played or engaged in activity around their home, in their neighborhood, or in local parks. As seen in figure 11.1, *a* through *d*, more than 60% of parents of young children express this fear about their child playing alone just outside their home; playing with a

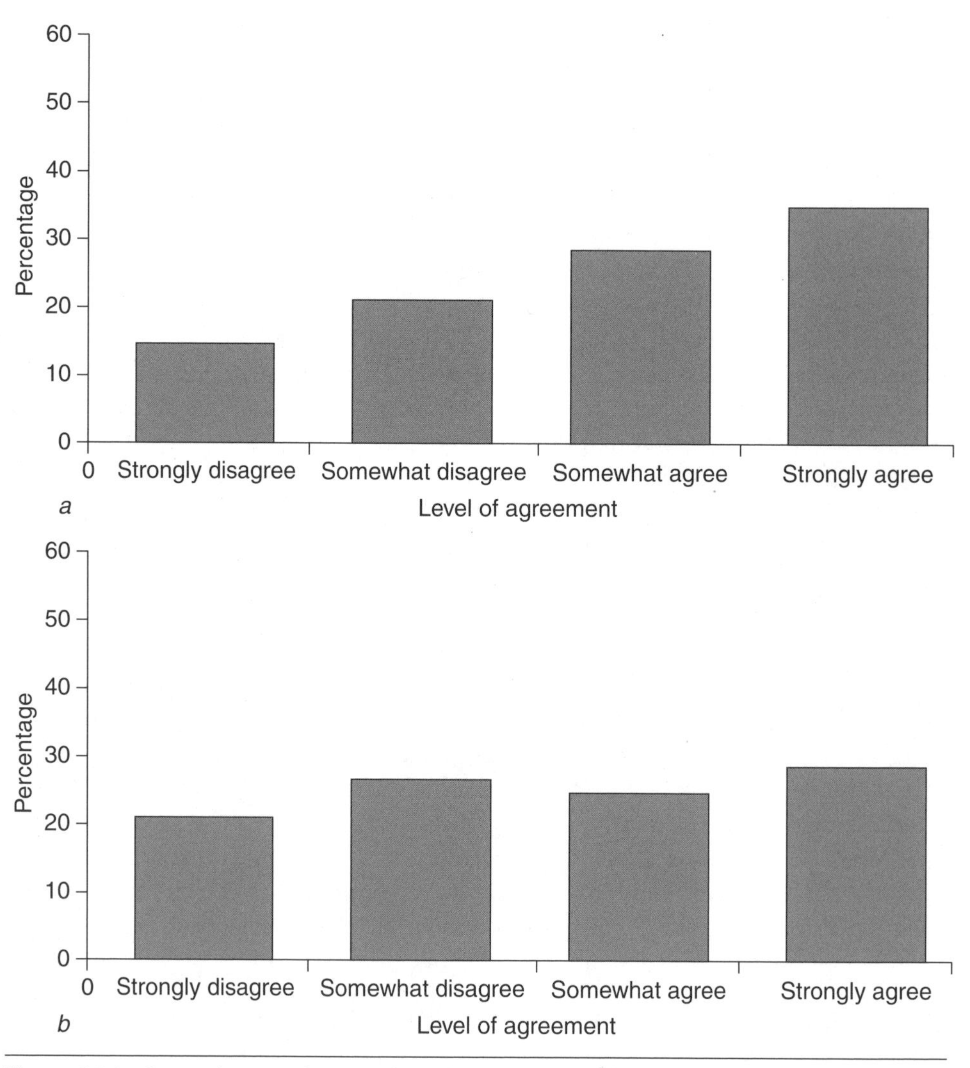

Figure 11.1 Parents' reported worry about their young child (<11 years old) being harmed or abducted by strangers *(a)* alone just outside the home (e.g., in the yard), *(b)* with a friend just outside the home, *(c)* alone or with a friend in the neighborhood, and *(d)* alone or with a friend in a local park.

friend rather than alone reduced this percentage only to just over 50%. Notably, 81.5% of parents of these young children expressed such a fear regarding their children playing alone or with a friend in a local park.

These concerns were somewhat lower for parents of older children; still, though, one-third of parents expressed worry about adolescents' safety just outside their home, and almost half expressed worry about their adolescent being harmed or abducted in a local park by a stranger (figure 11.2, *a-d*). Perhaps not surprisingly, a recent study among a sample of Mexican American adolescent girls showed associations between nonschool outdoor physical activity and both adolescents' perceived neighborhood safety (positive association) and the

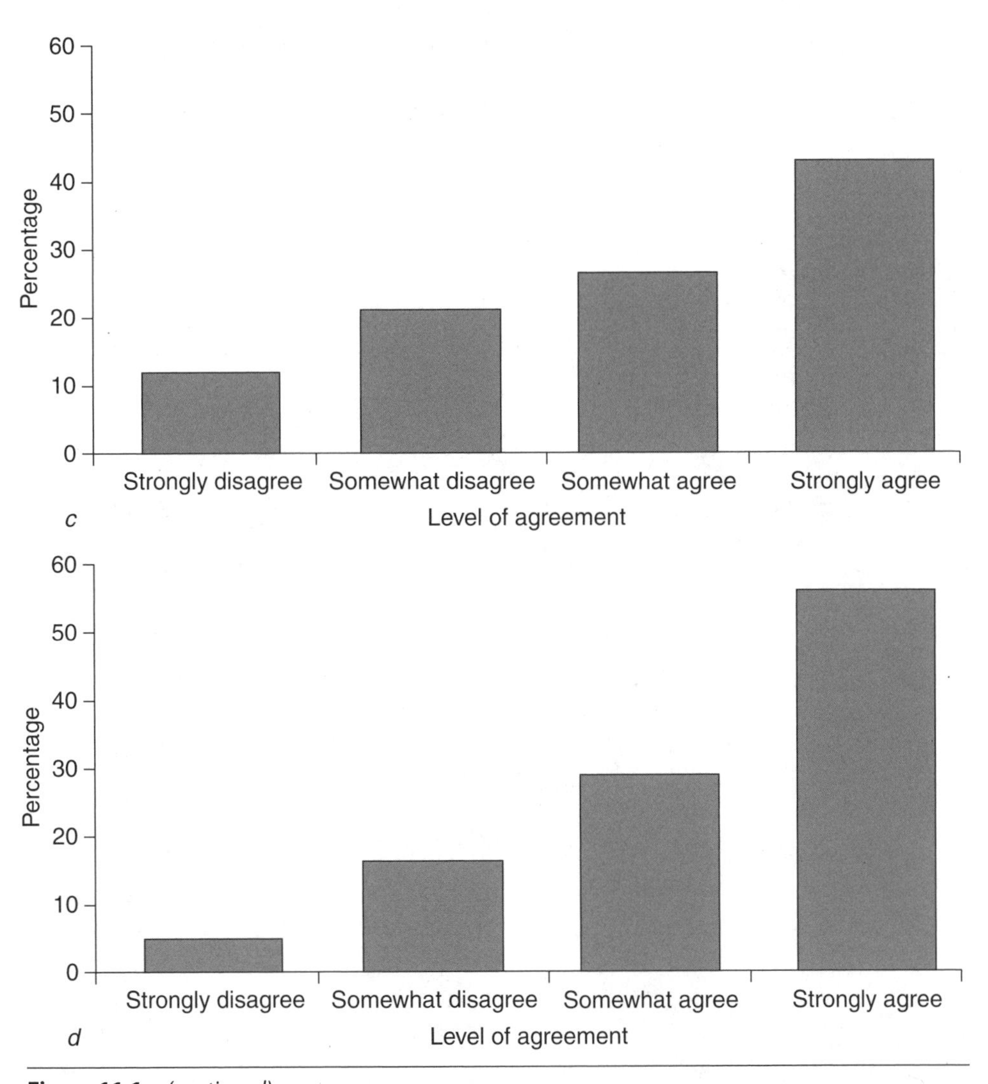

Figure 11.1 *(continued)*

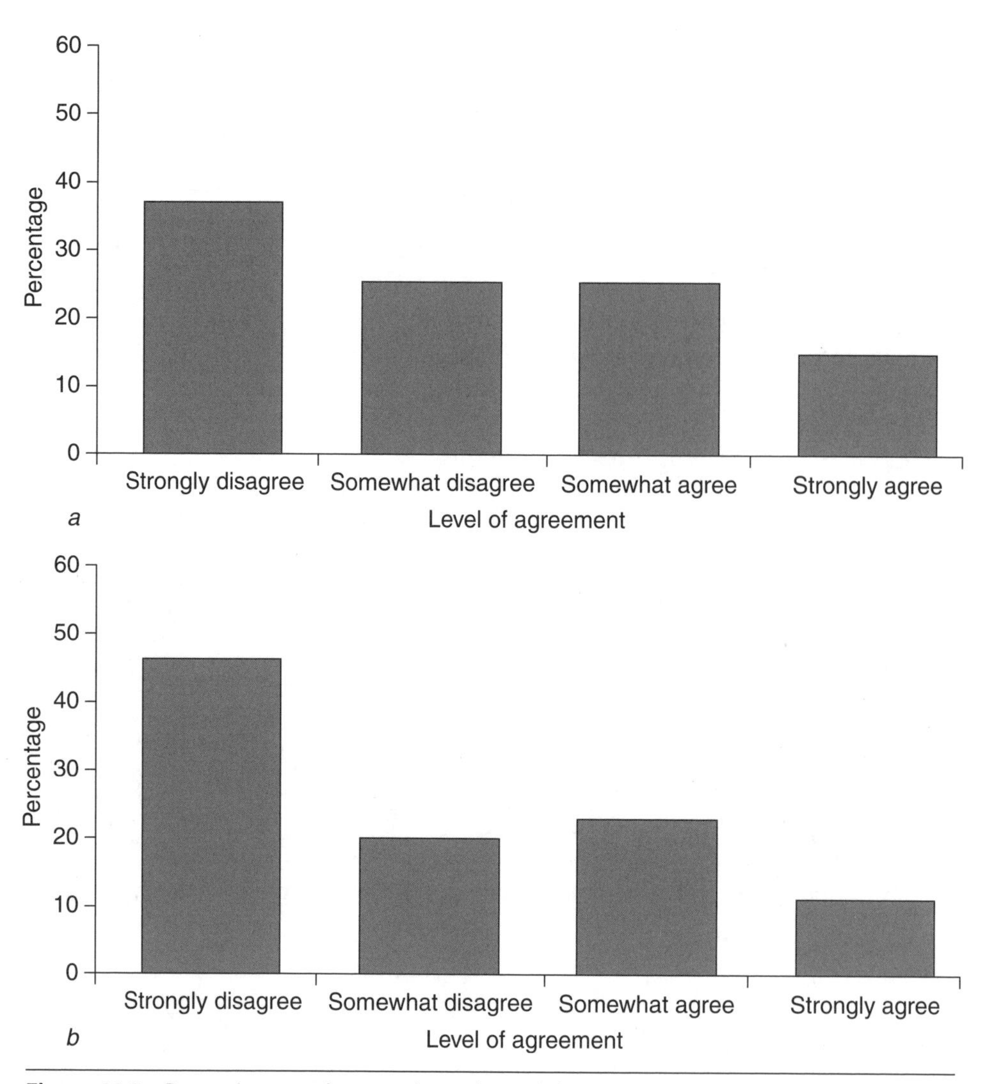

Figure 11.2 Parents' reported worry about their adolescent (≥12 years old) being harmed or abducted by strangers *(a)* alone just outside the home (e.g., in the yard), *(b)* with a friend just outside the home, *(c)* alone or with a friend in the neighborhood, and *(d)* alone or with a friend in a local park.

number of violent crimes in the surrounding neighborhood (negative association) (Gomez et al., 2004). Weir, Etelson, and Brand (2006) found a negative association between parent anxiety about neighborhood safety and overall physical activity of 5- to 10-year-old children. Parents who do not feel that their child is safe outside in their neighborhood perhaps inadvertently limit their child's overall physical activity. More research regarding safety is warranted.

For overall physical activity, the only factors for which the evidence is sufficient to allow one to draw conclusions, albeit tentative, are parent modeling and support. Recent findings regarding parent encouragement and support converge

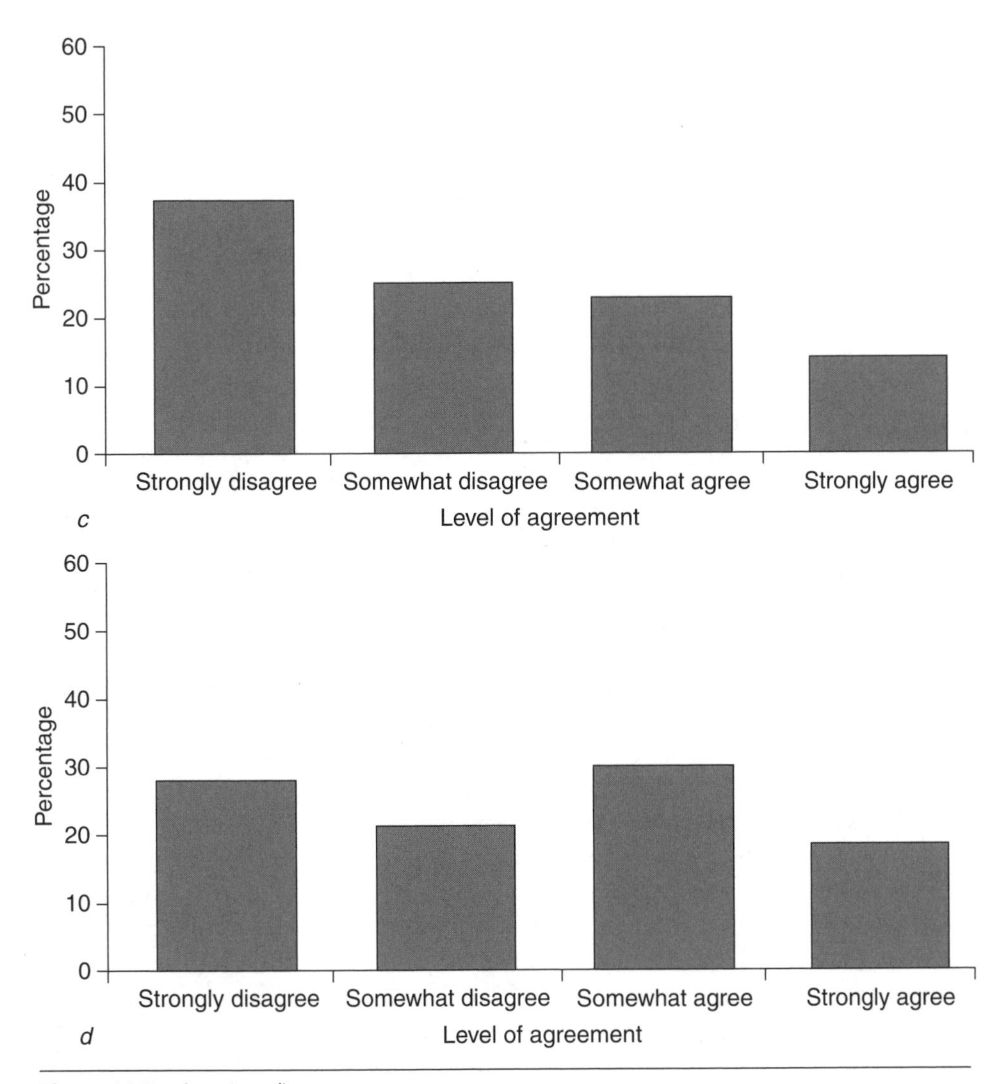

Figure 11.2 *(continued)*

with the evidence regarding parent modeling. It appears that children's overall physical activity increases if parents are physically active *with* the child but not if parents are physically active separately from the child. Likewise, more direct or active support (in the form of encouragement, being present during children's physical activity, and transportation) of children's physical activity is likely more consistently beneficial than passive factors (Duncan, Duncan, & Strycker, 2005; Gustafson & Rhodes, 2006), particularly for older children. A study by Biddle and Goudas (1996) highlights this distinction. These investigators found that adolescents' vigorous or strenuous physical activity and intentions to be active were significantly positively correlated with parental encouragement of physical activity, but minimally and insignificantly associated with parent self-reported physical activity.

Organized Physical Activity or Sport

Compared to the situation with overall physical activity, there has been less recent attention to whether parents influence children's organized activity or sport involvement. Organized physical activity or sport can be an important component of a child's overall level of physical activity. The family impact on children's participation in organized physical activity or sport can be quite substantial.

Parental Modeling

Recent evidence suggests a stronger relationship between parent behavior and child sport participation than between parent behavior and child overall physical activity levels (Saunders et al., 2004). For example, work has shown parent overall physical activity to be positively related to Australian youths' (ages 7 to 15 years) extracurricular sport participation (Cleland et al., 2005). In two U.S. studies, parent sport participation was directly associated with adolescents' sport participation (Nelson et al., 2005), and family support was positively related to sport participation among 8th-grade girls (Saunders et al., 2004). In a population-based investigation of approximately 3000 French 12-year-old children, Wagner and colleagues (2004) found that the likelihood of children engaging in structured physical activity outside of school was almost twice as high among children both of whose parents were involved in a sport compared to those with neither parent involved in sport.

In some countries, evidence of declining youth organized sport participation rates is mounting. Dollman, Norton, and Norton (2005) summarize such findings for Australian, Swedish, and U.K. youth. For instance, an Australian study documented a decrease of approximately 10% in the number of boys and girls participating in organized sport between 1985 and 1997 (Martin et al., 2005). Further, these investigators found that parents were also less likely to be involved in organized sport in 1997 than in 1985 and that there was decay in the association between parent modeling and child sport participation across this time period. These authors propose a changing social (e.g., changing role models) and familial climate as possibly contributing to the decreased association; that is, children are perhaps less likely to model their behavior (including physical activity) after their parents in a "society that is becoming increasingly complex, differentiated, and independent" (Martin et al., 2005, p. 75).

Other Parental Factors

There may be parental factors other than direct modeling that affect children's participation in organized sport. Sporting activities may be restricted by parental concerns. For example, one study suggested that parental concerns about injury and safety for children aged 5 to 12 years were sufficiently strong for parents to discourage children from participating in some sporting activities, but that the concern varied by child age and gender (Boufous, Finch, & Bauman, 2004). In a study surveying both children and parents, the child's perception of support for organized physical activity and the parent's reporting of support were both positively related to the child's participation in organized physical activity (Heitzler et al., 2006).

Active Commuting and Neighborhood Activity

Another context of children's activity is active commuting to and from school or other places. The literature on active commuting and neighborhood-based activities is relatively new (Saelens & Handy, in press), but is a potentially important component of the research on children's overall physical activity (Spinks et al., 2006).

Parental Modeling and Support

In contrast to the many studies on overall physical activity, there are only a few studies on the relationship between parental behavior or support and children's active commuting. Ziviani, Scott, and Wadley (2004) found that parent active commuting history and parent perception of the overall importance of physical activity were positively related to walking to school in 9-year-old Australian children. In another study, primary school children were more likely to walk or bike to school if their father walked or biked to school with them or if a parent walked or biked to work (Merom et al., 2005). One study based in the United States indicated that if no parent was at home after school, adolescents were more likely to walk or bike to school (Evenson et al., 2003). This may reflect unavailability of parents for transportation. Unexpectedly, Ommundsen and colleagues (2006) found that Norwegian parent support for physical activity was negatively related to children's active school commuting; but after age was accounted for (decrease in active commuting to school with increasing age), parent factors were not related to children's active commuting.

Other Family Factors

Walking and biking to school have decreased greatly over the past 30 years, and fewer children are allowed out alone (Dollman et al., 2005). This change could be partly due to an increase in parental concern for children's safety from traffic and crime. Yet violent crime rates and pedestrian fatalities have not increased over this time period; if anything, their incidence has decreased (Schieber & Vegega, 2002; Klaus & Rennison, 2002). The publicity about crime or pedestrian fatalities, however, may have increased, inflating parental concern about children's safety. Parents report traffic-related danger as a barrier to walking or biking by children of all ages (Centers for Disease Control and Prevention, 2005), and parent perceptions of heavy traffic are negatively related to adolescent boys' walking and girls' walking and biking for transportation (Carver et al., 2005).

Timperio and colleagues (2004) found that parent beliefs about crossing and intersection safety were related to walking or cycling in boys and girls 5 to 8 and 10 to 12 years old. In this study, parental perceptions of few other children in the neighborhood and no lights or crossings for their child were also related to less active commuting. In research that we conducted, parental concerns, including concern about crime, traffic, and distance to school, were more potent correlates (negative association) of walking to school than perceived and objective neighborhood built-environment factors commonly related to walkability (e.g., land use mix, street connectivity; Kerr et al., 2006). In addition, there was an interaction between neighborhood built environment (living in either a "walkable" or a "nonwalkable" neighborhood) and parental concerns. Among the children who lived in the more walkable neighborhoods, those children with parents reporting high concerns for their child's safety, as well as more logistical barriers to walking or biking (with the result that it was easier to drive), were significantly less likely to walk or bike to or from school. This suggests that educating parents about actual risk, as well as environmental change, is needed to improve children's activity levels in the realm of active commuting. Other authors have emphasized that it is important for future neighborhood activity studies to assess parent fear and control (Romero et al., 2001) and that parents' real and perceived concerns must be considered if active commuting to school is to be increased (Tudor-Locke et al., 2001). Studies must also address the factors that contribute to parental concerns, which in turn may affect youth behavior.

Sedentary Behavior—TV Viewing

The relationship between parent factors and child activities may be stronger for overall sedentary behavior than for children's overall physical activity. There are many potential reasons for this; for one, sedentary behavior is perhaps more easily shared simultaneously across generations (e.g., whole family watching television) than physical activity. One study that measured both physical activity and sedentary behavior showed that parent inactivity was strongly related to child inactivity (with the mother's influence stronger than the father's) but that parent physical activity showed only a weak link with child vigorous physical activity (Fogelholm et al., 1999).

Parental Modeling

Most studies of sedentary behavior focus specifically on TV watching. Gorely and colleagues (2004) reviewed 68 studies assessing the correlates of youth TV viewing and found that TV viewing was positively related to parent TV viewing habits as well as to having a TV in the bedroom (Gorely et al., 2004). Similar findings were reported by Salmon and colleagues (2005a), who found that watching TV together as a family was positively related to boys' TV watching time and that TV watching and rules of mothers and fathers were associated with boys' and girls' TV watching. In a study of 9- to 11-year-old girls, greater

TV viewing was related to greater time spent by parents viewing TV, parents viewing TV as a recreational activity, parents watching TV with the child, and no limits on TV watching (Davison, Francis, & Birch, 2005). In preschool and primary school children, children's TV viewing time has been related to sibling and parent TV viewing time (Songül Yalçin et al., 2002). Similarly, more than 2 h of TV viewing per day by parents of 12-year-olds has been related to greater youth TV watching time (Wagner et al., 2004).

Parental Rules

In our experience, parents are much more likely to establish specific rules for their children's sedentary behaviors than for physical activity. On the basis of child reports, Kennedy (2000) identified four parenting styles pertaining to TV rule enforcement: (1) laissez faire (no rules or guidance for children from parents regarding their TV viewing), (2) restrictive (parental limits set on child TV watching), (3) promotive (parental encouragement of child TV watching), and (4) selective (parent monitoring of time and content) styles. Others found that the laissez faire style of no parental limitations on TV time, few family dinners, and the presence of a TV in the child's bedroom were positively correlated with 6th and 7th graders' TV viewing time (Wiecha et al., 2001). Although there was no main effect of overall TV time restriction, Salmon and colleagues (2005b) found that parents who had rules prohibiting TV during mealtimes had children who were less likely to watch more than 2 h of TV per day. Parental rules for TV may be particularly important if they mediate other factors related to children's TV use. For instance, parents may limit the total amount of their children's TV watching time (i.e., specific parental rule), thus changing the potential effect of having a TV in the child's bedroom.

Studies of Other Family Factors

In addition to parent modeling, parent support, and parental rules regarding children's physical activity and sedentary behavior, there are other family influences affecting these behaviors that have received empirical attention. Such influences range from factors that could be affecting the whole family environment (such as socioeconomic status) to factors that specifically affect children's activity choices in the moment (such as whether a sibling is being active or sedentary).

Parent Education and Socioeconomic Status

Studies have examined familial sociodemographic variables in relation to child physical activity (Frenn et al., 2005; O'Loughlin et al., 1999; Pratt, Macera, & Blanton, 1999) and sedentary behaviors (Certain & Kahn, 2002; Christakis et al., 2004; Gordon-Larsen, McMurray, & Popkin, 2000). Higher parental education level may reflect a better understanding of the health benefits of physical activity and also be a proxy for income and other resources available to promote child physical activity. Household income may be particularly important for child activity levels, as organized activities often require equipment or fees. In addition, one-parent families may have lower incomes as well as less time available for the parent to support activity or enforce TV rules. However, Gutstafson and Rhodes (2006) reported that there were not enough studies to permit conclusions about the influence of single-parent families or socioeconomic status on child activity levels. The review by Ferreira and colleagues (2006) suggests that most studies of younger children (<12 years old) fail to find any association between physical activity and any household economic factor or parent education level variable, but that adolescents' physical activity is more consistently positively related to mothers' education and family income.

In contrast, parent education and household income are consistently and negatively related to children's television watching (Gorely, Marshall, & Biddle, 2004). In a recent study of Australian children in early and late primary school, parent education was the only significant correlate of children's television watching across age and gender (Hesketh et al., 2006).

Siblings

Sallis and colleagues (2000) noted in their review that sibling physical activity was correlated with adolescents' physical activity; however, few studies have specifically focused on the role of siblings in youth physical activity behavior (Ferreira et al., 2006). Two recent studies provide somewhat contradictory evidence regarding sibling support. One indicated that general sibling support was not an independent correlate of 10- to 14-year-old youths' physical activity. However, the more specific support conferred when either a sibling, parent, or friend watches a child who is engaged in activity was found to be positively related to that child's physical activity (Duncan et al., 2005). In contrast, Davison and Schmalz (2006) found that sibling support for physical activity was positively associated with 6th to 8th graders' (mean age of 12.7 years for boys and 12.5 years for girls) physical activity, though this association was higher for children at greater risk for being less physically active (with risk defined as including two of the following: being female, being overweight, having low perceived sport competence). In focus groups in the United States and Canada, parents mentioned that sibling activity or inactivity and coparticipation by the sibling would influence their child's physical activity and sedentary behavior (Irwin et al., 2005; Thompson et al., 2003).

In a large quantitative study, Brazilian adolescents' sedentary lifestyle was positively related to low birth order (a possible proxy for number of siblings), which may indicate that the presence of siblings provides opportunity for being more sedentary (Hallal et al., 2006). TV watching, in particular, was shown to be positively related to sibling viewing time in a sample of Turkish youth (Songül Yalçin et al., 2002). However, in another study, having a sibling was related to less sedentary behavior time (although TV watching time did not differ based on presence of siblings) (Hesketh et al., 2006), perhaps because there are more familial active play opportunities when one has more siblings.

Interventions for weight loss or increasing physical activity often target families as a whole, but the role of brothers or sisters has not been investigated to date in these types of studies. During one intervention, Belgian children increased their physical activity partly through sibling modeling (Deforche et al., 2004). This suggests that specific attention to the role of siblings in youth physical activity behavior may be fruitful.

Moderators of Family Influence

Given the numerous levels of influence from the child to family to neighborhood to policy levels, and the various factors within each of these levels that can affect children's activity, it is important to consider nonfamilial mediators and moderators of familial factors related to children's activity. It is also likely that family influences themselves are interrelated and therefore may have both direct and indirect effects on young people's activity. The testing of such interrelatedness is becoming more common in observational studies designed to find correlates of youth activity (Lewis et al., 2002).

Child characteristics, including age, gender, and ethnicity, may moderate familial influences on physical activity. Direct comparison of younger and older youth to assess age as a moderator of parental influence is rare, though parental influences may be specific to the type and context of youth activity. For example, parental modeling and participation may be more critical for younger children who play closer to home, whereas practical support and encouragement may be more influential for older children and adolescents participating in organized activities. For the domain of active commuting, parent concerns and rules are likely to vary by age and likely by other factors such as gender.

Gender is more frequently examined as a moderator of parent influence on youth physical activity. Both parent gender and child gender and their concordance or divergence may be important. Gustafson and Rhodes' (2006) review indicated that boys tend to be better supported and more influenced by parent physical activity, but overall there were positive correlations between mother's and daughter's physical activity and between father's and son's physical activity. For example, Martin and colleagues (2005) found that paternal physical activity was positively related to boys' sport participation, whereas maternal inactivity was related to lower girls' sport participation. Another study showed that fathers tended to provide explicit physical activity modeling to boys and logistic support to girls (Raudsepp, 2006).

Family inactivity has also been related to boys' inactivity (Wagner et al., 2004). Parental rules may play a greater role in girls' activity than in boys'. One study showed that mother's authoritative parenting style ("balanced between responsiveness and control") was positively related to physical activity and negatively related to sedentary habits in female adolescents (Schmitz et al., 2002). Ferreira and colleagues' (2006) review suggests a more complicated interaction between child age and parent gender, such that fathers' physical activity is more consistently positively related to children's (<12 years old) physical activity but that neither mothers' nor fathers' physical activity is related to adolescents' overall physical activity.

Consideration of gender and ethnicity together may be important, as ethnically and culturally based parental expectations and rules may differ by gender. One study exploring the role of gender and ethnicity indicated that parent encouragement was positively related to boys' activity in all ethnic groups except Hispanic boys. Further, parent TV time was positively associated with TV watching in Hispanic boys but inversely related to TV viewing in Black boys (McGuire et al., 2002). Parent concern about fitness was associated with less TV time in White girls, but more TV time in Black girls (McGuire et al., 2002). A recent study of adolescent boys in Australia showed higher associations between father physical activity support and youth physical activity among Vietnamese- compared to Anglo-Australians (Wilson & Dollman, 2007). Overall, ethnicity appears to be an important moderator of family influence on physical activity, likely in combination with gender. If observational studies are used to design and conduct interventions targeting increasing physical activity or decreasing sedentary behaviors or both, these important moderation effects need to be considered.

Parents may encourage youth physical activity by actively participating in activities themselves, but family influence on youth activity behavior is complex.

Fewer studies address mediators of the association of familial variables with child physical activity. In one study, children's perceived athletic competence was a cross-sectional mediator of the relation between parental encouragement and children's activity (Biddle & Goudas, 1996). Similarly, parent support has been shown to be related to physical activity directly and indirectly through child self-efficacy (Trost et al., 2003).

Intervention Studies

Intervention studies have also included and considered the efficacy of familial components for increasing child physical activity. Such interventions vary in the level and amount of familial involvement they incorporate, ranging from minimal direct involvement, as when family members are provided with information (e.g., through mailings), to instruction or training of parents on how to support and encourage children's greater physical activity, to more direct involvement in the form of parents' participation in physical activity with their children. Guidelines have encouraged parental participation across these varying levels of involvement, adding advocacy for and modeling of an active lifestyle (National Center for Chronic Disease Prevention and Health Promotion, Centers for Disease Control and Prevention, 1997). Few studies have contrasted different ways in which the family is included in interventions (e.g., comparing interventions that focus on parental vs. sibling involvement), but evidence across studies generally suggests that more comprehensive and direct familial involvement is related to greater efficacy.

School-Based Interventions

Many youth activity interventions have been based or initiated in school settings (see chapter 13), and most school-based studies have incorporated minimal or no familial involvement (Stone et al., 1998). One large trial, the Child and Adolescent Trial for Cardiovascular Health (CATCH), did evaluate the added efficacy of a family component. The investigators examined whether adding a minimal family-based component to the primarily school-based program was more effective than use of a school-based intervention alone for changing child diet and physical activity. There was no evidence of an added benefit of the family-based component, consisting primarily of "activity packets" for parents and children to complete at home and "family fun nights" (Luepker et al., 1996). Additional analyses suggested greater improvement in dietary and physical activity knowledge and attitudes with increased parental involvement in the CATCH familial component, though again there was little impact on actual child activity (Nader et al., 1996).

To augment the preschool components directed toward the children, a recent preschool-based obesity prevention program among ethnic minority children included weekly newsletters and parent homework assignments, as well as offering exercise classes for parents twice a week (Fitzgibbon et al., 2002). Whereas at follow-up, desired weight change differences existed between this program and a general health education condition (which focused on various aspects of health including dental health and immunizations and which controlled for the attention provided in the obesity prevention condition), there was no child physical activity difference between these conditions. Most other school-based studies showed no additional increase in children's physical activity level when a minimal family involvement component was introduced (Davis et al., 2003; Kahn et al., 2002). In contrast, more active involvement by parents in the context of primarily school-based programs, though difficult to achieve, may be beneficial. Adding individual tailoring and parent involvement (in the form of activity-monitoring assistance and encouragement of children's physical activity) to an existing school-based fitness and nutrition program was found to decrease boys' television watching further than in the core program (Burke et al., 1998).

There are other noteworthy exceptions to the general lack of efficacy seen among school-based programs that include a minimal parental component. Recent programs have specifically targeted reductions in children's sedentary activity instead of increases in physical activity. For instance, electronic television time manager devices and parent newsletters were part of an effective school-based program for reducing 3rd- and 4th-grade (mean age of 8.9 years) children's television watching and related screen time (Robinson, 1999). Television watching was also reduced among preschool-aged children in a combined day care- and home-based intervention focused on reducing children's screen time (Dennison et al., 2004). A primary care–based diet and physical activity change program for healthy adolescents that included computer-based assessment, provider counseling, and subsequent telephone and mail intervention provided written materials to parents as well (Patrick et al., 2006). This multicomponent program also focused on reducing sedentary activity. Relative to a sun-protection comparison program (which differed in content from the physical activity change program but provided a similar frequency and amount of intervention attention), it resulted in more significant decreases in sedentary behaviors among both boys and girls and a greater number of days of sufficient physical activity in boys. The specific impact of family was not evaluated.

Community-Oriented Interventions

Other community-oriented intervention programs suggest that a minimal family involvement model (e.g., sending newsletters about the program and physical activity to the home) fails to add meaningfully to increases in children's physical activity (Pate et al., 2003). Some community-oriented programs targeting child activity changes have focused on more direct parent involvement. For example, in the intervention arm of the San Diego Family Health project, both parents and children attended activity sessions and education sessions addressing ways to improve the whole family's eating and physical activity. Whereas this resulted in changes in dietary behaviors and knowledge about physical activity, the intervention did not increase physical activity among parents or children (Nader et al., 1989).

More recent trials have examined the efficacy of coparticipation of mothers and daughters in physical activity interventions. In a series of small sample studies, Ransdell and colleagues investigated whether community/clinic-based or home-based interventions (or both) were effective in increasing physical activity for either mother or daughter or both. Their initial structured activity and classroom-based trial failed to result in increased days of physical activity for either mother or daughter (Ransdell et al., 2001). Subsequent trials enrolling mothers and daughters in either a community- or a home-based program showed increased physical activity, including improvements in aerobic, flexibility, and strength parameters (Ransdell et al., 2004, 2003). This suggests that the setting of an intervention in which family members are coactive may be an important aspect of increasing physical activity.

More intense family involvement in interventions exclusively targeting physical activity change is rare. Although it is difficult to isolate the direct effect of parental involvement on child activity, such involvement is critical to obese children's success in clinic-based weight control interventions. Indeed, training parents in behavior modification strategies (e.g., monitoring, contingency management, environmental control) around children's activity and diet is more effective than no or minimal parental involvement (McLean et al., 2003), particularly when the children are younger (Golan & Crow, 2004). Many pediatric obesity trials have taken a family-based approach, not only targeting the child for behavior change, but also targeting change across all family members. Concurrent participation of the parent in his or her own physical activity and dietary behavior change improves the child's outcome over simply having the parent support and implement behavior change exclusively in the child (Epstein et al., 1994). Based on the correlational studies regarding parent factors, it would not be surprising if more complete participation by parents was needed in interventions in order to successfully increase children's physical activity.

Conclusions and Recommendations for Research

Despite the number of activity, inactivity, and obesity studies that include parent influence constructs, firm conclusions are difficult to draw from the current literature. This partly reflects the multiple and complicated factors that influence behavior as outlined in ecological models, even within one level of influence (e.g., family). While focus group research often suggests that the family is important to physical activity, sport involvement, and TV watching (Allison et al., 2005; Kennedy et al., 2002; Kubik, Lytle, & Fulkerson, 2005; Thompson et al., 2003), quantitative studies do not always support these findings.

The correlation and intervention evidence on child activity generally suggests that family influences that are more "active" or participatory (e.g., when family members are active or are not active with a child; when they do or do not verbally encourage a child's activity) are more consistently related to children's physical activity than passive factors (e.g., when parents themselves are active or are not). Further, though the data regarding the relationship between parents' physical activity and young people's overall physical activity are inconsistent or weak, studies targeting more specific youth activity behaviors, such as organized activity or sport, yield more consistent, stronger associations. Future research on parental behavior and youth activity should continue to differentiate activity domains and types. Along with this, more specific assessment of familial support is warranted. A recent Norwegian study showed that parental support was not related to children's school-based unstructured physical activity (Ommundsen et al., 2006). This is not surprising given the lack of presence of the parent at the school, but such specificity requires more complex assessments of both children's activity (and its contexts) and type of support.

In particular, studies should determine whether the parent participates with the child. And, since the obesity epidemic is affecting both adults and children, interventions that encourage both children and parents to become active and reduce sedentary behaviors, and that help both child and parent to be actively supportive, should be investigated. In this way, researchers could also examine the influence of child activity levels on parent activity levels, recognizing that the child–parent relationship is not unidirectional (i.e., child behavior can influence parent behavior).

Future studies should address how different types of support influence children of different ages and gender and across other demographic variables. In addition to considering various activity domains, investigators should look at the relationship between support and activity in different activity locations. In particular, it is important to understand parents' concerns about their children's engaging in activity in public locations.

It is also difficult to draw conclusions from self-report data, especially if parents report their child's physical activity and barriers to activity as well as their own perceptions. Additional studies adopting more objective and comprehensive assessments of child activity and activity type would augment the existing literature (e.g., see Robinson et al., 2006). Evaluation of both young people's and their parents' beliefs and perceived barriers to activity is needed, as parents and children do not always share similar beliefs or perceived barriers to activity (e.g., see Dunton, Jamner, & Cooper, 2003).

Family structure is changing, and more parents are relying on after-school programs or day care for their children than in the past. Researchers should investigate these environments and explore the impact of one-parent families and siblings (or lack of siblings) on youth physical activity and sedentary behavior. Family influences on children's activity are evolving and changing, requiring a broader examination of these influences and an exploration of their interaction with influences from other ecological levels (e.g., policy, built environment). It could be that such changes explain the weak findings regarding parental modeling and youth overall physical activity levels, though some of the support and context-specific findings suggest a continued important role for the family in shaping youth activity behavior.

APPLICATIONS FOR RESEARCHERS

The family can influence youth physical activity and sedentary behavior in many ways; more consistent associations have been found between active factors (e.g., direct support) and youth's overall physical activity than for passive factors (e.g., parent modeling), though parental modeling of sedentary behavior is often associated with children's sedentary behavior. There is growing evidence of context specificity; for example, parent modeling of sport participation is consistently and positively related to children's sport participation. More investigation of familial support (both from parents and from other family members such as siblings) should target context-specific influences. More research is needed to increase understanding of the mechanisms by which family demographic factors affect youth activity. Potential moderators of familial influence on youth activity are also understudied. Physical activity intervention research should seek and test strategies that can optimize positive family environments for youth physical activity.

APPLICATIONS FOR PROFESSIONALS

Evidence suggests that the family influences youth physical activity, but the association is complex. Parents need to be made aware that their behavior and attitudes can inhibit or facilitate their children's physical activity and sedentary behavior. This may vary, however, by age of child, type of activity, and context. Parents may influence youth by their own activity behaviors and degree of support for physical activity. In the home, clear TV rules and not modeling sedentary activity may reduce youth sedentary behavior. Outside of the home, logistical support may be critical for youth physical activity, as parent physical activity in itself does not always translate into youth physical activity. Also, professionals should address parent concern for child safety in local neighborhoods and parks. Finally, even early on, parents should be advised to consider the physical activity environment of available child care options, as this can get children off to an active start in life.

References

Adkins, S., Sherwood, N.E., Story, M., & Davis, M. (2004). Physical activity among African-American girls: The role of parents and the home environment. *Obesity Research, 12*, 38S-45S.

Allison, K.R., Dwyer, J.J., Goldenberg, E., Fein, A., Yoshida, K.K., & Boutilier, M. (2005). Male adolescents' reasons for participating in physical activity, barriers to participation, and suggestions for increasing participation. *Adolescence, 40*, 155-170.

Anderssen, N., Wold, B., & Torsheim, T. (2006). Are parental health habits transmitted to their children? An eight year longitudinal study of physical activity in adolescents and their parents. *Journal of Adolescence, 29*, 513-524.

Bandura, A. (1986). *Social foundations of thought and action: A social cognitive theory*. Englewood Cliffs, NJ: Prentice Hall.

Biddle, S.J.H., Gorely, T., & Stensel, D.J. (2004). Health-enhancing physical activity and sedentary behaviour in children and adolescents. *Journal of Sports Sciences, 22*, 679-701.

Biddle, S., & Goudas, M. (1996). Analysis of children's physical activity and its association with adult encouragement and social cognitive variables. *Journal of School Health, 66*, 75-78.

Boufous, S., Finch, C., & Bauman, A. (2004). Parental safety concerns—a barrier to sport and physical activity in children? *Australian and New Zealand Journal of Public Health, 28*, 482-486.

Burke, V., Milligan, R.A., Thompson, C., Taggart, A.C., Dunbar, D.L., Spencer, M.J., et al. (1998). A controlled trial of health promotion programs in 11-year-olds using physical activity "enrichment" for higher risk children. *Journal of Pediatrics, 132*, 840-848.

Campbell, P.T., Katzmarzyk, P.T., Malina, R.M., Rao, D.C., Perusse, L., & Bouchard, C. (2001). Prediction of physical activity and physical work capacity (PWC150) in young adulthood from childhood and adolescence with consideration of parental measures. *American Journal of Human Biology, 13*, 190-196.

Carver, A., Salmon, J., Campbell, K., Baur, L., Garnett, S., & Crawford, D. (2005). How do perceptions of local neighborhood relate to adolescents' walking and cycling? *American Journal of Health Promotion, 20*, 139-147.

Centers for Disease Control and Prevention. (2005). Barriers to children walking to or from school—United States, 2004. *Morbidity and Mortality Weekly Report, 54*, 949-952.

Certain, L.K., & Kahn, R.S. (2002). Prevalence, correlates, and trajectory of television viewing among infants and toddlers. *Pediatrics, 109*, 634-642.

Christakis, D.A., Ebel, B.E., Rivara, F.P., & Zimmerman, F.J. (2004). Television, video, and computer game usage in children under 11 years of age. *Journal of Pediatrics, 145*, 652-656.

Cleland, V., Venn, A., Fryer, J., Dwyer, T., & Blizzard, L. (2005). Parental exercise is associated with Australian children's extracurricular sports participation and cardiorespiratory fitness: A cross-sectional study. *International Journal of Behavioral Nutrition and Physical Activity, 2*, 3. doi:10.1186/1479-5868-2-3.

Davis, S.M., Clay, T., Smyth, M., Gittelsohn, J., Arviso, V., Flint-Wagner, H., et al. (2003). Pathways curriculum and family interventions to promote healthful eating and physical activity in American Indian schoolchildren. *Preventive Medicine, 37,* S24-S34.

Davison, K.K., & Birch, L.L. (2001). Childhood overweight: A contextual model and recommendations for future research. *Obesity Reviews, 2,* 159-171.

Davison, K.K., Cutting, T.M., & Birch, L.L. (2003). Parents' activity-related parenting practices predict girls' physical activity. *Medicine and Science in Sports and Exercise, 35,* 1589-1595.

Davison, K.K., Francis, L.A., & Birch, L.L. (2005). Links between parents' and girls' television viewing behaviors: A longitudinal examination. *Journal of Pediatrics, 147,* 436-442.

Davison, K.K., & Schmalz, D.L. (2006). Youth at risk of physical inactivity may benefit more from activity-related support than youth not at risk. *International Journal of Behavioral Nutrition and Physical Activity, 3,* 5. doi:10.1186?1479-5868-3-5.

Deforche, B., De Bourdeaudhuij, I., Tanghe, A., Hills, A.P., & De Bode, P. (2004). Changes in physical activity and psychosocial determinants of physical activity in children and adolescents treated for obesity. *Patient Education and Counseling, 55,* 407-415.

Dennison, B.A., Russo, T.J., Burdick, P.A., & Jenkins, P.L. (2004). An intervention to reduce television viewing by preschool children. *Archives of Pediatrics and Adolescent Medicine, 158,* 170-176.

Dollman, J., Norton, K., & Norton, L. (2005). Evidence for secular trends in children's physical activity behaviour. *British Journal of Sports Medicine, 39,* 892-897.

Duncan, S.C., Duncan, T.E., & Strycker, L.A. (2005). Sources and types of social support in youth physical activity. *Health Psychology, 24,* 3-10.

Duncan, S.C., Duncan, T.E., Strycker, L.A., & Chaumeton, N.R. (2004). A multilevel approach to youth physical activity research. *Exercise and Sport Sciences Reviews, 32,* 95-99.

Dunton, G.F., Jamner, M.S., & Cooper, D.M. (2003). Assessing the perceived environment among minimally active adolescent girls: Validity and relations to physical activity outcomes. *American Journal of Health Promotion, 18,* 70-73.

Epstein, L.H., Valoski, A., Wing, R.R., & McCurley, J. (1994). Ten-year outcomes of behavioral family-based treatment for childhood obesity. *Health Psychology, 13,* 373-383.

Evenson, K.R., Huston, S.L., McMillen, B.J., Bors, P., & Ward, D.S. (2003). Statewide prevalence and correlates of walking and bicycling to school. *Archives of Pediatrics and Adolescent Medicine, 157,* 887-892.

Ferreira, I., van der Horst, K., Wendel-Vos, W., Kremers, S., van Lenthe, F.J., & Brug, J. (2006). Environmental correlates of physical activity in youth: A review and update. *Obesity Reviews,* doi:10.1111/j.1467-789X.2006.00264.x.

Fitzgibbon, M.L., Stolley, M.R., Dyer, A.R., VanHorn, L., & KauferChristoffel, K. (2002). A community-based obesity prevention program for minority children: Rationale and study design for Hip-Hop to Health Jr. *Preventive Medicine, 34,* 289-297.

Fogelholm, M., Nuutinen, O., Pasanen, M., Myohanen, E., & Saatela, T. (1999). Parent-child relationship of physical activity patterns and obesity. *International Journal of Obesity, 23,* 1262-1268.

Frenn, M., Malin, S., Villarruel, A.M., Slaikeu, K., McCarthy, S., Freeman, J., et al. (2005). Determinants of physical activity and low-fat diet among low income African American and Hispanic middle school students. *Public Health Nursing, 22,* 89-97.

Giles-Corti, B., Timperio, A., Bull, F., & Pikora, T. (2005). Understanding physical activity environmental correlates: Increased specificity for ecological models. *Exercise and Sport Sciences Reviews, 33,* 175-181.

Golan, M., & Crow, S. (2004). Parents are key players in the prevention and treatment of weight-related problems. *Nutrition Reviews, 62,* 39-50.

Gomez, J.E., Johnson, B.A., Selva, M., & Sallis, J.F. (2004). Violent crime and outdoor physical activity among inner-city youth. *Preventive Medicine, 39,* 876-881.

Gordon-Larsen, P., McMurray, R.G., & Popkin, B.M. (2000). Determinants of adolescent physical activity and inactivity patterns. *Pediatrics, 105,* E83.

Gorely, T., Marshall, S.J., & Biddle, S.J. (2004). Couch kids: Correlates of television viewing among youth. *International Journal of Behavioral Medicine, 11,* 152-163.

Gustafson, S.L., & Rhodes, R.E. (2006). Parental correlates of physical activity in children and early adolescents. *Sports Medicine, 36,* 79-97.

Hallal, P.C., Wells, J.C.K., Reichert, F.F., Anselmi, L., & Victora, C.G. (2006). Early determinants of physical activity in adolescence: Prospective birth cohort study. *British Medical Journal, 332,* 1002-1007.

Heitzler, C.D., Martin, S.L., Duke, J., & Huhman, M. (2006). Correlates of physical activity in a national sample of children aged 9-13 years. *Preventive Medicine, 42,* 254-260.

Hesketh, K., Crawford, D., & Salmon, J. (2006). Children's television viewing and objectively measured physical activity: Associations with family circumstance. *International Journal of Behavioral Nutrition and Physical Activity, 3,* 36. doi:10.1186/1479-5868-3-36.

Hoefer, W.R., McKenzie, T.L., Sallis, J.F., Marshall, S.J., & Conway, T.L. (2001). Parental provision of transportation for adolescent physical activity. *American Journal of Preventive Medicine, 21,* 48-51.

Irwin, J.D., He, M., Bouck, L.M., Tucker, P., & Pollett, G.L. (2005). Preschoolers' physical activity behaviors: Parents' perspectives. *Canadian Journal of Public Health, 96,* 299-303.

Kahn, E.B., Ramsey, L.T., Brownson, R.C., Heath, G.W., Howze, E.H., Powell, K.E., et al. (2002). The effectiveness of interventions to increase physical activity: A systematic review. *American Journal of Preventive Medicine, 22,* 73-107.

Kennedy, C.M. (2000). Television and young Hispanic children's health behaviors. *Pediatric Nursing, 26,* 283-288, 292-294.

Kennedy, C.M., Strzempko, F., Danford, C., & Kools, S. (2002). Children's perceptions of TV and health behavior effects. *Journal of Nursing Scholarship, 34,* 289-294.

Kerr, J., Rosenberg, D., Sallis, J.F., Saelens, B.E., Frank, L.D., & Conway, T. (2006). Active commuting to school: Associations with built environment and parental concerns. *Medicine and Science in Sports and Exercise, 38,* 787-794.

Klaus, P., & Rennison, C.M. (2002). *Age patterns in violent victimization, 1976-2000.* Bureau of Justice Statistics Crime Data Brief: U.S. Department of Justice NCJ 190104. Available www.ojp.usdoj.gov/bjs/pub/pdf/apvv00.pdf. Accessed February 6, 2008. Washington, DC: U.S. Department of Justice.

Kohl, W.H., & Hobbs, K.E. (1998). Development of physical activity behaviors among children and adolescents. *Pediatrics, 101*, S549-S554.

Kubik, M.Y., Lytle, L., & Fulkerson, J.A. (2005). Fruits, vegetables, and football: Findings from focus groups with alternative high school students regarding eating and physical activity. *Journal of Adolescent Health, 36*, 494-500.

Lewis, B.A., Marcus, B.H., Pate, R.R., & Dunn, A.L. (2002). Psychosocial mediators of physical activity behavior among adults and children. *American Journal of Preventive Medicine, 23*, 26-35.

Luepker, R.V., Perry, C.L., McKinlay, S.M., Nader, P.R., Parcel, G.S., Stone, E.J., et al. (1996). Outcomes of a field trial to improve children's dietary patterns and physical activity: The Child and Adolescent Trial for Cardiovascular Health—CATCH collaborative group. *Journal of the American Medical Association, 275*, 768-776.

Martin, M., Dollman, J., Norton, K., & Robertson, I. (2005). A decrease in the association between the physical activity patterns of Australian parents and their children, 1985-1997. *Journal of Science and Medicine in Sport, 8*, 71-76.

McGuire, M.T., Hannan, P.J., Neumark-Sztainer, D., Cossrow, N.H., & Story, M. (2002). Parental correlates of physical activity in a racially/ethnically diverse adolescent sample. *Journal of Adolescent Health, 30*, 253-261.

McLean, N., Griffin, S., Toney, K., & Hardeman, W. (2003). Family involvement in weight control, weight maintenance and weight-loss interventions: A systematic review of randomised trials. *International Journal of Obesity, 27*, 987-1005.

Merom, D., Rissel, C., Mahmic, A., & Bauman, A. (2005). Process evaluation of the New South Wales Walk Safely to School Day. *Health Promotion Journal of Australia, 16*, 100-106.

Moore, L.L., Lombardi, D.A., White, M.J., Campbell, J.L., Oliveria, S.A., & Ellison, R.C. (1991). Influence of parents' physical activity levels on activity levels of young children. *Journal of Pediatrics, 118*, 215-219.

Nader, P.R., Sallis, J.F., Patterson, T.L., Abramson, I.S., Rupp, J.W., Senn, K.L., et al. (1989). A family approach to cardiovascular risk reduction: Results from the San Diego Family Health Project. *Health Education Quarterly, 16*, 229-244.

Nader, P.R., Sellers, D.E., Johnson, C.C., Perry, C.L., Stone, E.J., Cook, K.C., et al. (1996). The effect of adult participation in a school-based family intervention to improve children's diet and physical activity: The Child and Adolescent Trial for Cardiovascular Health. *Preventive Medicine, 25*, 455-464.

National Center for Chronic Disease Prevention and Health Promotion, Centers for Disease Control and Prevention. (1997). Guidelines for school and community programs to promote lifelong physical activity among young people. *Journal of School Health, 67*, 202-219.

Nelson, M.C., Gordon-Larsen, P., Adair, L.S., & Popkin, B.M. (2005). Adolescent physical activity and sedentary behavior: Patterning and long-term maintenance. *American Journal of Preventive Medicine, 28*, 259-266.

O'Loughlin, J., Paradis, G., Kishchuk, N., Barnett, T., & Renaud, L. (1999). Prevalence and correlates of physical activity behaviors among elementary schoolchildren in multiethnic, low income, inner-city neighborhoods in Montreal, Canada. *Annals of Epidemiology, 9*, 397-407.

Ommundsen, Y., Klasson-Heggebø, L., & Anderssen, S.A. (2006). Psycho-social and environmental correlates of location-specific physical activity among 9- and 15-year-old Norwegian boys and girls: The European Youth Heart Study. *International Journal of Behavioral Nutrition and Physical Activity, 3*, 32. doi:10.1186/1479-5868-3-32.

Osler, M., Clausen, J.O., Ibsen, K.K., & Jensen, G.B. (2001). Social influences and low leisure-time physical activity in young Danish adults. *European Journal of Public Health, 11*, 130-134.

Pate, R.R., Saunders, R.P., Ward, D.S., Felton, G., Trost, S.G., & Dowda, M. (2003). Evaluation of a community-based intervention to promote physical activity in youth: Lessons from Active Winners. *American Journal of Health Promotion, 17*, 171-182.

Patrick, K., Calfas, K.J., Norman, G.J., Zabinski, M.F., Sallis, J.F., Rupp, J., et al. (2006). Randomized controlled trial of a primary care and home-based intervention for physical activity and nutrition behaviors: PACE+ for adolescents. *Archives of Pediatrics and Adolescent Medicine, 160*, 128-136.

Pratt, M., Macera, C.A., & Blanton, C. (1999). Levels of physical activity and inactivity in children and adults in the United States: Current evidence and research issues. *Medicine and Science in Sports and Exercise, 31*, S526-S533.

Ransdell, L.B., Dratt, J., Kennedy, C., O'Neill, S., & DeVoe, D. (2001). Daughters and mothers exercising together (DAMET): A 12-week pilot project designed to improve physical self-perception and increase recreational physical activity. *Women and Health, 33*, 101-116.

Ransdell, L.B., Robertson, L., Ornes, L., & Moyer-Mileur, L. (2004). Generations exercising together to improve fitness (GET FIT): A pilot study designed to increase physical activity and improve health-related fitness in three generations of women. *Women and Health, 40*, 77-94.

Ransdell, L.B., Taylor, A., Oakland, D., Schmidt, J., Moyer-Mileur, L., & Shultz, B. (2003). Daughters and mothers exercising together: Effects of home- and community-based programs. *Medicine and Science in Sports and Exercise, 35*, 286-296.

Raudsepp, L. (2006). The relationship between socio-economic status, parental support and adolescent physical activity. *Acta Paediatrica, 95*, 93-98.

Robinson, J.L., Winiewicz, D.D., Fuerch, J.H., Roemmich, J.N., & Epstein, L.H. (2006). Relationship between parental estimate and an objective measure of child television watching. *International Journal of Behavioral Nutrition and Physical Activity, 3*, 43. doi:10.1186/1479-5868-3-43.

Robinson, T.N. (1999). Reducing children's television viewing to prevent obesity: A randomized controlled trial. *Journal of the American Medical Association, 282*, 1561-1567.

Romero, A.J., Robinson, T.N., Kraemer, H.C., Erickson, S.J., Haydel, K.F., Mendoza, F., et al. (2001). Are perceived neighborhood hazards a barrier to physical activity in children? *Archives of Pediatrics and Adolescent Medicine, 155*, 1143-1148.

Saelens, B.E., & Handy, S.L. (in press). Built environment correlates of walking: A review. *Medicine and Science in Sports and Exercise.*

Sallis, J.F., & Owen, N. (1997). Ecological models. In K. Glanz, F.M. Lewis, & B.K. Rimer (Eds.), *Health behavior and health education: Theory, research, and practice* (2nd ed., pp. 403-424). San Francisco: Jossey-Bass.

Sallis, J.F., Prochaska, J.J., & Taylor, W.C. (2000). A review of correlates of physical activity of children and adolescents. *Medicine and Science in Sports and Exercise, 32*, 963-975.

Salmon, J., Ball, K., Crawford, D., Booth, M., Telford, A., Hume, C., et al. (2005a). Reducing sedentary behaviour and increasing physical activity among 10-year-old children: Overview and process evaluation of the "Switch-Play" intervention. *Health Promotion International, 20*, 7-17.

Salmon, J., Timperio, A., Telford, A., Carver, A., & Crawford, D. (2005b). Association of family environment with children's television viewing and with low level of physical activity. *Obesity Research, 13*, 1939-1951.

Saunders, R.P., Motl, R.W., Dowda, M., Dishman, R.K., & Pate, R.R. (2004). Comparison of social variables for understanding physical activity in adolescent girls. *American Journal of Health Behavior, 28*, 426-436.

Schieber, R.A., & Vegega, M.E. (2002). Reducing childhood pedestrian injuries. *Injury Prevention, 8*, i3-i8.

Schmitz, K.H., Lytle, L.A., Phillips, G.A., Murray, D.M., Birnbaum, A.S., & Kubik, M.Y. (2002). Psychosocial correlates of physical activity and sedentary leisure habits in young adolescents: The Teens Eating for Energy and Nutrition at School study. *Preventive Medicine, 34*, 266-278.

Songül Yalçin, S., Tu rul, B., Naçar, N., Tuncer, M., & Yurdakök, K. (2002). Factors that affect television viewing time in preschool and primary schoolchildren. *Pediatrics International, 44*, 622-627.

Spinks, A., Macpherson, A., Bain, C., & McClure, R. (2006). Determinants of sufficient daily activity in Australian primary school children. *Journal of Paediatrics and Child Health, 42*, 674-679.

Stone, E.J., McKenzie, T.L., Welk, G.J., & Booth, M.L. (1998). Effects of physical activity interventions in youth: Review and synthesis. *American Journal of Preventive Medicine, 15*, 298-315.

Thompson, V.J., Baranowski, T., Cullen, K.W., Rittenberry, L., Baranowski, J., Taylor, W.C., et al. (2003). Influences on diet and physical activity among middle-class African American 8- to 10-year-old girls at risk of becoming obese. *Journal of Nutrition Education and Behavior, 35*, 115-123.

Timperio, A., Crawford, D., Telford, A., & Salmon, J. (2004). Perceptions about the local neighborhood and walking and cycling among children. *Preventive Medicine, 38*, 39-47.

Trost, S.G., Sallis, J.F., Pate, R.R., Freedson, P.S., Taylor, W.C., & Dowda, M. (2003). Evaluating a model of parental influence on youth physical activity. *American Journal of Preventive Medicine, 25*, 277-282.

Trudeau, F., Laurencelle, L., & Shephard, R.J. (2004). Tracking of physical activity from childhood to adulthood. *Medicine and Science in Sports and Exercise, 36*, 1937-1943.

Tudor-Locke, C., Ainsworth, B.E., & Popkin, B.M. (2001). Active commuting to school: An overlooked source of childrens' physical activity? *Sports Medicine, 31*, 309-313.

Wagner, A., Klein-Platat, C., Arveiler, D., Haan, M.C., Schlienger, J.L., & Simon, C. (2004). Parent-child physical activity relationships in 12-year old French students do not depend on family socioeconomic status. *Diabetes Metabolism, 30*, 359-366.

Weir, L.A., Etelson, D., & Brand, D.A. (2006). Parents' perceptions of neighborhood safety and children's physical activity. *Preventive Medicine, 43*, 212-217.

Wiecha, J.L., Sobol, A.M., Peterson, K.E., & Gortmaker, S.L. (2001). Household television access: Associations with screen time, reading, and homework among youth. *Ambulatory Pediatrics, 1*, 244-251.

Wilson, A.N., & Dollman, J. (2007). Social influences on physical activity in Anglo- and Vietnamese-Australian adolescent males in a single sex school. *Journal of Science and Medicine in Sport, 10*, 147-155.

Ziviani, J., Scott, J., & Wadley, D. (2004). Walking to school: Incidental physical activity in the daily occupations of Australian children. *Occupational Therapy International, 11*, 1-11.

CHAPTER

12

Peers

Alan L. Smith, PhD ▪ Meghan H. McDonough, PhD

As detailed in a number of chapters in this book, considerable effort has been devoted to understanding the significance of social agents to youth physical activity and sedentary behavior. Key individuals in young people's lives are believed to influence their physical activity behavior in a variety of ways, for example by transmitting attitudes or values, modeling active or inactive pursuits, affording access to environments that foster or inhibit active living, and providing psychological support or reinforcement for certain behaviors. A large base of correlational research provides support for these potential mechanisms of influence on youth physical activity (e.g., see Sallis, Prochaska, & Taylor, 2000). In turn, this has provided researchers and practitioners with a foundation for pursuing further research that addresses social influence on youth physical activity and sedentary behavior and for constructing physical activity interventions.

Understandably, much of the research on social influence on youth physical activity levels has targeted the role of parents. This is logical given the degree to which young people depend on their parents for instrumental and emotional support and the amount of time they are exposed to their parents, especially during childhood. This emphasis is also, in part, likely a function of the tendency for physical activity researchers and professionals to view young people as passive agents in the socialization process. Such a viewpoint places young people at the conceptual endpoint of the socialization process rather than acknowledging that they are active socialization agents—a prevalent theme in contemporary developmental theory and research (e.g., see Bugental & Grusec, 2006). A natural result is to overlook young people themselves and young people's peers as agents that foster or undermine physical activity behavior.

We maintain that researchers and practitioners would benefit by more closely examining the link between peers and youth physical activity and sedentary behavior. Research investigations can be found in the literature that specifically emphasize the role of peers in youth physical activity levels (e.g., Anderssen & Wold, 1992; Luszczynska et al., 2004; Prochaska, Rodgers, & Sallis, 2002; Smith, 1999; Voorhees et al., 2005). These studies are, however, few in number, largely correlational, and often limited to broad-based survey research that happens to include a peer variable. Thus, there is a considerable knowledge gap on peers and youth physical activity. This is a surprising state of affairs given that a good deal of young people's physically active and inactive behavior is pursued along with other young people.

Our aim for this chapter is to encourage greater attention to peers in youth physical activity and sedentary behavior research and practice. We first address key conceptual matters relative to the study of peers by defining what a peer is, presenting a brief synopsis of selected germane theoretical perspectives, and providing an overview of key peer-related constructs and variables that are examined in developmental and physical activity research. In the second section of this chapter we detail what is known about peers and youth physical activity, sharing key research findings and discussing their implications for theory and practice. In the final section we discuss several future research directions that hold promise for clarifying the link between peers and youth physical activity and sedentary behavior. Specifically, we argue that research targeting closeness of relationships, the interface of peers and other social agents in a young person's life, and the role of peers in successful physical activity promotion efforts can meaningfully extend the knowledge base. We supplement this discussion by sharing important considerations in pursuing such research.

Basic Concepts in Peer-Related Research

Fundamental to the study of peers and youth physical activity and sedentary behavior is a conceptual understanding of peers, from definitional, theoretical, and operational standpoints. The following sections provide the base upon which quality peer-related research may be conducted.

Definitions

Peer is rarely explicitly defined in developmental or exercise science research. A widely held assumption is that peers are individuals who are at or near the same age as the target(s) of study (see Smith, 2007). Thus, individuals in the same grade or classroom, youth on an organized sport team, or children using an age-specified play structure at a park would be considered peers. In certain activity settings, such as organized sport, it may be appropriate to define peers instead with reference to similarity in characteristics other than age, such as skillfulness (Smith, 2007). This can be useful when researchers wish to address questions pertaining to competence and motivation, though often in this circumstance skillfulness and age are confounded. Nonetheless, in the overwhelming bulk of youth physical activity research, a peer is considered someone of the same age or close to the same age as the young person of interest.

In youth physical activity research, the distinction of concern is generally that of peers versus significant adults in one's life such as parents or teachers. Peers are distinct from these adults not only in age, but also with reference to balance of power. Young people can be constrained by rules and expectations of significant adults as well as by their dependence on adults for some modes of transportation. In contrast, young people are in a stronger position to negotiate expectations and exert personal control with peers. Further, within the constraints of where one lives, goes to school, and participates in organized activities, youth have some opportunity to choose their same-age friends and close associates. For these reasons, the nature of young people's relationships with peers is considered distinct from the nature of their relationships with adults.

An important special case is that of siblings. Siblings constitute peers in that they are more similar to the young person in age and power relative to adults, though often age differences result in a power asymmetry favoring older siblings. At the same time, siblings represent part of a young person's family and, like parents, cannot be chosen. Siblings have certain common life experiences that are not shared with other peers, most notably the socialization influences of their parents. Thus, when siblings are examined in research on young people's physical activity they are often, appropriately, treated as nested within the family system (e.g., Duncan et al., 2004). Given the limited amount of physical activity research emphasizing siblings, particularly within a peer framework, we do not address siblings in this chapter.

Theoretical Perspectives

Various theoretical approaches to understanding the psychosocial importance of peers have shaped the direction of human development research, and to some degree physical activity research on peers. Furthermore, motivational frameworks that incorporate social influence constructs are useful guides for research on peers and youth physical activity. This is the case because choice is involved in the expression of physical activity and sedentary behaviors, physical activity is effortful, and persistent engagement in physical activity is required in order for one to accrue health benefits. In the following subsections we provide a brief description of selected developmental and motivational perspectives that are germane to the examination of peers and youth physical activity and sedentary behavior. These perspectives illuminate peer constructs of importance as well as specify potential pathways for peer influence on physical activity. We refer readers interested in an extended treatment of developmental theory addressing peers to Rubin, Bukowski, and Parker's (2006) review chapter in *Handbook of Child Psychology*. Further, we direct readers interested in an extended treatment of motivation theory, as applied to youth physical activity and sedentary behavior, to other chapters in this book (e.g., chapters 7 and 8).

Interpersonal Theory of Psychiatry

The interpersonal theory of psychiatry (Sullivan, 1953) has been an especially important guide to research on the psychosocial importance of peers to young people, for several reasons. First, the theory is developmental, addressing in detail interpersonal processes from infancy through late adolescence, as well as the ways in which shortcomings in youth development can translate to deficits in adult functioning. Second, Sullivan addressed both the peer group and specific friendships as salient developmental entities. He argued that both the peer group and friends help enable a young person's perspective of self to evolve from one that is egocentric to one that is understood in relation to others. In younger childhood, acceptance by the peer group is important to the development of perspectives on authority, competition, cooperation, and other matters. Moving into middle and later childhood, young people actively seek close same-sex friendships because these fulfill a budding need for interpersonal intimacy and validation. Thus, a third reason for the importance of this theory to the study of peers is that emphasis is placed on closeness or quality of associations with peers. In other words, peer *relationships* involving familiarity, exchanges, and interactions are of central importance to development. Framing this within the context of physical activity and sedentary behavior, it is not sufficient to assess whether peers participate in active or sedentary pursuits with a young person. One must understand how youth relate to one another when engaged in such activities. Young people who feel accepted by the group and validated by specific peers in physical activity settings in most cases would be expected to hold physical activity as a higher priority than those who do not feel accepted or able to fulfill needs for intimacy.

Finally, though Sullivan (1953) viewed the peer group and specific friendships as having distinct developmental roles, he viewed them as related entities that can serve as surrogates for one another when necessary. For example, developmental shortcomings stemming from rejection by the peer group could be overcome by the existence of a high-quality friendship. This extends to other types of relationships. According to this perspective, the potential negative impact of strained child–parent relationships, for example, can be offset to some degree by high-quality peer relationships. The implication of this idea is that assessing multiple markers of social relationships in youth physical activity research may be required for an accurate understanding of physical activity and sedentary behavior choices of young people. The constellation of relationships a young person experiences in physical activity settings may be the important predictor of her or his behavior. Overall, this perspective has been underutilized in youth physical activity research to date. It does, however, have much to offer in helping us understand the salience of peers to the physical activity and sedentary behavior of young people.

Social Cognitive Theory

In contrast to the interpersonal theory of psychiatry, Bandura's (1986) social cognitive theory has had a broad impact on the study of youth physical activity. Bandura suggests that personal factors (e.g., cognition), the environment, and behavior affect one another in reciprocal fashion. This process is referred to as triadic reciprocity. Social agents are conceptualized as features of an individual's environment that are involved in this process. Bandura's ideas stemmed from social learning theory traditions, and therefore social reinforcement is an important feature of his perspective. Constructs such as peer support for physical activity or sedentary behavior—when defined as perceived encouragement by peers—offer a basic representation of reinforcement by peers that is commonly employed in youth physical activity research.

Bandura (1986) additionally introduced the concept of vicarious learning, also termed modeling. This is a process whereby visual or verbal information contained in the actions of others is perceived, stored as a cognitive representation in memory, and then converted to one's own thoughts, feelings, and actions. The modeling process and resultant motor skill and psychological outcomes have been extensively studied in motor learning and performance settings (see McCullagh & Weiss, 2001, for a review). Like representations of reinforcement, a broad depiction of modeling is typically employed in youth physical activity research. The level of activity of friends or peers, often as perceived by the young research participant, is a customary representation of peer modeling. Overall, social cognitive theory has been significant in helping researchers conceptualize ways in which peers can influence young people's lives because it draws attention to social agents as models of behavior and sources of reinforcement.

Competence Motivation Theory

Harter's (1978, 1981, 1987) competence motivation theory and her related work on the development of the self have been extensively applied in youth organized sport research (see Weiss & Ferrer-Caja, 2002) and suggest pathways by which social agents foster motivation in particular achievement domains. Competence motivation theory is an extension of White's (1959) views on the sources of motivation underlying competence development. It specifies that mastery behaviors in specific domains (e.g., physical) increase or decrease intrinsic motivation for pursuing such behavior in the future as a function of the responses of significant others, affect, and self-perceptions. Reinforcement, modeling, and approval of mastery attempts by significant others are proposed to foster young people's internalization of rewards and goals, enhance their perceptions of competence and internal control, and generate positive affect. This pathway yields enhancement of intrinsic motivation. Alternatively, lack of reinforcement or modeling and expression of disapproval of mastery attempts result in child dependence on others for rewards and goals, increased perceptions of incompetence and external control, and negative affect. This pathway undermines intrinsic motivation. Thus, how significant others respond to a young person in the physical domain is critical to motivation for physical pursuits. When youth feel well regarded by others in physical activity settings they are more likely to seek challenges, exert effort, and persist in their physical activity pursuits. If they do not feel well regarded by others in physical activity settings, other settings where they do experience such validation may become relatively more attractive. Given the intensive social exchanges among peers in many youth physical activity and sedentary contexts, the pathways of social influence described by this framework are useful in advancing understanding of how peers may impact youth physical activity levels.

Other Conceptual Frameworks

The theories just outlined are only a sampling of conceptual frameworks that can be used to undergird peer-focused efforts to better understand and promote youth physical activity. Among other frameworks that can offer guidance to researchers and practitioners, several emphasize the goals that individuals possess (see chapter 8). Nicholls' (1984, 1989) perspective, for example, is that motivational challenges are associated with emphasizing normative comparison goals when one is seeking to demonstrate competence in achievement settings. Because only a select few can be "the best," individuals who do not maintain high perceptions of competence while possessing such goals are proposed to be motivationally at risk. Achievement goal perspectives also call attention to the role of the motivational climate that young people perceive as they formulate their goals (Ames, 1992). Motivational climate perceptions are shaped by the expectations held, rewards distributed, and degree of autonomy supported by significant others. Recent work in the organized youth sport setting provides evidence that peers are key agents in shaping motivational climates (Ntoumanis, Vazou, & Duda, 2007). Additionally, beyond the competence-related goals typically emphasized in this literature, there is interest in the agentic social goals (e.g., gaining the admiration of others) and communal social goals (e.g., affiliating with others) that individuals bring to achievement settings and the motivational implications of their degree of fulfillment (Ojanen, Grönroos, & Salmivalli, 2005; Urdan & Maehr, 1995).

Closely linked with the pursuit of social goals is the process of impression management, whereby individuals attempt to control the impressions that others form of them (Leary & Kowalski, 1990). Leary (1992) showcased a variety of ways in which impression management is salient in exercise and sport contexts, with a number of examples linked to physical activity motivation. For youth, physical activity pursuits such as organized sport can offer the means to fulfill social goals and make positive impressions on peers, as physical prowess represents valuable social currency among peers (Smith, 2007). At the same time, a strong desire to make a positive impression with peers, coupled with concerns about one's physical appearance or competence, could drive a young person to avoid physical activities and seek alternative pursuits that enable desired impressions to be made. Aligning with yet another theoretical orientation, perceived expectations of significant others regarding physical activity or sedentary pursuits, as well as one's desire to comply with those expectations, constitute the subjective norm construct specified in the theories of reasoned action and planned behavior (Ajzen, 1985; Fishbein & Ajzen, 1975; see also chapter 7). Overall, a host of conceptual frameworks suggest that, beyond what peers say and do, an individual's goals and social concerns are important considerations for understanding motivated behaviors such as physical activity.

Peer Constructs

Based on the theoretical perspectives we have outlined, a host of peer constructs emerge that can be examined in youth physical activity research. As noted in the previous section, peer support and peer modeling, albeit in relatively unrefined forms, have been the most frequently employed constructs in this research area. Typically they have been incorporated into larger-scale correlational investigations designed to illuminate potential determinants of physical activity. Other peer constructs warrant greater attention in such efforts, as well as in more fine-grained investigations that are theoretically grounded and that specifically target social influences on youth physical activity. We assemble these constructs into three general categories in table 12.1 and provide a description of both the construct and how it is typically assessed in human development or physical activity research or both. The first two categories of peer constructs, those pertaining to the peer group (e.g., peer network structure, peer acceptance) and specific peer relationships (e.g., friendship quality), are inspired by Sullivan's (1953) view that peer groups and friendships represent distinct agents of social influence. The third category, peer-referenced constructs, represents broad social goals or concerns that can shape a young person's affect, beliefs, and behavior in physical activity contexts as well as in contexts that involve sedentary behavior. While our list is not exhaustive, it clearly showcases a broad range of peer constructs with potential bearing on youth physical activity levels and sedentary behavior. In the next section of this chapter, we provide an overview

Table 12.1 Selected Peer Constructs of Potential Importance to Youth Physical Activity and Sedentary Behavior

Construct	Description and assessment
	PEER GROUP
Peer network	Often not explicitly defined, but characterized by relational connections among peers that can exhibit structure reflecting features such as density of peer connections, cliques, and position or centrality of particular members of a group. Self-report and other methods (e.g., mother report of child's playmates) can be employed to assess peer networks.[4]
Peer acceptance/ popularity	Being liked or accepted by peers.[3] Sociometric techniques whereby number of positive nominations or average liking ratings within a group of peers are used to assess popularity. Also, young people's perceptions of peer acceptance are assessed via self-report.
Peer rejection	Being overtly disliked by peers.[1] Often assessed by sociometric techniques whereby number of negative nominations or average liking ratings within a group of peers are calculated.

(continued)

Table 12.1 *(continued)*

Construct	Description and assessment
	PEER GROUP
Peer support	Social support stemming from the peer group. Social support is reflected in three interdependent features: structural (e.g., network composition), functional (e.g., tangible assistance), and perceptual (e.g., appraisals of received support).[10] Assessed by a range of self-report tools in physical activity research.
	SPECIFIC PEERS
Friendship	A dyadic relationship characterized as close and mutual.[3] Sociometric, parent/teacher report and self-report techniques are used to assess the existence of friendships; that is, whether a young person believes he or she has friends, whether friendship nominations are reciprocated, how many friends one has (a reflection of peer network), and the stability of friendships.
Friend characteristics	Behaviors, beliefs, and other qualities of a friend. Characteristics of interest will depend on the topic of study[6]. In physical activity research, physical activity and sedentary behavior attitudes and behaviors of one's friend are of interest.
Friendship quality	The nature of a given relationship, reflected in features such as closeness/warmth, reliable alliance, provision of support, and conflict.[6] Often assessed through interview or self-report methods.
Peer modeling	Patterning or acquiring thoughts, emotions, and behaviors in response to observing a peer (or multiple peers).[9] A variety of approaches can be used to assess psychological and behavioral effects of observing a peer, though peer modeling is most often represented by self-report of the model's behavior.
	PEER REFERENCED
Social goals	Goals linked to social motives or needs that can be broadly characterized as agentic (desire for admiration) or communal (desire for affiliation, closeness with another).[8] Assessed via self-report scales.
Impression management	The process by which an individual controls the impressions that other people form of her or him.[7] Self-report assessments can capture constructs related to this process such as impression motivation, identity, and social physique anxiety.
Subjective norm	Composed of perceptions of expectations of significant others regarding one's involvement in a behavior and the desire to comply with those expectations.[2,5] Traditionally assessed by self-report.

[1]Asher 1990; [2]Ajzen 1985; [3]Bukowski and Hoza 1989; [4]Cairns et al. 1998; [5]Fishbein and Ajzen 1979; [6]Leary and Kowalski 1990; [7]Ojanen et al. 2005; [8]Schunk 1998; [9]Vaux 1988.

of research that has explored how such peer constructs relate to youth physical activity participation as well as sedentary behavior.

Peer Research

A great deal of evidence suggests that peers are important social agents in the youth physical activity context. Qualitative studies investigating adolescents' perceptions of influences on their physical activity behavior consistently show that friends are cited as an important influence (e.g., Allison et al., 2005; Humbert et al., 2006; Spink et al., 2006; Vu et al., 2006). Predictive research on potential determinants of physical activity behavior also shows peer variables, in particular peer support and peer modeling, to be key correlates (e.g., Raudsepp & Viira, 2000; Sallis et al., 2002; Taylor et al., 2002). Likewise, with regard to peer relationships, greater perceived peer acceptance and close friendship are associated with more youth physical activity behavior (Smith, 1999). Some of these peer factors also may foster inactivity and sedentary behaviors (e.g., Strauss & Pollack, 2003), though research in this area is sparse. Finally, work on interactions of peer constructs (Vilhjalmsson & Thorlindsson, 1998) and efforts to determine the potential role of peers in physical activity interventions (e.g., Lieberman et al., 2000) represent promising avenues of research activity.

Physical Activity

One of the most frequently examined peer constructs in youth physical activity research is peer support. Research employing this construct is often couched within social cognitive theory, and typically incorporates peer support as a form of social reinforcement for activity. Overall, perceptions of peer support for physical activity, including encouragement, praise, watching, talking about activity, and being active together, predict physical activity behavior (Beets et al., 2006; De Bourdeaudhuij et al., 2005; Duncan, Duncan, & Strycker, 2005; Ommundsen, Klasson-Heggebø, & Anderssen, 2006; Voorhees et al., 2005). For girls, perceived direct help in physical activity from a best friend has also been linked to higher activity rates (Anderssen & Wold, 1992). This is consistent with qualitative work showing that adolescents view participation with friends as enabling of, and providing motivation for, physical activity (Coakley & White, 1992; Spink et al., 2006; Wilson et al., 2005). Finally, a link between peer support and youth physical activity has been demonstrated both in studies that include peer support among a host of potential physical activity determinants (Sallis et al., 2002; Taylor et al., 2002) and in studies that specifically target types and sources of social support for activity (Beets et al., 2006; Duncan et al., 2005).

Additionally, as specified in social cognitive theory, if peers model physical activity behavior, this should have an impact on the observing youth's activity levels. Qualitative work on adolescents indicates that participation of friends in physical activity is a reason why young people participate themselves, particularly in unstructured forms of physical activity (Spink et al., 2006). Some quantitative studies have shown that youths' perceptions of their friends' physical activity participation levels predict physical activity behavior (Anderssen & Wold, 1992; Wold & Anderssen, 1992). However, not all studies support this finding. For example, Vilhjalmsson and Kristjansdottir (2003) found that best friend physical activity levels did not predict youth activity level if sport club participation was controlled for. Voorhees and colleagues (2005) found that a composite score of the physical activity status (yes/no) of up to three best friends was not associated with physical activity in multivariate analyses. However, measurement shortcomings may have influenced their results. Overall, peer activity levels may play a role in youth physical activity, but measurement issues and the interaction between friend activity level and organized physical activity participation will need to be addressed before firm conclusions can be drawn.

A number of studies have combined peer support and peer modeling into a single construct, often termed peer influences. While this work does not enable us to explore the relative contributions of support and modeling to youth physical activity, the peer influences construct typically emerges as a significant predictor of activity (Sabiston & Crocker, 2008; Wu, Pender, & Noureddine, 2003), even when many other possible determinants of physical activity behavior are simultaneously considered (Sallis et al., 2002; Taylor et al., 2002). In contrast to this research approach, one study by Vilhjalmsson and Thorlindsson (1998) adopted a more fine-grained examination of peer support and modeling by exploring the interaction of these constructs in predicting youth physical activity. They found that friend activity level was a stronger predictor of one's activity level when friend support was higher, suggesting that the interaction of these variables is important.

A limited number of studies have examined the association between peers and youth physical activity using sociometric or peer network assessments (e.g., Schofield et al., 2007; Strauss & Pollack, 2003; Voorhees et al., 2005). For example, using sociometric methods, Strauss and Pollack (2003) demonstrated that receiving more friendship nominations from classmates, an indication of popularity, is a significant predictor of physical activity participation among adolescents aged 13 to 18 years. This suggests that researchers need to consider not only individual perceptions of peer acceptance but also popularity as identified by one's peers in order to fully understand the influence of peers on physical activity levels. Voorhees and colleagues (2005) used a peer network–type approach by having adolescent girls identify three of their closest friends. The girls then rated these friends on a variety of characteristics and behaviors including activity level, whether they were active together and how often, and whether they asked each other to be active together. In multivariate analyses the researchers found that the frequency of being active together, a form of peer support, was the only significant predictor of activity in this sample.

Few studies have explored whether elements of *relationships* with peers (e.g., closeness of friendships, feelings of acceptance by the peer group) contribute to the prediction of activity. Qualitative studies on adolescent boys (Allison et al., 2005) and on adolescents of high and low socioeconomic status (Humbert et al., 2006) suggest that meeting friends, spending time together, and building relationships are motivating factors for engaging in physical activities. Having opportunities for socializing, characterized by having more friends in the neighborhood and many same-age peers to hang out with, are predictors of walking and cycling in one's neighborhood among adolescents (Carver et al., 2005). Even possessing a relatively higher perceived ability to make new friends, or sociability, has been shown to predict activity behavior (Vilhjalmsson & Thorlindsson, 1998). In a study targeting peer relationships specifically, Smith (1999) found that perceptions of peer acceptance and close friendship in the activity context uniquely contributed to the prediction of physical activity motivation and behavior. Overall, this work suggests that it is not just peer characteristics or reinforcement that are associated with youth physical activity levels, but also the nature of a young person's relationships in the physical activity environment.

Peer-referenced constructs have also been examined relative to physical activity. In a physical activity and nutrition study of adolescents from four countries, Luszczynska and colleagues (2004) investigated both tendency to make social comparisons and perception of physical activity as normative in the peer group. Youth scoring higher on these constructs were more physically active. Wu and colleagues (2003) assessed subjective norm for physical activity among Taiwanese adolescents in terms of how much their peers expected them to participate in activity; however, these authors combined peer support, peer modeling, and subjective norm together as a peer construct. This composite both directly and indirectly, via self-efficacy, predicted physical activity behavior. Unfortunately, the relative contribution of peer subjective norm is not distinguishable in this study. Motl and colleagues (2004) found no relationship between subjective norm and moderate-to-vigorous physical activity of Black and White adolescent girls in the United States. However, their items tapped normative beliefs about the expectations of a variety of significant others (i.e., peers, teachers, parents, siblings) and therefore not exclusively peer-referenced subjective norm. This measurement issue may in part explain why, in general, findings for the subjective norm construct are not as robust as findings linked to other components of the theory of planned behavior (see chapter 7).

While research largely supports a link between peers and youth physical activity, with effect sizes of small to moderate magnitude, many studies fail to support this relationship. In most cases, an argument can be made that these findings are influenced by researchers combining health behavior or peer measures in ways that could mask the relationship. For example, health behaviors such as physical activity, eating behaviors, substance use, and safety-related behaviors have been examined in combination as a dependent variable, even though they are not highly correlated (Terre et al., 1992). While studies of this kind typically yield no association between peer constructs and health behaviors (Hazard &

Lee, 1999; McLellan et al., 1999; Terre et al., 1992), they obviously fall short of invalidating a possible link between peers and physical activity behavior specifically. Similarly, peer constructs are often combined with constructs connected to other social agents (e.g., parents, coaches) or become lost in measures that direct respondents to consider "others," a nebulous term that can bring a variety of social agents to mind. As a result, an association (Motl et al., 2004; Saunders et al., 1997), or lack of association (DiLorenzo et al., 1998; Haverly & Krahnstoever Davison, 2005), between the amalgamated social construct and physical activity is difficult to interpret relative to peers.

Other researchers have employed methods that may artificially minimize the relationship between peer constructs and youth physical activity. For example, Vilhjalmsson and Kristjansdottir (2003) entered sport club participation into their regression model before entering a block of social and other potential predictors of youth physical activity, including frequency of a best friend's physical activity participation. The best friend variable did significantly predict youth physical activity, but additional explanation of physical activity variance was only roughly 2% to 3% for the block of variables entered at that step of the regression analysis. Sport clubs are venues where friend activity behavior is likely salient, and therefore this outcome is not surprising. Other work concluding that peer factors (e.g., peer behaviors, beliefs, and approval regarding physical activity) do not predict physical activity failed to properly assess activity (Monge-Rojas et al., 2002). Specifically, the authors factor analyzed self-efficacy, body image, parental influence, social environment influence, and peer influence items that refer to activity, then formed conclusions about predictors of physical activity based on the magnitude of factor loadings. Overall, these studies do not convincingly counter the position that peers and physical activity are meaningfully linked.

While most studies have explored the roles of support and peer activity behavior separately or as additive predictors in regression-type models, there is some evidence that the interaction of peer support and peer modeling uniquely predicts youth physical activity. As noted previously, Vilhjalmsson and Thorlindsson (1998) found that friends' physical activity level is more strongly linked with activity in more emotionally supportive friendships. These findings further suggest that combining different peer constructs into a single subscale or latent variable (e.g., support, modeling and norms; Wu et al., 2003) can mask the nature of a possible relationship between peers and youth physical activity.

Inactivity or Sedentary Behavior

Compared to work emphasizing physical activity, much less research is available on peers and inactivity or sedentary behavior. Qualitative work suggests that boys' attitudes about and treatment of girls in physical activity contexts can contribute to their female peers' inactivity (Kunesh, Hasbrook, & Lewthwaite, 1992; Vu et al., 2006). Also, having inactive friends, experiencing social intimidation, having social invitations that conflict with workout time, lacking a training partner, and having negative experiences with peers within the activity context have been identified as potential barriers to physical activity, particularly among those transitioning out of high school into university (Gyurcsik et al., 2006). Research on sedentary behaviors among youth has been equivocal. Strauss and Pollack (2003) found that receiving more friendship nominations was associated with less television viewing in both normal-weight and overweight adolescents. In another study, teasing by peers was associated with overweight children's greater preference for sedentary versus physically active behaviors (Hayden-Wade et al., 2005). Conversely, Taylor and colleagues (2002) found that peer influences (i.e., friend's activity frequency, friend's encouragement of activity, and friend's participation in activity with the focal child) did not predict sedentary behaviors in normal-weight or overweight adolescents. Also, friends' concerns about fitness did not predict TV viewing in a study of junior and senior high school students (McGuire, Neumark-Sztainer, & Story, 2002). Given that engagement in sedentary pursuits does not necessarily correspond to lack of engagement in active pursuits (see chapter 1), this mix of findings is understandable. Variables such as friends' sedentary behavior frequency and participation of friends in TV viewing with a youngster, for example, are more logical to examine in research on peers and youth sedentary behavior.

Peer-Based Interventions

Finally, there is preliminary evidence that peer-based interventions may have promise for promoting physical activity among youth. One unique study using a peer tutoring intervention matched 10- to 12-year-old deaf students with hearing peers of the same age and gender who were trained to provide cues, modeling, assistance, and feedback in physical education classes (Lieberman et al., 2000). Both the deaf and the hearing students increased their moderate-to-vigorous physical activity from baseline to intervention phases, and effects were retained over a maintenance period. Thus, this type of intervention can have positive effects for both the targeted students and the tutors. Jelalian and Mehlenbeck (2002) added weekly peer-based skills training to a cognitive-behavioral weight management intervention that targeted diet, physical activity, and behavior modification in overweight adolescents. The peer-based skills were developed through outward bound–type activities (e.g., ropes courses) designed to enhance trust, social skills, and self-confidence. Significant changes in physical self-worth were obtained, but not in perceptions of peer rejection or loneliness. Case illustrations suggested that despite the peer rejection and loneliness findings, participants did become more outgoing and assertive with peers. This could influence the effectiveness of youth in obtaining peer support for their lifestyle (i.e., diet, physical activity) changes. Clearly more work is needed in the area of peer-based interventions to replicate existing findings, broaden the range of strategies employed, broaden the range of peer constructs used in evaluation of such interventions, and obtain better understanding of mechanisms underlying intervention success or failure.

Promising Future Research Directions

The extant literature suggests that attempts to better understand the role of peers in youth physical activity levels and sedentary behavior would be well worth researchers' time and effort. Peer constructs predict physical activity; however, at the same time a host of knowledge gaps are evident. Many of the peer constructs listed in table 12.1 have received limited or no attention, and research investigations that incorporate peer constructs have been minimally guided by developmental or motivation theory. In the present section we discuss three future research directions that hold promise for filling important knowledge gaps. Specifically, we encourage future peer-focused physical activity research that attends to closeness of peer relationships, constellations of peer and other social constructs, and peers as agents of change within youth physical activity promotion efforts. We end the section by sharing important considerations in peer research that are pertinent to these and other research directions.

Closeness of Peer Relationships

According to Rubin and colleagues (2006), a full understanding of the psychosocial relevance of peers to young people requires consideration of a hierarchical, integrated social complex. This complex includes characteristics of the individual, current interactions among individuals, patterns of interactions among individuals over time, and characteristics of the group. The reader is referred to Rubin and colleagues' work for extended description of these components. For our purposes, we point out that there are many layers at which peers have potential to influence young people. Assuming that peers are important contributors to youth physical activity, for example, merely knowing that a youngster has active peers may not sufficiently explain that youngster's own physical activity attitudes and behavior. Also of likely importance are *how* the peers interact with the youngster, the history the youngster has with those peers, and the broader psychosocial implications of being affiliated with a certain group of peers.

It is difficult to capture all features of this social complex within a particular investigation; therefore researchers often must make choices about how to explore the potential influence of peers on physical activity and sedentary behavior. We believe that researchers would make significant strides by choosing to examine *relationships* among peers in physical activity and sedentary pursuits, particularly as reflected in the closeness or quality of those relationships. Such work would yield several benefits. For one, it would increase the scope of peer constructs that are considered in the research and would encourage researchers to make use of Sullivan's (1953) interpersonal theory of psychiatry, a particularly promising yet underutilized developmental theoretical framework for understanding youth physical activity. In particular, research targeting friends as potential sources of influence on youth physical activity levels and sedentary behavior would become more thorough.

Hartup (1996) points out that the developmental significance of friends goes beyond simply having them. The characteristics of one's friends and the quality of relationships with friends are also very important. Settings where youth make friends with common interests who also help them feel well regarded, supported, and validated are settings we would expect youth to value and revisit. Although some research on physical activity behavior (e.g., Smith, 1999) and youth sport motivation (e.g., Ommundsen et al., 2005; Ullrich-French & Smith, 2006; Weiss & Smith, 2002) has considered friendship closeness or quality, this work is in the early stages. To this point friendship has been broadly defined in youth physical activity research, often making it difficult to distinguish from peer group constructs. As noted earlier in this chapter, Sullivan (1953) views friendships and the peer group as making distinct contributions to youth well-being.

Another benefit of targeting closeness of peer relationships in future research is that this will move us toward an emphasis on dyads and small groups, whereas the dominant research emphasis at present is on the individual (i.e., psychological vantage) or the broader population (i.e., public health vantage). Attending to dyads and small groups in future work could afford understanding of peer dynamics in youth physical activity settings that translate well to practice. Many professionals seeking to benefit from emerging knowledge on youth physical activity levels and sedentary behaviors—especially physical educators—manage exchanges among young people in such small-group settings. Finally, a related potential benefit of research emphasizing relationship closeness is that the knowledge obtained could shift the dominant practice orientation from promoting physical activity to promoting quality exchange among young people in physical activity settings. By bringing human exchange to the forefront of youth physical activity research and practice, we will make meaningful progress toward an ethical paradigm for physical activity promotion and away from objectifying young people in our zeal to make them healthy (see chapter 5).

Constellations of Peer and Other Social Constructs

As overviewed earlier in this chapter and presented in table 12.1, a host of peer constructs are worth continued or intensified consideration in youth physical activity and sedentary behavior research. There are also, of course, numerous other social agents who may influence young people's physical activity and sedentary behavior. Thus, many social constructs (e.g., parent modeling, teacher reinforcement) beyond those explicitly linked to peers require attention. A significant challenge that we believe has not been sufficiently met by researchers is how to combine these constructs in research designs to yield meaningful information about social influence on youth physical activity levels. The chief strategies used to this point fall into two categories. The first, as discussed previously in this chapter, has been to amalgamate social constructs (e.g., Anderssen & Wold, 1992; DiLorenzo et al., 1998; Stucky-Ropp & DiLorenzo, 1993). For example, perceived support from parents, peers, and others might be combined to produce a general social support assessment. There are obvious shortcomings of this approach, most notably that various sources of social support are presumed to be of equal importance to young people and that social support is an additive construct. Certain people may be more important to youth than others, and not all significant people in a young person's life will think and act similarly. What are the implications, for example, of having highly supportive parents along with unsupportive peers relative to physical activity involvement?

The second strategy, which is not mutually exclusive with the first, is to assess multiple social constructs and analyze them in parallel within regression-type models (e.g., Anderssen & Wold, 1992; Sallis et al., 2002; Stucky-Ropp & DiLorenzo, 1993). Typically, interest is in how much physical activity variance is explained by the respective constructs, with the predictor variable that explains the most variance considered to be of greatest importance. Given some degree of overlap among social constructs, naturally there are challenges in obtaining consistent patterns of associations across studies. Furthermore, and this is not an insignificant criticism despite the well-known challenges in obtaining sufficient statistical power, interactions of social constructs are regularly ignored in such investigations.

This is unfortunate because dependencies may exist among constructs, as demonstrated in Vilhjalmsson and Thorlindsson's (1998) work. These authors found that friend physical activity associated more strongly with youth physical activity in the context of relatively more supportive friendships. Social constructs deemed less important when analyzed in parallel may exhibit greater importance if they moderate the association of other social constructs with physical activity or sedentary behavior. For example, in Ullrich-French and Smith's (2006) sport motivation research that included parent–child and peer relationship predictor variables, main effects showed perceived acceptance by peers to be a dominant construct. However, significant interaction terms suggested that relatively high perceived acceptance by peers in sport may be insufficient for optimal motivation-related outcomes. The most adaptive outcomes were observed when such perceptions were coupled with relatively high perceptions of friendship quality or parent–child relationships in sport (or both). Thus, there is danger in prematurely dismissing constructs as unimportant when their interaction with other constructs has not been examined.

An alternative to examining interaction terms in regression models is to employ cluster analysis to uncover meaningful constellations of social constructs. By generating ideographic profiles from a chosen set of variables, this person-centered strategy offers a more holistic understanding of the individual than traditionally employed variable-centered research strategies (see Magnusson, 1998). Recent work, for example, has used cluster analysis to understand individual profiles of engagement in different forms of sedentary behavior, physical activity, or both (Marshall et al., 2002; Wang et al., 2006; Zabinski et al., 2007), showing that such engagement falls into a finite set of combinations.

Applied to social constructs, cluster analysis can illuminate the existing combinations of social experiences that individuals possess. Smith and colleagues (2006) cluster analyzed young (ages 10 to 14 years) sport camp participants' perceptions of peer acceptance, positive friendship quality, and friendship conflict in sport. Their analysis generated five peer relationship profiles that appear in figure 12.1. With use of a z-score criterion of ±0.5 in judging clusters as relatively high or low on a given peer construct, the first profile was labeled *isolate* to reflect individuals with low perceptions of peer acceptance and positive friendship quality along with a tendency toward low perceptions of friendship conflict. The second profile was labeled *reject* because the individuals in that group held undesirable peer acceptance and friendship perceptions. Those in the *survive* profile had sound friendship responses but perceived low peer acceptance. The fourth profile was labeled *thrive* because all perceptions were in adaptive directions. That is, these individuals simultaneously possessed high peer acceptance and low friendship conflict perceptions along with a tendency toward high positive friendship quality perceptions. The final group was labeled *alpha*, as individuals in this group held perceptions similar to those in

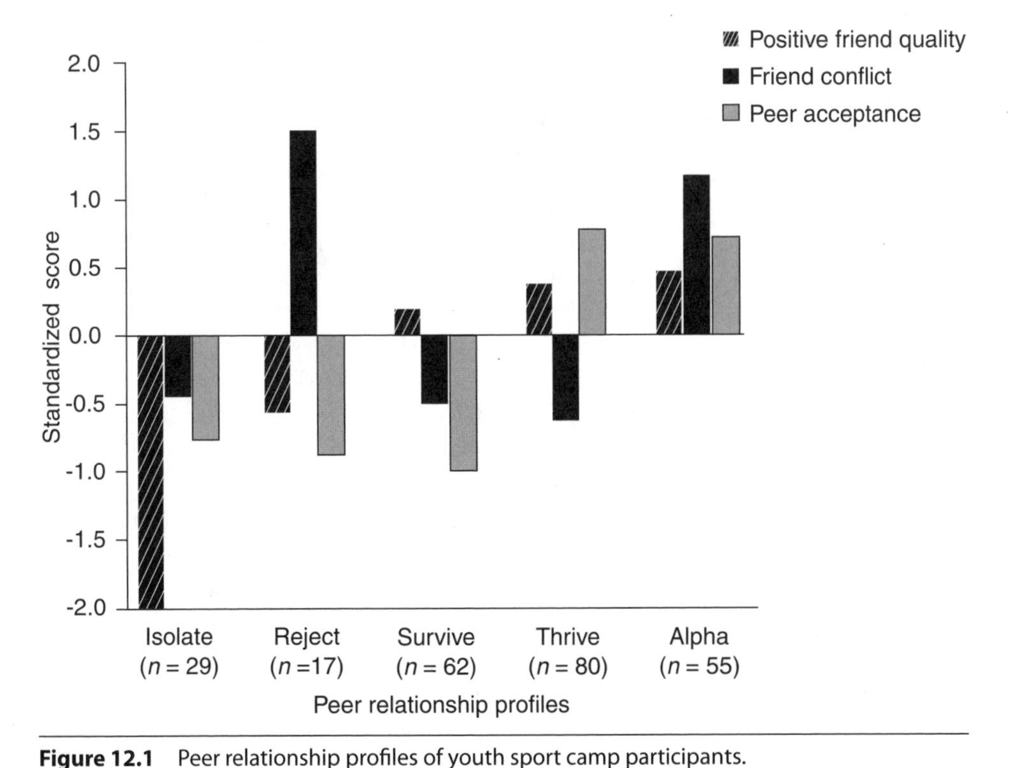

Figure 12.1 Peer relationship profiles of youth sport camp participants.

Reprinted, by permission, from A. L. Smith et al., 2006, "Peer relationship profiles and motivation in youth sport," *Journal of Sport and Exercise Psychology* 28: 373.

the thrive group for peer acceptance and positive friendship quality. However, they coupled this with high perceptions of friendship conflict. These profiles were motivationally distinct, as supported by profile group differences on key motivation-related constructs.

Yet to be published work on peer relationships and general physical activity attitudes and behavior of middle school students shows that these profiles are replicable, stable across a one-year time span, and salient to motivation-related constructs, including self-reported physical activity behavior (Smith, Ullrich-French, & Wisdom, 2005). Interestingly, this work shows that, while the peer relationship profiles are stable, more than half of youth will change profile membership over the year. This suggests that peer constructs are sufficiently malleable to justify the development of peer-based interventions for physical activity promotion.

Adopting research strategies that capture constellations of social constructs holds great promise for filling key knowledge gaps on youth physical activity and sedentary behavior. Such an approach would be wholly consistent with Sullivan's (1953) theoretical perspective as overviewed earlier in this chapter. It also may illuminate strategies for better modeling the interchange of psychological and social factors that can facilitate or impede active living. This would contribute to emerging efforts that are grounded in ecological models and target better understanding of extraindividual influences on physical activity behavior (see chapter 4; Sallis & Owen, 1997; Spence & Lee, 2003; Welk, 1999).

Peers and Physical Activity Promotion

Preliminary work suggests that there is value in incorporating peer tutoring, peer modeling, peer reinforcement, and other peer-based strategies in youth physical activity promotion efforts (Jelalian & Mehlenbeck, 2002; Lieberman et al., 2000). These approaches are largely inspired by social cognitive theory (Bandura, 1986), which emphasizes modeling and support mechanisms for change, and are in line with research showing that peer support and modeling correlate with youth physical activity. Other intervention approaches would appear to function, at least in part, through influence on peer constructs as well. For example, interventions based on playground markings (see chapter 13) may create change through modification of peer networks and peer interactions. This is so because markings can encourage certain configurations of youth to play games and can provide boundaries that afford previously marginalized youth on the playground more room to play or the opportunity to be more centrally involved in active recess pursuits.

Also, fostering climates that address youths' desire for social affiliation and validation, encouraging pursuit of cooperative goals and shared decision making, and limiting anxiety-producing normative comparisons (e.g., picking teams) that can marginalize some youth can promote more positive peer relationships in physical activity settings (see Weiss & Stuntz, 2004). Ideally this translates into more positive attitudes toward and involvement in physical activity. Overall, a range of potentially efficacious peer-based interventions can be generated based on existing theory and empirical work. The theoretical perspectives overviewed earlier in this chapter, including Sullivan's (1953) underutilized framework, are amenable to practical usage and should be considered in physical activity promotion efforts. As part of this process, future research is warranted that carefully assesses the degree to which peers may assist in facilitating, or at least not inhibiting, youth physical activity.

Because many interventions that are not explicitly peer based may, nonetheless, succeed or fail as a function of intervention effects on peer interactions, it would behoove researchers to consider the following questions. Are peers potential sources of modeling and reinforcement, for good or bad, in the intervention program? Does the intervention program foster supportiveness in ways that can enhance closeness of peer relationships within the physical activity setting? Does the intervention program foster normative comparison in ways that can undermine positive social perceptions in the physical activity setting? Answering such questions could lead to useful modifications of physical activity promotion programs, or at least direct researchers to important peer constructs to track when they are evaluating intervention efficacy and the retention of any effects. Given that many youth physical activity interventions take place in group settings (e.g., physical education) and that youth spend their school day and much after-school time with peers, it seems that much could be gained by considering peers in the evaluation of these interventions.

General Research Considerations

Important considerations in conducting research on peers in physical activity settings have been discussed in this chapter and also are described in detail elsewhere (see Smith, 2003, 2007; Weiss & Stuntz, 2004). These include addressing multiple levels of social complexity (e.g., individual and group), addressing the interface of peer and other social agents, and drawing from existing theory and research in constructing and evaluating peer-based interventions. Additional considerations include age, gender, onset of puberty, social transitions (e.g., moving from primary school to secondary school), and concerns about peers outside of the physical activity context.

For example, as youth move from childhood to adolescence, they become more concerned about loyalty and intimacy in their friendships—relationship features that are less tangible than simply sharing common characteristics or interests. They also go through puberty, which, as a function of the timing of onset relative to peers, can affect their physical self-perceptions (see chapter 6). Further, this period is often marked by transition to a new school and, therefore, concerns about "fitting in" with new peers who are coparticipants in physical activity and peers who are not physically active. In summary, there are a host of factors that likely moderate the peer–physical activity linkage. The Smith (2003,

2007) and the Weiss and Stuntz (2004) papers provide additional discussion of these and other issues that require consideration by researchers examining peers in physical activity and sedentary behavior contexts.

Conclusion

We believe that peers merit greater attention from youth physical activity researchers and practitioners because peers are closely and extensively engaged in physical activity and sedentary behaviors with young people and because they are of considerable psychosocial importance through childhood and adolescence. Theory and research provide some understanding of the link between peers and youth physical activity levels; however, substantial knowledge gaps exist that will require significant effort to fill. Attending to a greater breadth of peer constructs, exploring the intersection of peers and other agents within young people's social networks, and considering how peers complement or detract from physical activity promotion efforts can yield considerable research and practice advancements. We encourage readers to vigorously and thoughtfully take part in these endeavors.

Applications for Researchers

Peers are understudied by youth physical activity and sedentary behavior researchers despite theory and human development research highlighting peers as influential in the lives of young people. The extant physical activity research literature emphasizes peer support and modeling, in many cases showing positive associations with youth physical activity levels. To enhance the knowledge base, other peer constructs should be examined. Those capturing the closeness of a young person's relationships with peers are particularly important to target. Close relationships arguably have greater influence on young people's behavior than less close or marginal relationships. Beyond this, it is important to examine the constellations of social relationships in young people's lives. At present there is little understanding of the behavioral choices young people make when, for example, parents encourage physical activity and peers do not. Finally, assessing peer constructs when one is evaluating intervention programs could shed light on the effectiveness or ineffectiveness of youth physical activity promotion efforts.

APPLICATIONS FOR PROFESSIONALS

Much more research on the link between peers and physical activity is needed before confident recommendations can be made to practitioners. Nonetheless, preliminary evidence suggests that peer-based interventions for the promotion of physical activity hold great promise. Successful interventions will likely be those that enhance a young person's perceptions of closeness with peers in physical activity contexts and involve sustained reinforcement and modeling of physical activity by peers. When constructing a physical activity promotion program, practitioners should consider how peers may contribute to or detract from intervention goals, determine whether or not elements of the program promote supportiveness and cooperation among participants, and track peer constructs as a means of assessing program effectiveness. Practitioners also should be aware that age, gender, biological maturation, social transitions, and social goals or concerns can influence program effectiveness. Physical activity promotion programs that are tailored to the developmental needs and motives of young people are most likely to be efficacious.

References

Ajzen, I. (1985). From intentions to actions: A theory of planned behavior. In J. Kuhl & J. Beckmann (Eds.), *Action-control: From cognition to behavior* (pp. 11-39). Heidelberg: Springer.

Allison, K.R., Dwyer, J.J.M., Goldenberg, E., Fein, A., Yoshida, K.K., & Boutilier, M. (2005). Male adolescents' reasons for participating in physical activity, barriers to participation, and suggestions for increasing participation. *Adolescence, 40*, 155-170.

Ames, C. (1992). Classrooms: Goals, structures, and student motivation. *Journal of Educational Psychology, 84*, 261-271.

Anderssen, N., & Wold, B. (1992). Parental and peer influences on leisure-time physical activity in young adolescents. *Research Quarterly for Exercise and Sport, 63*, 341-348.

Asher, S.R. (1990). Recent advances in the study of peer rejection. In S.R. Asher & J.D. Coie (Eds.), *Peer rejection in childhood* (pp. 3-14). New York: Cambridge University Press.

Bandura, A. (1986). *Social foundations of thought and action: A social cognitive theory.* Englewood Cliffs, NJ: Prentice Hall.

Beets, M.W., Vogel, R., Forlaw, L., Pitetti, K.H., & Cardinal, B.J. (2006). Social support and youth physical activity: The role of provider and type. *American Journal of Health Behavior, 30*, 278-289.

Bugental, D.B., & Grusec, J.E. (2006). Socialization processes. In N. Eisenberg (Vol. Ed.), *Handbook of child psychology:* Vol. 3. *Social, emotional, and personality development* (6th ed., pp. 366-428). Hoboken, NJ: Wiley.

Bukowski, W.M., & Hoza, B. (1989). Popularity and friendship: Issues in theory, measurement, and outcome. In T.J. Berndt & G.W. Ladd (Eds.), *Peer relationships in child development* (pp. 15-45). New York: Wiley.

Cairns, R., Xie, H., & Leung, M. (1998). The popularity of friendship and the neglect of social networks: Toward a new balance. In W.M. Bukowski & A.H. Cillessen (Eds.), *Sociometry then and now: Building on six decades of measuring children's experiences with the peer group* (pp. 25-53). San Francisco: Jossey-Bass.

Carver, A., Salmon, J., Campbell, K., Baur, L., Garnett, S., & Crawford, D. (2005). How do perceptions of local neighborhood relate to adolescents' walking and cycling? *American Journal of Health Promotion, 20*, 139-147.

Coakley, J., & White, A. (1992). Making decisions: Gender and sport participation among British adolescents. *Sociology of Sport Journal, 9*, 20-35.

De Bourdeaudhuij, I., Lefevre, J., Deforche, B., Wijndaele, K., Matton, L., & Philippaerts, R. (2005). Physical activity and psychosocial correlates in normal weight and overweight 11 to 19 year olds. *Obesity Research, 13*, 1097-1105.

DiLorenzo, T.M., Stucky-Ropp, R.C., Vander Wal, J.S., & Gotham, H.J. (1998). Determinants of exercise among children. II. A longitudinal analysis. *Preventive Medicine, 27*, 470-477.

Duncan, S.C., Duncan, T.E., & Strycker, L.A. (2005). Sources and types of social support in youth physical activity. *Health Psychology, 24*, 3-10.

Duncan, S.C., Duncan, T.E., Strycker, L.A., & Chaumeton, N.R. (2004). A multilevel analysis of sibling physical activity. *Journal of Sport and Exercise Psychology, 26*, 57-68.

Fishbein, M., & Ajzen, I. (1975). *Belief, attitude, intention, and behavior: An introduction to theory and research.* Reading, MA: Addison-Wesley.

Gyurcsik, N.C., Spink, K.S., Bray, S.R., Chad, K., & Kwan, M. (2006). An ecologically based examination of barriers to physical activity in students from grade seven through first-year university. *Journal of Adolescent Health, 38*, 704-711.

Harter, S. (1978). Effectance motivation reconsidered. *Human Development, 21*, 34-64.

Harter, S. (1981). A model of intrinsic mastery motivation in children: Individual differences and developmental change. In W.A. Collins (Ed.), *Minnesota symposium on child psychology* (Vol. 14, pp. 215-255). Hillsdale, NJ: Erlbaum.

Harter, S. (1987). The determinants and mediational role of global self-worth in children. In N. Eisenberg (Ed.), *Contemporary topics in developmental psychology* (pp. 219-242). New York: Wiley.

Hartup, W.W. (1996). The company they keep: Friendships and their developmental significance. *Child Development, 67*, 1-13.

Haverly, K., & Krahnstoever Davison, K. (2005). Personal fulfillment motivates adolescents to be physically active. *Archives of Pediatrics and Adolescent Medicine, 159*, 1115-1120.

Hayden-Wade, H.A., Stein, R.I., Ghaderi, A., Saelens, B.E., Zabinski, M.F., & Wilfley, D.E. (2005). Prevalence, characteristics, and correlates of teasing experiences among overweight children vs. non-overweight peers. *Obesity Research, 13*, 1381-1392.

Hazard, B.P., & Lee, C-F. (1999). Understanding youth's health-compromising behaviors in Germany. *Youth and Society, 30*, 348-366.

Humbert, M.L., Chad, K.E., Spink, K.S., Muhajarine, N., Anderson, K.D., Bruner, M.W., et al. (2006). Factors that influence physical activity participation among high- and low-SES youth. *Qualitative Health Research, 16*, 467-483.

Jelalian, E., & Mehlenbeck, R. (2002). Peer-enhanced weight management treatment for overweight adolescents: Some preliminary findings. *Journal of Clinical Psychology in Medical Settings, 9*, 15-23.

Kunesh, M.A., Hasbrook, C.A., & Lewthwaite, R. (1992). Physical activity socialization: Peer interactions and affective responses among a sample of sixth grade girls. *Sociology of Sport Journal, 9*, 385-396.

Leary, M.R. (1992). Self-presentational processes in exercise and sport. *Journal of Sport and Exercise Psychology, 14*, 339-351.

Leary, M.R., & Kowalski, R.M. (1990). Impression management: A literature review and two-component model. *Psychological Bulletin, 107*, 34-47.

Lieberman, L.J., Dunn, J.M., van der Mars, H., & McCubbin, J. (2000). Peer tutors' effects on activity levels of deaf students in inclusive elementary physical education. *Adapted Physical Activity Quarterly, 17*, 20-39.

Luszczynska, A., Gibbons, F.X., Piko, B.F., & Tekozel, M. (2004). Self-regulatory cognitions, social comparison, and perceived peers' behaviors as predictors of nutrition and physical activity: A comparison among adolescents in Hungary, Poland, Turkey, and USA. *Psychology and Health, 19*, 577-593.

Magnusson, D. (1998). The logic and implications of a person-oriented approach. In R.B. Cairns, L.R. Bergman, & J. Kagan (Eds.), *Methods and models for studying the individual* (pp. 33-64). Thousand Oaks, CA: Sage.

Marshall, S.J., Biddle, S.J.H., Sallis, J.F., McKenzie, T.L., & Conway, T.L. (2002). Clustering of sedentary behaviors and physical activity among youth: A cross-national study. *Pediatric Exercise Science, 14*, 401-417.

McCullagh, P., & Weiss, M.R. (2001). Modeling: Considerations for motor skill performance and psychological responses. In R.N. Singer, H.A. Hausenblas, & C.M. Janelle (Eds.), *Handbook of sport psychology* (2nd ed., pp. 205-238). New York: Wiley.

McGuire, M.T., Neumark-Sztainer, D.R., & Story, M. (2002). Correlates of time spent in physical activity and television viewing in a multi-racial sample of adolescents. *Pediatric Exercise Science, 14*, 75-86.

McLellan, L., Rissel, C., Donnelly, N., & Bauman, A. (1999). Health behaviour and the school environment in New South Wales, Australia. *Social Science and Medicine, 49*, 611-619.

Monge-Rojas, R., Nuñez, H.P., Garita, C., & Chen-Mok, M. (2002). Psychosocial aspects of Costa Rican adolescents' eating and physical activity patterns. *Journal of Adolescent Health, 31*, 212-219.

Motl, R.W., Dishman, R.K., Saunders, R.P., Dowda, M., & Pate, R.R. (2004). Measuring social provisions for physical activity among adolescent black and white girls. *Educational and Psychological Measurement, 64*, 682-706.

Nicholls, J.G. (1984). Achievement motivation: Conceptions of ability, subjective experience, task choice, and performance. *Psychological Review, 91*, 328-346.

Nicholls, J.G. (1989). *The competitive ethos and democratic education.* Cambridge, MA: Harvard University Press.

Ntoumanis, N., Vazou, S., & Duda, J.L. (2007). Peer-created motivational climate. In S. Jowett & D. Lavallee (Eds.), *Social psychology in sport* (pp. 145-156). Champaign, IL: Human Kinetics.

Ojanen, T., Grönroos, M., & Salmivalli, C. (2005). An interpersonal circumplex model of children's social goals: Links with peer-reported behavior and sociometric status. *Developmental Psychology, 41*, 699-710.

Ommundsen, Y., Klasson-Heggebø, L., & Anderssen, S.A. (2006). Psycho-social and environmental correlates of location-specific physical activity among 9- and 15-year-old Norwegian boys and girls: The European Youth Heart Study. *International Journal of Behavioral Nutrition and Physical Activity, 3*, 32.

Ommundsen, Y., Roberts, G.C., Lemyre, P., & Miller, B.W. (2005). Peer relationships in adolescent competitive soccer: Associations to perceived motivational climate, achievement goals and perfectionism. *Journal of Sports Sciences, 23*, 977-989.

Prochaska, J.J., Rodgers, M.W., & Sallis, J.F. (2002). Association of parent and peer support with adolescent physical activity. *Research Quarterly for Exercise and Sport, 73*, 206-210.

Raudsepp, L., & Viira, R. (2000). Sociocultural correlates of physical activity in adolescents. *Pediatric Exercise Science, 12*, 51-60.

Rubin, K.H., Bukowski, W.M., & Parker, J.G. (2006). Peer interactions, relationships, and groups. In N. Eisenberg (Vol. Ed.), *Handbook of child psychology:* Vol. 3. *Social, emotional, and personality development* (6th ed., pp. 571-645). Hoboken, NJ: Wiley.

Sabiston, C.M., & Crocker, P.R.E. (2008). Exploring self-perceptions and social influences as correlates of adolescent leisure-time physical activity. *Journal of Sport and Exercise Psychology, 30*, 3-22.

Sallis, J.F., & Owen, N. (1997). Ecological models. In K. Glanz, F.M. Lewis, & B.K. Rimer (Eds.), *Health behavior and health education: Theory, research, and practice* (2nd ed., pp. 403-424). San Francisco: Jossey-Bass.

Sallis, J.F., Prochaska, J.J., & Taylor, W.C. (2000). A review of correlates of physical activity of children and adolescents. *Medicine and Science in Sports and Exercise, 32*, 963-975.

Sallis, J.F., Taylor, W.C., Dowda, M., Freedson, P.S., & Pate, R.R. (2002). Correlates of vigorous physical activity for children in grades 1 through 12: Comparing parent-reported and objectively measured physical activity. *Pediatric Exercise Science, 14*, 30-44.

Saunders, R.P., Pate, R.R., Felton, G., Dowda, M., Weinrich, M.C., Ward, D.S., et al. (1997). Development of questionnaires to measure psychosocial influences on children's physical activity. *Preventive Medicine, 26*, 241-247.

Schofield, L., Mummery, W.K., Schofield, G., & Hopkins, W. (2007). The association of objectively determined physical activity behavior among adolescent female friends. *Research Quarterly for Exercise and Sport, 78*, 9-15.

Schunk, D.H. (1998). Peer modeling. In K. Topping & S. Ehly (Eds.), *Peer-assisted learning* (pp. 185-202). Mahwah, NJ: Erlbaum.

Smith, A.L. (1999). Perceptions of peer relationships and physical activity participation in early adolescence. *Journal of Sport and Exercise Psychology, 21*, 329-350.

Smith, A.L. (2003). Peer relationships in physical activity contexts: A road less traveled in youth sport and exercise psychology research. *Psychology of Sport and Exercise, 4*, 25-39.

Smith, A.L. (2007). Youth peer relationships in sport. In S. Jowett & D. Lavallee (Eds.), *Social psychology in sport* (pp. 41-54). Champaign, IL: Human Kinetics.

Smith, A.L., Ullrich-French, S., Walker, E., & Hurley, K.S. (2006). Peer relationship profiles and motivation in youth sport. *Journal of Sport and Exercise Psychology, 28*, 362-382.

Smith, A.L., Ullrich-French, S., & Wisdom, S.A. (2005). Stability of youth peer relationship profiles in the physical domain. *Journal of Sport and Exercise Psychology, 27*, S142-S143. [Abstract]

Spence, J.C., & Lee, R.E. (2003). Toward a comprehensive model of physical activity. *Psychology of Sport and Exercise, 4*, 7-24.

Spink, K.S., Shields, C.A., Chad, K., Odnokon, P., Muhajarine, N., & Humbert, L. (2006). Correlates of structured and unstructured activity among sufficiently active youth and adolescents: A new approach to understanding physical activity. *Pediatric Exercise Science, 18*, 203-215.

Strauss, R.S., & Pollack, H.A. (2003). Social marginalization of overweight children. *Archives of Pediatrics and Adolescent Medicine, 157*, 746-752.

Stucky-Ropp, R.C., & DiLorenzo, T.M. (1993). Determinants of exercise in children. *Preventive Medicine, 22*, 880-889.

Sullivan, H.S. (1953). *The interpersonal theory of psychiatry.* New York: Norton.

Taylor, W.C., Sallis, J.F., Dowda, M., Freedson, P.S., Eason, K., & Pate, R.R. (2002). Activity patterns and correlates among youth: Differences by weight status. *Pediatric Exercise Science, 14*, 418-431.

Terre, L., Drabman, R.S., Meydrech, E.F., & Hsu, H.S.H. (1992). Relationship between peer status and health behaviors. *Adolescence, 27*, 595-602.

Ullrich-French, S., & Smith, A.L. (2006). Perceptions of relationships with parents and peers in youth sport: Independent and combined prediction of motivational outcomes. *Psychology of Sport and Exercise, 7*, 193-214.

Urdan, T.C., & Maehr, M.L. (1995). Beyond a two-goal theory of motivation and achievement: A case for social goals. *Review of Educational Research, 65*, 213-243.

Vaux, A. (1988). *Social support: Theory, research, and intervention.* New York: Praeger.

Vilhjalmsson, R., & Kristjansdottir, G. (2003). Gender differences in physical activity in older children and adolescents: The central role of organized sport. *Social Science and Medicine, 56*, 363-374.

Vilhjalmsson, R., & Thorlindsson, T. (1998). Factors related to physical activity: A study of adolescents. *Social Science and Medicine, 47*, 665-675.

Voorhees, C.C., Murray, D., Welk, G., Birnbaum, A., Ribisl, K.M., Johnson, C.C., et al. (2005). The role of peer social network factors and physical activity in adolescent girls. *American Journal of Health Behavior, 29*, 183-190.

Vu, M.B., Murrie, D., Gonzales, V., & Jobe, J.B. (2006). Listening to girls and boys talk about girls' physical activity behaviors. *Health Education and Behavior, 33*, 81-96.

Wang, C.K.J., Chia, Y.H.M., Quek, J.J., & Liu, W.C. (2006). Patterns of physical activity, sedentary behaviors, and psychological determinants of physical activity among Singaporean school children. *International Journal of Sport and Exercise Psychology, 4*, 227-249.

Weiss, M.R., & Ferrer-Caja, E. (2002). Motivational orientations and sport behavior. In T.S. Horn (Ed.), *Advances in sport psychology* (2nd ed., pp. 101-183). Champaign, IL: Human Kinetics.

Weiss, M.R., & Smith, A.L. (2002). Friendship quality in youth sport: Relationship to age, gender, and motivation variables. *Journal of Sport and Exercise Psychology, 24*, 420-437.

Weiss, M.R., & Stuntz, C.P. (2004). A little friendly competition: Peer relationships and psychosocial development in youth sport and physical activity contexts. In M.R. Weiss (Ed.), *Developmental sport and exercise psychology: A lifespan perspective* (pp. 165-196). Morgantown, WV: Fitness Information Technology.

Welk, G.J. (1999). The youth physical activity promotion model: A conceptual bridge between theory and practice. *Quest, 51*, 5-23.

White, R.W. (1959). Motivation reconsidered: The concept of competence. *Psychological Review, 66*, 297-333.

Wilson, D.K., Williams, J., Evans, A., Mixon, G., & Rheaume, C. (2005). A qualitative study of gender preferences and motivational factors for physical activity in underserved adolescents. *Journal of Pediatric Psychology, 30*, 293-297.

Wold, B., & Anderssen, N. (1992). Health promotion aspects of family and peer influences on sport participation. *International Journal of Sport Psychology, 23*, 343-359.

Wu, T-Y., Pender, N., & Noureddine, S. (2003). Gender differences in the psychosocial and cognitive correlates of physical activity among Taiwanese adolescents: A structural equation modeling approach. *International Journal of Behavioral Medicine, 10*, 93-105.

Zabinski, M.F., Norman, G.J., Sallis, J.F., Calfas, K.J., & Patrick, K. (2007). Patterns of sedentary behavior among adolescents. *Health Psychology, 26*, 113-120.

CHAPTER

13

Physical Activity Levels During the School Day

Gareth Stratton, PhD ▪ Stuart J. Fairclough, PhD ▪ Nicola D. Ridgers, PhD

In 2000, the U.S. Department of Health and Human Services recommended the school environment for both physical activity promotion and interventions aimed at increasing daily activity. School provides opportunities and clues about how to be physically active, and provides messages to young people about acceptable or unacceptable behaviors. As youth spend 40% to 45% of their waking hours at school, this context may play a critical role in the development of physical activity and health-related behaviors (Department of Health, 2004; Strong et al., 2005). Furthermore, the rationale for considering the school as a significant player in children's physical activity is clear. First, the full socioeconomic and ethnic spectrum of youth attend school. Second, youth attend school for the majority of the calendar year. Third, schools are centers of learning where students are exposed to the full smorgasbord of educational curricula including messages promoting health and physical activity. Finally, after-school programs allow students to access further physical activities, the majority of which are organized, competitive sports.

The increased emphasis on physical activity promotion at school by physical educators is interesting given the fact that recent changes in Western education systems have increased pressure on academic achievement at the expense of physical education and physical activity (Hardman, 2000). Further, active transport, physical education, time in recess, and after-school sport have been declining; and external factors such as school policy, parental rules in relation to safety and convenience, and environmental factors have all served to promote inactivity and create an educational environment where inactive messages outweigh active ones (Dollman, Norton, & Norton, 2005).

While schools can exert a strong influence on youths' physical activity, the effect of specific school environments on young people's physical activity is relatively unexplored in the literature (Fein et al., 2004). Chen and Zhu (2005) analyzed the data from the U.S. Early Childhood Longitudinal study and found that school and home variables, including the number of physical education classes per week, teacher experiences of teaching physical education, and neighborhood safety, significantly affected children's interest in physical activity. Fein and colleagues (2004) reported that with regard to school, neighborhood, and home domains, only the perceived importance of the school environment was significantly and positively related to physical activity in youth. Furthermore, Timperio and colleagues (2004) found that whole-school approaches using curriculum, policy, and environmental strategies for increasing physical activity were more effective than curriculum-only approaches. More recent studies have measured the relative contribution of each part of the school day to overall habitual activity. Trost and colleagues (2002) found that 39% and 12% of youth activity was accounted for by the "school day" and "lunchtime and out-of-school activities," respectively (see figure 13.1), whereas 4% of daily activity was achieved through structured exercise programs.

Gavarry and colleagues (1998) used heart rate reserve to calculate habitual physical activity in a group of French schoolchildren. They reported that mean heart rate was the greatest during physical education lessons (128 ± 11 beats/min), recreation (113 ± 15 beats/min), and lunch break (108 ± 12 beats/min) whereas the most inactive part of the day was evening time, when the mean heart rate was 94 ± 10 beats/min. Mean heart rate was higher during lunch break only on school days when compared to nonschool days. Moreover, Riddoch and colleagues (1991) reported that 33% of children took part in vigorous physical activity (VPA) only during physical education lessons. Physical education has a historic role in the promotion of young people's physical activity (Williams, 1988); and Stratton (1999) has promoted the merits of school recess as a context within which daily physical activity should be encouraged, while Tudor-Locke and colleagues (2001) identified travel to and from school as an obvious context within which children could engage in moderate-intensity physical activity.

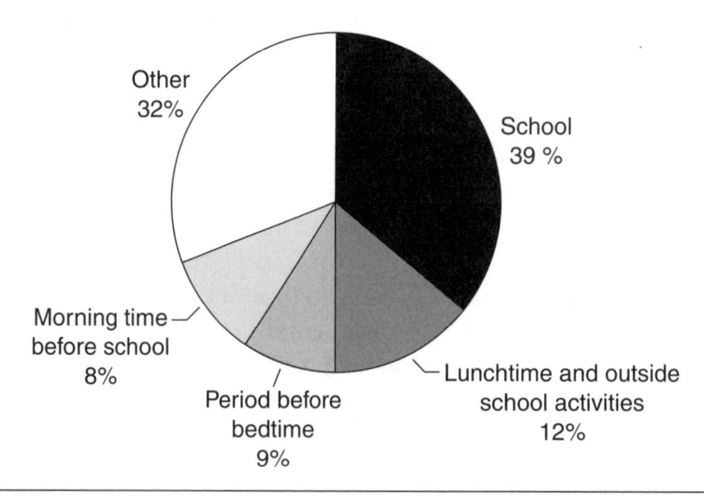

Figure 13.1 The percentage of children's physical activity accumulated across different sections of the school day.

Based on Trost et al. 2002.

It stands to reason, then, that public health policy tends to focus on three main contexts for physical activity promotion during the school day, namely active transport to and from school, recess, and physical education. Given the plethora of school physical activity issues that are available to consider, this chapter examines the empirical evidence on physical activity and inactivity during active transport to and from school, recess, and physical education lessons and analyzes these findings within the context of Welk's Youth Physical Activity Promotion Model (1999). This model proposes that enabling, predisposing, reinforcing, and sociodemographic factors affect the promotion of youth physical activity.

Active Transportation

Environmental influences on children's physical activity are particularly notable in travel to and from school (Tudor-Locke et al., 2001). Increases in travel to school by car and perceived parental worries about safety are implicated in depressing children's physical activity immediately before and after school. As a result, this readily available source of physical activity is underutilized by today's youth. Described by Ziviani and colleagues (2004) as "incidental physical activity," travel to school can be categorized into two main areas: *inactive transport*, for example commuting by bus, car, or train, and *active transport*, such as walking, skating, or bicycling. However, some youngsters engage in both inactive transport (taking the bus) and active transport (walking from the bus stop to school) on their journey to and from school. Thus categorizing children as active or inactive travelers may oversimplify their activity behavior. Notwithstanding these limitations in categorization, active transportation to school (ATS) can make a significant contribution to the 60 min of daily physical activity recommended for children (Cavill, Biddle, & Sallis, 2001; Strong et al., 2005). For example, if a child walked 1 mile (about 1.6 km) to and from school at an average speed of 3 miles per hour (4.8 km/h), this would be equivalent to 20 min of moderate-to-vigorous physical activity (MVPA) per day, which is about a third of the recommended amount of daily MVPA.

Because of the potential that ATS has in helping children attain 60 min of physical activity per day, policies and strategies that aim to provide greater opportunities for children to walk, cycle, or skate (rare) to school are in vogue. "Walk to school week," "walking buses" (children walk together in orderly lines) and "walking taxis" (an adult or another child collects a child from home and walks with the child to school), bicycle storage areas on school grounds, safe cycle pathways, and environmental adaptations (such as better lighting and traffic-calming measures around schools) are all examples of policies and strategies that aim to promote active transportation. Further, the U.S. *Healthy People 2010* objectives (U.S. Department of Health and Human Services [USDHHS], 2000) include increasing the proportion of trips made by walking to and from school by 50% for children who live within 1 mile (about 1.6 km) of school. Whether these initiatives have any immediate or long-term effect on youths' physical activity is open to question.

Research

There are few if any data on the effect of walking or cycling to school on any health outcome in young people; however, there is some evidence that active transport can improve health in adults. For example, Andersen and colleagues (2000), after controlling for leisure-time physical activity, found a 40% lower risk of all-cause mortality in adults who bicycled to work. Hayashi and colleagues (1999) also reported a decreased risk of hypertension in Japanese men who walked to work. As these studies have shown lower morbidity and mortality in adults who actively transport themselves to and from work, one can hypothesize that youth may also accrue positive health benefits if they regularly engage in ATS.

Over the past two to three decades there have been significant changes in the way youth travel to and from school. In the United Kingdom, walking to school dropped by a fifth between 1970 and 1991 (Hillman, 1993). In the mid-1980s, 67% of children in the United Kingdom aged 5 to 10 years walked to school, and more than 6% of children aged 11 to 16 years cycled. By the late 1990s, these percentages had fallen to 56% of 5- to 10-year-olds walking and less than 2% of 11- to 16-year-olds cycling (Noble, 1999). Decrements of 25% and 75% in walking and cycling to school, respectively, between 1985 and 2001 have been reported for Australian 9- to 13-year-olds (Salmon et al., 2005). According to McCann (2000), active transportation in the United States declined by over a third between 1977 and 1995.

Determinants of active walking and bicycling to school, such as perceived and actual neighborhood safety, gender, neighborhood design, child peer groups, rural or urban dwelling, distance between home and school, age, and parental mode of transport, have been reported. Gielen and colleagues (2004) found that while mothers and fathers taught their children how to cross the road safely, parental concerns about the physical and social environment were negatively related to the socioeconomic status of areas. Further studies of adult perceptions of neighborhood safety, traffic, strangers, and road safety also suggest that parents' perceptions of local neighborhoods influence youths' physical activity (Timperio et al., 2004; Ziviani et al., 2004). Parental support of youths' ATS was related to the importance that the parents placed on physical activity, their individual history of transport, and distance from home to school (Ziviani et al., 2004). Carver and colleagues (2005) found that girls' and boys' ATS was related to the number of peers or friends they had. Moreover, girls who perceived local roads as safe were more likely to take an active transport option during the week. Intuitively, the number of routes from home to school may also affect active transportation, as a relatively higher number of routes would make travel varied, interesting, and sometimes more convenient. Conversely, Braza and colleagues (2004) reported that active transportation rates (walking and biking) were positively related to population density and school size but were not related to the number of street intersections per mile.

Distance from school, age, and gender also influence the prevalence of ATS. Sleap and Warburton (1996) reported that half of 4- to 11-year-old English children were driven less than a mile to school whereas the Early Bird study in England showed that two-thirds of 5-year-olds walked 0.4 miles (about 0.6 km) to school for a duration of between 5 and 10 min (Metcalf et al., 2004). Interestingly, children in the Early Bird study who opted for ATS accumulated only 2% more physical activity across the school day than children who were driven to school, and there were no differences in body mass index (BMI) or skinfolds between active and inactive commuters. Young children are the most active group in society, and it is perhaps unsurprising that ATS represents only a small proportion of their daily physical activity.

In a study of 13- to 14-year-old youth, those who actively traveled to *and* from school, to *or* from school, or neither engaged in MVPA for 123.1, 113.1, and 97.2 min per weekday, respectively (Alexander et al., 2005). Furthermore, children who walked at least one way to school were more active at every other time of the school day than those children who did not. Cooper and colleagues (2005) also found that Danish children who chose ATS (mainly cycling) were more active during all times of the school day when active choices were available compared to children who chose inactive transportation. This study suggests that children who choose ATS are more predisposed to physical activity, or value the positive social milieu provided by activity opportunities before and after school, or both. Mackett and colleagues (2005), analyzing activity across weekdays and weekend days, reported that walking came second only to physical education lessons and ball games in terms of energy expenditure, and also that walking to school used more calories than either structured or unstructured out-of-home

Active transport to school offers opportunity for young people to attain recommended daily physical activity levels.

events. From these data they concluded that ATS used more energy in one week than physical education alone.

Perhaps the largest study on ATS was a cross-sectional survey in the United States involving 1395 adult-child pairs; the children were in school grades 4 through 12. The survey, undertaken in 1996, showed that the majority of youth traveled to school by either bus (41.2%) or car (34%), whereas 9.4% drove themselves, 1.3% traveled on public transport, and only 14% took an active transport option (11.4% walked; 2.6% bicycled). The prevalence of ATS was different between boys (16.6%) and girls (11.1%) and by age (20.5%, 17.4%, and 8% in grades 4-6, 7-9, and 10-12, respectively). There were no differences in ATS by race, gender, family structure (i.e., one vs. two parents in home), or neighborhood safety in the corrected model (Fulton et al., 2005). Fulton and colleagues concluded that less availability of pavements (sidewalks), greater distance to school, and lower child perception of safety were related to less ATS. Moreover, they reported that few teenage girls chose ATS as their preferred mode of travel, suggesting that intervention programs with this group were required.

In summary, while evidence on the effects of ATS is limited, the majority of studies suggest that youth who engage in ATS tend to be more active in other parts of the school day than those who choose inactive forms of transport to school. Moreover, girls and adolescents seem less inclined to take nonmotorized transport to school compared to boys and younger children, respectively. Active transport to school is more prevalent in safe environments and, when walking or cycling to school is chosen as the preferred form of transport, children can expend more energy in ATS than they may in structured activity such as physical education lessons.

Evaluation of Interventions

Few studies have addressed the effect of school active transport programs on the physical activity and health of youth. Heelan and colleagues (2005) assessed the effect of a five-month active commuting program on the BMI of 9- to 11-year-old children from eight rural schools in the United States and reported that active commuting did not provide sufficient physical activity to attenuate BMI. Boarnet and colleagues (2005) evaluated the California Safe Routes to School (SR2S) program by using a parental survey methodology to assess the relationship between urban form changes and ATS in 3rd- to 5th-grade youth in 10 schools. The SR2S program provided funding for construction projects near schools, with the intent of increasing pedestrian and bicyclist safety and improving the environment for active transportation to and from school. Two groups were created based on whether parents stated that their children would travel a route that would be affected by the SR2S program on the way to school (group 1) or not (group 2). Children who used SR2S program routes were more likely to show increases in ATS than children who did not (15% vs. 4%).

Furthermore, half of the programs had a positive impact on perceived safety and safety-related behaviors.

Developing safe walking routes to schools has become popular in many countries. In the United Kingdom, surveys indicate that cycling levels have increased by more than 50% in pilot schools since 1996 and that walking levels have increased by 20% (SUSTRANS, 2001). Hotz and colleagues (2004) have also reported improved safety knowledge in a "Walk Safe" school-based pedestrian program delivered to nearly 6500 children in 16 elementary schools in a high-risk district in Miami. However, they did not report increased prevalence of walking to school.

Overall, the limited evidence to date suggests that ATS intervention programs have a positive effect on physical activity and perception of neighborhood safety. While no program included evaluation of whether behavior change had been sustained, no reports exist of any negative impact of ATS programs.

Implications for Future Research and Practice

In summary, studies of ATS are relatively new, and more research is required before the effects of cycling, walking, or skating to school on the health and well-being of youngsters in the short and long term are better understood (Tudor-Locke et al., 2001). While one cross-sectional study sampled across age and gender (Fulton et al., 2005), there are no longitudinal studies on ATS. Thus the effects of ATS on health and well-being are yet to be fully elucidated. Clearly, better-designed research projects that track changes in ATS over time across school-age groups are needed. Informed by the results from basic research, intervention projects will subsequently be required to investigate the effectiveness of ATS programs on youth health, well-being, and academic achievement. The *Healthy People 2010* objectives include a target that 50% of youth have active transport as their preferred means of travel to and from school by 2010 (USDHHS, 2000). While this is a useful target, few studies have used ATS as a physical activity intervention. Thus, meaningful data allowing cause-and-effect statements are not available.

Recess

Recess (called *playtime* or *breaktime* in the United Kingdom) refers to the regularly scheduled, noncurriculum time allocated by schools between lessons for children to engage in leisure activities and free play. Recess not only provides opportunities for children to engage in physically active play behaviors but also enables them to develop social, emotional, and cognitive skills, sometimes through inactive behaviors (Pellegrini & Bohn, 2005). During recess, children have numerous choices for interacting with their peers through active or sedentary behaviors (Jarrett & Duckett-Hedgebeth, 2003). Concern has been expressed that the time allocated to recess is being reduced (Blatchford &

Sumpner, 1998) or removed altogether (National Association of Early Childhood Specialists in State Departments of Education, 2002). The reduction is in response to worries that recess detracts from curriculum time and facilitates negative behavior and aggression on the playground (Pellegrini & Smith, 1993). However, recess not only complements curricular time by providing activity and social interaction opportunities, but also contributes to daily physical activity and young people's physical health (Pellegrini & Bohn, 2005; Ridgers, Stratton, & Fairclough, 2006).

In the United Kingdom, children may experience up to 600 recess periods a year (Stratton, 1999), accounting for nearly a quarter of the elementary school day (Boulton, 1992). Comparatively, in the United States there is no consistency in the way in which recess is implemented (Pellegrini & Bohn, 2005). In the United Kingdom, the total daily average recess time provided for elementary children is 83 min (Blatchford & Sumpner, 1998). Internationally, a recent review reported that daily recess duration ranged from 14 to 46 min (Ridgers et al., 2006). Despite the differences globally, recess represents one of the main opportunities for children to be physically active on a day-to-day basis in school (Sarkin, McKenzie, & Sallis, 1997). However, little empirical research has focused on the physical activity levels of children within a recess context. Furthermore, there is a paucity of research centered on the impact of recess-based interventions on youth physical activity and the utility of recess as a health promotion setting.

Research

Physical activity research has primarily focused on quantifying recess activity levels using a range of objective measures, while a small number of studies have also addressed youths' behavior during recess using direct observation (Kraft, 1989; McKenzie et al., 2000). With the exception of a number of studies in Portuguese schools (Mota et al., 2005; Mota & Stratton, 2003), research has shown that boys are significantly more active than girls during recess (Ridgers, Stratton, & Fairclough, 2005; Sarkin et al., 1997).

The reasons underlying gender differences in recess activity levels have not been established in the literature. However, links have been made to behavioral research in which recess as a social context has been evaluated in more detail (Ridgers et al., 2006). For example, Blatchford and colleagues (2003) investigated the activities that children engage in during recess and found that boys tend to play more active games and ball games than girls whereas girls tend to engage in more inactive play and social behaviors, such as passive and verbal games, conversation, and socializing. There is concern that girls' lower activity levels may be attributed to the domination of available space by boys who are playing vigorous ball games during recess (Evans, 1996; Renold, 1997), suggesting that low amounts of active behavior have not necessarily been chosen by girls. In circumstances in which the majority of the available play space is dominated by vigorous games, the remaining children tend to situate themselves around the perimeter and engage in inactive behaviors (Renold, 1997). It has not, however, been clearly established whether girls have access to the same opportunities and choose less active games or whether this access is restricted due to the reported dominance of boys in the playground (Blatchford et al., 2003; Epstein et al., 2001; Pellegrini et al., 2004).

Seasonal factors have been found to influence children's habitual physical activity engagement. However, some studies have shown that youth are more active in the summer months compared to the winter months (Hagger, Cale, & Almond, 1997; Loucaides, Chezdoy, & Bennett, 2003), while others have indicated a trend for physical activity to be higher in winter than summer (Ridgers et al., 2006; Stratton, 1999). While these seasonal differences in recess physical activity levels were not significant, the seasonal variation may be attributable to a thermoregulatory need to keep warm during the colder winter weather (Stratton, 1999), as children in the United Kingdom, where these studies were conducted, do not have the option of playing inside on cold days (Thomson, 2004). These studies give insight into the potential impact of seasonality on recess activity, though it has been recommended that future studies using larger samples of children from different geographical regions be conducted to investigate this issue (Ridgers et al., 2006). In addition, there is a need to establish what play behaviors children engage in across the seasons in order to unravel seasonal impacts on children's activity.

Recently researchers have attempted to evaluate the contribution that recess makes to current daily physical activity recommendations. Recess has been found to contribute from 4.7% to 40% and 4.5% to 30.7% toward optimal daily physical activity recommendations for boys and girls, respectively (Ridgers et al., 2006). This suggests that some young people are effective in using their recess time to engage in physical activity, though there is potential for interventions to be implemented in this context to enhance activity levels further. This is particularly important for children who engage in little or no activity during recess, as research indicates that they do not increase their activity levels outside of school hours to accommodate for this (Dale, Corbin, & Dale, 2000).

Empirical research has generally focused on determining recess physical activity levels or children's behavior in this context. Surprisingly little research has dealt with the relationship between physical activity levels of children and their play behavior during recess. Research is required to ascertain the determinants of children's physical activity during recess in conjunction with their play behaviors and activity and inactivity choices. There is a need to know what behaviors children engage in when they are being active or sedentary, as well as to consider the social effects of the playground culture on children's behavior. In addition, understanding why children choose active or sedentary behaviors during recess would help inform suitable recess interventions that aim to increase physical activity.

Current cross-sectional investigations into recess physical activity are limited by small sample sizes, inconsistent methodologies, and cultural differences in recess among countries (Ridgers et al., 2006). A number of studies have reported activity levels using one recess period (Kraft, 1989; Stratton, 1999; Zask et al., 2001) or a combination of recess periods on one day (Ridgers et al., 2005). Larger population-based studies using similar methods and longer monitoring periods on girls and boys in all school years would enable researchers to better understand the overall contribution of recess to physical activity during the school day.

Evaluation of Interventions

Ridgers and colleagues (2006) reported that individual schools often initiate recess changes, with the provision of equipment being a common strategy, though they rarely rigorously evaluate or describe any effects of the changes. It is only recently that the effects of recess-based interventions to increase young people's physical activity, using playground markings, fitness breaks, and games and equipment, for example, have been reported in the empirical literature (table 13.1).

Multicolor playground markings have been used as a recess intervention with the intention of increasing elementary school children's physical activity levels by providing an environment that supports and facilitates physically active behaviors (Wechsler et al., 2000). Playground markings consist of games (e.g., soccer lines, snakes and ladders), activities (e.g., targets, hopscotch), and objects (e.g., number squares, clock faces) that are generally painted on the school playground to stimulate physical activity and play behaviors. These can provide opportunities and cues to a large population of children regarding how to be active (Giles-Corti & Donovan, 2002). Several studies have shown positive short-term increases in children's recess physical activity levels following an intervention using playground markings. Stratton (2000) reported an increase in MVPA of 11.1% in the four weeks following an intervention in a single school. This equated to an additional 5.9 min a day (9.8%) toward optimal daily activity guidelines. In a second study that used a larger sample of children from four schools (two experimental, two control), Stratton and Mullan (2005) reported similar increases in MVPA (13.6%) four weeks postintervention. In this study, the experimental children spent over half of recess engaged in MVPA (50.3%) following the intervention. Boys and girls experienced similar increases in their recess activity levels.

Stratton and Leonard (2002) reported increases in total energy expenditure (TEE) and rate of energy expenditure (REE) of 35% and 6%, respectively, following a playground markings intervention. Interestingly, boys and girls ages 5 to 7 years experienced similar increases in TEE and REE, which is consistent with the physical activity intervention studies. This may be attributable to the involvement of children in the design of the play area, as prior to the intervention they were consulted concerning what markings they would like in order

Table 13.1 Summary of Intervention-Based Studies Investigating Children's Physical Activity (PA) During Recess

Author	Participants	Intervention	Design and PA assessment	Key findings (F) and critique (C)
Scruggs et al. (2003)	10 boys, 17 girls, 5th grade U.S.	Fitness breaks	HR telemetry and pedometers Morning playtime, lunch, and FB monitored for 3 consecutive days Repeated measures ANOVAs	(F) Intervention benefited both boys' and girls' PA Girls—lower enjoyment of FB (C) Limited generalizability Short-term intervention Possible novelty effects Small sample size
Connolly and McKenzie (1995)	56 elementary children U.S.	Games playtime	Self-report, accelerometry, direct observation Each day: one games playtime, one standard playtime Paired t-tests	(F) Benefited boys and girls Enjoyment similar for the two playtime types (C) Possible novelty effects Short-term intervention Games played not detailed
Stratton (2000)	Exp: 18 boys, 18 girls Con: 12 boys, 12 girls Years R-2 England	Playground markings	Three children per school monitored each day HR telemetry 4-week follow-up ANCOVA	(F) PA increased after painting Playtime duration increased (C) Short-term follow-up Possible novelty effects
Stratton and Leonard (2002)	32 boys, 30 girls aged 5-7 years England	Playground markings	1 week pre- and 1 week postpainting measures taken HR telemetry RM ANCOVA (covariates: BM and play duration)	(F) Intervention benefited both sexes (C) Short-term follow-up Most effective markings unknown Potential novelty effects
Stratton and Mullan (2005)	Exp: 60 boys, 60 girls (four schools) Con: 60 boys, 60 girls (four schools) Wales, England	Playground markings	Baseline: 4 weeks prepainting Intervention: 4 weeks postpainting HR telemetry 3 ANCOVAs (covariates: BM and play duration)	(F) Painting benefited both sexes (C) Possible seasonal effects on control groups' PA levels Increases in MVPA may link to novelty effects Short-term follow-up

Table ordered by sample size (smallest to largest) according to intervention; ANOVA = analysis of variance; ANCOVA = analysis of covariance; BM = body mass; Con = control group; Exp = experimental group; FB = fitness breaks; HR = heart rate; MVPA = moderate-to-vigorous physical activity; PA = physical activity; RM = repeated measures.

Adapted and reprinted, by permission, from N.D. Ridgers, G. Stratton, & S.J. Fairclough, 2006, "Physical activity levels of children during school playtime: A review," *Sports Medicine* 36: 367.

to enable them to take ownership of and recognize the activities represented by the markings. Furthermore, the spacing and types of markings across the playground potentially open up the play space to children who may in the past have been pushed to the perimeter by dominant games (Stratton & Leonard, 2002). Children also had the opportunity to choose which games to play during the recess period, as their decisions were governed by their enjoyment of particular activities (Stratton, 2000).

Structured fitness breaks have been implemented as an environmental intervention alternative to playground markings. Fitness breaks aim to engage children in significantly more MVPA compared to traditional unstructured recess periods. Scruggs and colleagues (2003) employed a continuous 400 m obstacle course across three sections of a school play area. Fifth-grade children participated in activities such as dance, jump rope, zigzag runs, and crawling over objects for 15 min. The fitness break was implemented in the last hour of the school day. Children had already participated in two regularly scheduled recess periods earlier in the day. In comparison to recess, the fitness break stimulated higher MVPA, VPA, and steps per minute in a small sample of both boys and girls. While boys and girls engaged in similar amounts of MVPA during the fitness break, the step count and VPA for boys were higher than for girls. A unique aspect of Scruggs and colleagues' research design was the measurement of enjoyment. Although there was no significant difference in boys' and girls' enjoyment levels for recess, girls' enjoyment of the fitness break was significantly lower than the boys' and girls' enjoyment of unstructured recess periods (Scruggs et al., 2003). There was no difference between the enjoyment of recess and that of the fitness break for boys. Since enjoyment is a central facet of activity participation, this project highlights the inherent problem of overemphasis on fitness at the expense of intrinsic factors for girls.

A third recess intervention investigated the effects of games implemented by playground supervisors at one elementary school on students' voluntary recess physical activity levels (Connolly & McKenzie, 1995). The children were significantly more active during the games recess than the traditional recess period; and in contrast to the findings of Scruggs and colleagues (2003), there were no gender differences in enjoyment or physical activity during the intervention. Additionally, girls and boys reported similar enjoyment levels of the games and standard recess, again in contrast to the findings of Scruggs and colleagues (2003) for the girls. While the types of games and activities used in the intervention were not detailed, the results suggest that a games approach is effective in increasing recess physical activity levels.

The effect of games equipment on elementary children's recess physical activity has recently been investigated by Verstraete and colleagues (2006). Children in the intervention group engaged in significantly more MVPA during morning recess and in more MVPA and VPA during lunch recess compared to the control group. However, positive effects of the intervention were noted only for lunch recess, during which boys had greater increases in moderate activity compared to girls while girls had greater increases in VPA compared to boys. During morning recess, boys' MVPA and VPA levels decreased and girls' VPA decreased. This may be explained by recess duration. The organization of equipment and games in physical education lessons can account for up to a quarter of lesson time (McKenzie et al., 1997). Applying this information to recess, the organization of equipment may not have a large impact on children's physical activity behavior during lunch recess as there is more time to play games once the organization has occurred compared to morning recess, therefore, explaining the greater increases observed. It appears that a games equipment intervention is effective in increasing recess physical activity when recess is of sufficient duration.

Overall, the results from recess interventions indicate that physical activity levels increase when environmental modifications and strategies are implemented. Changing the playground environment can create opportunities for physical activity engagement and provide messages to the recipients about acceptable or unacceptable behaviors within the specific milieu (Cohen, Scribner, & Farley, 2000; Sallis & Owen, 1999). Drawing on ecological models of health promotion, recess interventions enable children to be active by creating space on the playground, providing relevant equipment, and improving access to activities within a safe environment (Welk, 1999). Overall, the results indicate that playground markings, fitness breaks, and games and equipment interventions create such an environment.

In its position statement, the National Association of Early Childhood Specialists in State Departments of Education (2002) acknowledges that recess should afford children the opportunity to make choices and plan their activity, with schools facilitating these choices through adequate supervision and playground environments. The intervention studies indicate that playground markings, fitness breaks, and games recess periods have similar effects on boys' and girls' activity, showing that all children will take the opportunity to be active in suitable environments. However, it could be argued that interventions need to focus on increasing girls' activity levels during recess, as they are less active at baseline and some approaches generate different levels of enjoyment for boys and girls. In addition, the relative effectiveness of various types of markings or activities for increasing children's recess activity is not known (Ridgers et al., 2006).

Implications for Future Research and Practice

Research into children's recess physical activity levels is still in its infancy. There are key issues, such as the use of different methodologies across studies, small sample sizes, and short-term intervention studies, that limit our knowledge to date and warrant acknowledgement. Current research indicates that children enjoy recess and take advantage of enriched recess environments that use playground markings, obstacle courses, and games and equipment to be physically active in this context. However, the long-term sustainability of these approaches is not known, as research has focused on the short- to medium-term effects of interventions. The extent of novelty effects on children's recess requires detailed investigation. In addition, there is a need for empirical research employing multidisciplinary methods such as direct observation, coupled with physical activity monitoring using accelerometers and heart rate telemetry, for example, to establish children's activity levels and play behavior (with a particular emphasis on determining the longitudinal effects of recess-based interactions) to increase knowledge of physical activity during recess.

Physical Education

The potential of physical education to affect young people's physical activity is apparent from two standpoints. First, high-quality, active physical education can contribute to recommended physical activity levels for youth (Cavill et al., 2001; Corbin & Pangrazi, 2003). Second, positive physical education experiences may be key in the promotion of physical activity beyond the curriculum and outside of school (Cavill et al., 2001; Corbin & Pangrazi, 2003).

Context and Climate

An international survey of the state of physical education concluded that the subject is under threat in many countries because it is commonly perceived as a low-status curriculum area (Hardman, 2000). Trost (2004) applied the RE-AIM public health evaluation framework (Glasgow, Vogt, & Boles, 1999) to demonstrate how this view of physical education may be damaging from a public health perspective. Taking U.S. high school physical education as an example, he estimated that the subject initially has the potential to reach 100% of high school students. When actual enrollment rates and estimated implementation of quality programs are factored in, however, less than 1% of students would attend daily physical education through each high school grade (Trost, 2004). On this basis, the probable public health benefits of physical education appear to be limited. While concerning, this estimation is unsurprising when one considers that only half of students are enrolled in physical education during high school and that this figure drops dramatically from grade 9 onward (Kann, 2001). Although daily physical education is included as a public health goal in *Healthy People 2010* (USDHHS, 2000), less than 10% of students in all grades are exposed to daily physical education for the entire school year (Burgeson et al., 2003).

The amount of weekly time allocated to physical education is also a factor in determining students' activity engagement. In the United States, this ranges dramatically from 30 min to 5 h (Morrow, Jackson, & Payne, 1999), whereas in the United Kingdom, weekly physical education time for students aged 5 to 14 years ranges from 96 to 124 min (Qualifications and Curriculum Authority,

2005). Other determinants of high-quality physical education include teacher expertise and curriculum content. The lack of specialist physical educators in primary schools is common in most countries (Warburton, 2000), and this can compromise the quality of physical education provision (McKenzie et al., 1997).

In terms of physical education curricula, students should experience as wide a range of activities as possible, but curricula tend to focus on competitive team sports at the expense of other more lifetime-oriented activities (Fairclough, Stratton, & Baldwin, 2002; Napper-Owen et al., 1999). This is problematic because the nature of competitive team sports dictates that it is usually the fittest and most able students who experience success. For those students who prefer to engage more in lifetime activities, physical education generally provides limited opportunities. The worst-case scenario is that these students become alienated from the subject and possibly from physical activity participation outside of school and later in life.

Research on Physical Activity Levels

The influence of physical education on youth physical activity is difficult to assess. The increased prevalence of childhood overweight and obesity is linked at least in part to a decline in habitual physical activity levels, suggesting that the influence of physical education beyond the curriculum may be minimal (Trost, 2004). On the other hand, it is possible that youths' physical activity levels would be even lower were they not exposed to any physical education during some point in their schooling. For example, a study involving 859 English schoolchildren (aged 10-11 years) showed that students with the highest levels of physical education enjoyment and perceived competence engaged in the most physical activity out of school (Carroll & Loumidis, 2001). Although the findings were descriptive, they support the notion that when students' perceived competence and enjoyment in physical education are optimized, they will be more motivated to participate in out-of-school physical activity (Wallhead & Buckworth, 2004). Other examples associating physical education participation with increases in out-of-school physical activity come predominantly from intervention studies that are discussed later in this chapter.

A more deliberate method of physical education–based physical activity promotion is through engaging youth in appropriate amounts of activity during lesson time (Simons-Morton, 1994). Though some argue that physical activity during physical education cannot make a meaningful difference to overall physical activity, as children spend less than 1% of their waking hours in that setting (Fox, Cooper, & McKenna, 2004), others espouse the importance of the activity accumulated during physical education (Corbin & Pangrazi, 2003). Indeed, *Healthy People 2010* objective 22.10 recommends increasing the number of students who are physically active for at least 50% of physical education lesson time (USDHHS, 2000). Comparing physical activity levels across the physical education curriculum is problematic, as data are confounded by variations in lesson content, teaching personnel, environment, motivation, skill ability, class size, and so forth. Investigations that have compared physical activity in physical education classes across grade levels have generally noted an increase in activity as students move through the elementary grades (Levin et al., 2001; Luepker et al., 1996). A review of elementary physical education activity levels reveals moderate and small effects for increases in activity between grades 3 and 4 (*ES* = 0.61) and grades 4 and 5 (*ES* = 0.31), respectively (Fairclough & Stratton, 2006). One could speculate that increases in activity as children move through the elementary years are related to increased skill acquisition and skill-related fitness, which better enable students to actively participate in structured physical activity (Gallahue & Cleland-Donnelly, 2003).

In middle and high school physical education, there is limited intergrade variation in activity levels, possibly because more students are taught by physical education specialists, who are more able than elementary school nonspecialists to differentiate lessons to accommodate wide ranges in abilities. Furthermore, as middle and high school physical education curricula often include a greater range of activities than are taught in elementary school, variations in class activity levels are often related more to the activities being taught. For example, the nature and requirements of team invasion games and fitness activities generally elicit greater physical activity levels (~40-50% of lesson time in MVPA) than dance, gymnastics, and striking games (Fairclough & Stratton, 2005b; Stratton, 1996). These patterns may also be apparent during elementary school physical education; but as few studies have reported physical activity in relation to activity type in elementary school (Fairclough & Stratton, 2006), it is difficult to form conclusions at this time.

Girls' and boys' physical activity levels are generally comparable during physical education (Fairclough & Stratton, 2005b, 2006), possibly because students in coeducational physical education receive the same stimuli from the teacher and work within the same environment. However, though equitable physical activity opportunities appear to be available to all students, boys and girls have been observed to respond differently in their physical activity behaviors (McKenzie et al., 2004a). Where boys and girls are taught separately, it is likely that the type of activity explains differences in their activity levels. In schools that deliver physical education in single-sex lessons, it is common for boys' activities to include a greater proportion of team games whereas girls' activities often consist more of movement activities such as gymnastics and dance (Fairclough & Stratton, 2005c).

Evaluation of Interventions

Most physical education–based interventions have involved elementary grades and have employed a physical education component as part of a larger intervention. The Go for Health physical education intervention component consisted of a modified curriculum designed to encourage enjoyable MVPA, as well as physical activity knowledge and self-efficacy (Simons-Morton et al., 1991). Significant increases in MVPA during physical education were achieved as well as improvements in knowledge and self-efficacy, but there were no improvements in habitual physical activity (Simons-Morton et al., 1991). Physical activity during physical education was also increased by 16% in the Nebraska School Study following an intervention to maximize physical activity of grades 3 through 5 students during physical education classes and recess periods (Donnelly et al., 1996). However, a reduction in habitual physical activity among the intervention students of almost 16% was also reported, which the authors offered no explanation for. Conversely, the Cardiovascular Health in Children (CHIC) study employed physical education lessons that emphasized fun aerobic activities to reduce cardiovascular disease risk factors among 3rd and 4th graders (Harrell et al., 1996). Results indicated a 23% increase in self-reported physical activity in intervention schools, compared to 15% improvement among the control schools (Harrell et al., 1996).

The Child and Adolescent Trial for Cardiovascular Health (CATCH) project (McKenzie et al., 1996) and the Sports, Play and Active Recreation for Kids (SPARK) project (McKenzie et al., 1997) were two of the most prominent randomized controlled, sustained, multicenter trials. CATCH involved 96 schools and aimed to promote enjoyment of and participation in MVPA during physical education, as well as to provide skills to promote participation in out-of-school physical activity. The physical education intervention component involved a CATCH-PE curriculum, which consisted of developmentally appropriate activities organized into units of instruction. Additionally, the generalist class teachers received in-service training on physical activity–promoting pedagogy as well as subsequent site visits and support. SPARK was implemented with 4th and 5th graders in seven schools and also aimed to improve physical activity during physical education and out of school. The intervention included modified physical education curricula that emphasized health-related physical activities and sport skills. SPARK also incorporated a behavior change element using a self-management program to enhance students' out-of-school physical activity.

Both the CATCH and SPARK projects consistently engaged students in MVPA for around 50% of physical education class time. Furthermore, in both programs, physical education specialists or classroom teachers who had received in-service training (or both) engaged students in significantly more MVPA than classroom teachers in control schools (McKenzie et al., 1997, 2001). During these lessons, physical education specialists employed more efficient instructional methods and activity-promoting tasks (e.g., increased time for fitness activities and dynamic skill drills; McKenzie et al., 2001). Both interventions had little impact on the general out-of-school physical activity engaged in by the students in the experimental schools (McKenzie et al., 1996; Sallis et al., 1997). In SPARK, the self-management intervention component was not implemented as effectively as the others (Sallis et al., 1997); and as a result the knowledge and attitudes of the students may not have changed sufficiently for a significant change in physical activity behaviors to be recorded (Marcoux et al., 1999).

Fewer middle and high school physical education interventions have been conducted. Project Active Teens targeted 9th graders who attended two weekly lessons of conceptual physical education for one school year (Dale, Corbin, & Cuddihy, 1998). The lessons focused on physical activity knowledge, behavioral skills, and physical skills. Two to three years later, significantly fewer students reported sedentary behaviors compared to students who had followed a traditional sport-based physical education program. An alternative approach was used in Project Planet Health, in which modified sessions were delivered to students in grades 6 and 7 that promoted increased physical activity, improved diet, and reduced TV viewing (Gortmaker et al., 1999). The sessions were implemented for two years in physical education classes and four other subject areas. A reduced prevalence of obesity was reported among girls, but not boys; however, because the intervention was implemented across multiple curriculum areas, the effectiveness of the physical education component was not known (Gortmaker et al., 1999).

The Lifestyle Education for Activity Program (LEAP) was a physical education–specific intervention for grade 9 girls that deemphasized competition, was choice based, was gender sensitive and segregated, and optimized small-group interaction (Felton et al., 2005). The main findings were improved self-efficacy and enjoyment, which resulted in increased physical activity (Dishman et al., 2004, 2005). The Middle School Physical Activity and Nutrition project (M-SPAN) was a two-year middle school intervention involving students in grades 6 through 8 from 24 San Diego middle schools (McKenzie et al., 2004b). The study aimed to increase the students' energy expenditure and improve their nutritional behavior while at school. Physical education teachers were subject specialists who received voluntary in-service training to increase their awareness of the need for active physical education classes.

Unlike the teachers in CATCH and SPARK, teachers in the M-SPAN study did not receive a purpose-made curriculum to teach. Instead they were helped to modify existing curricula and teaching strategies to increase students' MVPA during lessons. Other intervention elements were regular on-site consultation and support visits by the M-SPAN trainers as well as social marketing techniques, such as newsletters detailing the changes to the curricula and promoting positive physical activity messages (Sallis et al., 2003). During physical education, students in intervention schools increased their MVPA from 48% to around 52% of lesson time, which was about 4% (or 3 min) greater than in the control schools (McKenzie et al., 2004b). Furthermore, boys significantly increased their physical activity equally during physical education and away from physical education (e.g., during recess, extracurricular programs). Girls also increased their physical activity, but this was mainly through physical education classes (Sallis et al., 2003).

Physical education is often cited as a key environment for promoting youth physical activity. The available evidence suggests that students generally engage in modest amounts of activity during class time, though this varies depending on the activities being taught. Physical education–based interventions have successfully increased students' class-time physical activity, as well as psychological variables such as enjoyment and self-efficacy. Few studies, however, have reported improvements in students' habitual physical activity, which raises a question about the extent of impact that physical education may have on youth physical activity.

Physical Inactivity During Physical Education

Most studies have shown that students spend less than 50% of physical education class time in MVPA and that for the majority of the time they are engaged in light activity or periods of stillness (Fairclough & Stratton, 2005b, 2006). Objective physical activity measures, such as those obtained with accelerometers and heart rate monitors, can extract when episodes of inactivity occur, but they do not detect physical activity behaviors such as walking, running, and jumping or contextual factors (i.e., skill practice, warm-up, management, etc.). Conversely, systematic observation systems (e.g., SOFIT; McKenzie, Sallis, & Nader, 1991) are able to provide descriptions of what students are doing and what the lesson context is during sampling intervals. For example, the proportion of lesson time that students spent lying, sitting, and standing was found to be similar between elementary school lessons taught by generalists (54%) and specialists (49%; McKenzie et al., 1993) and middle school physical education delivered by specialists (51%; McKenzie et al., 2004b). Moreover, during these lessons, over one-third of the time was devoted to lesson contexts such as management, delivery of general physical education knowledge, and static skill drills (McKenzie et al., 1993, 2004b), which contribute to student inactivity. These inactive lesson phases are a necessary part of physical education and its potential to foster long-term positive physical activity attitudes and skills. Active and inactive physical education through enjoyable and motivating learning environments is required to promote physical activity knowledge, understanding, and perceived competence (Goudas & Biddle, 1994; Goudas et al., 1995).

Thus, it is important to note that some periods of low activity or inactivity should be expected during physical education, as the emphasis is justifiably on learning *through* and *about* physical activity, not exclusively on physical activity participation itself. Through considered lesson planning, teachers can deliver educationally and psychologically beneficial physical education classes while optimizing MVPA within the context of the activity that is being taught. If lessons are planned with physical activity as a formal objective, and delivered efficiently in terms of student groupings, teaching styles, and use of resources, it is possible for physical education to engage students in optimal activity levels within purposeful and fun learning environments (Fairclough & Stratton, 2005a; Martin & Hodges-Kulinna, 2004).

Through careful lesson planning, physical education instructors can help students attain physical activity as well as educational and skill-based objectives.

Implications for Future Research and Practice

Physical education studies should gather data relating to the type of activity being taught and the ways in which the teaching and learning environment affects students' physical activity opportunities. Such data would provide greater clarity about what determines key mediators of physical activity like enjoyment, perceived competence, and self-efficacy. Moreover, longitudinal tracking of pupils over a number of physical education lessons would illustrate typical responses to lessons as well as changes in physical activity with age and between different activities.

Curriculum interventions are required that draw on established good practice from previous studies. Elements of effective interventions have included in-service training and support for teachers (especially generalists); modified curricula focused on enhanced physical activity, enjoyment, and self-efficacy; use of social marketing to promote the intervention messages to teachers, parents, and significant others; and employment of whole-school approaches to emphasize physically active behaviors (as well as other healthful messages, such as healthy eating) before, during, and after the school day.

As physical activity during physical education does not occur in a vacuum, we need to gain a better understanding of the interrelationships between physical education variables such as physical activity, psychological responses, social dynamics, and cognitive learning. This latter area may be particularly appealing to education administrators and teachers who may be more concerned with academic attainment than physical activity. Research on the relationship between school-based physical activity and academic attainment has generally yielded equivocal results (Shephard, 1997). Though not all studies have described improved academic performance following additional physical activity time, a consistent conclusion is that even when extra physical activity time is provided at the expense of classroom time, academic performance is not negatively affected (Sallis et al, 1999; Shephard, 1997). Moreover, a recent study showed associations between the amount of vigorous activity (but not moderate activity) performed during physical education and academic achievement (Podulka-Coe et al., 2006), while others have suggested that physical activity may be indirectly related to academic attainment through improvements to students' physical health and self-esteem (Tremblay, Inman, & Willms, 2000).

While it is appealing to link physical education activity with academic performance, more work is required in this area before firm associations can be conclusively drawn. For a better understanding of how physical education variables and outcomes are related to physical activity, combined methodological approaches are particularly attractive because they afford a more rounded picture of the physical education environment and its influence on activity levels.

To effectively promote physical activity within lessons, physical educators should consciously plan physical activity objectives within appropriate curriculum activities. By focusing on active participation, teachers will need to modify instructional and organizational strategies to maximize students' *active* learning. At the same time, physical educators should recognize that physical inactivity can be positive in specific teaching and learning contexts (e.g., during gymnastics lessons that aim to develop balance and stillness). Teachers also need to move away from traditional team sport–based curriculum models to embrace more inclusive, student-centered curricula that place greater emphasis on lifetime physical activity. Furthermore, physical education teachers should lead other colleagues to implement whole-school approaches that make physical activity promotion a collective responsibility for teachers, students, parents, coaches, and youth leaders.

Summary

Physical activity in relation to school has been considered in three contexts: ATS, recess, and physical education. Recess and ATS can provide daily opportunities to engage in moderate and vigorous physical activity that is unstructured. Inactive transport is largely a result of parent perception of neighborhood safety, choice of transport to school in younger children, and convenience in older teens. The prevalence of inactive transport is increasing, and this worrying shift has only recently been opposed by major programs promoting ATS as a sustainable effort toward health promotion in schoolchildren. Recess can also provide a sustainable context for activity promotion, and the limited number of intervention programs undertaken has increased physical activity in the short term. Recently, some play spaces have also been designed to include relaxation areas as an approach to reduce stress and allow youngsters to engage in positive social, but inactive, behavior. Recess also promotes vigorous activity that is more likely to stimulate changes in aerobic fitness.

Physical education has many goals, only one of which is physical activity. All these goals are normally delivered in less than 2 h per week; therefore physical education cannot be responsible for increasing the fitness of students, as there is simply not enough curriculum time. Furthermore, inactivity during physical education, mainly a result of teacher instruction and management time, is essential if the key principles of long-term activity and skill promotion are to be effectively delivered. Interventions into school physical education have successfully increased MVPA during lesson time, though there is little evidence to suggest that these changes have carried over into habitual physical activity. Youths' physical activity is also mediated by the quality of teaching and by whether lessons are delivered by qualified physical educators. Physical inactivity is more common in lessons developed by nonspecialist teachers compared to physical education teachers.

Finally, there has been limited research into ATS, recess, and, to a degree, physical education from an activity-promoting perspective. Limited evidence to date supports the effective promotion of physical activity through the school day, and further examination of inactive and active behaviors is required within these contexts.

APPLICATIONS FOR RESEARCHERS

Although physical activity measurement limitations affect research on ATS, recess, and physical education, the empirical database is improving. On the other hand, many studies of physical activity and sedentary behavior during the school day do not report behavioral or contextual information, and this needs to be addressed. Furthermore, basic data on physical activity using multiple methods and disciplines are lacking, so it is difficult to fully understand the impact of school on youth physical activity. More descriptive research is required to help develop new research tools and inform the design of intervention studies targeting ATS, recess, and physical education. Evidence suggests that interventions in physical education and recess can increase physical activity in the short to medium term, though there is limited evidence on whether these increases positively affect health, well-being, and academic performance. Thus, significantly more research is needed to improve the quality and quantity of this database. Finally, mixed-method, multidisciplinary designs are required before the contribution that school makes to habitual physical activity can be elucidated.

APPLICATIONS FOR PROFESSIONALS

Promoting ATS may increase activity, reduce congestion around schools, and improve air quality. More than half of students should take an active transport option, and late-teen girls particularly should be targeted when this option is promoted. Youngsters are more active in stimulating recess environments and should have the opportunity to engage in outdoor unstructured play three times per day. Moderate-to-vigorous physical activity accumulated during recess can significantly contribute to recommended amounts of daily physical activity. Playgrounds should be organized to allow equal access to activity for girls and boys of all abilities and ages. Specialist physical educators who plan to integrate appropriate amounts of physical activity tend to have more active lessons and are more effective in delivering positive physical activity messages than nonspecialists. Physical education curricula should embrace inclusive, health-related activity principles. Finally, whole-school approaches that integrate ATS, recess, and physical education should be designed to promote a culture of lifelong physical activity.

References

Alexander, L.M., Inchley, J., Todd, J., Currie, D., Cooper, A.R., & Currie, C. (2005). The broader impact of walking to school among adolescents: Seven day accelerometry based study. *British Medical Journal, 331*, 1061-1062.

Andersen, L.B., Schnohr, P., Schroll, M., & Hein, H.O. (2000). All-cause mortality associated with physical activity during leisure time, work, sports, and cycling to work. *Archives of Internal Medicine, 160*, 1621-1628.

Blatchford, P., Baines, E., & Pellegrini, A.D. (2003). The social context of school playground games: Sex and ethnic difference, and changes over time after entry to junior school. *British Journal of Developmental Psychology, 21*, 481-505.

Blatchford, P., & Sumpner, C. (1998). What do we know about breaktime? Results from a national survey of breaktime and lunchtime in primary and secondary schools. *British Educational Research Journal, 24*, 79-94.

Boarnet, M., Anderson, C., Day, K., McMillan, T., & Alfonzo, M. (2005). Evaluation of the California Safe Routes to School legislation. Urban form changes and children's active transportation to school. *American Journal of Preventive Medicine, 28*(2S2), 134-140.

Boulton, M.J. (1992). Participation in playground activities at middle school. *Educational Research, 34*, 167-181.

Braza, M., Shoemaker, W., & Seeley, A. (2004). Neighborhood design and rates of walking and biking to elementary school in 34 California communities. *American Journal of Health Promotion, 19*, 128-136.

Burgeson, C.R., Wechsler, H., Brener, N.D., Young, J.C., & Spain, C.G. (2003). Physical education and activity: Results from the School Health Policies and Programs Study 2000. *Journal of Physical Education, Recreation and Dance, 74*(1), 20-36.

Carroll, B., & Loumidis, J. (2001). Children's perceived competence and enjoyment in physical education and physical activity outside school. *European Physical Education Review, 7*, 24-43.

Carver, A., Salmon, J., Campbell, K., Baur, L.A., Garnett, S., & Crawford, D. (2005). How do perceptions of local neighborhood relate to adolescents' walking and cycling? *American Journal of Health Promotion, 20*, 139-147.

Cavill, N., Biddle, S.J.H., & Sallis, J.F. (2001). Health enhancing physical activity for young people: Statement of the United Kingdom expert consensus conference. *Pediatric Exercise Science, 13*, 12-25.

Chen, A., & Zhu, W. (2005). Young children's intuitive interest in physical activity: Personal, school, and home factors. *Journal of Physical Activity and Health, 2*, 1-15.

Cohen, D., Scribner, R., & Farley, T. (2000). A structural model of health behavior: A pragmatic approach to explain and influence health behaviors at the population level. *Preventive Medicine, 30*, 146-154.

Connolly, P., & McKenzie, T.L. (1995). Effects of a games intervention on the physical activity levels of children at recess. *Research Quarterly for Exercise and Sport, 66*(Suppl.), A60.

Cooper, A.R., Andersen, L.B., Wedderkop, N., Page, A.S., & Froberg, K. (2005). Physical activity levels of children who walk, cycle, or are driven to school. *American Journal of Preventive Medicine, 29*, 179-184.

Corbin, C.B., & Pangrazi, R.P. (2003). *Guidelines for appropriate physical activity for elementary school children. 2003 update.* Reston, VA: NASPE.

Dale, D., Corbin, C.B., & Cuddihy, T.F. (1998). Can conceptual physical education promote physically active lifestyles? *Pediatric Exercise Science, 10*, 97-109.

Dale, D., Corbin, C.B., & Dale, K.S. (2000). Restricting opportunities to be active during school time: Do children compensate by increasing physical activity levels after school? *Research Quarterly for Exercise and Sport, 71*, 240-248.

Department of Health. (2004). *Choosing health. Making healthier choices easier.* London: HMSO.

Dishman, R.K., Motl, R., Saunders, R., Felton, G., Ward, D.S., Dowda, M., et al. (2004). Self-efficacy partially mediates the effect of a school-based physical-activity intervention among adolescent girls. *Preventive Medicine, 38*, 628-636.

Dishman, R.K., Motl, R.W., Saunders, R., Felton, G., Ward, D.S., Dowda, M., et al. (2005). Enjoyment mediates effects of a school-based physical activity intervention. *Medicine and Science in Sports and Exercise, 37*, 478-487.

Dollman, J., Norton, K., & Norton, L. (2005). Evidence of secular trends in children's physical activity behaviour. *British Journal of Sports Medicine, 39*, 892-897.

Donnelly, J., Jacobsen, D., Whatley, J., Hill, J., Swift, L., Cherrington, A., et al. (1996). Nutrition and physical activity program to attenuate obesity and promote physical activity and metabolic fitness in elementary school children. *Obesity Research, 4*, 229-243.

Epstein, D., Kehily, M., Mac an Ghaill, M., & Redman, P. (2001). Boys and girls come out to play. *Men and Masculinities, 4*, 158-172.

Evans, J. (1996). Children's attitudes to recess and changes taking place in Australian primary schools. *Research in Education, 56*, 49-61.

Fairclough, S.J., & Stratton, G. (2005a). Improving health-enhancing physical activity in girls' physical education. *Health Education Research, 20*, 448-457.

Fairclough, S.J., & Stratton, G. (2005b). Physical activity levels in middle and high school physical education: A review. *Pediatric Exercise Science, 17*, 217-236.

Fairclough, S.J., & Stratton, G. (2005c). "Physical education makes you fit and healthy." Physical education's contribution to young people's physical activity levels. *Health Education Research, 20*, 14-23.

Fairclough, S.J., & Stratton, G. (2006). A review of physical activity levels during elementary school physical education. *Journal of Teaching in Physical Education, 25*, 240-258.

Fairclough, S.J., Stratton, G., & Baldwin, G. (2002). The contribution of secondary school physical education to lifetime physical activity. *European Physical Education Review, 8*, 69-84.

Fein, A.J., Plotnikoff, R.C., Wild, T.C., & Spence, J.C. (2004). Perceived environment and physical activity in youth. *International Journal of Behavioral Medicine, 11*, 135-142.

Felton, G., Saunders, R., Ward, D., Dishman, R.K., Dowda, M., & Pate, R.R. (2005). Promoting physical activity in girls: A case study of one school's success. *Journal of School Health, 75*, 57-62.

Fox, K., Cooper, A., & McKenna, J. (2004). The school and promotion of children's health-enhancing physical activity: Perspectives from the United Kingdom. *Journal of Teaching in Physical Education, 23*, 336-355.

Fulton, J.E., Shisler, J.L., Yore, M.M., & Caspersen, C.J. (2005). Active transportation to school: Findings from a national survey. *Research Quarterly for Exercise and Sport, 76*, 352-357.

Gallahue, D.L., & Cleland-Donnelly, F.C. (2003). *Developmental physical education for all children.* Champaign, IL: Human Kinetics.

Gavarry, O., Bernard, T., Giacomoni, M., Seymat, M., Euzet, J., & Falgairette, G. (1998). Continuous heart rate monitoring over 1 week in teenagers aged 11-16 years. *European Journal of Applied Physiology, 77*, 125-132.

Gielen, A.C., DeFrancesco, S., Bishai, D., Mahoney, P., Ho, S., & Guyer, B. (2004). Child pedestrians: The role of parental beliefs and practices in promoting safe walking in urban neighborhoods. *Journal of Urban Health: Bulletin of the New York Academy of Medicine, 81*, 545-555.

Giles-Corti, B., & Donovan, R.J. (2002). The relative influence of individual, social, and physical environment determinants of physical activity. *Social Science and Medicine, 54*, 1793-1812.

Glasgow, R.E., Vogt, T.M., & Boles, S.M. (1999). Evaluating the public health impact of health promotion interventions: The RE-AIM framework. *American Journal of Public Health, 89*, 1322-1327.

Gortmaker, S., Peterson, K., Wiecha, J., Sobol, A., Dixit, S., Fox, M., et al. (1999). Reducing obesity via a school-based interdisciplinary intervention among youth. *Archives of Pediatric and Adolescent Medicine, 153*, 409-418.

Goudas, M., & Biddle, S. (1994). Intrinsic motivation in physical education: Theoretical foundations and contemporary research. *Education and Child Psychology, 11*, 68-76.

Goudas, M., Biddle, S., Fox, K., & Underwood, M. (1995). It ain't what you do, it's the way that you do it! Teaching style affects children's motivation in track and field lessons. *Sport Psychologist, 9*, 254-264.

Hagger, M., Cale, L., & Almond, L. (1997). Children's physical activity levels and attitudes towards physical activity. *European Physical Education Review, 3*, 144-164.

Hardman, K. (2000). The state and status of physical education in schools in international context. *European Physical Education Review, 6*, 203-229.

Harrell, J.S., McMurray, R.G., Bangdiwala, S., Frauman, A., Gansky, S., & Bradley, C. (1996). Effects of a school-based intervention to reduce cardiovascular risk factors in elementary-school children: The Cardiovascular Health in Children (CHIC) Study. *Journal of Pediatrics, 128*, 797-805.

Hayashi, T., Tsumura, K., Suematsu, C., Okada, F., Fujii, S., & Endo, G. (1999). Walking to work and the risk for hypertension in men: The Osaka Health Survey. *Annals of Internal Medicine, 131*, 21-26.

Heelan, K.A., Donnelly, J.E., Jacobsen, D.J., Mayo, M.S., Washburn, R., & Greene, L. (2005). Active commuting to and from school and BMI in elementary school children—preliminary data. *Child: Care, Health and Development, 31*, 341-349.

Hillman, M. (1993). *Children, transport and the quality of life.* PSI research report 716. London: Policy Studies Institute.

Hotz, G., Cohn, S., Castelblanco, A., Colston, S., Thomas, M., Weiss, A., et al. (2004). WalkSafe: A school-based pedestrian safety intervention program. *Traffic Injury Prevention, 5*, 382-389.

Jarrett, O.S., & Duckett-Hedgebeth, M. (2003). Recess in middle school: What do the students do? In D.E. Lytle (Ed.), *Play and educational theory and practice* (pp. 227-241). Westport, CT: Greenwood Publishing Group.

Kann, L. (2001). The Youth Risk Behavior Surveillance System: Measuring health-risk behaviors. *American Journal of Health Behavior, 25*, 272-277.

Kraft, R.E. (1989). Children at play: Behaviour of children at recess. *Journal of Physical Education, Recreation and Dance, 60*, 21-24.

Levin, S., McKenzie, T.L., Hussey, J., Kelder, S.H., & Lytle, L. (2001). Variability of physical activity during physical education lesson across school grades. *Measurement in Physical Education and Exercise Science, 5*, 207-218.

Loucaides, C.A., Chedzoy, S.M., & Bennett, N. (2003). Pedometer-assessed physical activity in Cypriot children. *European Physical Education Review, 9*, 43-55.

Luepker, R.V., Perry, C.L., McKinlay, S.M., Nader, P.R., Parcel, G.S., Stone, E.J., et al. (1996). Outcomes of a field trial to improve children's dietary patterns and physical activity. *Journal of the American Medical Association, 275*, 768-776.

Mackett, R.L., Lucas, L.L., Paskins, J., & Turbin, J. (2005). The therapeutic value of children's everyday travel. *Transportation Research, 39*, 205-219.

Marcoux, M.F., Sallis, J.F., McKenzie, T.L., Marshall, S., Armstrong, C., & Goggin, K.J. (1999). Process evaluation of a physical activity self-management program for children: SPARK. *Psychology and Health, 14*, 659-677.

Martin, J.J., & Hodges-Kulinna, P.H. (2004). Self-efficacy theory and the theory of planned behavior: Teaching physically active physical education classes. *Research Quarterly for Exercise and Sport, 75*, 288-297.

McCann, B. (2000). *Driven to spend. The impact of sprawl on household transportation expenses: A Transportation and Quality of Life publication.* Retrieved May 4, 2006, from www.transact.org/PDFs/DriventoSpend.pdf.

McKenzie, T.L., Marshall, S., Sallis, J.F., & Conway, T.L. (2000). Leisure-time physical activity in school environments: An observational study using SOPLAY. *Preventive Medicine, 30*, 70-77.

McKenzie, T.L., Nader, P.R., Strikmiller, P., Yang, M., Stone, E., Perry, C., et al. (1996). School physical education: Effect of the Child and Adolescent Trial for Cardiovascular Health. *Preventive Medicine, 25*, 423-431.

McKenzie, T.L., Prochaska, J.J., Sallis, J.F., & LaMaster, K.J. (2004a). Coeducational and single-sex physical education in middle schools: Impact on physical activity. *Research Quarterly for Exercise and Sport, 75*, 446-449.

McKenzie, T.L., Sallis, J.F., Faucette, N., Roby, J.J., & Kolody, B. (1993). Effects of a curriculum and inservice program on the quantity and quality of elementary physical education classes. *Research Quarterly for Exercise and Sport, 64*, 178-187.

McKenzie, T.L., Sallis, J.F., Kolody, B., & Faucette, F. (1997). Long term effects of a physical education curriculum and staff development program: SPARK. *Research Quarterly for Exercise and Sport, 68*, 280-291.

McKenzie, T.L., Sallis, J.F., & Nader, P.R. (1991). SOFIT: System for Observing Fitness Instruction Time. *Journal of Teaching in Physical Education, 11*, 195-205.

McKenzie, T.L., Sallis, J.F., Prochaska, J.J., Conway, T.L., Marshall, S.J., & Rosengard, P. (2004b). Evaluation of a two year middle school physical education intervention: M-SPAN. *Medicine and Science in Sports and Exercise, 36*, 1382-1388.

McKenzie, T.L., Stone, E.J., Feldman, H.A., Epping, J., Yang, M., Strikmiller, P., et al. (2001). Effects of the CATCH physical education intervention: Teacher type and lesson location. *American Journal of Preventive Medicine, 21*, 101-109.

Metcalf, B., Voss, L., Jeffrey, A., Perkins, J., & Wilkin, T. (2004). Physical activity cost of the school run: Impact on schoolchildren of being driven to school (EarlyBird 22). *British Medical Journal, 329*, 832-833.

Morrow, J.R., Jackson, A.W., & Payne, V.P. (1999). Physical activity promotion and school physical education. *President's Council on Physical Fitness and Sports Research Digest, 3*(7), 1-8.

Mota, J., Silva, P., Santos, M., Ribeiro, J.C., Oliveira, J., & Duarte, J. (2005). Physical activity and school recess time: Differences between the sexes and the relationship between children's playground physical activity and habitual physical activity. *Journal of Sports Sciences, 23*, 269-275.

Mota, J., & Stratton, G. (2003). Gender differences in physical activity during recess in Portuguese primary schools. *Revista Portuguesa de Ciências do Desporto, 3*, 150.

Napper-Owen, G.E., Kovar, S.K., Ermler, K.L., & Mehrhof, J.H. (1999). Curricula equity in required ninth-grade physical education. *Journal of Teaching in Physical Education, 19*, 2-21.

National Association of Early Childhood Specialists in State Departments of Education. (2002). *Recess and the importance of play. A position statement on young children and recess.* Retrieved April 12, 2004, from naecs.crc.uiuc.edu/position/recessplay.pdf.

Noble, B. (1999). *Review of the National Travel Survey.* Department for the Environment, Transport and the Regions. Retrieved May 4, 2006, from www.statistics.gov.uk/methods_quality/quality_review/downloads/NTSQA_2000.pdf.

Pellegrini, A.D., Blatchford, P., Kato, K., & Baines, E. (2004). A short-term longitudinal study of children's playground games in primary school: Implications for adjustment to school and social adjustment in the USA and the UK. *Social Development, 13,* 107-123.

Pellegrini, A.D., & Bohn, C.M. (2005). The role of recess in children's cognitive performance and school adjustment. *Educational Researcher, 34,* 13-19.

Pellegrini, A.D., & Smith, P.K. (1993). School recess: Implications for education and development. *Review of Educational Research, 63,* 51-67.

Podulka-Coe, D., Pivarnik, J.M., Womack, C.J., Reeves, M.J., & Malina, R.M. (2006). Effect of physical education and activity levels on academic achievement in children. *Medicine and Science in Sports and Exercise, 38,* 1515-1519.

Qualifications and Curriculum Authority. (2005). *Physical education. 2004/5 annual report on curriculum and assessment.* London: QCA.

Renold, R. (1997). "All they've got on their brains is football." Sport, masculinity and gendered practices of playground relations. *Sport, Education and Society, 2,* 5-23.

Riddoch, C., Savage, J.M., Murphy, N., Cran, G.W., & Boreham, C. (1991). Long term health implications of fitness and physical activity patterns. *Archives of Disease in Childhood, 66,* 1426-1433.

Ridgers, N.D., Stratton, G., & Fairclough, S.J. (2005). Assessing physical activity during recess using accelerometry. *Preventive Medicine, 41,* 102-107.

Ridgers, N.D., Stratton, G., & Fairclough, S.J. (2006). Physical activity levels of children during school playtime. *Sports Medicine, 36,* 359-371.

Sallis, J.F., McKenzie, T.L., Alcaraz, J.E., Kolody, B., Faucette, N., & Hovell, M.F. (1997). The effects of a 2 year physical education programme (SPARK) on physical activity and fitness in elementary school students. *American Journal of Public Health, 87,* 1328-1334.

Sallis, J.F., McKenzie, T.L., Conway, T., Elder, J., Prochaska, J.J., Brown, M., et al. (2003). Environmental interventions for eating and physical activity. A randomized control trial in middle school. *American Journal of Preventive Medicine, 24,* 209-217.

Sallis, J.F., McKenzie, T.L., Kolody, B., Lewis, S., Marshall, S.J., & Rosengard, P. (1999). Effects of health-related physical education on academic achievement: Project SPARK. *Research Quarterly for Exercise and Sport, 70,* 127-134.

Sallis, J.F., & Owen, N. (1999). *Physical activity and behavioral medicine.* Thousand Oaks, CA: Sage.

Salmon, J., Timperio, A., Cleland, V., & Venn, A. (2005). Trends in children's physical activity and weight status in high and low socio-economic status areas of Melbourne, Victoria, 1985-2001. *Australian and New Zealand Journal of Public Health, 29,* 337-342.

Sarkin, J., McKenzie, T.L., & Sallis, J.F. (1997). Gender differences in physical activity during fifth-grade physical education and recess periods. *Journal of Teaching in Physical Education, 17*, 99-106.

Scruggs, P.W., Beveridge, S.K., & Watson, D.L. (2003). Increasing children's school time physical activity using structured fitness breaks. *Pediatric Exercise Science, 15*, 156-169.

Shephard, R.J. (1997). Curricular physical activity and academic performance. *Pediatric Exercise Science, 9*, 113-126.

Simons-Morton, B.G. (1994). Implementing health-related physical education. In R.R. Pate & R.C. Hohn (Eds.), *Health and fitness through physical education* (pp. 137-146). Champaign, IL: Human Kinetics.

Simons-Morton, B.G., Parcel, G.S., Baranowski, T., Forthofer, R., & O'Hara, N.M. (1991). Promoting physical activity and healthful diet among children: Results of a school-based intervention study. *American Journal of Public Health, 81*, 986-991.

Sleap, M., & Warburton, P. (1996). Physical activity levels of 5-11 year old children in England: Cumulative evidence from three direct observation studies. *International Journal of Sports Medicine, 17*, 248-253.

Stratton, G. (1996). Children's heart rates during physical education lessons: A review. *Pediatric Exercise Science, 8*, 215-233.

Stratton, G. (1999). A preliminary study of children's physical activity in one urban primary school playground. Differences by sex and season. *Journal of Sport Pedagogy, 5*, 71-81.

Stratton, G. (2000). Promoting children's physical activity in primary school: An intervention study using playground markings. *Ergonomics, 43*, 1538-1546.

Stratton, G., & Leonard, J. (2002). The effects of playground markings on the energy expenditure of 5–7-year-old children. *Pediatric Exercise Science, 14*, 170-180.

Stratton, G., & Mullan, E. (2005). The effect of multicolor playground markings on children's physical activity level during recess. *Preventive Medicine, 41*, 828-833.

Strong, W.B., Malina, R.M., Blimkie, C.J.R., Daniels, S.R., Dishman, R.K., Gutin, B., et al. (2005). Evidence based physical activity for school-age youth. *Journal of Pediatrics, 146*, 732-737.

SUSTRANS. (2001). *Healthy and active travel. Information sheet FH01.* Bristol, UK: SUSTRANS.

Thomson, S. (2004). Just another classroom? Observations of primary school playgrounds. In P. Vertinsky & J. Bale (Eds.), *Sites of sport: Space, place, experience* (pp. 73-84). London: Routledge.

Timperio, A., Crawford, D., Telford, A., & Salmon, J. (2004). Perceptions about the local neighborhood and walking and cycling among children. *Preventive Medicine, 38*, 39-47.

Tremblay, M.S., Inman, J.W., & Willms, J.D. (2000). The relationship between physical activity, self-esteem, and academic achievement in 12-year-old children. *Pediatric Exercise Science, 12*, 312-323.

Trost, S.G. (2004). School physical education in the post-report era: An analysis from public health. *Journal of Teaching in Physical Education, 23*, 318-337.

Trost, S.G., Pate, R.R., Sallis, J.F., Freedson, P.S., Taylor, W.C., Dowda, M., et al. (2002). Age and gender differences in objectively measured physical activity in youth. *Medicine and Science in Sports and Exercise, 34*, 350-355.

Tudor-Locke, C., Ainsworth, B., & Popkin, B.M. (2001). Active commuting to school. An overlooked source of children's physical activity? *Sports Medicine, 31*, 309-313.

U.S. Department of Health and Human Services. (2000). *Healthy people 2010: Understanding and improving health*. Washington, DC: DHHS.

Verstraete, S.J.M., Cardon, G.M., De Clercq, D.L.R., & De Bourdeaudhuij, I. (2006). Increasing children's physical activity levels during recess periods in elementary schools: The effects of providing games equipment. *European Journal of Public Health, 16*, 415-419.

Wallhead, T., & Buckworth, J. (2004). The role of physical education in the promotion of youth physical activity. *Quest, 56*, 285-301.

Warburton, P. (2000). Initial teacher training—the preparation of primary teachers in physical education. *British Journal of Teaching Physical Education, 31*(4), 6-9.

Wechsler, H., Devereaux, R., Davis, M., & Collins, J. (2000). Using the school environment to promote physical activity and healthy eating. *Preventive Medicine, 31*, S121-S137.

Welk, G.J. (1999). The youth physical activity promotion model: A conceptual bridge between theory and practice. *Quest, 51*, 5-23.

Williams, A. (1988). The historiography of health and fitness in physical education. *British Journal of Physical Education Research Supplement 3*, 1-4.

Zask, A., van Beurden, E., Barnett, L., Brooks, L., & Dietrich, U. (2001). Active school playgrounds—myth or reality? Results from the "Move It Groove It" project. *Preventive Medicine, 33*, 402-408.

Ziviani, J., Scott, J., & Wadley, D. (2004). Walking to school: Incidental physical activity in the daily occupations of Australian children. *Occupational Therapy International, 11*, 1-11.

CHAPTER

14

Organized Sport and Physical Activity Promotion

Robert J. Brustad, PhD ▪ Runar Vilhjalmsson, PhD ▪ Antonio Manuel Fonseca, PhD

It is both important and timely to consider youth sport as a domain for the promotion of physical activity. Because physical activity is so strongly related to favorable physical, mental, and emotional health outcomes, it is fundamentally important to try to understand how early experiences in the physical domain may contribute to the disposition to be physically active throughout the life span. The topic is also timely because organized sport involvement has a strong and increasing presence in most modern industrialized nations (DeKnop et al., 1996). Furthermore, a recent trend in many industrialized countries, most notably the United States, has been for organized youth sport involvement to supplant more traditional forms of physical play for youth, such as self-organized physical games and modified forms of traditional sports (Centers for Disease Control and Prevention, 2003; Coakley, 2004). To the extent that this occurs, organized youth sport does not necessarily add opportunities for youngsters to be physically active but instead replaces traditional forms of youth physical activity. Thus, we need to take a close look at the role of organized youth sport in promoting a physically active lifestyle in youth in both the present and the future.

By "organized sport" we are referring to all types of adult-structured competitive sport opportunities that are provided for children and adolescents. The specific nature of this involvement can vary on a continuum ranging from recreational to extremely competitive and intensive. In each circumstance, however, the sport has been organized by adults, and adults thus assume the primary role in determining the philosophy, goals, and structure of the sport experience. Furthermore, it is adults who explicitly or implicitly make other important decisions, such as who will have access to participation and who will be excluded from participation. Influential adults also are responsible for making decisions about the length of the season, the numbers of competitions and practice sessions, and the specific structure and goals of each practice or competition. Since adults establish the value structure within which youth sport takes place, we cannot assume that youth sport involvement by itself will inherently foster favorable physical activity attitudes and behavioral practices among child and adolescent participants.

This chapter addresses current knowledge about the ways in which organized sport involvement can favorably or unfavorably influence youngsters' current and lifelong patterns of physical activity. Of particular importance is how sport involvement during childhood is likely to affect motivation to be physically active across the adolescent and adult years. Because little research has systematically explored the sustained or long-term physical, social, and psychological effects of organized sport involvement during childhood and adolescence, this chapter centers primarily on logical, theoretically based links between sport participation and lifelong physical activity involvement. Systematic research lines are needed for a fuller understanding of the possible role of sport involvement in the promotion of healthy physical activity practices. As researchers and practitioners in the youth sport domain, we seek to develop strategies that maximize the long-term benefits of this involvement.

Organized Sport Participation and Children's Well-Being

There is good general support for the notion that participation in structured youth sport has beneficial effects on the physical activity behaviors, lifestyle practices, and psychological well-being of youngsters during the years corresponding with their involvement. Specifically, sport participants are more likely to engage in regular strenuous physical activity than are nonparticipants and are also more likely to consume healthy foods and to abstain from use of cigarettes and illicit drugs (Pate et al., 2000).

A study of European middle adolescents by Vilhjalmsson and Thorlindsson (1992) showed that involvement in organized sport clubs was more strongly associated with health-related behaviors and psychological well-being than was involvement in other leisure-time sport settings, and this relationship was attributed to the high physical performance demands of participation and the high levels of social integration present within the sport clubs. A large-scale investigation conducted with 5200 Portuguese adolescents and young adults, ages 13 to 20 years, examined participants' level of involvement in sport and exercise activities (e.g., type of activity, frequency and duration of usual practice) in relation to their satisfaction with life (Alves et al., 2004). Findings revealed that individuals who had moderate levels of involvement in sport and exercise had greater life satisfaction than did those individuals with lower levels, and also that young people with the highest levels of involvement in sport and exercise had the highest levels of life satisfaction.

It seems that youngsters who participate in organized youth sport demonstrate many desirable physical and mental health characteristics. However, it is logical to presume that youngsters who self-select into organized youth sport are also more likely to be physically fit and more motivated to engage in physical activity to begin with than are youngsters who do not participate in youth sport. Thus, without controlling for other relevant lifestyle variables it is not possible to accurately estimate the beneficial consequences of organized sport relative to the simultaneous influences of current physical fitness status and physical activity motivation.

In a study that provides some insight on this issue, Tomson and colleagues (2003) examined the relative influence of nonschool sport involvement, overall physical activity level, and health-related fitness status on the depressive symptomology of 933 American children aged 8 to 12 years. Their findings indicated that those children who participated in organized sport were 1.3 to 2.4 times (according to gender; larger values for boys) less likely to suffer from depressive symptomology than were children who did not participate in sport. Whereas sport participation had a positive effect, sport participation was not as strongly linked to reduced depressive symptomology as overall physical activity level was. Active children were 2.8 to 3.4 times less likely to suffer depressive symptomology than were inactive children. Furthermore, children who had favorable health-related fitness profiles with respect to aerobic fitness and body fat levels were 1.5 to 4.0 times less likely to suffer from depressive symptomology than were those youngsters with unfavorable health-related fitness profiles. Thus, though organized sport participation was linked to reduced depressive symptomology in children, a physically active lifestyle and favorable physical fitness status were more strongly associated with this outcome than was mere sport participation.

Youth Sport Value Structures

It is also fair to ask, "Why should we expect organized youth sport involvement to contribute to physical activity promotion?" and "Where does physical activity promotion reside in the typical value structure of organized youth sport programs?" These are difficult questions, since no known systematic research exists about the value structures of organized youth sport programs; it is also difficult to generalize across programs in different countries, for different sports, and at different age levels that also vary in their recreational or competitive orientation. It is safe to say, however, that youth sport programs have tended to focus much more on physical skill development and competitive strategies than on physical activity promotion, and this focus is reflected by a strong orientation toward competitive outcomes and the exclusion of those who are less talented. Furthermore, recent trends point in the direction of greater sport specialization for youngsters at earlier ages (Wiersma, 2000) as well as the greater privatization of youth sport programs, with restrictive and for-profit programs having increasingly replaced community-based and inclusive programs (Coakley, 2004). Each of these trends reflects a predominant value orientation toward competitive results rather than physical activity promotion or personal growth.

Organized youth sport programs also tend to be very much bound by tradition. Sport programs for youth almost invariably follow the same rules and competitive structures as their professional sport counterparts (Brustad, Babkes, & Smith, 2001). Coaches typically are hired because of their experience with the skills, techniques, and competitive strategies of the sport and not for their knowledge about or interest in promoting physical activity for youth through sport. The tradition-bound nature of organized youth sport programs logically makes these programs less responsive to current social trends and needs, such as the need to promote a physically active lifestyle in the face of increasing public health issues. Without trying to overgeneralize about the orientations of youth sport programs, we suggest that one of the main reasons that organized sport is not always an optimal venue for the promotion of physical activity in youth is simply that the value structure of organized sport may not be compatible with such a mission.

Theoretical Perspectives

To more concretely address the link between youth sport participation and concurrent and subsequent physical activity practices, it seems important to examine relevant theoretical perspectives that could shed light on this issue. In this regard, developmental, motivational, and social theories are all relevant to our understanding.

In trying to understand the nature of the relationship between sport involvement during childhood and adolescence and subsequent patterns of participation in physical activity, it is helpful to apply a life span developmental perspective. This is the case because our interest is in ascertaining how an individual's experiences at one phase of the life cycle are likely to influence the same individual's attitudes, motivation, and behavior at another phase. Furthermore, our approach to understanding this issue must be grounded in relevant motivational theory that takes into account how individuals maintain, or change, attitudes and behaviors as a consequence of personal experiences and psychological and social influences. The combination of appropriate life span developmental and motivational theoretical perspectives can help us to better understand how sport and physical activity participation is likely to be sustained or discontinued at differing points in the life span. In addition, it may help us to identify specific developmental phases that may represent "sensitive" or "critical" phases in the formation of attitudes toward physical activity and thus may be more strongly linked to physical activity involvement during adulthood. Such an understanding may help us to design and implement youth sport programs that facilitate continued involvement in sport and physical activity.

The need for a strong theoretical framework from which to address this issue is bolstered by the fact that well-designed longitudinal research does not exist on this topic. In addition, the available cross-sectional research is not always sufficiently insightful to allow projections about the role of youth sport in affecting current and future physical activity patterns. Our discussion will start with a consideration of the role of social influence in shaping orientations toward sport and physical activity. Subsequently, we propose a general life span developmental perspective on involvement using continuity theory (Atchley, 1989, 1993) and apply motivational theory in the form of self-determination theory (Deci & Ryan, 1985, 1991; Ryan & Deci, 2000) and achievement goal theory (Nicholls, 1984) to assist in our understanding of this question. Social influence, developmental considerations, and motivational patterns are all highly interrelated, of course. Thus, we will present a conceptual overview of some of the major forms of influence and seek integration of these perspectives within our overall understanding of the likely influence of youth sport participation on physical activity involvement.

Social Influence

Youth sport involvement takes place within a social milieu. How do we understand the ways in which this social environment has both current and enduring effects on youngsters' physical activity practices? Bronfenbrenner's (1993) ecological systems theory proposes that an individual develops attitudes, beliefs, and cognitions within a social environment that consists of multiple overlapping spheres of social influence, including family, peers, and other significant social groups. We can examine the meaning of youth sport involvement in relation to these social influences through Bronfenbrenner's framework to understand social influence in youth sport. What are the meanings (e.g., anticipated achievement, affiliation, and personal growth outcomes), values, and expectations that shape this involvement? Youth sport involvement is strongly influenced by peer social relationships as well, including those with peers generally and those with specific friends (Smith, 2003; Weiss & Stuntz, 2004). Thus, we need to consider peer-based forms of social influence as they may affect an individual's involvement. Finally, the child's involvement in sport takes place within a larger frame of influence that has been structured primarily in accordance with adult values—specifically the values of coaches, parents, and administrators—and that reflects the broader cultural values within which participation is embedded. We will discuss family and peer forms of influence and also address how cultural attitudes and practices are likely to be influential.

During early and middle childhood, the family is usually the primary form of social influence on children's sport and physical activity involvement (see chapter 11). Research indicates that strong relationships exist between the physical activity levels of parents and the physical activity levels of their children (Anderssen & Wold, 1992; Freedson & Evenson, 1991; Wold & Anderssen, 1992), and that parental beliefs and socialization practices have important effects upon youngsters' attraction to physical activity (Brustad, 1996). It is not clear whether fathers and mothers generally exert the same type of influence or whether the same-sex or opposite-sex parent exerts greater influence on these relationships (Bois et al., 2005; Seabra et al., 2004). Research also indicates that siblings tend to have highly similar levels of physical activity (Duncan et al., 2004; Sallis et al., 1999), suggesting that a family's activity patterns are strongly shaped by parental values and practices. A study carried out by Wold and Anderssen (1992) with children and adolescents between 11 and 15 years of age showed that when best friends, parents, and siblings were involved in physical activity, 84% of the male participants and 71% of the female participants engaged in physical activity at least two times per week. Conversely, when these three important significant others did not engage in physical activity, only 52% of the males and 30% of the females were involved in physical activity at least twice per week.

As children approach later childhood and adolescence, the peer relationships and friendships that they have established become highly important forms of social influence on their sport and physical activity involvement (see chapter 12). In this phase of development, individuals are likely to look for a "reflected appraisal" from others that provides them with additional insight into how they are regarded by others. Such feedback also indicates the extent to which similar others support, or fail to support, young people's continued involvement in sport or any other domain of achievement (Horn & Weiss, 1991; Veroff, 1969).

It is also essential to recognize that one of the primary motives for individuals of any age to become and stay involved in sport and physical activity is the opportunity that the physical domain provides for establishing and maintaining favorable social relationships. The affiliation motive has consistently been recognized as a fundamental precursor to interest in youth sport, as youngsters seek to make new friends and acquaintances and to strengthen existing friendships in this context (Weiss & Petlichkoff, 1989). In a large-scale longitudinal study conducted in Scotland by Hendry and colleagues (1993) across different adolescent age cohorts (ages 13-14 and 15-16 years) and types of sport involvement (competitive or recreational), more than 70% of the males and more than 50% of the females reported that they participated in competitive sport at least in part because their friends participated, and more than 75% of the male and female recreational sport participants did so for the same reasons. Other researchers (e.g., Raudsepp & Viira, 2000; Zeijl et al., 2000) have also found that youngsters are much more likely to participate in sport when their best friends also participate. For example, a large-scale study conducted with more than 7000 Portuguese children and adolescents (Corte-Real et al., 2004) showed that children and adolescents who reported that their best friend participated in sport were twice as likely as other youngsters to participate in sport. Research has also linked physical competence with social acceptance in the physical domain (Chase & Dummer, 1992; Weiss & Duncan, 1992), suggesting that youngsters desire to develop physical skills in order to enhance the likelihood of gaining social acceptance from their peers.

As noted in chapter 12, both friends and peers can have important effects on children's and adolescents' physical activity practices. Friendship refers to a close, reciprocal relationship between two individuals that contains a strong emotional component (Bukowski & Hoza, 1989). The peer group can be considered a form of influence independent from one's friends in that the peer group represents "generalized others" and provides a generalized form of acceptance by similar others (Bukowski & Hoza, 1989). In his study of male and female adolescents, Smith (1999) found that both the peer group and friends affected youngsters' desire to be involved in physical activity as well as their actual physical activity behavior. Specific findings indicated that stronger perceptions of friendship in the physical domain were associated with higher levels of attraction to physical activity, and that more favorable perceptions of peer acceptance were associated with stronger perceptions of physical self-worth for the participants.

More favorable relational experiences for children within the social environment of youth sport result in more favorable affective experiences and memories that will enhance continued interest in the physical domain (Brustad, 1996). Conversely, experiences of rejection by others will deter such continued involvement. It is important to recognize that one of the fundamental reasons individuals seek to become and stay involved in sport and physical activity at any age is the positive social relationships they may encounter in this domain (Brodkin & Weiss, 1990; Whaley & Ebbeck, 1997). We will continue to consider social influence throughout the chapter, but next direct primary attention to developmental change across the life span as an influence on sustained interest in physical activity involvement.

Continuity Theory

A fundamental life span development principle is that a person's current and anticipated patterns of behavior reflect the cumulative and sequential effects of the individual's life history and experiences (Langley & Knight, 1999; McPherson, 1994). In continuity theory (Atchley, 1989, 1993), a life span developmental framework, it is proposed that individuals seek to maintain a consistent sense of self across the life span that is reflected in their thought processes, social networks, and activity choices. The fundamental tenet of this theoretical perspective is that humans have an inherent desire to maintain a stable sense of self that is reflected by self-related cognitions, such as their identity and sense of self, interests, goals, and perceptions of competence across various domains of activity (Langley & Knight, 1999). Atchley (1993) proposed that individuals proceed through a selective investment of their resources so as to devote time and energy to those activities that are most likely to enable continuity of personal and social characteristics. It is anticipated that the motive for continuity is particularly strong toward activities that have resulted in goal attainment in the past. Given this presumed desire to maintain customary patterns of thought and behavior throughout the life span, we would generally anticipate individuals to be motivated to maintain similar interests and activity involvement across developmental phases.

Atchley (1989) proposed two distinct forms of continuity, which he referred to as internal continuity and external continuity. Internal continuity refers to the person's own cognitive and psychological patterns and characteristics, such as his or her emotional response to various activities and his or her values, attitudes, and belief system. These personal characteristics are considered to be relatively stable across the life span. External continuity refers to the desire to maintain a reasonably stable social network and social environment. Individuals are thus believed to be predisposed to want to maintain consistent patterns of social relationships with others who share common interests, beliefs, and values. However, external continuity is more difficult to maintain than internal continuity given changes in social networks across time as a consequence of life events.

As a life span developmental perspective, continuity theory is entirely suitable for examining the relationship between childhood sport involvement and physical activity participation during all phases of the life cycle. To date, however, the theory has been used primarily to understand activity choices made by middle-aged and older adults (i.e., Langley & Knight, 1999). From the perspective of continuity theory, we need to understand the role of sport in relation to the internal continuity of the individual across time. In this regard we would ask, "How important is sport and physical activity participation to the child's sense of self?" and "What is the meaning of participatory involvement to the young person?" In addition, we need to consider the factors that might affect external continuity, particularly in relation to the quality of the social relationships that the person experienced while previously involved and the perceived likelihood that favorable social relationships in physical activity are possible in the future.

If youth sport involvement is to have a positive effect upon physical activity involvement throughout adolescence and into adulthood, we would expect to see a consistency in the motives, personal meanings, and patterns of social interaction that underlie participation. However, it is important to recognize that the motives and meanings regarding sport participation vary considerably in accordance with the individual's developmental status. Thus, although continuity theory provides a general framework for understanding continuity of involvement, it is also important to recognize that individuals undergo a great deal of change with development, particularly during adolescence.

In one of the few studies of its kind, Brodkin and Weiss (1990) examined the motives for involvement in competitive swimming for individuals aged 6 years to more than 60 years. Although all individuals participated within the same general structured competitive sport context, self-reported motives for participation varied widely in accordance with developmental status. The youngest swimmers, including those in both age-groups 6-9 and 10-14 years, cited participation motives strongly related to significant others (i.e., parents and friends want me to participate). Adolescent swimmers (ages 15-18 years) and college-age swimmers (ages 19-22 years) were much more inclined to cite as important to them motives related to gaining social status. Younger and middle adults rated health and fitness motives as highly important, and older adults rated fun most highly. Affiliation motives were rated highly across all age-groups. These findings suggest the importance of recognizing that continuity of physical activity across the life span may be affected by changes in the personal meanings and motives for involvement.

Developmental and Motivational Considerations

If individuals are to be motivated to engage in physical activity from childhood through adulthood, the adolescent developmental period is obviously an essential link in the chain. In order to try to facilitate continued sport and physical activity involvement during this period, we need to understand general characteristics of adolescent participatory involvement while maintaining a perspective that is attentive to adolescent developmental change and motivational issues during these changes. Research indicates that physical activity and sport involvement declines for young people across the adolescent years and that this decline is particularly apparent for girls (Baranowski et al., 1997; Pratt, Macera, & Blanton, 1999; Vilhjalmsson & Kristjansdottir, 2003). Adolescence appears to be a very important time frame during which individuals make decisions about which activities they will continue to pursue and which former interests will be discontinued (Fredricks et al., 2002).

Developmental Perspectives

Logically, adolescents' issues related to their own identity and "coming of age" may strengthen or weaken the desire to maintain sport and physical activity participation. Coakley and White (1992) identified five common issues cited by 13- to 18-year-olds concerning how they made decisions about whether or not to engage in regular physical activity and sport. It is important to consider adolescent sport involvement within the broader context of these common adolescent developmental concerns, which are as follows:

- Making the transition to adulthood
- Perceived opportunities to demonstrate physical competence and autonomy to others
- Young people's perceptions about the extent of the social support they will receive from parents and same-sex friends
- The social constraints imposed by finances, parents, and opposite-sex friends
- Young people's own personal recollections of their sport and physical education experiences

In a related line of research, Eccles and Barber (1999) examined correlates of adolescents' extracurricular activity participation (sport, performing arts, academic clubs, and involvement in prosocial groups). They found that involvement in these activities was strongly associated with issues related to the adolescents' processes of identity formation, as well as to the extent of peer group support that they received for their involvement. In a related study addressing why adolescents desire to participate in extracurricular achievement activities, Fredricks and colleagues (2002) found that participants in extracurricular activities cited high perceived ability and the opportunity to be with friends as fundamental contributors to their interest. Similarly, Adler, Kless, and Adler (1992) reported that adolescents' satisfaction with their role identity in their peer group culture was an important contributor to their desire to participate in various achievement domains. Since the adolescent peer group provides a form of reflected appraisal for a youngster, it is not surprising that research has also revealed that identity and peer relationship issues can strengthen or undermine commitment to various free-choice activities.

Eccles and Midgley (1989, 1990) have argued for the consideration of a fit between developmental stage and environment in seeking a better understanding of continuity of behavioral patterns from childhood through adulthood. This perspective holds that if the social environment provides a good "fit" with an individual's needs at a particular developmental period, favorable motivational consequences should ensue. However, if the transition to a subsequent developmental phase (e.g., from childhood to adolescence or from adolescence to adulthood) is not met with an appropriate "match" between opportunities and developmental needs, it is logical that motivational levels will be undermined. Thus, if continuity of participation in sport and physical activity is to be maintained through the life span, we need to consider how well the involvement meets the unique developmental needs of individuals during childhood, adolescence, and adulthood.

How does this knowledge about adolescent participatory involvement patterns and adolescents' social and psychological development contribute to the purpose of our chapter? It supports the idea that if we are to develop the types of sport and physical activity programs that will attract adolescent participants, these programs need to meet their developmental needs. At the present time, we are clearly not meeting these needs, or we would not be witnessing such a decline in sport participation during the adolescent years. Some of the key factors that affect adolescents in making decisions to participate or not participate in any domain of achievement involve

- maintaining personal autonomy or control over their involvement,
- opportunities to be with friends,
- social identity concerns,
- perceptions about the extent to which they will be able to demonstrate competence, and
- the perceived level of social support that they anticipate for their involvement.

To what extent do youth sport programs address these needs? Should we not anticipate a decline in adolescent involvement in sport for the mere reason that programs typically are structured around a traditional sport model that takes little account of the developmental needs and goals of the participants? A highly unfortunate consequence of adolescent dropout from sport is that youngsters may lose their best link to a physically active lifestyle.

Motivational Perspectives

Individuals' continued involvement in any domain, such as sport and physical activity, is dependent on motivational processes. It is highly unlikely that people can maintain their involvement in sport and physical activity across developmental phases without significant motivation, most notably intrinsic motivation. Thus, our conceptual framework for understanding sport and physical activity across the life span must consider the role of intrinsic motivation in the process.

Self-determination theory (Deci & Ryan, 1985, 1991; Ryan & Deci, 2000) is an appropriate theoretical approach to understanding sport and physical activity involvement patterns throughout the life span because it directs primary attention to intrinsic, self-determined forms of motivation that are essential to maintaining involvement in free-choice activities such as sport and physical activity. Self-determination theorists propose three important general motivational profiles that differentiate people with regard to their motivation in any given context: intrinsic motivation, extrinsic motivation, and amotivation (see chapter 8).

Intrinsic motivation is the most desirable type of motivation because this form is self-determined, which means that it is under the control of the individual. Intrinsic motivation is expressed in the physical domain when individuals take part in sport or physical activity because of the inherent pleasure and satisfaction they experience through their involvement. Amotivation and extrinsic motivation are much less desirable motivational patterns. Amotivation is a lack of motivation to engage in a given achievement context as the individual has neither intrinsic nor extrinsic reasons to be engaged. People are extrinsically motivated when their participation revolves around the rewards or tangible benefits that they receive rather than their own intrinsic interest and enjoyment. Although extrinsic motivation can have some benefits for a youngster's motivation in the present, this form of motivation should have limited long-term benefits since many extrinsic sources of motivation (e.g., social status, rewards such as trophies, and parental reinforcement) are less likely to be available to them in later years.

Self-determination theorists propose that motivation is more likely to be self-determined, or intrinsic, when three fundamental needs are met: autonomy, competence, and social relatedness. Autonomy is the feeling that one is the originator and regulator of one's own actions. Competence refers to feelings that one has the necessary skills and abilities to realize desired outcomes in a given

Parents who express strong negative reactions to umpires' decisions, children's mistakes, or the success of opposing teams can undermine children's intrinsic motivation for sport involvement.

achievement context. Social relatedness refers to feelings that one can attain and maintain positive social relationships with other individuals in a particular achievement setting. Baumeister and Leary (1995) have provided compelling arguments that the desire to belong to social groups, to be accepted by other social members, and to establish strong social networks with others reflects a fundamental need underlying human motivational processes. Self-determination theory is one of the few theories that include social needs in their explanations of motivational processes.

To address the issue of how sport involvement during childhood and adolescence is likely to facilitate physical activity involvement, it is important to examine how well the needs for autonomy, competence, and relatedness are met in the youth sport domain.

Studies on Autonomy

In accordance with contemporary motivational perspectives, in relation to autonomy needs we should attempt to promote feelings of self-regulation and perceptions of control in youngsters during their sport experiences. The reality is that adult coaches, parents, and administrators typically organize and control such a large part of the organized youth sport experience that it becomes unlikely that many youngsters experience true feelings of autonomy in this context.

Coakley and White's (1992) research indicated that the extent to which adolescents perceive they will be able to retain autonomy over their involvement is instrumental to their sustaining or discontinuing participation in their sport. In a physical education setting, Goudas and colleagues (1995) examined the influence of an autonomy-supportive environment on youngsters' intrinsic motivation to play sports. The authors compared two different teaching methods, involving either teacher-controlled or student-controlled decisions, related to choice of track and field activities. As anticipated, students who participated in an environment in which they had greater autonomy exhibited greater intrinsic motivation toward track and field than did those youngsters who participated in the environment in which the teacher made all of the decisions. Similarly, Hagger and colleagues (2003) found that perceived autonomy support in physical education classes had a positive association with intrinsic motivation and contributed to students' intentions to participate in leisure-time physical activity, as well as to their subsequent leisure-time activity. Using a large national sample of British adolescents, Wang and Biddle (2001) found that perceived autonomy in the sport and physical education context was significantly related to adolescents' actual levels of physical activity involvement, which suggests that providing autonomy in educational settings can translate to greater intrinsic motivation for individuals outside of the school context.

In a sport-related intervention study, Pelletier and colleagues (2001) provided workshops intended to facilitate autonomy support in swim coaches to determine if increased autonomy support would have favorable motivational consequences for young swimmers. They found that coaches who received the autonomy-support training intervention had a dramatically lower rate of dropout among their young swimmers than control coaches. In addition, swimmers of coaches who received the autonomy-support training demonstrated increases in intrinsic motivation levels over the course of the swim season in comparison to the control group swimmers, whose intrinsic motivation remained unchanged. Sarrazin and coworkers (2002) found that female adolescents who discontinued their team handball participation reported lower feelings of personal autonomy than did those who maintained involvement. In a unique study that used athletes' diaries over a period of four weeks, Gagné, Ryan, and Bargmann (2003) found that coach and parental autonomy support resulted in greater self-determined motivation for adolescent female gymnasts.

Research also indicates that the perception of low autonomy can result in feelings of emotional exhaustion or burnout, particularly in highly invested adolescent athletes. Coakley (1992) found that adolescent athletes who had parents with a strong investment in their sport were likely to perceive that they had little input into the decisions affecting their involvement and subsequently were more likely to experience burnout from their sport. Similarly, Raedeke (1997) reported that adolescent swimmers who felt "trapped" or controlled by the expectations of others regarding their sport involvement were more likely to experience the unfortunate emotional state of burnout. Thus, the well-intended investment of resources by significant adults such as parents and coaches in a young athlete's sport involvement can actually backfire and have a less than favorable effect by reducing the youngster's sense of autonomy.

Autonomy is likely to be an important contributor to physical activity involvement as well, though this has been rarely explored. Parental autonomy support may be particularly important in facilitating the physical activity involvement of young people. A recent study by Marlatt (2005) showed that higher levels of maternal autonomy support and combined parental autonomy support were significantly related to higher levels of free-time physical activity among 11- to 14-year-old children. If adult coaches, parents, and administrators were more willing to structure youth sport and physical activity experiences in ways that encouraged autonomy in youngsters, it is likely that youngsters would feel more control over their free-time physical activity experiences and have greater motivation to be physically active.

Studies on Competence

Competence perceptions are also fundamental contributors to motivation from the perspective of self-determination theory and other principal contemporary motivational perspectives, such as self-efficacy and achievement goal theories (e.g., Bandura, 1986; Nicholls, 1984). A large body of knowledge has indicated that individuals' perceptions of competence are likely to affect their subsequent involvement in the physical domain. For instance, Gould and colleagues (1982) found that adolescent swimmers who dropped out of swimming frequently cited being "not good enough" as a reason for discontinuing their involvement. Wang and Biddle (2001) found that perceived physical competence was significantly related to physical activity involvement in their adolescent sample. As reported by Coakley and White (1992), adolescents feel that they need to be able to demonstrate competence if they are to maintain their sport involvement. In Sarrazin and colleagues' (2002) study of female handball players, individuals who held more favorable perceptions of competence were more likely to maintain their involvement relative to individuals with less favorable competence perceptions.

From a developmental perspective, research would suggest that the competence need becomes particularly important during adolescence. Wigfield (1994) concluded that children can be motivated to participate in activities in which they don't feel competent, but that adolescents need to feel competent in an activity in order to maintain their involvement. Harter's research on competence perceptions among students in the elementary and middle school years indicates that most youngsters demonstrate a decline in perceived competence from the ages of 6 to 12 years (Harter, 1998). This decline in perceived competence is attributed to the developmental capacity to utilize social comparison information more effectively (Eccles & Midgley, 1989). Logic suggests that such declines in perceived competence would reduce intrinsic motivation to participate in physical activity and sport.

In addition to the developmentally related factors that may dispose youngsters to hold lower perceptions of competence with age, the climate in which youth sport involvement takes place may exert a major influence on perceptions of competence. Research indicates that coaches are very important in affecting young athletes' perceptions of competence. Specifically, coaches who provide ideal forms of feedback by way of technical instruction, contingent praise, and encouragement are likely to strengthen young athletes' competence perceptions (Allen & Howe, 1998; Black & Weiss, 1992).

In line with achievement goal theory (Nicholls, 1984) and as discussed in chapter 8, when people use predominantly self-referenced or mastery criteria in assessing their competence, there is no logical reason for them to perceive that they have low competence as long as they are exerting effort toward the realization of their improvement goal. However, it is likely that individuals predominantly holding a social comparison standard of success and competence will experience a reduction in perceived competence with age, because they are comparing their abilities to those of others in a sport system in which maintenance of involvement requires greater relative ability each year.

Since organized youth sport follows a hierarchical "pyramid" system in which sport involvement increasingly depends upon ability as children age, less able athletes tend to be selected out each year, reducing overall participation numbers. Youngsters adopting a social comparison criterion of competence will accordingly be more prone to perceive that they have low competence as they compare themselves to an increasingly more rigorous standard. Perceptions of personal competence during adolescence thus become even more important because youngsters are aware that their continued participation depends on the demonstration of high ability relative to their peers. Compounding the problem is the fact that youngsters with later maturation relative to their peers may prematurely develop negative competence perceptions and quit a given sport or sport in general. Significant adults, particularly coaches, have an important role in this regard because they help to establish the motivational climate operating within a given sport context. If a coach has a strong social comparison orientation toward success, athletes are likely to view success and failure in the same way. Conversely, a motivational style oriented toward personal mastery will encourage a different view of success in young participants and help them to use self-referenced criteria in their appraisals of success. Such an orientation will prove useful in sport and physical activity contexts because this competence orientation falls under the individual's personal control.

Studies on Relatedness

Relatedness is the third principal contributor to intrinsic motivation from the standpoint of self-determination theory and is essential to youth sport motivation. In a comprehensive review of the research on motives for participation in youth sport, Weiss and Petlichkoff (1989) identified affiliation and relatedness goals as fundamental contributors to youngsters' desires to become involved, and stay involved, in sport. Brodkin and Weiss (1990) found that affiliation motives were highly important motives for involvement in competitive swimming across the entire age span of participation. Sarrazin and colleagues (2002) also reported that differences in feelings of relatedness explained differences in sport involvement or discontinuance. Ullrich-French and Smith (2006) found in their study of 10- to 14-year-old youth soccer players that stronger self-determined motivation was predicted by higher peer acceptance, friendship quality, and more positive parent–child relationships. Thus, research supports the proposition that relatedness is an important contributor to self-determined motivation, whether these feelings of relatedness involve peers, friends, or parents.

Gender and Socioeconomic Influences

It is also necessary to consider the role of gender and socioeconomic influences in shaping the relationship between childhood sport involvement and subsequent physical activity participation. We have seen in previous chapters that, overall, adolescents' participation in organized sport drops with age (Baranowski et al., 1997) and that this change is more profound for girls than for boys (Duke, Huhman, & Heitzler, 2003). We have also seen a number of reasons suggested for this disparity, including the following:

- Gender stereotypes in which competitive sports have been associated with traits considered "masculine," such as being tough, aggressive, competitive, and physically strong (Coakley, 2004; Washington & Karen, 2001)
- The predominance of males among administrators and coaches of organized sport programs (Vilhjalmsson & Kristjansdottir, 2003; Washington & Karen, 2001)
- An emphasis in organized programs on sports that may be more popular among boys
- Greater concern among girls' parents about neighborhood and transportation safety issues as barriers to their child's participation (Duke et al., 2003)

Individuals with limited education or low income are less likely to have sport or exercise facilities in their residential area and are more likely to be confronted by neighborhood safety and transportation problems as well as prohibitive

Courtesy of T.H. Farris

This young teen overcame barriers such as gender discrimination to continue playing a sport she loves.

costs of participation (Duke et al., 2003). Studies also show that lower-income parents are less likely to have time to participate with, and support, their child's physical activity involvement (Duke et al., 2003). Other factors pertaining to family context and lifestyle may also hamper sport involvement of youth from lower social strata. A study by Vilhjalmsson and Thorlindsson (1998) showed that students of lower socioeconomic status engaged in more TV viewing and paid work during the school year and had parents who tended to be less involved in physical activity, which helped to explain lower rates of sport and physical activity involvement among these students. Additionally, adults in lower income or education brackets are less likely to participate in organized physical activity (Canadian Fitness and Lifestyle Research Institute, 2005). Chapters 6, 11, 16, and 18 also discuss the influence of lower parental socioeconomic status on the activity levels of children.

Even if organized sport programs are formally open to all individuals in a given geographical area, their actual operation can result in differential participation by individuals and groups. In this context, Merton (1957) distinguished between *manifest functions* of institutions, having to do with official purposes and publicly expected outcomes, and *latent functions* pertaining to the unintended consequences of human action. As Vilhjalmsson and Kristjansdottir (2003) observed, there are three ways in which organized sport programs can contribute to disparities in sport and physical activity involvement. First, *differential enrollment* may result from the fact that the particular sports and events that are offered cater unequally to individuals and groups, or that facilities, training times, or trainers or coaches are not equally accessible. Second, *differential withdrawal* may result from negative experiences of some individuals already enrolled, leading them to withdraw from the sport program. Finally, *differential involvement* may be observed among those enrolled, as a result of unequal opportunities to train or compete within the programs (differential coaching, unequal access to equipment or facilities, or other variations in training or competition opportunities). Further research is needed to increase understanding of how organized sport programs shape involvement in sport and physical activity and contribute to disparities between individuals and social groups.

Programming to Facilitate Lifelong Physical Activity

Understanding the role of organized sport involvement during childhood in subsequent physical activity behavior is a timely issue. We have acknowledged that direct empirical tests of this relationship are lacking. However, we can make some inferences based on our knowledge about how youth sport programs are structured in relation to developmental, motivational, and social influences.

Youth sport participation occurs within a social context that is heavily influenced by adults, specifically parents, coaches, and league administrators. Adult influence is quite powerful due to the developmental status of the young participants, who typically enter youth sport programs at young ages and who therefore look to adults to form attitudes and judgments about their sport involvement. Although virtually all adults involved in youth sport programs have good intentions, they do not necessarily provide a sport experience for youngsters that will facilitate their continued physical activity involvement.

Addressing Motivation

It is undoubtedly with regard to autonomy that the greatest changes in youth sport programs are needed. Currently, youngsters have little involvement in or input into most of the major decisions related to their engagement in sport. From a developmental standpoint, this is not appropriate, as perceived autonomy has been identified as a key factor affecting adolescents' willingness to maintain participation in extracurricular activities. It would be beneficial to youngsters' sustained involvement to allow them greater influence over many of the decisions that relate to their sport practice. Adults should provide youngsters with greater opportunities to structure some of their practices—for example, to choose the types of drills and learning opportunities that they find most interesting and challenging, select the order and sequence of drills, and organize small-sided drills and scrimmages. Furthermore, youngsters should have the opportunity to modify some of the rules of the game (e.g., substitutions) to allow them greater autonomy over their involvement. Related research conducted in sport and physical education contexts (e.g., Goudas et al., 1995; Wang & Biddle, 2001) clearly indicates that the provision of autonomy support to youngsters will increase their self-determined motivation.

A secondary strategy to facilitate lifelong involvement would be to enhance feelings of social relatedness in youngsters. The desire to develop and maintain favorable peer relationships is one of the primary reasons why youngsters seek to become involved in youth sport in the first place (Weiss & Petlichkoff, 1989). Moreover, a major reason for discontinuing their organized sport involvement pertains to its effects on their social relational patterns (Adler et al., 1992; Coakley & White, 1992; Eccles & Barber, 1999). Research indicates that more favorable relational patterns with peers and parents are associated with higher self-determined motivation in young athletes (Ullrich-French & Smith, 2006). Although organized youth sport programs typically provide youngsters with many opportunities to develop and maintain positive social relationships, these opportunities could be increased. For example, youngsters could be given more opportunities to remain with the same set of friends and teammates over a longer period of time if they were allowed to have input into team selection. Allowing youngsters more autonomy over their involvement, as already suggested, would encourage more social interaction and collective problem solving within the group, which should enhance relatedness. Finally, more opportunities for positive social interaction with team members away from the sporting fields and courts will be of additional benefit in fulfilling relatedness needs.

Promoting perceptions of competence is the third prong of strategies based on a self-determination perspective. Competence is a particularly important consideration because we know that youngsters typically show diminished perceptions of competence with age. In many cases, coaches need to develop greater expertise in providing feedback to young participants so that these individuals recognize increases in personal ability. Furthermore, youth sport coaches should be encouraged to foster mastery orientations toward success in which perceptions of competence are developed in line with personal improvement criteria.

Addressing Gender and Socioeconomic Status

There are many additional forms of social influence to consider, particularly in relation to gender and socioeconomic status. Further efforts need to be made to retain girls in organized sport during the adolescent years. Because many organized sport activities are still gender stereotyped as "male" activities, and because male coaches and administrators are prominent in the delivery of such programs, efforts to offer a greater variety of sports and recruit more female coaches and administrators could facilitate girls' involvement in youth sport and their subsequent physical activity.

Currently, as a consequence of socioeconomic differences, organized youth sport programs are not equally available to all youngsters. An unfortunate trend in the United States has been for privatized youth sport programs to replace community and recreational programs (Coakley, 2004). One consequence of this trend has been that many lower-income families cannot enroll their children in youth sport programs and do not have alternatives to these programs. A second consequence has been that for-profit, privatized programs tend to be highly specialized and to have a strong corresponding tendency toward near-total adult control. If one of the purposes of organized sport involvement for youth is to promote lifelong physical activity, we need to be more inclusive in the development of our sport programs so that youngsters have equal opportunities for participation. Achieving greater inclusiveness entails addressing differential enrollment, involvement, and withdrawal by offering a variety of subsidized or free sports and activities catering to a broader group of participants; safe and affordable transportation; and equal opportunities for training, engagement, and competition within the programs.

Involvement in organized youth sport during childhood is likely to influence individuals' attitudes and orientations toward physical activity in many different ways. Although we do not yet have a complete understanding of the relationship between organized sport involvement and youngsters' current and future activity patterns, we can make strong inferences about this relationship in accordance with developmental and motivational theories. Because of the great social and economic importance of physical activity, we need to design youth sport programs with the intention of facilitating, simultaneously, the immediate and lifelong physical activity practices of the participants.

APPLICATIONS FOR RESEARCHERS

Considerable attention has been devoted to the psychological and social consequences of organized sport involvement for youth over the past few decades. However, attention to the physical health consequences of this involvement is highly warranted given public health issues associated with physical inactivity. Research is needed to examine both the short-term and the long-term effects of sport involvement on physical activity motivation and behavioral patterns. Examination of the long-term effects should be the higher priority, and such research must be structured so that it is consistent with current perspectives on motivation and developmental issues that affect individuals' sport and physical activity interests.

APPLICATIONS FOR PROFESSIONALS

If youth sport programs are to effectively promote physical activity, a change will be required in the prevailing orientations of most youth sport organizations. Influential adults, such as youth sport administrators, coaches, and parents, must help to establish a climate in which lifelong health goals are valued. This would entail a corresponding deemphasis on the competitive goals that typically dominate the youth sport environment. Furthermore, in accordance with contemporary motivational theory, we need to increase youngsters' feelings of autonomy, social relatedness, and competence. In addition, youth sport programs need to become more inclusive to allow for greater participation among youth with different sport interests, and with varying sport skills and abilities, to more proactively facilitate the participation of girls and to include more youth from families with disadvantaged backgrounds.

References

Adler, P.A., Kless, S.J., & Adler, P. (1992). Socialization to gender roles: Popularity among elementary school boys and girls. *Sociology of Education, 65*, 169-187.

Allen, J.B., & Howe, B. (1998). Player ability, coach feedback, and female adolescent athletes' perceived competence and satisfaction. *Journal of Sport and Exercise Psychology, 20*, 280-299.

Alves, J.R., Corte-Real, N.J., Corredeira, R., Brustad, R., Balaguer, I., & Fonseca, A.M. (2004). The involvement of young people in sport and exercise and their satisfaction with life: Is there any relationship? *Annual conference of the British Psychological Society: Health psychology—a positive perspective* (p. 52). Edinburgh: Queen Margaret University College.

Anderssen, N., & Wold, B. (1992). Parental and peer influences on leisure-time physical activity in young adolescents. *Research Quarterly for Exercise and Sport, 63*, 341-348.

Atchley, R.C. (1989). A continuity theory of normal aging. *Gerontologist, 29*, 183-190.

Atchley, R.C. (1993). Continuity theory and the evolution of activity in later adulthood. In J. Kelly (Ed.), *Activity and aging: Staying involved in later life* (pp. 5-16). Newbury Park, CA: Sage.

Bandura, A. (1986). *Social foundations of thought and action: A social cognitive theory*. Englewood Cliffs, NJ: Prentice Hall.

Baranowski, T., Bar-Or, O., Blair, S., Corbin, C., Dowda, M., et al. (1997). Guidelines for school and community programs to promote lifelong physical activity among young people. *Morbidity and Mortality Weekly Report, 46*(RR-6), 1-36.

Baumeister, R.F., & Leary, M.R. (1995). The need to belong: Desire for interpersonal attachments as fundamental human motivation. *Psychological Bulletin, 117*, 497-529.

Black, S.J., & Weiss, M.R. (1992). The relationship among perceived coaching behaviors, perceptions of ability, and motivation in competitive age-group swimmers. *Journal of Sport and Exercise Psychology, 14*, 309-325.

Bois, J.E., Sarrazin, P.G., Brustad, R.J., Cury, F., & Trouilloud, D.O. (2005). Elementary schoolchildren's perceived competence and physical activity involvement: The influence of parents' role modeling behaviors and perceptions of their child's competence. *Psychology of Sport and Exercise, 6*, 381-397.

Brodkin, P., & Weiss, M.R. (1990). Developmental differences in motivation for participating in competitive swimming. *Journal of Sport and Exercise Psychology, 12*, 248-263.

Bronfenbrenner, U. (1993). The ecology of cognitive development: Research models and fugitive findings. In R.H. Wozniak & K.W. Fischer (Eds.), *Development in context: Acting and thinking in specific environments* (pp. 3-44). Hillsdale, NJ: Erlbaum.

Brustad, R.J. (1996). Attraction to physical activity in urban schoolchildren: Parental socialization and gender influences. *Research Quarterly for Exercise and Sport, 67*, 316-323.

Brustad, R.J., Babkes, M.L., & Smith, A.L. (2001). Youth in sport: Psychological considerations. In R.N. Singer, H.A. Hausenblas, & C.M. Janelle (Eds.), *Handbook of sport psychology* (2nd ed., pp. 604-635). New York: Wiley.

Bukowski, W.H., & Hoza, B. (1989). Popularity and friendship: Issues in theory, measurement, and outcome. In T.J. Berndt & G.W. Ladd (Eds.), *Peer relationships in child development* (pp. 15-45). New York: Wiley.

Canadian Fitness and Lifestyle Research Institute. (2005). *2004 physical activity monitor.* Retrieved January 31, 2006, from www.cflri.ca/cflri/resources/pub.php.

Centers for Disease Control and Prevention. (2003). Physical activity levels among children aged 9-13 years: United States, 2002. *Mortality and Morbidity Weekly Report, 52*, 785-788.

Chase, M.A., & Dummer, G.M. (1992). The role of sports as a social status determinant for children. *Research Quarterly for Exercise and Sport, 63*, 418-424.

Coakley, J. (1992). Burnout among adolescent athletes: A personal failure or social problem? *Sociology of Sport Journal, 9*, 271-285.

Coakley, J. (2004). *Sports in society: Issues and controversies* (8th ed.). New York: McGraw-Hill.

Coakley, J., & White, A. (1992). Making decisions: Gender and sport participation among British adolescents. *Sociology of Sport Journal, 9*, 20-35.

Corte-Real, N., Alves, J.R., Corredeira, R., Balaguer, I., Brustad, R., & Fonseca, A.M. (2004). Um olhar sobre os estilos de vida dos adolescentes portugueses: A actividade física, os consumos e a importância das relações com os amigos. [An examination of lifestyle practices among Portuguese adolescents: Physical activity, consumption practices and the importance of friendships.] *Revista Portuguesa de Ciências do Desporto, 4*(2), S138.

Deci, E.L., & Ryan, R.M. (1985). *Intrinsic motivation and self-determination in human behavior.* New York: Plenum Press.

Deci, E.L., & Ryan, R.M. (1991). A motivational approach to self: Integration in personality. In R.A. Dientsbier (Ed.), *Nebraska symposium on motivation: Perspectives on motivation* (Vol. 38, pp. 237-288). Lincoln, NE: University of Nebraska.

DeKnop, P., Engstrom, L.M., Skirstad, B., & Weiss, M.R. (1996). *Worldwide trends in youth sport.* Champaign, IL: Human Kinetics.

Duke, J., Huhman, M., & Heitzler, C. (2003). Physical activity levels among children aged 9-13 years: United States, 2002. *Morbidity and Mortality Weekly Report, 52*(33), 785-788.

Duncan, S.C., Duncan, T.E., Strycker, L.A., & Chaumeton, N. (2004). A multilevel analysis of sibling physical activity. *Journal of Sport and Exercise Psychology, 26,* 57-68.

Eccles, J.S., & Barber, B.L. (1999). Student council, volunteering, basketball, or marching band? What kind of extracurricular involvement matters? *Journal of Adolescent Research, 14,* 10-43.

Eccles, J.S., & Midgley, C. (1989). Stage/environment fit: Developmentally appropriate classrooms for early adolescents. In R. Ames & C. Ames (Eds.), *Research on motivation in education* (Vol. 3, pp. 139-181). San Diego: Academic Press.

Eccles, J.S., & Midgley, C. (1990). Changes in academic motivation and self-perception during early adolescence. In R. Montemayor, G.R. Adams, & T.P. Gullotta (Eds.), *From childhood to adolescence: A transitional period?* (pp. 134-155). Newbury Park, CA: Sage.

Fredricks, J.A., Alfeld-Liro, C.J., Hruda, L.Z., Eccles, J.S., Patrick, H., & Ryan, A.M. (2002). A qualitative examination of adolescents' commitment to athletics and the arts. *Journal of Adolescent Research, 17,* 68-97.

Freedson, P.S., & Evenson, S. (1991). Familial aggregation in physical activity. *Research Quarterly for Exercise and Sport, 62,* 384-389.

Gagné, M., Ryan, R.M., & Bargmann, K. (2003). Autonomy support and need satisfaction in the motivation and well-being of gymnasts. *Journal of Applied Sport Psychology, 15,* 372-390.

Goudas, M., Biddle, S., Fox, K., & Underwood, M. (1995). It ain't what you do, it's the way that you do it! Teaching style affects children's motivation in track and field lessons. *Sport Psychologist, 9,* 254-264.

Gould, D., Feltz, D., Horn, T., & Weiss, M. (1982). Reasons for attrition in competitive youth swimming. *Journal of Sport Behavior, 5,* 155-165.

Hagger, M.S., Chatzisarantis, N.L.D., Culverhouse, T., & Biddle, S.J.H. (2003). The processes by which perceived autonomy support in physical education promotes leisure-time physical activity intentions and behavior: A trans-contextual model. *Journal of Educational Psychology, 95,* 784-795.

Harter, S. (1998). The development of self-representations. In W. Damon (Series Ed.) & N. Eisenberg (Vol. Ed.), *Handbook of child psychology:* Vol. 3. *Social, emotional, and personality development* (5th ed., pp. 553-617). New York: Wiley.

Hendry, L.B., Shucksmith, J., Love, J.L., & Glendinning, A. (1993). *Young peoples' leisure and lifestyles.* London: Routledge.

Horn, T.S., & Weiss, M.R. (1991). A developmental analysis of children's self-ability judgments in the physical domain. *Pediatric Exercise Science, 3,* 310-326.

Langley, D.J., & Knight, S.M. (1999). Continuity in sport participation as an adaptive strategy in the aging process: A lifespan narrative. *Journal of Aging and Physical Activity, 7,* 32-54.

Marlatt, H. (2005). *The relationship between children's perception of personal autonomy in organized sport and time spent in unstructured physical activity.* Unpublished master's thesis, University of Northern Colorado, Greeley.

McPherson, B. (1994). Sociocultural perspectives on aging and physical activity. *Journal of Aging and Physical Activity, 2,* 329-353.

Merton, R.K. (1957). *Social theory and social structure.* Glencoe, IL: Free Press.

Nicholls, J. (1984). Achievement motivation: Conceptions of ability, subjective experience, task choice, and performance. *Psychological Review, 91,* 328-346.

Pate, R.R., Trost, S.G., Levin, S., & Dowda, M. (2000). Sports participation and health-related behaviors among US youth. *Archives of Pediatric and Adolescent Medicine, 154,* 904-911.

Pelletier, L.G., Fortier, M.S., Vallerand, R.J., & Briere, N.M. (2001). Associations among perceived autonomy support, forms of self-regulation, and persistence: A prospective study. *Motivation and Emotion, 25,* 279-306.

Pratt, M., Macera, C.A., & Blanton, C. (1999). Levels of physical activity and inactivity in children and adults in the United States: Current evidence and research issues. *Medicine and Science in Sports and Exercise, 31*(11, Suppl.), S526-S533.

Raedeke, T.D. (1997). Is athlete burnout more than just stress? A sport commitment perspective. *Journal of Sport and Exercise Psychology, 19,* 396-417.

Raudsepp, L., & Viira, R. (2000). Influence of parents' and siblings´ physical activity on activity levels of adolescents. *European Journal of Physical Education, 5,* 169-178.

Ryan, R.M., & Deci, E.L. (2000). Self-determination and the facilitation of intrinsic motivation, social development, and well-being. *American Psychologist, 55,* 68-78.

Sallis, J., Alcaraz, J., McKenzie, T., & Hovell, M. (1999): Predictors of change in children's physical activity over 20 months. Variations by gender and level of adiposity. *American Journal of Preventive Medicine, 16*(3), 222-229.

Sarrazin, P., Vallerand, R., Guillet, E., Pelletier, L., & Cury, F. (2002). Motivation and dropout in female handballers: A 21-month prospective study. *European Journal of Social Psychology, 32,* 395-418.

Seabra, A., Mendonça, D., Maia, J., & Garganta, R. (2004). Influência de determinantes demográfico-biológicas e sócio-culturais nos níveis de actividade física de crianças e jovens. [Influence of biodemographic and sociocultural variables on levels of physical activity in children and youth.] *Revista Brasileira de Cineantropometria & Desempenho Humano, 6*(2), 67-72.

Smith, A.L. (1999). Perceptions of peer relationships and physical activity participation in early adolescence. *Journal of Sport and Exercise Psychology, 21,* 329-350.

Smith, A.L. (2003). Peer relationships in physical activity contexts: A road less traveled in youth sport and exercise psychology research. *Psychology of Sport and Exercise, 4,* 25-39.

Tomson, L.M., Pangrazi, R.P., Friedman, G., & Hutchison, N. (2003). Childhood depressive symptoms, physical activity and health related fitness. *Journal of Sport and Exercise Psychology, 25,* 419-439.

Ullrich-French, S., & Smith, A.L. (2006). Perceptions of relationships with parents and peers in youth sport: Independent and combined prediction of motivational outcomes. *Psychology of Sport and Exercise, 7,* 193-214.

Veroff, J. (1969). Social comparison and the development of achievement motivation. In C.P. Smith (Ed.), *Achievement-related motives in children* (pp. 46-101). New York: Russell Sage Foundation.

Vilhjalmsson, R., & Kristjansdottir, G. (2003). Gender differences in physical activity in older children and adolescents: The central role of organized sport. *Social Science and Medicine, 56*, 363-374.

Vilhjalmsson, R., & Thorlindsson, T. (1992). The integrative and physiological effects of sport participation: A study of adolescents. *Sociological Quarterly, 33*, 637-647.

Vilhjalmsson, R., & Thorlindsson, T. (1998). Factors related to physical activity: A study of adolescents. *Social Science and Medicine, 47*, 665-675.

Wang, C.K.J., & Biddle, S.J.H. (2001). Young people's motivational profiles in physical activity: A cluster analysis. *Journal of Sport and Exercise Psychology, 23*, 1-22.

Washington, R.E., & Karen, D. (2001). Sport and society. *Annual Review of Sociology, 27*, 187-212.

Weiss, M.R., & Duncan, S.C. (1992). The relationship between physical competence and peer acceptance in the context of children's sport participation. *Journal of Sport and Exercise Psychology, 14*, 177-191.

Weiss, M.R., & Petlichkoff, L.M. (1989). Children's motivation for participation in and withdrawal from sport: Identifying the missing links. *Pediatric Exercise Science, 1*, 195-211.

Weiss, M.R., & Stuntz, C.P. (2004). A little friendly competition: Peer relationships and psychosocial development in youth sport and physical activity contexts. In M.R. Weiss (Ed.), *Developmental sport and exercise psychology: A lifespan perspective* (pp. 165-196). Morgantown, WV: Fitness Information Technology.

Whaley, D.E., & Ebbeck, V. (1997). Older adults' constraints to participation in structured exercise classes. *Journal of Aging and Physical Activity, 5*, 190-212.

Wiersma, L.D. (2000). Risks and benefits of youth sport specialization: Perspectives and recommendations. *Pediatric Exercise Science, 12*, 13-22.

Wigfield, A. (1994). Expectancy-value theory of achievement motivation: A developmental perspective. *Educational Psychology, 6*, 49-78.

Wold, B., & Anderssen, N. (1992). Health promotion aspects of family and peer influences on sport participation. *International Journal of Sport Psychology, 23*, 343-359.

Zeijl, E., Poel, Y., Ravesloot, J., & Meulman, J. (2000). The role of parents and peers in the leisure activities of young adolescents. *Journal of Leisure Research, 32*(3), 281-302.

CHAPTER

15

Community Out-of-School Physical Activity Promotion

David A. Dzewaltowski, PhD

Out-of-school time waking hours provide an opportunity for participation in community settings that promote youth development. This chapter examines whether such programs do in fact provide youth with an opportunity for physical activity, and if they do not, the reasons for this failure. Suggestions for improvement of such programs are also offered. The chapter is limited to organized community opportunities that are not part of the school curriculum, such as after-school programs and clubs. These organized community settings offer a unique chance for studying voluntary physical activity. To a large extent, early in children's development, adults control the activities that youth participate in outside of the school day. Later in development, many youth control their out-of-school choices. Thus, health and fitness professionals working to promote physical activity in the community before or after school hours must view parents and youth as consumers. These professionals must provide a physical activity promotion experience that is effective, attractive to parents of younger children, and attractive to children themselves. This situation offers researchers the chance to study how to engage youth at different stages of development that may not be addressed in studies of mandatory programs.

After-School Programs in the United States

Increasingly, youth and families are seeking organized community opportunities for supervised activities during out-of-school time. The number of youth enrolled in after-school programs in the United States increased from 1.7 million in 1991 to 6.7 million in 1997 (Capizzano, Tout, & Adams, 2000). And the U.S. Department of Education estimates that 37.7% of all children in grades

The material in this chapter is based upon work supported by the Cooperative State Research, Education, and Extension Service, U.S. Department of Agriculture, under Award No. U.S.D.A. 2005-35215-15418.

K through 8 participated in some form of structured after-school activity at least once per week in 2001 (Wirt et al., 2004). It is likely that this trend will increase. Sturm (2005) summarized U.S. nationally representative data from University of Michigan surveys in 1981 and 1997 that used 24 h time diaries of children aged 3 to 12 years. The data suggested a 12% decline in discretionary time from 1981 to 1997 among children of these ages (Hofferth & Sandberg, 2001). A major contributor to the decline was the amount of time youth spent in organized activities.

Promotion of moderate and vigorous physical activity (MVPA) in youth does not appear to be a primary goal of many organized community settings. More common reasons for families' seeking out-of-school programs are to provide adult supervision for young children and to promote academic achievement. That supervision for young children is a concern, at least among U.S. parents, is supported by the U.S. National Survey of America's Families. This survey indicated that only 7% of 6- to 9-year-olds are regularly left home alone, whereas 26% of 10- to 12-year-olds and 47% of 14-year-olds are left unsupervised (Vandivere et al., 2003). The focus on academic achievement is suggested by an evaluation of the U.S. Department of Education's 21st Century Community Learning Centers Program (James-Burdumy et al., 2005), in which community funding for delivery of after-school programs is contingent on the inclusion of instructional enrichment, tutoring, or homework assistance. The Massachusetts After-School Research Study (MARS) examined 78 after-school programs and showed that in addition to providing child care, the prototypical after-school program focused on decreasing gaps in educational test scores. Promotion of physical activity and reduction of sedentary behavior were not among the top 10 goals of these programs. In summary, with the increasing trend of youth participation in organized out-of-school settings, as well as the failure of these programs to consistently and effectively provide opportunities for physical activities, public health practitioners involved in promoting healthful physical activity behaviors would probably do well to target them.

This chapter examines two interrelated hypotheses suggesting that after-school programs and other community out-of-school settings, when structured appropriately, provide opportunities for MVPA and decreased sedentary behavior in youth:

- Youth in communities that provide organized opportunities including physical activity will exhibit greater MVPA and less sedentary behavior than youth in communities without these opportunities.
- Youth participating in organized opportunities that include quality physical activity promotion programs will increase in personal assets, increase in MVPA, and decrease in sedentary behavior compared to youth who do not participate. A quality physical activity promotion setting includes a daily physical activity session providing moderate- to vigorous-intensity activity and a weekly physical activity regulation skill session structured for the promotion of positive youth development.

The first part of the chapter presents the Organized Community Opportunities Model. In the second section of the chapter, this model frames a review of existing research that pertains to the two hypotheses just presented. The chapter closes with conclusions and recommendations for future research and practice.

The Organized Community Opportunities Model

The Organized Community Opportunities Model for physical activity promotion is informed by developmental ecological systems theory and the RE-AIM (Reach, Efficacy, Adoption, Implementation, and Maintenance) framework (www.re-aim.org; Dzewaltowski, Estabrooks, & Glasgow, 2004; Glasgow, Vogt, & Boles, 1999). The model provides a tool for planning and evaluating the public health impact of organized community opportunities designed to promote physical activity in youth.

Youth grow up and interact in numerous environments that influence their development and behavior. For study of the environmental variables central to these interactions, the Organized Community Opportunities Model for physical activity promotion places environmental variables into community, delivery setting, and behavior setting categories. Definitions of key terms of the model are provided in table 15.1.

Table 15.1 Definitions of the Organized Community Opportunities Model

Term	Definition
Community	A group of youth and adults residing in a geographic neighborhood or multi-neighborhood area, no matter how they relate to one another (Ferguson & Dickens, 1999; Sampson, 2002)
Delivery setting	A location for physical activity promotion that includes the target audience and is bounded in space and time to provide the social structure and context for planning, implementing, and evaluating interventions (Green et al., 2000)
Behavior setting	The social and physical environment where youth behavior occurs
Organized activities	Activities that are regularly scheduled, offer some level of supervision by adults, and emphasize skill building (Mahoney, Larson, Eccles, & Lord, 2005)
Organized community opportunity	A delivery setting that implements regularly scheduled activities, offers some level of supervision by adults, and is organized around a broad goal or purpose that provides an opportunity for physical activity promotion
Physical activity promotion program	A protocol of intervention activities that is designed to reach a target audience (youth) to increase physical activity and decrease sedentary behavior

A community is a group residing in a geographic neighborhood or multi-neighborhood area, no matter how the members of the group relate to one another (Ferguson & Dickens, 1999; Sampson, 2002). Communities result from the impact of the landscape comprising culture, institutions, and policies on economic, social, natural, and built physical resources, as well as governmental institutional infrastructure (Heinrich & Lynn, 2000). Communities vary in their resources for organized physical activity opportunities. Some communities have numerous built environmental resources, such as indoor gym facilities. Other communities have fewer built environmental resources but more natural environmental resources, such as mountains and lakes. The challenge for research teams is to discover the opportunities for physical activity presented by the economic, social, natural, and built physical environment, as well as by the governmental and institutional infrastructure of communities. The challenge for health fitness professionals is to enhance the resources for physical activity within communities and to capitalize on existing resources to promote physical activity.

Communities include delivery settings that provide a location for physical activity promotion. Delivery settings include the target audience and provide the social and physical environment for planning, implementing, and evaluating interventions (Green, Poland, & Rootman, 2000). A delivery setting that implements regularly scheduled activity opportunities, offers supervision by adults, and emphasizes skill building provides opportunity for youth development (Mahoney et al., 2005). These youth development settings are defined as an organized community opportunity for youth physical activity promotion. Organized community opportunities are delivered by numerous youth-serving systems. Publicly sponsored delivery agencies, in addition to schools, include parks and recreation sites, libraries, and even zoos. In Australia, the Active After-School Communities program targets sporting organizations, sporting clubs, and private providers as well as Australian primary school and child care benefit–approved care services out of school hours. In the United Kingdom, after-school clubs have been used to promote healthful eating (Hyland et al., 2006) and are also a target for physical activity promotion. In the United States, examples of national community-based youth-serving organizations include Boys and Girls Clubs, Boy Scouts and Girl Scouts, YMCA and YWCA, Big Brothers and Big Sisters, and 4-H. Organized community opportunities can also be delivered by faith-based organizations and for-profit companies.

In the Organized Community Opportunities Model, delivery settings are distinguished from what Barker (1968) described as "behavior settings." Behavior settings are the physical and social environments where behavior occurs; for example, a Boys and Girls Club may conduct an after-school program at a school. Within that program, children may have access to gym, classroom, and outdoor field spaces. In each of these physical environments, the after-school program offers activities. These may be competing for this space with school and other community opportunities. Behavior settings comprise a physical environment and the standard social pattern of behavior that goes with that environment during a specific time period. For example, the snack time of an after-school program in the gym is a behavior setting. At 3:30 p.m. after school, the gym may be set up with tables and food and the time designated as snack time by the adult leadership and youth. This is a behavior setting that provides an after-school snack option in which physical activity is not encouraged. At 4:00, the behavior setting may be free play. Now the tables are moved to the side of the gym, physical activity equipment is provided, and the adult leaders and youth promote physical activity.

Figure 15.1 illustrates a community with two delivery settings: after-school program and home. Settings may include features that increase or decrease minutes of sedentary behavior, as illustrated on the left y-axis. As the right y-axis illustrates, settings also may include features that increase or decrease minutes of physical activity. Although these two health behaviors may be negatively correlated, settings may have features that influence both.

The after-school program meets the definition of an organized community opportunity for youth. The program is regularly scheduled to meet from 3:30 p.m. to 6:00 p.m. every day, is supervised by adults, and is organized around a broad goal of academic enrichment. At home and in the after-school program, there are behavior settings. Some behavior settings are structured to promote physical activity (e.g., parent-supervised play), and others are structured to promote sedentary behavior (e.g., computer lab).

Figure 15.1 also illustrates that for all delivery settings, outside community influences may foster physical activity or sedentary behavior. Influences that promote sedentary behavior may include media promoting such behavior and funding for after-school academic programs. Physical activity–promoting influences may include funded parent MVPA promotion programs, after-school program MVPA policies, and after-school program quality improvement efforts. Thus, delivery settings can be structured to include a quality physical activity promotion program—a protocol of intervention activities designed to reach a target audience (youth) to increase physical activity and decrease sedentary behavior. Within the delivery settings, intervention activities can be structured to create physical activity–promoting behavior settings.

Although youth contribute to the structure of behavior settings, key adults often control environmental opportunities and constraints. And, rather than letting the structure of the after-school or home or behavior setting emerge naturally, adult leaders, including parents, can intentionally develop behavior settings. Key adults are leaders in and developers of behavior settings and link behavior settings in the model (see figure 15.1). Key adults in the after-school staff determine the after-school program activities offered and the structure of the social and physical environment. They also link to the home environment by interacting with parents through face-to-face exchanges or through media such as newsletters. Parents determine a child's home environment. For example, the permissive parent may provide a child with a bedroom that includes

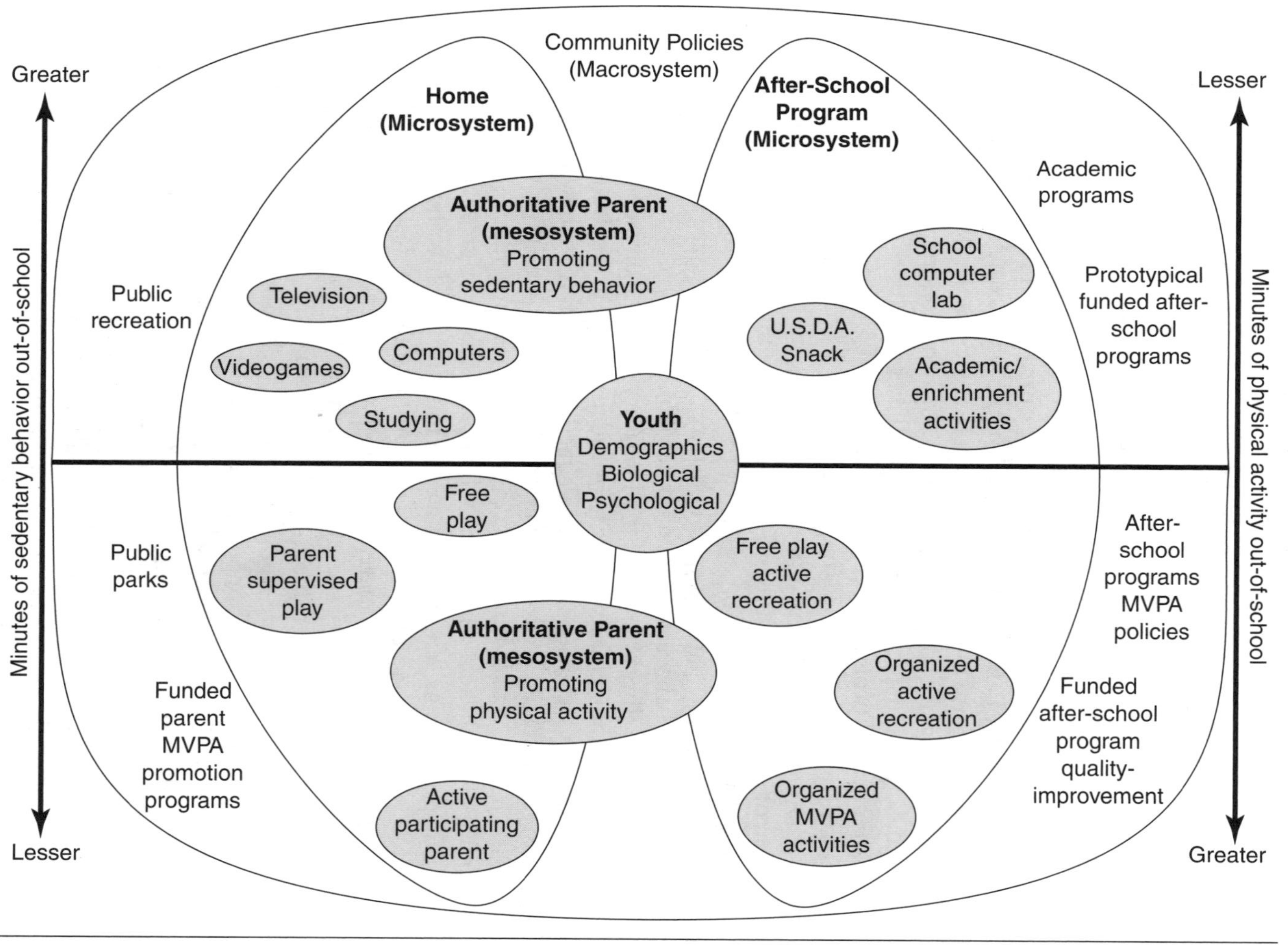

Figure 15.1 The Organized Community Opportunities Model for physical activity.

a television, video games, and other sedentary behavior–promoting activities. The social environment at home may be such that the child is allowed to have several hours of uninterrupted sedentary play in his or her room without adult intervention. Alternatively, parents can adopt an authoritative style for physical activity promotion and "monitor and impart clear standards for their children's conduct" (Baumrind, 1991, p. 62). Parents can place their children into community opportunities that include physical activity–promoting behavior settings, or they can construct physical activity–promoting behavior settings at home.

Developmental Ecological Systems Theory

The Organized Community Opportunities Model is consistent with developmental ecological systems theory (Hawkins, Catalano, & Arthur, 2002; Lerner & Castellino, 2002), which adopts many of the themes of other ecological models in the public health field (Booth et al., 2001; Dzewaltowski, 1997; McLeroy et al., 1988; Sallis & Owen, 2002). The term *ecological systems* refers to the contention that sustained participation in a behavior is embedded within an integrated matrix of variables derived from multiple levels of organization. These levels of organization can be thought of as a group of nested structures, each inside the next like a set of Russian dolls (Bronfenbrenner, 1979). *Developmental* refers to the time element missing from other ecological models. The study of physical activity promotion from a developmental ecological systems perspective necessitates examination of dynamic relations among youth and behavior within this multitiered matrix over time.

The key challenge in studying the impact of the ecological system is to define how levels of influence interact to determine physical activity or sedentary behavior. Bronfenbrenner (1979) labeled the outermost ring of the ecological system the macrosystem. The macrosystem includes the overarching pattern of stability, at the level of the subculture or culture as a whole, in forms of social organization, associated beliefs, and lifestyles. The Organized Community Opportunities Model defines community as a macrosystem expressed by residents within a geographic area.

Within a community or the macrosystem is the exosystem. The exosystem is defined as a setting that does not include the developing person but as one in which events occur that affect the setting containing the person (Bronfenbrenner, 1979). Although not depicted in figure 15.1, the exosystem as defined by the Organized Community Opportunities Model comprises settings that influence youth physical activity opportunities. Examples of exosystem settings include a community coalition, an after-school staff training program, and a school board. In these settings, decisions are made that affect other systems. Decisions that may positively influence MVPA include funding for after-school programs, policy making that affects after-school program MVPA offerings, and after-school program quality improvement efforts that may focus on MVPA.

Developmental ecological systems theory defines the next level as a mesosystem, a system of microsystems or the connections between people in the settings. Youth-serving systems are formalized mesosystems providing delivery settings. But a mesosystem can also include connections between home, school, and neighborhood peer group. The Organized Community Opportunities Model in figure 15.1 focuses on key adult leaders as the link between two or more delivery settings that contain the developing youth (e.g., the relation between home and the after-school program through parents and after-school staff). These adult leaders, therefore, are targets for intervention activities.

The next level of developmental ecology systems theory is the microsystem. "A microsystem is a pattern of activities, roles, and interpersonal relations experienced by the developing person in a given face-to-face setting with particular physical and material features and containing other persons with distinctive characteristics of temperament, personality and systems of belief" (Bronfenbrenner, 1992, p. 148). Within a delivery setting, such as an after-school program, are behavior settings. These are the sessions that young people move through as they participate in a delivery setting such as an after-school program. For example, youth may attend a snack session that is a strong influence on sedentary behavior and then move to a physical activity session that is a strong influence on MVPA. Therefore, the Organized Community Opportunities Model distinguishes between the delivery setting and the behavior setting. By targeting multiple behavior settings within a delivery setting, the model examines the microsystem processes of social interaction between a developing youth and others in the social environment, as well as processes of engagement on the part of a young person through her or his participation in progressively more complex activities.

The Organized Community Opportunities Model is consistent with common social psychological models of behavior change, which can be nested within a developmental ecological systems theory approach at the level of the microsystems. Lewin (1951) coined the term ecological psychology to describe the microsystem. The microsystem is the relation between the person and the outside environment. In Lewin's well-known equation, B = f(P,E), the person (P) and the environment (E) are hypothesized to interact and result in behavior (in this case physical activity or sedentary behavior). Bronfenbrenner (1992) made developmental outcomes a primary goal by adding the dimension of time to Lewin's equation, D = f(P,E), with "D" referring not to development but to its outcome at a specific point in time. Therefore, microsystems influence immediate behavior but also have an impact on the development of youth assets for future behavioral decisions. Focusing on only immediate behavior and not on youth development is not likely to lead to sustained physical activity across the life span.

In a report, the Committee on Community-Level Programs for Youth of the National Research Council (NRC) and the Institute of Medicine (IOM) evaluated current science of adolescent health and development to provide recommendations that meet the demands of community intervention program design, implementation, and evaluation. The committee defined environmental variables as features of positive developmental settings and defined development variables as personal and social assets. The hypothesized features of positive developmental settings included physical and psychological safety; appropriate structure; supportive relationships; opportunities to belong; positive social norms; support for efficacy; opportunities for skill building; and integration of family, school, and community youth development efforts (NRC-IOM, 2002). Personal and social assets comprised physical development (including good health behaviors), intellectual development, psychological and emotional development, and social development (NRC-IOM, 2002).

Ecological systems theory proposed that development was a function of the person and the environment. "P" represents the personal and perceived social assets that interact with the environment (features of positive developmental settings) to determine development. Several personal and social assets that facilitate positive youth development correspond to the correlates and mediators of MVPA in young people (Sallis, Prochaska, & Taylor, 2000). For example, perceived psychological development (including good coping skills and confidence in one's personal efficacy) and perceived social development (including connectedness and trust with regard to parents, other adults, and peers and attachment to prosocial institutions such as after-school programs) are key assets. Therefore, quality youth microsystem settings will result in developmental outcomes that build youths' personal and social assets and also influence MVPA.

In summary, *developmental ecological systems theory* suggests that sustained participation in physical activity is embedded within an integrated matrix of variables derived from multiple levels of organization. The Organized Community Opportunities Model for physical activity promotion draws on ecological systems theory and compartmentalizes environmental variables into community, delivery setting, and behavior setting levels. By placing variables into these categories, researchers can study, describe, and influence relationships between community and delivery setting, between delivery setting and behavior setting, and between best program practice and youth development. The model draws on the social psychology literature at the behavior setting level and includes other environmental levels of influence. Youth variables are defined as personal and social assets. Therefore, an organized community opportunity for physical activity promotion would intervene on the multiple levels of environment to provide a physical activity opportunity with proximal environmental features or critical elements that will immediately influence physical activity and also develop the assets of youth for lifelong physical activity. Identifying important variables in the multilevel systems and targeting these variables with multilevel interventions require that research and evaluation models adopt a multilevel approach such as the RE-AIM framework as well.

Evaluating the Community Opportunities Model with RE-AIM

Current discussion among public health researchers and practitioners centers on expanding evaluation criteria to best estimate the impact of intervention programs on community population outcomes. The RE-AIM framework, which has been proposed to achieve this end, emphasizes five dimensions important for evaluating the potential public health impact of intervention activities (www.re-aim.org; Dzewaltowski et al., 2004; Glasgow et al., 1999, 2006). The framework includes individual-level and community setting indicators. Individual-level indicators within RE-AIM include reach (R) and efficacy (E), whereas delivery setting–level indicators include adoption (A) and implementation (I). Maintenance (M) is assessed at both an individual and setting level (see table 15.2).

Reach is defined as the percent of eligible youth in the target population that are exposed to an intervention, as well as the extent to which those youth are representative of the target population. The community, for example, can be defined as the neighborhoods feeding the school program delivered on a school site. The target audience, therefore, is all youth within the geographic neighborhood or multineighborhood area attending the school. Reach is important to measure because an after-school program servicing youth at a school site may be very effective at promoting physical activity with a small group of children but not reach a large number. In addition, some programs have large reach in terms of participation and attendance but have participants who are not representative of the school as a whole. The participating students may be ethnically homogeneous, economically advantaged, more active, or less overweight compared to nonparticipants. For public health impact, an evidence-based program should have large reach. Achieving representativeness of the target population may require greater focus on those most in need who may not be as receptive to physical activity promotion programs.

Efficacy or **effectiveness** is the impact of an intervention program on important positive outcomes, such as increased physical activity, decreased sedentary behavior, or both. In the evaluation of impact it is also necessary to examine possible negative, unintended consequences and the economic costs to the individual and the community. Programs that are most effective at promoting physical activity could have unintended consequences, such as cutting into time for academic enrichment and leading to negative academic outcomes or negatively affecting self-esteem or body image. Further, the most effective programs promoting physical activity may be too expensive to permit widespread implementation.

Adoption is the percent of eligible organized community entities (delivery settings, staff) providing the physical activity promotion program. Thus, adoption is parallel to reach, but at the delivery setting level. Adoption of the physical activity promotion program can be characterized by (a) the absolute number of settings, (b) setting participation rate, and (c) representativeness of

Table 15.2 RE-AIM Dimension Questions for Evaluating Organized Community Opportunities for Physical Activity Promotion

Dimension (level)	Question	Example
Reach (individual)	What percentage of eligible youth participated and how representative were they of the larger target population in the community?	What percentage of elementary school students participated in an on-site after-school program, and were they similar in gender, ethnicity, and socioeconomic status to students enrolled at the school?
Efficacy/ Effectiveness (individual)	What impact did the intervention have on participants' physical activity and sedentary behavior and what were negative, unintended consequences?	Did elementary school students attending the summer school physical activity promotion program participate in more physical activity and less sedentary behavior than students participating in the summer school academic enrichment program? Did their involvement limit their academic progress?
Adoption (setting)	What percentage of delivery settings and physical activity promotion leaders within these settings agreed to deliver the program and how representative were they?	What percentage of 4th-grade teachers within the school district agreed to deliver the physical activity program, and were these teachers similar in years of experience to all 4th-grade teachers in the school district?
Implementation (setting)	To what extent were the critical elements of the physical activity promotion intervention implemented as intended?	Did nutrition education teachers delivering the obesity prevention program deliver both the six nutrition education lessons and the six physical activity promotion lessons during the fall semester?
Maintenance (setting)	To what extent did the various physical activity promotion program critical elements continue or become institutionalized or part of routine organizational practices and policies after the initial program or grant period expired?	Did the after-school program continue to offer daily physical activity sessions 6 months after the conclusion of the federal grant to prevent youth obesity?
(Individual)	What were the long-term effects of the promotion program on physical activity 6 or more months after completion of the intervention?	Did 4th graders enroll in the after-school physical activity promotion program for their 5th-grade school year?

the sample of settings and intervention staff. An intervention program that has wide adoption would have a more robust impact on the promotion of physical activity in youth than an intervention with limited adoption. Communities vary in the youth-serving organizations that provide delivery settings for youth. Also, these delivery settings differ in size, staff number, staff training, staff demographics, and other community resources for offering physical activity promotion programs. Some interventions may be too costly or complex to be adopted by poorly funded settings with untrained, low-paid staff.

Implementation refers to the quantity and quality of delivery of the intended intervention. Central to evidence-based practice is delivering intervention activities that have been shown through research to be effective. As stated earlier, an "organized" community opportunity includes implemented activities that are regularly scheduled, involve some level of supervision by adults, and are organized around a broad goal or purpose (Mahoney et al., 2005). For young children, the primary goal of the after-school site may be the safety that comes with supervised child care. In addition, most youth development programs for young children and adolescents have some larger goal for participation. The goal may be focused on developing some particular skill for achieving in areas such as academics, arts, clubs (e.g., 4-H, scouting), religious activities, and sport. If sport or physical activity is a primary objective, programs where every youngster is given an equal opportunity to participate regardless of physical ability would have greater reach, but not necessarily be better implemented to increase physical activity of the participants. Thus, an organized community opportunity for physical activity promotion includes a defined protocol or program of critical intervention elements that has physical activity as a key goal. In sum, implementation is defined as the extent to which the critical elements of the physical activity promotion intervention were carried out as intended.

Maintenance includes individual-level and setting-level elements. At the individual level, maintenance is the longer-term efficacy or effectiveness of the intervention. For example, outcomes six months or more postintervention reflect longer-term individual maintenance. The setting-level definition of maintenance refers to the institutionalization or sustainability of a program and is assessed by the percent of community entities that continue the intervention program upon completion of the formal program period, or after grant funds have been expended, or both.

The Organized Community Opportunities Model informed by the RE-AIM framework provides a structure for the planning and evaluation of physical activity promotion programs and is also a tool for policy makers and program adopters attempting to make difficult decisions about where to focus limited community resources. A program that is effective may not be the best choice unless the program will have widespread adoption in numerous settings, reach a large proportion of the target audience, be implemented by staff, and be sustained in the community. The next section draws on elements from the RE-AIM framework to review studies that address community opportunities for physical activity during out-of-school time.

Research Based on the Organized Community Model

This section uses the Organized Community Opportunities Model as a way of integrating existing literature pertaining to the two hypotheses presented at the beginning of the chapter. Throughout this review, elements of the RE-AIM framework are selectively addressed to direct attention to the potential public health impact of intervention studies.

Providing Physical Activity Opportunities

Hypothesis 1 states, "Youth in communities that provide organized opportunities including physical activity will exhibit greater MVPA and less sedentary behavior than youth in communities without these opportunities." There is some evidence that youth will be more active if communities provide facilities. In one study, greater perceived access to various places to play and exercise was associated with greater leisure-time physical activity in preadolescents and adolescents (Garcia et al., 1995). Data from the 1996 National Longitudinal Study of Adolescents showed that use of community recreation centers was significantly associated with engagement in MVPA (Gordon-Larsen, McMurray, & Popkin, 2000).

Beyond providing facilities, there is some evidence that community delivery organizations that provide equipment and supervision to promote physical activity will have greater youth physical activity participation. Sallis and colleagues (2001) examined potential indoor and outdoor activity spaces at 24 middle schools that were available before school, at lunch, and after school. Environments that had both physical improvements (e.g., basketball hoops and courts, baseball backstops, volleyball nets) and adult supervision had fourfold greater participation for boys and fivefold greater participation for girls compared to environments lacking in both. In a subsequent experimental study, Sallis and colleagues (2003) randomly assigned 12 middle schools to an intervention and 12 to a control condition. Baseline data were collected, and the intervention was implemented across two academic years. All intervention schools targeted increasing supervision, equipment, and organized activities before, during, and after school. Intervention school staff submitted plans for how money would be used to provide a more healthful environment to students, and schools received $2000 USD for physical activity programs or equipment. Physical activity increased at intervention schools at a greater rate over time compared to control schools. Subsequent analyses showed that intervention school physical activity among boys increased compared to changes in control schools, but the increase in physical activity at intervention schools among girls was not significantly different compared to control schools. An analysis of the changes contributing to the overall intervention effect revealed that boys increased in MVPA during both physical education class and out-of-school time and that girls increased their MVPA mainly in physical education class.

Another study examined the impact of a community's provision of citywide after-school programs at three school sites for elementary children from low-income households. Mahoney, Lord, and Carryl (2005) followed 439 elementary school students enrolled in grades 1 to 3 across a two-year period. Most of the students lived in poverty and were Hispanic or African American. The students had access to programs at their school every day between 3:00 p.m. and 6:00 p.m. The programs provided time for a snack, homework, enrichment learning (e.g., computers, visitors, musical instruments), supervised recreation (e.g., kickball, basketball, board games), and art. Although the percentage of children categorized as overweight increased from 22% to 29% over the two-year study, there were fewer overweight children among those who participated in the after-school program three or more days per week (21%) compared to nonparticipants (33%).

In a study of the prevention of obesity, the Georgia FitKid After-school Program was provided free of charge to all 3rd graders attending nine intervention schools beginning in 2003-2004 and through their 2005-2006 school year (5th-grade year). The nine schools were randomly assigned to receive the intervention, and nine schools served as controls. The schools were selected from a school district where 63% of youth were African American, 34% were White, and 67% were of low socioeconomic status and eligible for free or reduced-price lunch (Yin et al., 2005a, 2005b). The program ran for 2 h, beginning with a 40 min academic enrichment period including a healthy snack that met U.S. Department of Agriculture guidelines. An 80 min exercise intervention followed that included 40 min of activity at a moderate to vigorous intensity, muscle strengthening, stretching, and motor skill activities. First-year results revealed that 3rd-grade intervention school children who attended at least 40% of the after-school sessions showed a relative reduction of percentage body fat and improved cardiovascular fitness compared to children attending control schools (Yin et al., 2005a).

The study's reach was poor, however. At the intervention schools, all students were invited to participate (n = 603). Of those students, 54% consented to participate, 312 were measured at baseline, and 275 were measured at follow-up (Yin et al., 2005a). Participants were representative of the school district in ethnicity, and there were no differences between those who participated in both the baseline data and postintervention data collection and those lost at follow-up. However, of the participating students, only 182 participated in at least 40% of the free sessions. A subsequent study showed that the greater the attendance, the greater the change in percent body fat and fat mass and cardiovascular fitness (Yin et al., 2005c). Thus, communities need to provide organized opportunities for physical activity promotion, and these opportunities must be attractive enough and sufficiently high in quality to reach youth.

In sum, there is some evidence that youth in communities providing accessible organized community opportunities will exhibit greater physical activity than youth in communities that do not provide these opportunities. However, no studies were identified that directly examined this issue on a community-wide basis. Subsequent analysis would involve examining whether increasing the adoption rate of physical activity promotion programs by these delivery settings would lead to an increase in youth physical activity.

Researchers and practitioners can gain some insight for future work in relation to hypothesis 1 from the U.S. National Evaluation of the 21st Century Community Learning Centers Program (James-Burdumy et al., 2005). As an after-school program, 21st Century was appropriated $1 billion in 2002. Results indicated limited developmental outcomes, little academic impact, and no improvements in safety and health behavior. A key implementation finding was that the federal funding enabled programs to spend about $1000 for each student enrolled during the school year. Even with this level of economic input into local community systems, student participation was low (elementary attendance was two to three days per week, and middle school attendance was one day per week), as was involvement of trained and committed staff (turnover of program coordinators and staff was high because many were trained teachers who initially took on the position for supplemental income). Furthermore, programs did not collaborate with community organizations to deliver a coordinated, sustainable youth development experience of high quality. Rather, centers contracted with agencies to provide specific services that were not sustainable without continued funding. These findings reinforce the idea that to increase physical activity and decrease sedentary behavior, it is necessary for communities (a) to develop a sustainable community infrastructure for out-of-school opportunities and also (b) to provide quality physical activity promotion efforts that reach youth, are effective, and are adopted, implemented, and maintained by numerous delivery settings.

Improving the Quality of Physical Activity Opportunities

Hypothesis 2 states, "Youth participating in organized opportunities that include quality physical activity promotion programs will increase in personal assets, increase in MVPA, and decrease in sedentary behavior compared to youth who do not participate." The studies discussed next deal not only with that hypothesis, but also with the effects on program attendance when issues of reach and implementation are addressed.

Skill-Building Experiences

Three experimental studies offer some evidence that the quality of an opportunity can be improved if a physical activity promotion program, beyond providing physical activity, includes skill-building experiences that develop the personal assets necessary for maintaining physical activity across the life span.

Girlfriends for KEEPS (Keys to Eating, Exercising, Playing and Sharing) was a twice-a-week, 1 h after-school "club meeting" delivered at elementary schools (Story et al., 2003). African American girls, 8 to 10 years old, participated in a 12-week program that focused on the prevention of obesity through the achievement of several goals, including the promotion of physical activity and the reduction of time spent in sedentary activities. The program provided a positive developmental setting that offered peer support, role models, social reinforcement, and physical activity opportunities. The goal was to build youth development assets that included knowledge, values, and self-efficacy, as well as behavioral (practice, goal setting) and resiliency skills. In this preliminary study of 26 girls who were randomized to receive the intervention and 28 girls in a control condition, the intervention group showed greater preference for physical activity following the intervention than the control group. Although no effect was found for body mass index (BMI), there was a nonsignificant trend for physical activity such that girls receiving the program showed greater objectively monitored physical activity over three days than those not receiving the program. Parents reported that sedentary behavior (television watching) did not change. Therefore, there is some evidence of the potential to build skills in girls to promote physical activity. Future research on Girlfriends for KEEPS will need to determine whether the reach of the program is maintained with a larger population. The small volunteer sample of girls reported liking the club meetings (92%) and on average attended 21 of 24 sessions.

Go Girls Faith-Based Group Behavioral Skills Class was another skill-based approach to physical activity promotion. Go Girls targeted 12- to 16-year-old African American middle- and upper-income girls who were overweight. The study compared a high-intensity program of weekly group behavior sessions, conducted over six months (20 to 26 sessions) at five churches, to a six-session program delivered at five churches. Target behaviors for the intervention included increased fruit and vegetable intake, decreased fat intake, decreased fast food intake, decreased sedentary behavior, and increased physical activity. Two trained staff with graduate degrees led group sessions. Each session included a behavior skills activity; at least 30 min of MVPA; and preparation and/or consumption of low-fat, portion-controlled meals or snacks. All girls in the high-intensity session received a paging device, and messages developed by the girls were sent to them at key times throughout the day to prompt MVPA or healthful eating. They also received six calls from counselors using motivational interviewing techniques across the six months. In addition, parents were invited and encouraged to attend every other session. Changes in youth developmental assets were not reported in the primary outcome paper. There was a net difference between groups of 0.5 BMI units, which was not statistically significant. However, girls who attended more than three-quarters of the weekly sessions of the high-intensity program were lower in BMI and percent body fat at six months and at one-year follow-up compared to girls attending the six-session program. The skills-based approach may be effective, but reach and participation may be a challenge. Girls attended 57% of the sessions.

The Food, Fun and Fitness program targeted the prevention of obesity in 8-year-old African American girls through promoting physical activity, fruit and vegetable consumption, and increased water intake (Baranowski et al., 2003). Girls attended a four-week summer day camp where they built skills (youth assets) and participated in daily physical activity. Specific components included developing a girls' buddy system, asking girls to ask their parents to be physically active with them after camp or in the evening, exposing girls to fun physical activities, teaching them physical activity skills, and providing them with pedometers to self-monitor their activity. The day camp was followed by an eight-week home Internet intervention. In this preliminary study, the 19 girls randomized to receive the intervention were not significantly different in physical activity preference, self-reported physical activity, or objective monitoring of physical activity by accelerometer compared to 16 girls randomized to a control condition. The girls attending the daily camp that received the intervention program did have excellent attendance (91.5%), and there was a trend toward greater self-reported physical activity in the intervention group compared to the control group. Heavier girls at baseline who received the program exhibited a trend toward lower BMI compared to heavier girls in the regular camp program. The pilot study demonstrated the potential for promoting physical activity through a summer camp in that it appears that reach and good daily attendance are possible. However, overall, the three studies discussed in this section suggest that a relatively brief-duration, skills-based approach may be insufficient to promote physical activity.

In each of these studies, the intervention program included a skill-building component and various levels of frequency of physical activity opportunities. None of the programs offered a frequent physical activity opportunity across the length of the study. It may be necessary to provide skill training as well as a behavior setting that offers a physical activity opportunity across development.

Skill-Building Experiences Plus Sustained Physical Activity

The four studies reviewed next deal with intervention programs that included both skill-building and sustained opportunities for physical activity.

The Active Winners Summer and After School Program study tested an intervention with four components: Active Kids (after-school and summer program), Active Home, Active School, and Active Community (Pate et al., 2003). Active Kids was an organized community opportunity offered to students over two weeks for 5 h a day in the summer after their 5th-grade year, offered after school every day for 2 h during 6th grade, and offered over four weeks for 5 h per day in the summer after 6th grade. A transition program was provided in the fall for students in their 7th-grade year. Active Kids included Fit for

Life (a fitness activity), Be a Sport (physical activity skills), Social Rap (social skills), and Brain Games (academic skills). The Active Home, Active School, and Active Community components were designed to positively influence the social and physical environment and to provide external cues for physical activity. Evaluation of Active Winners showed no effect on MVPA when compared to that of children in another similar community over the program years. This may be attributable to poor reach. Of the target group of 255 students, 82% had at least one exposure to the program, but only 5% attended over half of the total sessions offered. The process evaluation showed that students did not attend because friends did not attend and because of the presence of "problem" students. All the staff time became dedicated to implementing the Active Kids program, so the larger social environmental intervention changes did not occur. This illustrates that all components of the RE-AIM model may be necessary to produce successful intervention effects.

CATCH Kids Club After School Program targeted physical activity promotion and healthful eating in grades K through 2 and grades 3 through 5 by providing a 15-lesson 15 to 30 min education component, a physical activity component, and a snack component (Kelder et al., 2005). The five-module education component was designed to build knowledge, skills, self-efficacy, and intentions to make healthy dietary and physical activity decisions. The physical activity component provided 30 min of physical activity with the goal that children would be moving at a moderate or vigorous intensity 40% of the time. Program sites were compared to reference sites for students in grades 3 through 5. Focus groups with staff documented that the educational component was not well received. A youth asset, self-efficacy for physical activity, did not improve due to the intervention. Sedentary behavior at home also did not differ between the children in program and reference sites. However, the program was effective at increasing physical activity during the physical activity session after school. Thus, it appears that an after-school program can provide a behavior setting that offers physical activity and that the quality of physical activity offered in after-school time can be improved.

Wilson and colleagues (2005) examined the influence of an after-school program that offered a skill-building behavior setting and a physical activity behavior setting to promote physical activity. The physical activity promotion program was implemented three days per week for 2 h after school, over four weeks, to primarily African American (83%), low-income 10- to 12-year-old boys and girls. The student-centered program had three main components: a homework/snack session (30 min), a physical activity session that the students selected each week (60 min), and a behavior skill session that included motivational strategies to increase their physical activity with friends and at home (30 min). The student-centered intervention drew on self-determination theory: "(a) allowing the students to develop positive coping strategies for making lifestyle changes using a strategic self-presentation (videotaped interview) and (b) allowing the students to participate in program development by selecting a variety of the physical activities offered weekly, developing a program name and motto, and developing ideas for promoting MVPA to friends and peers" (Wilson et al., 2005, p. 120). The quasi-experimental design showed that 6th-grade students receiving the program (n = 28), compared to matched comparison school students (n = 20), had greater increases in physical activity from baseline to week 4 as assessed by accelerometer monitoring. The student-centered program also produced greater increases in physical activity motivation and physical activity self-concept. Thus, combining a skill session with a behavior setting that offers physical activity may be an effective route to promoting physical activity during out-of-school time.

A final study tested the offering of a specific, enjoyable physical activity option that reached youth and included skill-based training. The GEMS Jewels After School Dance classes targeted obesity prevention through physical activity promotion and reduction of television, videotape, and video game use and were delivered through community centers and after-school programs (Robinson et al., 2003). The target audience was 8- to 10-year-old African American girls who had a BMI greater than the 50th percentile for age or at least one overweight parent or guardian. Girls were encouraged to attend the dance classes as often as possible across a three-month period. Each daily session occurred from 3:30 p.m. to 6:00 p.m. and started with a healthful snack, which was followed by an hour-long homework period. Then girls participated in 45 to 60 min of MVPA dance. The sessions concluded with a 30 min discussion of the importance of dance in the African American community and culture. In addition to the after-school component, a START (Sisters Taking Action to Reduce Television) intervention of five lessons was delivered over 12 weeks during home visits with participating families. Girls randomized to receive the intervention compared to girls in an active control condition increased in after-school physical activity during the program, and there was a trend in increased physical activity after the conclusion of the 12-week program. Girls in the program also reported a 23% reduction in television, videotape, and video game use and 10% fewer meals with the television on. Thus, there is intriguing preliminary evidence that a quality physical activity promotion program includes a daily physical activity session providing moderate- to vigorous-intensity activity and physical activity regulation skill sessions.

In the GEMS Jewels After School Dance program study, success was dependent on program reach and implementation. At one site to which the city provided daily after-school buses from the schools, 70% of the girls attended at least two days per week. At another site to which the city did not provide transportation, only 33% of the girls attended at least two days per week. It is unlikely that promising results from the program would have been seen if reach and implementation had not been achieved through the provision of transportation to one of the program sites. Thus, program quality requires addressing all elements of the RE-AIM model.

Conclusions and Future Directions

Communities and parents are seeking organized community opportunities out of school for their youth. As suggested by existing evidence, communities providing organized opportunities that include MVPA-promoting behavior settings will have youth who are more physically active than youth in communities without these opportunities. Also, there is evidence that youth participating in an organized community opportunity structured for the promotion of positive youth development and MVPA will increase in personal assets, increase in MVPA, and decrease in sedentary behavior compared to youth who do not participate.

The findings supporting these conclusions clearly have limitations. Only a few studies illustrate that these effects are possible. And, where positive effects were seen, the effect sizes were not impressive. However, there is enough preliminary evidence to encourage researchers and practitioners to move forward in targeting organized community opportunities as a route to promoting physical activity and decreasing sedentary behavior in youth.

Researchers need to further address several questions. First, what are the resources that communities need to obtain in order to provide organized community opportunities? Without a delivery setting that reaches youth, there is no sustained channel for producing public health outcomes. Second, what are the processes that determine that delivery settings offer physical activity promotion programs? Once a system that reaches youth is in place, how can the quality of this system be improved?

The research reviewed for this chapter documents that many community opportunities for youth did not include physical activity promotion as one of their primary goals. Why was this so? Researchers need to identify what variables influence these priorities. It may be that administrators making programming decisions do not value physical activity and are not trained in physical activity. Or, it may be that programs are funded to deliver other outcomes, such as academic improvement or substance abuse prevention. Alternatively, the primary influence on physical activity opportunities could simply reflect a parental demand that children complete homework before going home. The specific question is, What is necessary in order for organized community opportunities to adopt physical activity–promoting, evidence-based programs that reach youth and are implemented and maintained sufficiently to have a public health impact?

The central message from the existing literature is that health and fitness professionals need to view themselves as community developers and trainers of delivery agents, rather than simply as delivery agents (Dzewaltowski et al., 2002, 2004). The first challenge is to work with community organizations to increase the availability of physical activity opportunities. The second challenge is to work with existing opportunities to increase the quality of these programs. This chapter illustrates that quality programs may need three components: a community delivery system that provides community opportunities for physical activity addressing all of the RE-AIM elements, a consistent opportunity for physical activity weekly (preferably daily), and skill-building sessions for lifelong physical activity. Literature reviews show that a comprehensive evaluation of hypotheses 1 and 2, addressing all elements of the RE-AIM framework, has not been conducted (Dzewaltowski et al., 2004).

For both researchers and practitioners, much of the focus turns to understanding the role of adult leaders, both parents and after-school staff. These are the key agents who build the environments young people live within. They also provide links between these environments. Simply, the best after-school program will not affect youth unless parents allow their children to participate, and, taking a more proactive stance, seek such opportunities for their children. Further, adult leaders must help young people as they develop and become progressively more involved in choosing and building their own environments. Human beings are not passive responders to environmental stimuli; they are active builders of their circumstances. Fortunately, there is preliminary evidence that systematic efforts can positively influence the recent trend toward increased childhood obesity. Community leaders and parents can build social and physical environments that promote physical activity and decrease the sedentary behavior of youth.

APPLICATIONS FOR RESEARCHERS

According to preliminary evidence, youth in communities that provide accessible organized out-of-school community opportunities—opportunities that reach children and offer daily physical activity and skill-building experiences to develop the personal assets necessary for maintaining physical activity across the life span—will be more physically active than youth in communities that do not provide these opportunities. To build on this preliminary evidence, researchers need to identify community, delivery setting, and behavior setting variables that determine youth development and behavior and to target these variables for intervention evaluation. Communities vary in the resources necessary for providing physical activity opportunities. Delivery settings also vary in the resources necessary for providing physical activity promotion programs. Further research should examine physical activity promotion programs designed to meet existing resources of communities and to provide social and physical environments (behavior settings) that include daily opportunities for physical activity and for the development of lifelong physical activity behavioral skills.

Applications for Professionals

Given the current public health crises of physical inactivity and sedentary behavior, health and fitness professionals must not wait for the best evidence but must instead immediately draw on the best available evidence in their physical activity promotion efforts. First, professionals need to go beyond a focus on the delivery of physical activity programs and work to provide community opportunities that will reach a large segment of youth in their target population. This requires practitioners to become community leaders and work to develop community resources and delivery settings for the provision of evidence-based programs. Second, health and fitness professionals need to work with existing community delivery systems to offer evidence-based physical activity promotion programs. Preliminary evidence suggests that quality physical activity promotion programs need to reach youth, offer a consistent opportunity for physical activity daily, and build youth skills and assets for lifelong physical activity.

References

Baranowski, T., Baranowski, J.C., Cullen, K.W., Thompson, D.I., Nicklas, T., Zakeri, I.E., et al. (2003). The Fun, Food, and Fitness Project (FFFP): The Baylor GEMS pilot study. *Ethnicity and Disease, 13*(Suppl. 1), S30-39.

Barker, R.G. (1968). *Ecological psychology.* Stanford, CA: Stanford University Press.

Baumrind, D. (1991). The influence of parenting style on adolescent competence and substance use. *Journal of Early Adolescence, 11,* 56-95.

Booth, S.L., Sallis, J.F., Ritenbaugh, C., Hill, J.O., Birch, L.L., Frank, L.D., et al. (2001). Environmental and societal factors affect food choice and physical activity: Rationale, influences, and leverage points. *Nutrition Reviews, 59,* S21-S39.

Bronfenbrenner, U. (1979). *The ecology of human development.* Cambridge, MA: Harvard University Press.

Bronfenbrenner, U. (1992). Ecological systems theory. In U. Bronfenbrenner (Ed.), *Making human beings human: Bioecological perspectives on human development* (pp. 106-173). Thousand Oaks, CA: Sage. Reprinted from R. Vast (Ed.), (1992). *Six theories of child development: Revised formulations and current issues* (pp. 187-249). London: Jessica Kingsley.

Capizzano, J., Tout, K., & Adams, G. (2000). *Child care patterns of school-age children with employed mothers.* Washington, DC: Urban Institute.

Department of Health. (2004). *At least five days a week: Evidence on the impact of physical activity and its relationship to health—a report from the Chief Medical Officer.* London: Department of Health.

Dzewaltowski, D.A. (1997). The ecology of physical activity and sport: Merging science and practice. *Journal of Applied Sport Psychology, 9,* 254-276.

Dzewaltowski, D.A., Estabrooks, P.A., & Glasgow, R.E. (2004). The future of physical activity behavior change research: What is needed to improve translation of research into health promotion practice? *Exercise and Sport Sciences Reviews, 32*, 57-63.

Dzewaltowski, D.A., Estabrooks, P.A., Johnston, J.A., & Gyurscik, N. (2002). Promoting physical activity through community development. In J.L. Van Raalte & B.W. Brewer (Eds.), *Exploring sport and exercise psychology* (2nd ed., pp. 209-223). Washington, DC: American Psychological Association.

Ferguson, R.F., & Dickens, W.T. (1999). *Urban problems and community development.* Washington, DC: Brookings Institution Press.

Garcia, A.W., Broda, M.A., Frenn, M., Coviak, C., Pender, N.J., & Ronis, D.L. (1995). Gender and developmental differences in exercise beliefs among youth and prediction of their exercise behavior. *Journal of School Health, 65*, 213-219.

Glasgow, R.E., Klesges, L.M., Dzewaltowski, D.A., Estabrooks, P.A., & Vogt, T.M. (2006). Evaluating the impact of health promotion programs: Using the RE-AIM framework to form summary measures for decision making involving complex issues. *Health Education Research, 21*, 688-694.

Glasgow, R.E., Vogt, T.M., & Boles, S.M. (1999). Evaluating the public health impact of health promotion interventions: The RE-AIM framework. *American Journal of Public Health, 89*, 1322-1327.

Gordon-Larsen, P., McMurray, R.G., & Popkin, B.M. (2000). Determinants of adolescent physical activity and inactivity patterns. *Pediatrics, 105*, 1327-1328.

Green, L.W., Poland, B.D., & Rootman, I. (2000). The settings approach to health promotion. In B.D. Poland, L.W. Green, & I. Rootman (Eds.), *Settings for health promotion: Linking theory and practice* (pp. 1-43). Thousand Oaks, CA: Sage.

Hawkins, J.D., Catalano, R.F., & Arthur, M.W. (2002). Promoting science-based prevention in communities. *Addictive Behavior, 27*, 951-976.

Heinrich, C.J., & Lynn, L.E. (2000). *Governance and performance: New perspectives.* Washington, DC: Georgetown University Press.

Hofferth, S., & Sandberg, J.F. (2001). Changes in American children's time, 1981-1997. In T. Owens & S. Hofferth (Eds.), *Children in the millennium: Where have we come from, where are we going? Advances in life course research* (pp. 193-229). New York: Elsevier Science.

Hyland, R., Stacy, R., Adamson, A., & Moynihan, P. (2006). Nutrition-related health promotion through an after-school project: The responses of children and their families. *Social Science and Medicine, 62*, 758-768.

James-Burdumy, S., Dynarski, M., Moore, M., Deke, J., Mansfield, W., Pistorino, C., et al. (2005). *When schools stay open late: The national evaluation of the 21st Century Community Learning Centers Program: Final report.* Washington, DC: U.S. Department of Education, National Center for Educational Evaluation and Regional Assistance. Available at www.mathematica-mpr.com/publications/redirect_pubsdb.asp?strSite=pdfs/21stfinal.pdf.

Kelder, S., Hoelscher, D.M., Barroso, C.S., Walker, J.L., Cribb, P., & Hu, S. (2005). The CATCH Kids Club: A pilot after-school study for improving elementary students' nutrition and physical activity. *Public Health Nutrition, 8*, 133-140.

Lerner, R.M., & Castellino, D.R. (2002). Contemporary developmental theory and adolescence: Developmental systems and applied developmental science. *Journal of Adolescent Health, 31*(6 Suppl.), 122-135.

Lewin, K. (1951). *Field theory in social science: Selected theoretical papers*, D. Cartwright (Ed.). New York: Harper & Row.

Mahoney, J.L., Larson, R.W., Eccles, J.S., & Lord, H. (2005). Organized activities as developmental contexts for children and adolescents. In J.L. Mahoney, R.W. Larson, & J.S. Eccles (Eds.), *Organized activities as contexts of development: Extracurricular activities, after-school and community programs* (pp. 3-22). Mahwah, NJ: Erlbaum.

Mahoney, J.L., Lord, H., & Carryl, E. (2005). Afterschool program participation and the development of child obesity and peer acceptance. *Applied Developmental Science, 9*, 202-215.

McLeroy, K.R., Bibeau, D., Steckler, A., & Glanz, K. (1988). An ecological perspective on health promotion programs. *Health Education Quarterly, 15*, 351-377.

National Research Council and Institute of Medicine. (2002). *Community programs to promote positive youth development*, J. Eccles and J.A. Gootman (Eds.). Committee on Community-Level Programs for Youth, Board on Children Youth and Families, Division of Behavioral and Social Sciences and Education. Washington, DC: National Academies Press.

Pate, R.R., Saunders, R.P., Ward, D.S., Felton, G., Trost, S.G., & Dowda, M. (2003). Evaluation of a community-based intervention to promote physical activity in youth: Lessons from Active Winners. *American Journal of Health Promotion, 17*, 171-182.

Robinson, T.N., Killen, J.D., Kraemer, H.C., Wilson, D.M., Matheson, D.M., Haskell, W.L., et al. (2003). Dance and reducing television viewing to prevent weight gain in African-American girls: The Stanford GEMS pilot study. *Ethnicity and Disease, 13*(Suppl. 1), S65-77.

Sallis, J.F., Conway, T.L., Prochaska, J.J., McKenzie, T.L., Marshall, S.J., & Brown, M. (2001). The association of school environments with youth physical activity. *American Journal of Public Health, 91*, 618-620.

Sallis, J.F., McKenzie, T.L., Conway, T.L., Elder, J.P., Prochaska, J.J., Brown, M., et al. (2003). Environmental interventions for eating and physical activity: A randomized controlled trial in middle schools. *American Journal of Preventive Medicine, 24*, 209-217.

Sallis, J.F., & Owen, N. (2002). Ecological models of health behavior. In K. Glanz, B.K. Rimer, & F.M. Lewis (Eds.). *Health behavior and health education: Theory, research, and practice* (3rd ed., pp. 403-424). San Francisco: Jossey-Bass.

Sallis, J.F., Prochaska, J.J., & Taylor, W.C. (2000). A review of correlates of physical activity of children and adolescents. *Medicine and Science in Sports and Exercise, 32*, 963-975.

Sampson, R.J. (2002). Transcending tradition: New directions in community research, Chicago style. *Criminology, 40*, 213-230.

Story, M., Sherwood, N.E., Himes, J.H., Davis, M., Jacobs, D.R., Jr., Cartwright, Y., et al. (2003). An after-school obesity prevention program for African-American girls: The Minnesota GEMS pilot study. *Ethnicity and Disease, 13*(Suppl. 1), S54-64.

Sturm, R. (2005). Childhood obesity—what we can learn from existing data on societal trends. In *Preventing chronic disease: public health research, practice, and policy, 2*, 1-9.

Vandivere, S., Tout, K., Zaslow, M., Calkins, J., & Capizzano, J. (2003). *Unsupervised time: Family and child factors associated with self-care.* Washington, DC: Urban Institute.

Wilson, D.K., Evans, A.E., Williams, J., Mixon, G., Sirard, J.R., & Pate, R. (2005). A preliminary test of a student-centered intervention on increasing physical activity in underserved adolescents. *Annals of Behavioral Medicine, 30,* 119-124.

Wirt, J., Choy, S., Rooney, P., Provasnik, S., Sen, A., & Tobin, R. (2004). The condition of education 2004 (NCES 2004-077). U.S. Department of Education, Washington, DC: U.S. Government Printing Office.

Yin, Z., Gutin, B., Johnson, M.H., Hanes, J., Jr., Moore, J.B., Cavnar, M., et al. (2005a). An environmental approach to obesity prevention in children: Medical College of Georgia FitKid Project year 1 results. *Obesity Research, 13,* 2153-2161.

Yin, Z., Hanes, J., Jr., Moore, J.B., Humbles, P., Barbeau, P., & Gutin, B. (2005b). An after-school physical activity program for obesity prevention in children: The Medical College of Georgia FitKid Project. *Evaluation and the Health Professions, 28,* 67-89.

Yin, Z., Moore, J.B., Johnson, M.H., Barbeau, P., Cavnar, M., Thornburg, J., et al. (2005c). The Medical College of Georgia Fitkid project: The relations between program attendance and changes in outcomes in year 1. *International Journal of Obesity, 29*(Suppl. 2), S40-45.

CHAPTER

16

Living Environments

Jo Salmon, PhD ▪ John C. Spence, PhD ▪ Anna Timperio, PhD ▪ Nicoleta Cutumisu, MSc

This chapter explores the ways in which the living environment of young people influences their levels of physical activity and sedentary behavior. Living environments may be defined as the social, built, and natural physical structures in which everyday life occurs. Social environments may include proximal factors, such as family (e.g., single or dual parents, siblings, extended family) and peers, and distal factors, such as strong social networks in the neighborhood, perceptions of trust of people in the community, crime (against persons or property), and incivilities (e.g., graffiti, vandalism). Physical environments may include topography (e.g., hills or mountains, coasts, forests or bush), the built environment (e.g., buildings and roads, public open spaces and parks, home and yard size), and weather and climate. The living environment or perceived living environment is likely to be reciprocally related (Bandura, 1986) to young people's individual characteristics (e.g., motivation, preference, perceived barriers) and may also directly influence active or sedentary behavioral choices.

Depending upon where they live and the time of year, youth may participate in different organized and nonorganized physical activities and to a varying extent. For instance, during the winter months in the northern hemisphere Canadian children are more likely than Australian children to participate in ice hockey and alpine skiing. At the same time, Australian children and youth may be surfing or playing rugby. Within these countries, proximity to the mountains and oceans influences the extent to which particular physical activities occur (e.g., skiing and surfing). Also, between and within countries there are likely to be variations in the physical environment that may influence children's free play and walking and cycling for transport. Evidence of relationships between environmental design and adult physical activity is emerging (Humpel, Owen,

Anna Timperio is supported by a Public Health Research Fellowship from the Victorian Health Promotion Foundation. Jo Salmon is supported by a National Heart Foundation of Australia Career Development Award.

& Leslie, 2002; Owen et al., 2004); however, much less is known about how these features relate to child and adolescent physical activity. Even less is known about the influence of the physical home environment (e.g., the increased size of housing "footprints" and reduced yard or garden size, the increased access to sedentary recreation in the home such as electronic games and Internet) and the broader social environment on child and adolescent activity. Thus a more thorough discussion of the topic is warranted.

The objectives of this chapter are to describe several theoretical frameworks relevant to the role of living environments in youth physical activity and sedentary behavior, and to provide a selected overview of observational and intervention studies of the relationship between living environments and youth physical activity and sedentary behavior. In particular, we focus on the physical and broader social environments in the home and neighborhood. We consider how studies have operationalized, defined, and assessed young people's physical activity environments. On the basis of this overview, we identify potential research directions to further develop theory regarding youth physical activity and sedentary behavior in relation to home and the neighborhood. Potential strategies to decrease sedentary time and promote physical activity are also identified.

Conceptual and Theoretical Perspectives

Theories and models of health behavior can frame and enhance our understanding of the effects of living environments on youth physical activity or inactivity and sedentary behavior. For the purposes of this chapter, we can classify such theories based upon the extent to which the environment influences behavior. King and colleagues (2002) provided such a classification of theoretical frameworks ranging from intrapersonal theories to more extrapersonal theories (macro-environment theories). Intrapersonal theories (e.g., theory of planned behavior) incorporate an array of individual factors (e.g., attitudes) related to behavior motivation that operate at the level of the individual. Extrapersonal theories encompass individual factors but also address the complex of factors external to the individual such as the physical and social environments. Another distinction made by King and colleagues is that intrapersonal theories imply that behavior is more *choice driven* while extrapersonal theories emphasize the role of the environment for *enabling choice*. The influence of the environment on behavior, if any, within intrapersonal theories is mediated by individual-level constructs such as attitudes and norms. Conversely, the environment has a direct influence on behavior in extrapersonal theories. Because of their emphasis on interactions among person, environment, and behavior, theories such as social cognitive theory (SCT; Bandura, 1986), behavioral choice theory (BCT; Epstein, 1998; Rachlin, 1989), the Youth Physical Activity Promotion Model (Welk, 1999), and ecological systems theory (EST; Bronfenbrenner, 1979) are potentially useful for exploring the influence of environment on physical activity in young people. These theories would fall somewhere in the middle of the intrapersonal-extrapersonal theoretical continuum.

In SCT, behavior is conceptualized as a triadic interaction between person, behavior, and environment, which all exert varying degrees of bidirectional influence on one another (Bandura, 1986). Thus, behavior is not simply the result of a person's cognitions and the environment, but rather the interplay among all three factors. At the heart of this reciprocal determinism process lies self-efficacy, the sense of personal agency about one's ability to perform a specific behavior, which supposedly affects the amount of effort invested in a given task and the level of success experienced. According to Bandura (1986), self-efficacy is the most important prerequisite of behavioral change. The findings of a recent meta-analysis revealed a medium-sized correlation ($r = .28$) between self-efficacy and physical activity among children and adolescents (Spence et al., 2006). Primary sources of self-efficacy include mastery experience and vicarious learning through modeling. Thus, children can learn to do a particular physical activity by participating in the activity and experiencing some success, or by watching other children, parents, or teachers doing the same task successfully, or in both ways. Other important constructs of SCT include outcome expectations and incentives. These are the value-expectancy components of the model through which a person considers whether he or she sees incentives in the anticipated outcome (expectations) of a particular behavior.

The BCT (Rachlin, 1989) or behavioral economic theory (Epstein, 1998) is a theoretical approach that attempts to understand how people make choices in terms of the costs associated with the potential options provided by the environment. The BCT accounts for the roles of environmental barriers, preference for the behavior, and influences on reinforcement value (Vuchinich & Tucker, 1988). According to Epstein (1998), BCT posits that choices among alternatives are based on individual preferences and a certain attractiveness of choice, the availability and accessibility of various choices, the reinforcing (positive or negative) value of each choice for the person, the behavioral costs of obtaining the choice, and the degree to which the person likes the given activity. On the basis of BCT, it is possible to increase physical activity not simply by restricting access to sedentary behaviors, but also by emphasizing the access to physical activities and increasing the reinforcing value of being physically active while decreasing the reinforcing nature of being sedentary. For instance, Saelens and Epstein (1998) showed that when asked to choose between a lower-preference sedentary behavior and a higher-preference sedentary behavior that was contingent upon an active behavior, children still chose the higher-preference sedentary behavior with its associated cost.

Social ecological models encompass individual factors as well as the complex factors external to the individual such as the physical and social environments (Bronfenbrenner, 1979; Sallis & Owen, 1997). The Youth Physical Activity Promotion Model (Welk, 1999) is informed by ecological models (McLeroy et al., 1988) and SCT (Bandura, 1986), and posits that a multitude of intrapersonal and extrapersonal (sociocultural and environmental) factors interact to influence the physical activity of youth. In developing the Youth Physical Activity Promotion Model, Welk (1999) categorized a select number of demographic, biological, psychological, sociocultural, and environmental correlates of physical activity in youth as predisposing, reinforcing, or enabling factors. The decision about which variables to include in the model was based upon the strength of the relationship between the variable and physical activity as well as the variable's potential for change. The predisposing factors include two categories: self-evaluative constructs (perception of control over personal behavior, perception of competence, self-efficacy, physical self-worth) and cognitive assessment constructs (cost-benefit, enjoyment, beliefs, attitudes). The reinforcing factors include the influence of families, peers, teachers, and coaches on young people (Welk, Wood, & Morss, 2003). The enabling factors include both personal (e.g., level of fitness, motor skills) and environmental (e.g., access to facilities, playground equipment) elements. All three sets of factors are moderated by the personal demographics of the child or adolescent.

Ecological systems theory highlights the importance of the "ecological niche" in which children grow and develop (Davison & Birch, 2001). This "ecological niche" incorporates personal attributes, family and peer influences, and school and community environments. All these "layers" of influence interact. For example, the child's individual characteristics interact with the family, school, and community environments, and all these environments also interact. It has been argued (Davison & Birch, 2001) that this theory provides a useful framework for understanding the development of obesity in children by identifying the context in which dietary, physical activity, and sedentary behavior patterns evolve from a young age. Spence and Lee (2003) have proposed an ecological model of physical activity that, while not specific to children, can be used to better understand how the environment influences physical activity. This model hypothesizes that biological factors are likely to moderate relationships between extraindividual factors and physical activity while other intraindividual factors, such as psychological factors, both directly influence behavior and mediate relationships between the environment and physical activity. The model also highlights the proximal and distal nature of environmental influences on behavior (e.g., from micro-level influences such as high-quality playground facilities to macro-level influences such as societal values about physical activity).

In summary, all four models recognize the influence of the environment on behavior. For SCT (Bandura, 1986) and BCT (Rachlin, 1989), environment indirectly influences behavior through cognitions and beliefs. In SCT, environment is important to the extent that it allows or provides opportunities for individuals to overcome perceived barriers to meet their outcome expectations. Alternatively, the environment serves as the main generator of potential choices in BCT and thus may play a more prominent role than in SCT. As SCT and BCT focus on indirect influences of the environment, the Youth Physical Activity Promotion Model (Welk, 1999) and EST (Bronfenbrenner, 1979) model may be more appropriate frameworks for understanding the role of the environment in youth physical activity and sedentary behavior because they focus on the environment as a direct influence on behavior.

The Home Environment

Children and youth who live in highly sedentary home environments may be more at risk of physical inactivity, though research findings are mixed. For example, sedentary opportunities within the home (e.g., number of television sets, Internet access, a personal computer, pay television, electronic games consoles) that are readily accessible may be important correlates of children's sedentary behavior and inactivity. Among preschool children (Dennison, Erb, & Jenkins, 2002) and early adolescents (Wiecha et al., 2001), having a television set in the bedroom has been found to be associated with higher levels of television viewing. In contrast, a recent Australian study of children's home environments failed to show any association between having sedentary opportunities at home and children's television viewing or activity levels (Salmon et al., 2005). Additionally, that study failed to find a relationship between having a television set in the bedroom and children's television viewing, which is also consistent with a longitudinal study of children in the United States (Saelens et al., 2002). In the Australian study, access to pay television in the home was associated with higher levels of physical activity among boys (Salmon et al., 2005). Having pay television in the home may be an indicator of a family's socioeconomic circumstances. Although the analyses adjusted for maternal education levels, socioeconomic circumstances may be a possible explanation for the positive relationship with physical activity. It could also be that television viewing and other sedentary behaviors cluster with physical activity (Marshall et al., 2002). However, evidence of an inverse relationship between television viewing and physical activity in the literature is inconsistent (Biddle et al., 2004; Marshall et al., 2004).

Other features of the home physical environment that may be important influences on physical activity include play or sporting equipment (e.g., bicycles, balls), facilities at home (e.g., trampoline, basketball ring), and yard space. A study of physical activity determinants in obese and nonobese children showed that, irrespective of weight status of the child, access to physical activity and exercise equipment in the home was positively associated with activity levels (Trost et al., 2001). A review of 54 studies of 3- to 12-year-olds published between 1976 and 1999 found that time spent outdoors was a consistent correlate of physical activity in that age-group (Sallis, Prochaska, & Taylor, 2000). However, whether time spent outdoors is associated with yard size or space available outside the home is not known. In qualitative research exploring parents' perceptions of factors influencing where their child plays (Veitch et al., 2006), many parents reported that their child played mainly in the yard at home. Fewer reported that their child usually played in public open spaces (e.g., parks, playgrounds). Interestingly, a consistent theme among families who lived in a cul de sac (or court) was that their child usually played outside in the street rather than in their own yard at home or at the park. The reasons provided by most parents were the social elements of the court (i.e., many other children to play with) and safety (Veitch et al., 2006).

While many elements inside and outside the home may influence young people's physical activity and sedentary behaviors, evidence for such influence is not well established. Further research is needed to identify (a) how elements of the home environment interact with the surrounding neighborhood environment and affect child and adolescent physical activity, and (b) whether parents' own beliefs and concerns about their child's safety and ability to be independently mobile interact with opportunities to be active at home, in the neighborhood, or both. An emerging literature has examined the relationship between the neighborhood social and physical environment and youth physical activity. A selected portion of this literature is reviewed in the following section.

The Neighborhood Environment

In this section we describe the methods used to search and review the literature relating to physical activity and the neighborhood social and physical environment. We summarize how the environment has been operationalized in the research literature and then overview the trends and specific findings of this research.

Concepts, Constructs, and Operational Definitions

Conceptually, physical activity research on the neighborhood social and physical environment is not well developed; and though some studies have employed a theoretical framework, many have not. Using key words and phrases such as "child," "adolescent," "physical activity," "walk," "television," "environment," "neighborhood," "public open space," "safety," and "urban design," we searched the following databases: MEDLINE, PsycARTICLES, SPORTDiscus, Sociological Abstracts, ERIC, and Cambridge Scientific Abstracts. Reference lists from published articles and reports were also searched. As a first step in reviewing the evidence of relationships between the environment and youth physical activity, we summarized the many definitions of environment that have been employed, and provide an overview of the manner in which these definitions have been operationalized (see tables 16.1 and 16.2). It must be noted, however, that for consistency we have defined some items as neighborhood social environment constructs whereas the authors may have defined them as physical environment constructs.

The Social Environment

Elements of the neighborhood social environment that have been examined in relation to child and youth physical activity can be grouped into four general constructs: neighborhood safety, social disorder, hazards or neighborhood problems, and social connectedness. The neighborhood safety construct has been operationalized in several ways in the youth physical activity literature: as crime at the population level, perception of safety from strangers, and general neighborhood safety.

As shown in table 16.1, these constructs have been assessed in various ways, with definitions ranging from perceptions of feeling unsafe when certain adults or older children are around to concerns about roaming dogs. Social disorder has been assessed based on observation of peer groups with gangs, public intoxication, prostitution, and adults loitering or congregating, fighting or arguing, selling drugs, or consuming alcohol in public. Neighborhood hazards have been measured by the perceived presence of litter, traffic, noise, crime, drugs, prejudice, and a lack of access to parks. Although access to parks is better placed with the neighborhood physical environment constructs, previous studies have incorporated this element within a broader neighborhood hazards scale (Romero, 2005; Romero et al., 2001). Social connectedness has typically been assessed as sense of community, neighborhood relationships, having friends or peers or other children living nearby, and neighborhood social cohesion.

The Physical Environment

Table 16.2 shows that previous studies of youth physical activity have included elements of the neighborhood physical environment that can be conceptualized into nine different constructs: quality of facilities, perception of safety of facilities, access and availability, neighborhood design, population density, physical disorder, the road or traffic environment, topography, and aesthetics. Quality of facilities have been operationalized as the quality of built facilities and equipment, and opening hours, and also as a more global measure of whether there are good facilities in the neighborhood. Perceived safety of facilities incorporates whether the facility itself is safe, is located in a safe area, or both. The perceived safety of facilities focuses on safety of the physical or built environment; this is distinct from neighborhood safety, identified in table 16.1, which focuses on crime, strangers, or general safety in the neighborhood.

Access and availability has usually been operationalized as the perceived availability of places for sport or play, the availability of good sport facilities, the availability of take-away or fast food stores, the accessibility of facilities (cost and transport), and the availability of public transport, as well as an objective assessment of whether the route to school is within walking distance for a child. Although the terms access and availability are not strictly interchangeable (e.g., a facility may be available as it is located close by, but it may not be accessible due to restricted opening hours), for simplicity and because the definitions are often inconsistent we have combined studies that have used these terms. Neighborhood design can be further defined as street connectivity, urban form, and housing type. Street connectivity refers to the interconnectedness of street networks and is thought to be a particularly important predictor of walking and cycling for transport because of its potential impact on travel time when a more direct route is possible (Saelens, Sallis, & Frank, 2003). In studies of children,

Table 16.1 Conceptual and Operational Definitions of the Neighborhood Social Environment

Environmental construct	Operational definition
Neighborhood safety	**Crime** Recorded incidents and locations of serious crime/100,000 population[3,4] Parent report: fear of crime[9] Youth report: crime rate makes it unsafe/unpleasant to walk in neighborhood[7] **Safety from strangers** Youth report: concerns about strangers[2,12,13] Parent report: concerns about strangers[12,13] Youth report: concerns about older children[2] Youth report: feel unsafe when certain adults around[10] **General safety** Youth report: safe to play outside[1] Community-level measure: neighborhood unsafe to play in[6] Parent report: safe place to walk/ride bike[2] Youth report: concerns about roaming dogs[2] Youth report: neighborhood safety (combined safe to walk/jog alone during day, ease of walking/jogging due to traffic, sidewalks, dogs, gangs)[8]
Social disorder	Videotape and observation of adults loitering/congregating; adults fighting/arguing in public; alcohol in public; peer groups with gangs; public intoxication; people selling drugs; prostitution; community perceptions of whether adults watch out for children[6]
Hazards/ neighborhood problems	Child report: how much of a problem various hazards are in the neighborhood—litter, traffic, crime, noise, lack of access to parks, drugs, prejudice[10,11]
Social connectedness	Mother report: frequency of play with peers; sense of community and neighborhood relationships[9] Youth report: neighborhood relationships—know some neighbors well; wave or talk to neighbors most days[2] Youth report: social connections—have many friends in neighborhood; have friends close to house; lots of boys/girls of same age to hang out with[2] Parent report: few other children in neighborhood child can play with[12] Children's neighborhood mind maps of important places: number of opportunities for social interaction in neighborhood[5] Youth report: observation of others being active in neighborhood[7]

[1]Adkins et al. 2004; [2]Carver et al. 2005; [3]Gomez et al. 2004; [4]Gordon-Laren et al. 2000; [5]Hume et al. 2005; [6]Molnar et al. 2004; [7]Mota et al. 2005; [8]Motl et al. 2005; [9]Prezza et al. 2001; [10]Romero 2005; [11]Romero et al. 2001; [12]Timperio et al. 2006; [13]Timperio et al. 2004.

street connectivity has been operationalized as the number of intersections per street mile within a specified radius and as the directness of the shortest possible route to school. Urban form incorporates the layout of neighborhood street design and the presence and characteristics of sidewalks. Housing type has been operationalized as the presence of condominiums, houses with private street entrance, and proximity to parks.

Population density has rarely been incorporated into studies of children's physical activity. When it has been, it is operationalized as the number of residents within a 0.5-mile (about 0.8 km) radius of school. As is the case for social disorder, physical disorder has been assessed by observation of environmental features. These include graffiti, cigarettes and cigars, empty beer bottles, abandoned cars, condoms, and needles and syringes. Some of these measures, however, are the result of social activity (e.g., drug use). The road/traffic environment has been assessed as perceived traffic density and level of danger regarding traffic, the presence of so much traffic that it is difficult or unpleasant to walk, and concerns about road safety. Specific to active commuting, exposure to traffic in the road environment has also been operationalized as whether the shortest possible route to school is alongside or crosses a "busy" road. Topography has been objectively assessed as the steepness of hills on children's routes to school, and aesthetics has been operationalized as "tidiness" and as the presence of interesting things to look at during walking.

From this brief review of operational definitions, it is apparent that there is substantial overlap between some of the constructs and the ways in which some elements of the constructs have been operationalized. Nevertheless, while acknowledging these limitations, it is important to identify current evidence of the relationship between the neighborhood social and physical environment and physical activity of young people. As noted earlier, we reviewed 17 published studies for this chapter that examined the relationship between the neighborhood social or physical environment (or both) and some aspect of youth physical activity. Types and measures of physical activity reported in the studies were diverse, including the following:

- Self-reported moderate- to vigorous-intensity physical activity duration (MVPA; Fein et al., 2004; Gordon-Larsen, McMurray, & Popkin, 2000; Motl et al., 2005; Romero et al., 2001)
- Proxy-reported independent mobility and autonomy (Prezza et al., 2001)
- Self-reported frequency of vigorous exercise (Romero et al., 2001) or physical activity (Henning Brodersen et al., 2005; Romero, 2005)
- Proxy-reported (Timperio et al., 2004, 2006) and self-reported (Braza, Shoemaker, & Seeley, 2004; Carver et al., 2005) frequency and duration of walking and cycling
- Proxy-reported playing with peers (Prezza et al., 2001)

- Self-reported MVPA outside of school hours (Gomez et al., 2004; Mota et al., 2005)
- Proxy-reported hours spent in a recreational program (Molnar et al., 2004)
- Movement counts (Hume, Salmon, & Ball, 2005), minutes of light activity (Jago et al., 2005), and MVPA (Adkins et al., 2004; Jago et al., 2005) assessed by accelerometry

Two studies also dealt with associations between the environment and youth fitness (Fridlund Dunton, Schneider Jamner, & Cooper, 2003; Romero et al., 2001). Sample sizes ranged from 52 to greater than 17,000. Ten studies were performed in the United States (Adkins et al., 2004; Braza et al., 2004; Fridlund Dunton et al., 2003; Gomez et al., 2004; Gordon-Larsen et al., 2000; Jago et al., 2005; Molnar et al., 2004; Motl et al., 2005; Romero, 2005; Romero et al., 2001); three in Australia (Carver et al., 2005; Hume et al., 2005; Timperio et al., 2004, 2006); and one each in the United Kingdom (Henning Brodersen et al., 2005), Canada (Fein et al, 2004), Italy (Prezza et al., 2001), and Portugal (Mota et al., 2005). One of the U.S. studies was a longitudinal study (Motl et al., 2005).

Physical Activity and the Social Environment

Several studies have addressed aspects of the neighborhood social environment (see table 16.1) in relation to youth physical activity. Most dealt with the relationship of neighborhood safety with children's physical activity and were cross-sectional; and most included various physical activity outcomes, making it difficult to synthesize the findings. For example, among adolescents in the United States, incidents of serious crime were inversely associated with self-reported physical activity (Gordon-Larsen et al., 2000) and with self-reported physical activities outside of school (Gomez et al., 2004). However, fear of crime was not associated with Italian children's proxy-reported independent mobility (Prezza et al., 2001), and Portuguese adolescents' perceptions that crime makes their neighborhood unsafe for walking was unrelated to physical activity level (Mota et al., 2005). Apart from one study (Romero, 2005), perceptions of safety from strangers were not found to be associated with children's (Timperio et al., 2004) or adolescents' (Carver et al., 2005) walking or cycling in the neighborhood or to or from school (Timperio et al., 2006). General perceptions of the neighborhood's being a safe place to play or to walk or ride a bicycle were not associated with young girls' objectively assessed physical activity (Adkins et al., 2004), or with adolescents' self-reported walking or cycling in the neighborhood (Carver et al., 2005). The latter study did, however, show an inverse relationship between concerns about roaming dogs and cycling for recreation or walking for exercise (Carver et al., 2005). Motl and colleagues (2005) found no cross-sectional or longitudinal relationship between their measure of neighborhood safety and adolescents' physical activity.

Other social environment measures, such as social disorder, were found to be inversely associated with young people's recreational activity (Molnar et al., 2004). In contrast, a study of 4th-grade children in the United States revealed that children who perceived hazards in the neighborhood (e.g., litter, traffic, crime, noise) engaged in higher levels of overall physical activity (Romero et al., 2001). That study also assessed fitness (20 m shuttle run test), but no associations between perceived hazards and children's fitness were identified. When the same perceived hazards measure was used in a different study with a much smaller sample (n = 74), no association with frequency of vigorous exercise was reported (Romero, 2005).

Positive relationships between various aspects of social connectedness (e.g., child's frequency of interaction with peers, neighborhood relationships, other children living close by in the neighborhood) and child independent mobility and adolescent walking and cycling in the neighborhood were identified in Italian (Prezza et al., 2001) and Australian (Carver et al., 2005) studies, respectively. Further, one study indicated that children were less likely to actively commute to school if their parents perceived that there were few other children nearby (Timperio et al., 2006). In contrast, an Australian study showed no association between the number of opportunities for social interaction identified in environmental mind maps drawn by children and their objectively assessed physical activity (Hume et al., 2005). In that study, however, children were instructed to draw only "places and things important to them," not places where they could be physically active or interact with others. Seeing others being active in the neighborhood was unrelated to physical activity among adolescents (Mota et al., 2005).

In summary, evidence of the importance of the neighborhood social environment to youth physical activity appears to be unsupportive. Safety has been studied more often than other social constructs such as neighborhood hazards or social disorder, but the evidence is not strongly supportive for a relationship between any of these constructs and physical activity. Three of the five studies on social connectedness yielded significant associations, suggesting that the evidence is mixed. Thus, although relationships appear to differ depending on the physical activity outcome, the overall evidence of physical activity associations with the neighborhood social environment is not supportive.

Physical Activity and the Physical Environment

Sixteen cross-sectional studies were identified that explored some aspect of the physical environment (see table 16.2) and youth physical activity.

Spaces and Facilities

Perceived quality of built facilities and the presence of good places in the neighborhood in which to be active were positively associated with youth vigorous exercise (Romero, 2005) and cycling in the neighborhood (Carver et al., 2005). Perceptions of safety of facilities (e.g., the nearest park is safe during the day and

Table 16.2 Conceptual and Operational Definitions of the Neighborhood Physical Environment

Environmental construct	Operational definition
Quality of facilities	Youth report: quality of built facilities, equipment, opening hours (e.g., sports complexes, gymnasiums)[13] Parent report: good sport facilities in neighborhood; good places where child can be active[3]
Perception of safety of facilities	Parent report: child has no place to play but the street; closest park/playground is well kept; park safe during day; park safe at night[9] Youth report: facilities located in areas where feel safe; feel safe if have to walk to facilities[13]
Access and availability	Child, youth, or parent report: availability/accessibility of various places for sports or play[1,5,10,13,14,15] Youth report: equipment availability (combination of enough supplies and pieces of sport equipment at home and playgrounds; parks or gyms close to home that can get to easily)[11] Youth report: availability of take-away/fast food stores, convenience stores,[3] stores within walking distance[10] Recorded number of sports pitches/fields; investment in leisure facilities and open spaces per head of population within neighborhood[6] Parent report: availability of public transport[9,14] Youth report: ease of travel by bicycle[3] Youth report: ease of walking to transit stop[10] Children's neighborhood mind maps of important places: number of places to be physically active, number of parks, number of roads, number of food locations[7] Youth report: audit of convenience of facilities, availability of space (e.g., roads, sidewalks), perceived safety in neighborhood, and importance of environmental resources in neighborhood[4] Measured shortest possible route between school and home within walking distance (<800 m)[14] Measured shortest possible route to school includes steep hill[14] Observed walking/cycling ease[8]
Neighborhood design	**Street connectivity** Measured number of intersections per street mile within 0.5-mile radius of school[2] Measured route directness to school (shortest possible route between school and homedivided bycrow-fly distance)[14] Youth report: many four-way intersections[10] Observed street access and condition (score)[8] **Urban form** Parent report: availability of traffic lights/pedestrian crossings; need to cross roads to reach play areas[14,15] **Housing type** Youth report: sidewalks on most streets[10] Observed sidewalk characteristics[8] Recorded urban area: buildings with condominium courtyards; age of neighborhood; buildings with entrance on private streets; buildings adjacent to parks[12]

Environmental construct	Operational definition
Population density	Recorded number of residents within 0.5 miles of school[2]
Physical disorder	Videotape and observation of tagging graffiti; graffiti painted over; gang graffiti; political graffiti; cigarettes/cigars; empty beer bottles; abandoned cars; condoms; needles; syringes[9]
Road/traffic environment	Mother[12] report: level of danger regarding traffic Parent[14,15] and youth[3,14,15] report: concerns about road safety Parent and youth[14,15] report: heavy traffic in local streets Parent[3] or youth[10] report: so much traffic that it is difficult/unpleasant to walk Measured shortest possible route to school alongside a "busy" road; crosses a "busy" road[14]
Topography	Measured shortest possible route to school includes a steep hill[14]
Aesthetics	Youth report: many interesting things to look at while walking[10] Observed "tidiness"[8]

[1]Adkins et al. (2004); [2]Braza et al. (2004); [3]Carver et al. (2005); [4]Fein et al. (2004); [5]Fridlund Dunton et al. (2003); [6]Henning Brodersen et al. (2005); [7]Hume et al. (2005); [8]Jago et al. (2005); [9]Molnar et al. (2004); [10]Mota et al. (2005); [11]Motl et al. (2005); [12]Prezza et al. (2001); [13]Romero (2005); [14]Timperio et al. (2006); [15]Timperio et al. (2004).

is well kept) were also positively associated with youth physical activity (Molnar et al., 2004; Romero, 2005). Access to and availability of facilities was positively related to fitness in one study (Fridlund Dunton et al., 2003). Access and availability of facilities were positively related to child and adolescent physical activity in six studies (Henning Brodersen et al., 2005; Hume et al., 2005; Mota et al., 2005; Motl et al., 2005; Timperio et al., 2004, 2006), inversely related in one study (Carver et al., 2005), and unrelated in four studies (Adkins et al., 2004; Fein et al., 2004; Fridlund Dunton et al., 2003; Romero, 2005). The four studies that showed no relationships all incorporated total measures of physical activity (including accelerometry), and it may be that access is more likely to be related to context-specific types of physical activity such as organized sports or play activities. Another possible reason for the null finding in the Canadian study (Fein et al., 2004) may have been the use of a composite measure to assess availability of space that also included perceptions of safety. This may have diminished the ability to detect a relationship between the availability within the physical environment and physical activity. In one of few studies to include environmental constructs specific to the physical activity behavior of interest, Timperio and colleagues (2006) found a strong association between active commuting to school and a route to school shorter than 800 m (one-half mile).

A study of Australian adolescents showed that girls who reported having a convenience store near home walked less frequently in their neighborhood compared to those who did not (Carver et al., 2005). It was argued that convenience stores may not be destinations to which girls walk or cycle. However, this does not fully explain why adolescent girls were less likely to walk if the store was nearby. It may be that perceived access to specific destinations is an indicator of other aspects of the environment that are inversely related to walking, such as poor walkability or connectivity. Or perhaps girls who do not frequently walk in their neighborhood are less able to estimate accessibility to destinations with any accuracy as they are less aware of and have less knowledge about facilities and destinations in their neighborhood compared to those who frequently walk.

Neighborhood Design

Associations between aspects of neighborhood design (e.g., street connectivity, urban form, and housing type) and young people's physical activity were examined in six studies. An investigation of 34 elementary schools in the United States showed that the number of residents living within a 0.5-mile (about 0.8 km) radius of school was positively associated with the proportion of children walking or cycling to school (Braza et al., 2004). That study, however, also indicated that street connectivity was only weakly related to the proportion of children walking to those schools (Braza et al., 2004), but did not consider how far the children lived from school. Among adolescents, Mota and colleagues (2005) found that perceiving many four-way intersections in their area was unrelated to physical activity. Similarly, Jago and colleagues (2005) reported that scores for street access and condition were unrelated to adolescent boys' objectively measured physical activity.

In contrast to these investigations, a study that objectively examined the shortest possible route to school of each child in the sample showed that a more direct route (i.e., better connectivity) was associated with a *lower* likelihood of active commuting to school (Timperio et al., 2006). It may be that living in an area with good street connectivity is associated with higher levels of traffic exposure and thus parents do not allow their child to walk to school due to concerns about traffic safety. On the other hand, living in an area with poor connectivity, which has been operationalized as the number of cul de sacs or courts within an area (Timperio et al., 2006), may be important for other aspects of children's physical activity such as active free play. For example, a qualitative study of parents showed that families who lived in or near a court or cul de sac considered this to be a positive influence on their child's spending time outside playing (Veitch et al., 2006). Therefore, connectivity may relate quite differently to children's physical activity compared with that of adults. Further studies are needed to clarify relationships between connectivity and different types of children's physical activity (i.e., active transport, active free play), as well as underlying reasons for these relationships (e.g., parents' concerns about traffic safety).

In an Australian study, a perceived lack of traffic lights and pedestrian crossings, as well as a need to cross several roads to reach play areas, was negatively associated with children's walking and cycling in the neighborhood (Timperio et al., 2004). Evidence regarding the influence of sidewalks on youth physical activity is mixed; one study showed no relationship (Mota et al., 2005) while another showed that sidewalk characteristics were associated with light physical activity measured by accelerometry (Jago et al., 2005). Regarding housing type, a study of children in Italy indicated that living in a building with condominium courtyards, living in a new neighborhood, living in a building with an entrance onto private streets, and living in a building adjacent to parks were all positively associated with children's independent mobility (Prezza et al., 2001). This may be due to reduced exposure to traffic and easy access to open space for play.

Road and Traffic Environment

Associations between the road and traffic environment and children's physical activity have been assessed in five studies.

Four of the five studies assessed elements of traffic safety (Carver et al., 2005; Prezza et al., 2001; Timperio et al., 2004, 2006). An Italian study showed no association between perceived level of danger regarding traffic in the area and children's independent mobility (Prezza et al., 2001). This is consistent with findings reported by Timperio and colleagues, who found no association

© PA Archive/PA photos

Studies involving active transport address the link between traffic and youth physical activity.

between parents' and children's perceptions of road safety in their area and active transport in the neighborhood (2004) or walking or cycling to school (2006). In contrast, in a study of Australian adolescents, those who agreed that road safety was a concern were less likely to cycle or walk in their neighborhood (Carver et al., 2005). Although it is unexpected that three of the four studies showed no association between traffic safety and children's independent mobility or active transport, it is important to note that high proportions (~80%) of parents report being concerned about this issue (Timperio et al., 2004). It may be that the measures used in these studies are unable to differentiate between parents who are highly concerned about this issue, and therefore restrict these types of physical activity, and those who are highly concerned about safety but still allow their children to engage in these activities.

Four of the five studies assessed concerns about both traffic density and physical activity. Parental perceptions of heavy traffic were positively associated with 5- to 6-year-old boys' walking or cycling to destinations in their neighborhood in one study (Timperio et al., 2004). As that study was cross-sectional, it is not known whether the positive relationship was due to a higher level of awareness of traffic in the neighborhood among those whose children walk or cycle in their neighborhood compared to those whose children do not, or possibly due to heavy traffic as an indicator of a potentially important correlate of children's active transport such as urban form. However, in that study no such associations were found among older children, nor were there associations in further analyses between perceptions of heavy traffic and active transport to school (Timperio et al., 2006).

Carver and colleagues (2005) found an inverse association between the perception that there is "so much traffic in the neighborhood that it is difficult/unpleasant to walk" and adolescents' walking or cycling in the neighborhood; and Mota and colleagues (2005) found no relationship with physical activity. In addition, a study of 5- to 6- and 10- to 12-year-old youth showed that those whose shortest possible route to school crossed a "busy" road were less likely to walk or cycle to school than those whose route did not cross a busy road (Timperio et al., 2006). However, in that same study, a route alongside a "busy" road was unrelated to active transport to school. Thus, the evidence regarding the road and traffic environment is mixed, varying according to age of the child, how the road and traffic environment has been operationalized, and the type of activity being assessed.

Other Factors

One study identified an inverse relationship between physical disorder in the neighborhood and the time youth spend in active recreation (Molnar et al., 2004). The study also indicated an inverse relationship between social disorder and physical activity. These are novel measures, and few studies have assessed these aspects of the neighborhood in relation to physical activity among children and youth. Further research is required, however, to more clearly delineate social and physical disorder, and these definitions will likely vary between and possibly within countries.

Finally, few studies have dealt with the association of topography and aesthetics with young people's physical activity. In one study, a steep hill en route to school was associated with a lower likelihood of active travel to school among 5- to 6-year-old but not 10- to 12-year-old children (Timperio et al., 2006), suggesting that topography may be important for some groups. Two studies considered perceived aesthetics of the neighborhood. In an investigation by Mota and colleagues (2005), adolescents who perceived that there were many interesting things to look at while walking in their neighborhood were 30% more likely to be classified as "active." In contrast, Jago and colleagues (2005) found no relationship between observed "tidiness" and light activity or MVPA.

In summary, most available evidence on the relationship between the neighborhood physical environment and children's physical activity relates to access and availability and to neighborhood design. Studies that showed null associations with children's physical activity generally employed measures of overall activity, such as accelerometry, rather than context-specific measures such as walking or organized sport. While road safety is one of the most frequently cited reasons parents give for not allowing their child to walk or cycle in the neighborhood, few researchers have looked at this issue, and the existing evidence is mixed. A deeper understanding of the aspects of the neighborhood environment that influence youth physical activity is required.

Interventions

Interventions to reduce children's sedentary behavior have incorporated environmental change in the home through the use of devices such as the TV Allowance monitor, which is attached to the television set and limits the number of hours each day the television can be watched (Ford et al., 2002; Robinson, 1999). In combination with other strategies, this approach was successful in reducing children's television viewing (Ford et al., 2002; Robinson, 1999) and preventing weight gain (Robinson, 1999) compared with findings for children in the control group. Other environmental strategies in the home that are yet to be tested in an intervention, but may be effective given the observational associations with sedentary behavior (Saelens et al., 2002) and low levels of physical activity (Salmon et al., 2005), include removing the television set from children's bedrooms, switching off the television during mealtimes, and removing electronic games from children's bedrooms (particularly for boys).

In a successful neighborhood-level intervention, Boarnet and colleagues found that changes in urban form (e.g., improvements to sidewalks and bicycle paths; availability of, or improvements to, traffic lights/pedestrian crossings) were positively associated with proxy-reported changes in walking/cycling to school among California children (Boarnet et al., 2005). Of the 10 schools included in the evaluation, the majority were within one-fourth mile (about 400 m) of the environmental improvements. However, conclusions about the effectiveness of the intervention are limited by a cross-sectional evaluation after completion of the project that used retrospective measures, as well as the lack of a control or comparison condition (analysis of the sample involved comparing those who would pass the environmental improvements on their usual route to and from school with those who would not).

Implications for Research and Practice

It is apparent from the literature reviewed that much of the current evidence for relationships of the social and physical environment with youth physical activity and sedentary behavior is inconsistent and therefore preliminary.

Several factors may help explain the lack of consistency in the results, including the fact that many measures were not well conceptualized. Global measures of safety, for example perceptions of whether it is "safe to play outside" (Adkins et al., 2004) or whether the neighborhood is a "safe place to walk/ride bike" (Carver et al., 2005), were unrelated to physical activity, possibly because of a lack of construct specificity on particular aspects of safety (e.g., traffic safety or fear of harm from strangers). Accessibility and availability of facilities for physical activity have also been operationalized in different ways, from investment in leisure facilities to number of sport venues within the neighborhood. Thus, much more work is required to conceptualize which aspects of the social and physical environment may be relevant, as well as to define, operationalize, and develop reliable scales to assess these constructs.

The small number of studies on the influence of the home environment suggest that the presence of physical activity equipment in the home may be important for children's activity levels, and that aspects of the sedentary home environment, such as having a television set in the bedroom, may be related to higher levels of television viewing. Studies of neighborhood social and physical environments and physical activity indicate that neighborhood social safety, social connectedness, quality of facilities, and neighborhood design may be important constructs. But the findings are so mixed that any conclusions drawn are equivocal and can be only preliminary at best.

Research

It is clear, then, that more and better-conceived and better-conducted research is needed before reliable conclusions can be reached. Next we present a number of suggestions that we believe would improve the reliability of research results if they were to be implemented.

We propose that future studies apply theoretical frameworks such as the SCT (Bandura, 1986), BCT (Epstein, 1998; Rachlin, 1989), Youth Physical

Activity Promotion Model (Welk, 1999), and EST (Bronfenbrenner, 1979) to investigate the direct and indirect effect of the social and physical neighborhood environment on young people's physical activity. Adopting such models would ensure that the investigation of constructs is not random but instead based on established criteria. As King and colleagues (2002) suggest, by applying such models we should gain a better understanding of the spatial factors that influence personal choice and activity-driven choice at all levels of the environment.

While environment is an important construct in each of the models discussed, none of the models offer good descriptions of the environment. Thus, we suggest incorporating environment-specific frameworks such as those used in the following studies:

- Moudon and Lee's model (2003) operationalizes environmental variables into three classes: spatiophysical (physical aspects of environment), spatiobehavioral (types and intensity of human uses), and spatiopsychosocial (people's internal response to environment).
- Pikora and colleagues (2003) identify four features of the local environment that may influence walking and cycling (functional features, safety, aesthetic features, and destinations); the elements that compose each feature; and specific items that can be directly measured or modified to improve each element.
- Krizek, Birnbaum, and Levinson (2004) focus on community design and its influences on the time youth spend in different activities. Specifically, destination locations and types of activities, as well as travel between destinations, are highlighted as the key elements that influence active (e.g., walking, cycling) or sedentary (e.g., driving) time spent by youth.

The models proposed by Moudon and Lee and by Pikora and colleagues were originally used to identify and structure the types of variables to be examined to assess the physical environment for bicycling and walking, and could be adapted for children's living environments. The usefulness of Krizek and colleagues' schematic is suggested by the following example: The total number of daily trips taken by U.S. children has increased since 1977 (Sturm, 2005b) and children are more likely to be involved in organized activities such as sport (Sturm, 2005a), but free time has substantially declined and unstructured playtime has decreased (Sturm, 2005a). Thus, it is possible that the commuting time and preparation associated with participation in organized sport have resulted in less opportunity for children to engage in active play. Krizek and colleagues' schematic could assist with identifying areas of interventions to increase active time among young people.

Another useful suggestion for researchers is that of Giles-Corti and colleagues (2005), who have argued that behavior-specific measures of the environment should be used to predict context-specific behaviors. Such behaviors are those that take place within the context under study, such as active free play, or walking or cycling *within the neighborhood.* Such activities are more likely to be related to social and physical environmental factors within that context (i.e., the neighborhood) than activities that take place in multiple settings and contexts (e.g., global physical activity, which may take place at school and within the home as well as in the neighborhood). Conceptually, aspects of the environment that are relevant to these different behaviors may also be very different (e.g., access to sport facilities may be important for participation in organized sport, while access to age-appropriate playground equipment might be most relevant to active play), and these relationships may vary depending on the child's age and gender.

The findings of our review show that a multitude of outcome measures are employed in varying environments, possibly explaining the conflicting relationships observed between aspects of the social and physical environment and different physical activity outcomes. Further, studies of the home sedentary environment have focused primarily on television viewing, which is just one sedentary behavior that occurs in the home. Influences in the home environment on other sedentary behaviors that may be less developmentally valuable than reading or homework (e.g., Internet or computer use, electronic games use) need to be identified. Future researchers seeking to examine the influence of the environment on youth physical activity need to consider these important distinctions. In addition, a focus on context-specific behaviors for various age-groups, analyzed from a temporal perspective, may elucidate which factors are most influential at various developmental stages in children. This would shed light on how to modify these factors to reduce time spent in sedentary behavior and promote a physically active lifestyle from the early stages of development.

Given that existing studies often revealed important differences in parent and child perceptions of heavy traffic, road safety, and strangers, it is also important to consider how perceptions of the social and neighborhood environments are assessed. These differences in perceptions may vary, and may have more or less of an influence, depending on the age of the child or the child's level of independence. Ideally, perceptions of the environment should be measured for both the parent and child and compared with objective measures of the environment. Future researchers could employ Geographic Information Systems (GIS) to facilitate greater precision in studying relationships of the perceived and objective environment with youth physical activity; this would ensure that consistent definitions of the "environment" are applied for all participants and neighborhoods.

Practice

As evidenced in our review, physical activity equipment in the home, social connectedness, access to facilities, and quality and safety of facilities have been identified as factors that show stronger associations with youth physical activity than other environmental factors. It is important to keep in mind, however, that much of the evidence for associations between the social and physical environment and youth physical activity is cross-sectional; very few environmental interventions promoting physical activity in youth have been published. Therefore, associations based on cross-sectional evidence may not be in the anticipated direction. For example, those who are already physically active may be more likely to have exercise equipment in the home. Although no definite conclusions can be drawn from these associations, making practical recommendations with regard to these findings is still worthwhile. In addition, well-designed interventions can be based on these preliminary findings and continue to inform policy and practice at the local and national levels.

Community initiatives to reduce physical and social disorder and to strengthen neighborhood ties, such as after-school programs, may provide more physical activity opportunities for children. To increase the feeling of security and safety (e.g., personal safety, traffic safety) and to decrease young people's exposure to various forms of risk in their neighborhoods, community upgrading programs need to be implemented. For example, "Safewalk" and "Neighborhood Watch" initiatives could be linked to programs that ensure lower fences and better street lighting to increase visibility or that ensure safer crosswalks and sidewalks (better signage, monitoring teams).

Creating environmental support for physical activity that particularly influences play, sport, and active transportation in neighborhoods should be a priority. Design issues, such as the quality and aesthetics of the built neighborhood environment, population density, mixed land use in neighborhoods, the quality of building and urban furniture materials, and the quality of cycling infrastructure should be a focus. Similarly, ensuring equity of facility distribution, periodic maintenance and upgrading of facilities, and increased access to facilities for physical activity is essential to promoting active lifestyles of youth within communities.

Parents should consider the potential effects of the home environment on their children's active or sedentary choices. If there are many sedentary opportunities within the home (e.g., numerous television sets, a television set in the child's bedroom, Internet and computer access, pay television, electronic games consoles), children may be more inclined to make sedentary rather than active choices; this is particularly important during daylight hours. The importance of providing children with access to physical activity equipment at home, which does not need to be expensive (e.g., balls, bats, skipping ropes), is an important message that can be conveyed to parents. For those who live in disadvantaged areas with a poor physical activity neighborhood environment, access to physical activity alternatives at home may be even more important for young people's activity levels.

In summary, partnerships between families, practitioners, and communities need to be developed to help incorporate physical activity into the daily routines of young people. Comprehensive programs and policies targeting family, school, and neighborhood, as well as broader cultural and societal factors, may help build physically friendly communities for children and adolescents.

APPLICATIONS FOR RESEARCHERS

Neighborhood social safety, social connectedness, access to facilities, and design are related to youth physical activity, though the strength of these findings is mixed. It is important to design theoretically based multilevel studies that incorporate individual, social, and physical environment factors. Additionally, future research requires better conceptualization of aspects of the social and physical environment within young people's homes and neighborhoods that are relevant to their physical activity and sedentary behavior; development of reliable scales and measurement items; the study of context-specific behaviors (i.e., occurring within particular environments such as the neighborhood or home); and behavior-specific features of the environment (i.e., aspects relevant to particular physical activity behaviors such as walking or organized sport). Future research also requires prospective studies to establish temporal relationships between aspects of the environment and youth physical activity, as well as new theoretical models that better describe environmental influences on youth physical activity and sedentary behavior.

APPLICATIONS FOR PROFESSIONALS

Evidence for environmental influences on youth physical activity is unclear; however, it appears that neighborhood social safety, social connectedness, access to facilities, and design are key features for practitioners to consider. Although physical activity and sedentary behaviors can coexist, parents should be encouraged to provide children with a home environment that supports active rather than sedentary choices, particularly in daylight hours after school or on weekends. Recommendations include devising programs and initiatives that are theoretically informed and that involve joint consideration of social and physical environments. Programs focusing on specific behaviors and specific age-groups in different settings are also needed. Professionals from multiple fields (e.g., health, sport, urban planning, transport, justice) who may influence physical activity opportunities need to create collaborative frameworks that facilitate cross-disciplinary efforts targeting youth physical activity promotion. Finally, better dissemination of best practice is necessary, as well as better monitoring of physical activity programs in natural settings.

References

Adkins, S., Sherwood, N.E., Story, M., & Davis, M. (2004). Physical activity among African-American girls: The role of parents and the home environment. *Obesity Research, 12*, 38-45.

Bandura, A. (1986). *Social foundations of thought and action: A social cognitive theory*. Englewood Cliffs, NJ: Prentice Hall.

Biddle, S.J.H., Gorely, T., Marshall, S.J., Murdey, I., & Cameron, N. (2004). Physical activity and sedentary behaviours in youth: Issues and controversies. *Journal of the Royal Society for the Promotion of Health, 124*, 29-33.

Boarnet, M.G., Anderson, C.L., Day, K., McMillan, T., & Alfonzo, M. (2005). Evaluation of the California Safe Routes to School legislation: Urban form changes and children's active transportation to school. *American Journal of Preventive Medicine, 28*, 134-140.

Braza, M., Shoemaker, W., & Seeley, A. (2004). Neighborhood design and rates of walking and biking to elementary school in 34 California communities. *American Journal of Health Promotion, 19*, 128-136.

Bronfenbrenner, U. (1979). *The ecology of human development*. Cambridge, MA: Harvard University Press.

Carver, A., Salmon, J., Campbell, K., Baur, L., Garnett, S., & Crawford, D. (2005). How do perceptions of local neighborhood relate to adolescents' walking and cycling? *American Journal of Health Promotion, 20*, 139-147.

Davison, K.K., & Birch, L.L. (2001). Childhood overweight: A contextual model and recommendations for future research. *Obesity Reviews, 2*, 159-171.

Dennison, B.A., Erb, T.A., & Jenkins, P.L. (2002). Television viewing and television in bedroom associated with overweight risk among low-income preschool children. *Pediatrics, 109*, 1028-1035.

Epstein, L. (1998). Integrating theoretical approaches to promote physical activity. *American Journal of Preventive Medicine, 15*, 257-265.

Fein, A.J., Plotnikoff, R.C., Wild, T.C., & Spence, J.C. (2004). Perceived environment and physical activity in youth. *International Journal of Behavioral Medicine, 11*, 135-142.

Ford, B.S., McDonald, T.E., Owens, A.S., & Robinson, T.N. (2002). Primary care interventions to reduce television viewing in African-American children. *American Journal of Preventive Medicine, 22*, 106-109.

Fridlund Dunton, G., Schneider Jamner, M., & Cooper, D.M. (2003). Assessing the perceived environment among minimally active adolescent girls: Validity and relations to physical activity outcomes. *American Journal of Health Promotion, 18*, 70-73.

Giles-Corti, B., Timperio, A., Bull, F., & Pikora, T. (2005). Understanding physical activity environmental correlates: Increased specificity for ecological models. *Exercise and Sports Science Reviews, 33*, 175-181.

Gomez, J.E., Johnson, B.A., Selva, M., & Sallis, J.F. (2004). Violent crime and outdoor physical activity among inner city youth. *Preventive Medicine, 39*, 876-881.

Gordon-Larsen, P., McMurray, R.G., & Popkin, B.M. (2000). Determinants of adolescent physical activity and inactivity patterns. *Pediatrics, 105*, 1-8.

Henning Brodersen, N., Steptoe, A., Williamson, S., & Wardle, J. (2005). Sociodemographic, developmental, environmental, and psychological correlates of physical activity and sedentary behavior at age 11 to 12. *Annals of Behavioral Medicine, 29*, 2-11.

Hume, C., Salmon, J., & Ball, K. (2005). Children's perceptions of their home and neighborhood environments, and their association with objectively measured physical activity: A qualitative and quantitative study. *Health Education Research, 20*, 1-13.

Humpel, N., Owen, N., & Leslie, E. (2002). Environmental factors associated with adults' participation in physical activity. A review. *American Journal of Preventive Medicine, 22*, 188-199.

Jago, R., Baranowski, T., Zakeri, I., & Harris, M. (2005). Observed environmental features and the physical activity of adolescent males. *American Journal of Preventive Medicine, 29*, 98-104.

King, A.C., Stokols, D., Talen, E., Brassington, G.S., & Killingsworth, R. (2002). Theoretical approaches to the promotion of physical activity: Forging a transdisciplinary paradigm. *American Journal of Preventive Medicine, 23*, S15-S25.

Krizek, K.J., Birnbaum, A.S., & Levinson, D.M. (2004). A schematic for focusing on youth in investigations of community design and physical activity. *American Journal of Health Promotion, 19*, 33-38.

Marshall, S.J., Biddle, S.J.H., Gorely, T., Cameron, N., & Murdey, I. (2004). Relationships between media use, body fatness and physical activity in children and youth: A meta-analysis. *International Journal of Obesity, 28*, 1238-1246.

Marshall, S.J., Biddle, S.J.H., Sallis, J.F., McKenzie, T.L., & Conway, T.L. (2002). Clustering of sedentary behaviours and physical activity among youth: A cross-national study. *Pediatric Exercise Science, 14*, 401-417.

McLeroy, K.R., Bibeau, D., Steckler, A., & Glanz, K. (1988). An ecological perspective on health promotion programs. *Health Education Quarterly, 15*, 351-377.

Molnar, B.E., Gortmaker, S.L., Bull, F.C., & Buka, S.L. (2004). Unsafe to play? Neighborhood disorder and lack of safety predict reduced physical activity among urban children and adolescents. *American Journal of Health Promotion, 18*, 378-386.

Mota, J., Almeida, M., Santos, P., & Ribeiro, J.C. (2005). Perceived neighborhood environments and physical activity in adolescents. *Preventive Medicine, 41*, 834-836.

Motl, R.W., Dishman, R.K., Ward, D.S., Saunders, R.P., Dowda, M., Felton, G., et al. (2005). Perceived physical environment and physical activity across one year among adolescent girls: Self-efficacy as a possible mediator? *Journal of Adolescent Health, 37*, 403-408.

Moudon, A.V., & Lee, C. (2003). Walking and bicycling: An evaluation of environmental audit instruments. *American Journal of Health Promotion, 18*, 21-37.

Owen, N., Humpel, N., Salmon, J., & Oja, P. (2004). Environmental influences on physical activity. In P. Oja & J. Borms (Eds.), *Health enhancing physical activity. Perspectives—the multidisciplinary series of physical education and sport science* (Vol. 6, pp. 393-426). Oxford: Meyer and Meyer Sport.

Pikora, T., Giles-Corti, B., Bull, F., Jamrozik, K., & Donovan, R. (2003). Developing a framework for assessment of the environmental determinants of walking and cycling. *Social Science and Medicine, 56*, 1693-1703.

Prezza, M., Pilloni, S., Morabito, C., Sersante, C., Alparone, F.R., & Giuliani, M.V. (2001). The influence of psychosocial and environmental factors on children's independent mobility and relationship to peer frequentation. *Journal of Community and Applied Social Psychology, 11*, 435-450.

Rachlin, H. (1989). *Judgement, decision, and choice: A cognitive/behavioral synthesis.* New York: Freeman.

Robinson, T. (1999). Reducing children's television viewing to prevent obesity: A randomized controlled trial. *Journal of the American Medical Association, 282*, 1561-1566.

Romero, A.J. (2005). Low-income neighbourhood barriers and resources for adolescents' physical activity. *Journal of Adolescent Health, 36*, 253-259.

Romero, A.J., Robinson, T.N., Kraemer, H.C., Erickson, S.J., Haydel, F., Mendoza, F., et al. (2001). Are perceived neighborhood hazards a barrier to physical activity in children? *Archives of Pediatrics and Adolescent Medicine, 155*, 1143-1148.

Saelens, B., & Epstein, L. (1998). Behavioural engineering of activity choice in obese children. *International Journal of Obesity, 22*, 275-277.

Saelens, B.E., Sallis, J.F., & Frank, L.D. (2003). Environmental correlates of walking and cycling: Findings from the transportation, urban design, and planning literatures. *Annals of Behavioral Medicine, 25*, 80-91.

Saelens, B., Sallis, J., Nader, P., Broyles, S., Berry, C., & Taras, H. (2002). Home environmental influences on children's television watching from early to middle childhood. *Journal of Developmental and Behavioural Pediatrics, 23*, 127-132.

Sallis, J.F., & Owen, N. (1997). Ecological models. In K. Glanz (ed.), *Health behavior and health education: Theory, research and practice* (2nd ed., pp. 403-424). San Francisco: Jossey-Bass.

Sallis, J.F., Prochaska, J.J., & Taylor, W.C. (2000). A review of correlates of physical activity of children and adolescents. *Medicine and Science in Sports and Exercise, 32*, 963-975.

Salmon, J., Timperio, A., Telford, A., Carver, A., & Crawford, D. (2005). Family environment and children's television viewing and low level physical activity. *Obesity Research, 13*, 1939-1951.

Spence, J.C., Burgess, J.A., Cutumisu, N., Lee, J.G., Moylan, B., Taylor, L., et al. (2006). Self-efficacy and physical activity: A quantitative review. *Journal of Sport and Exercise Psychology, 28*, S172-S173. [Abstract]

Spence, J.C., & Lee, R.E. (2003). Toward a comprehensive model of physical activity. *Psychology of Sport and Exercise, 4*, 7-24.

Sturm, R. (2005a). Childhood obesity—what we can learn from existing data on societal trends, part 1. *Preventing Chronic Disease, 2*(1), A12.

Sturm, R. (2005b). Childhood obesity—what we can learn from existing data on societal trends, part 2. *Preventing Chronic Disease, 2*(2), A20.

Timperio, A., Ball, K., Salmon, J., Roberts, R., Giles-Corti, B., Simmons, D., et al. (2006). Personal, family, social and environmental correlates of active commuting to school. *American Journal of Preventive Medicine, 30*, 45-51.

Timperio, A., Crawford, D., Telford, A., & Salmon, J. (2004). Perceptions about the local neighborhood and walking and cycling among children. *Preventive Medicine, 38*, 39-47.

Trost, S.G., Kerr, L.M., Ward, D.S., & Pate, R.R. (2001). Physical activity and determinants of physical activity in obese and non-obese children. *International Journal of Obesity*, *25*, 822-829.

Veitch, J., Ball, K., Robinson, S., & Salmon, J. (2006). Where do children play? A qualitative study of parents' perceptions of influences on children's active free-play. *Health and Place*, *12*, 383-393.

Vuchinich, R., & Tucker, J. (1988). Contributions from behavioural theories of choice to an analysis of alcohol abuse. *Journal of Abnormal Psychology*, *97*, 181-195.

Welk, G.J. (1999). The youth physical activity promotion model: A conceptual bridge between theory and practice. *Quest*, *51*, 5-23.

Welk, G.J., Wood, K., & Morss, G. (2003). Parental influences on physical activity in children: An exploration of potential mechanisms. *Pediatric Exercise Science*, *15*, 19-33.

Wiecha, J.L., Sobol, A.M., Peterson, K.E., & Gortmaker, S.L. (2001). Household television access: Associations with screen time, reading, and homework among youth. *Ambulatory Pediatrics*, *1*, 244-251.

CHAPTER

17

Economic Principles

Chad D. Meyerhoefer, PhD

When people think of economic issues related to physical activity, the costs faced by those who engage in various sports or activities often come to mind: the parent's fuel cost for travel soccer league, for example, or the cost of skiing or dance lessons. People also commonly mention the cost of infrastructure necessary to support physical activity, such as playing fields, stadiums, or public parks. While issues of cost play an important role in economic analysis, the economic framework can be applied to a broad array of decisions related to physical activity. The reason is that economics is, fundamentally, the study of how people behave in an environment where they must allocate scarce resources in order to maximize their well-being.

In economics, people's well-being is measured through the pleasure they derive from the consumption of goods and services, including the ability to participate in physical activities or sedentary forms of leisure. Thus economists attempt to describe how physical activity levels are determined in the context of all other consumption decisions while taking into account the monetary and nonmonetary resources available to the decision maker. Arguably the most important resource required for individuals to engage in physical activity is time, and this is also a resource with many competing demands. Therefore, time allocation decisions by both parents and their children play a key role in the analysis of youth physical activity levels.

Like many other academic disciplines, economics has a long history with a rich theoretical basis and many analytical and empirical findings. A thorough discussion of the theoretical axioms and mathematical constructs underlying the economic principles of physical activity is beyond the scope of this chapter. Instead, I present a general economic framework of individual physical activity

The author thanks John Cawley for helpful comments and suggestions on earlier versoins of this chapter.

The views expressed in this chapter are those of the author, and no official endorsement by the Agency for Healthcare Research and Quality, or the Department of Health and Human Services is intended or should be inferred.

participation decisions and describe their place in the context of the broader economy. Under ideal conditions, the unconstrained behavior of self-interested individuals has some very desirable implications for social welfare. However, departures from these conditions result in cases in which governments and other institutions can improve economic efficiency through their ability to influence people's opportunities to engage in physical activity. Consequently, the economic justification and analysis of public policies related to youth physical activity are taken up in the second part of the chapter.

Individual Choices and the Market Economy

Markets are nothing more than the collective representation of individual behavior as it relates to the production and consumption of various goods, services, and available resources. In this section I introduce a model of individual behavior that incorporates physical activity decisions and show how it relates to the demand and supply of goods and individuals' time at the market level. Because markets are a reflection of individual behavior, market-based economies have some very desirable properties that I describe before moving on to the ways in which markets can fail.

The SLOTH Model

A particularly useful framework for the analysis of physical activity decisions, developed by John Cawley (2004), is called the SLOTH model. Like all economic models of individual choice, it assumes that individuals' well-being or "utility" is generated through the consumption of various goods and services and the use of available resources. The process by which these factors affect utility is displayed diagrammatically in figure 17.1. The main inputs to utility shown at the bottom of the diagram are food, the individual's time, and all other goods and services.

The model gets its name from the exhaustive list of different ways individuals can spend their time. These mutually exclusive categories include sleeping (S); participation in various (active and sedentary) leisure activities (L); time at one's occupation or time spent in school (O); time spent in transportation (T); and time devoted to household chores, meal preparation, and similar productive activities (H). Time spent in various activities and the consumption of goods and services has a direct and differential effect on a person's well-being (as denoted by the solid arrows) as well as an indirect effect (denoted by the dashed arrows) through the production of health and body weight. For example, a child derives utility not only from the consumption of tasty and nutritious foods, but also from the feeling of good health that this consumption engenders.

By allocating different amounts of time to various activities and choosing different levels of goods consumption, individuals are able to maximize their well-being within the confines of several resource constraints. First among these is the requirement that people (or families) not spend more than their total

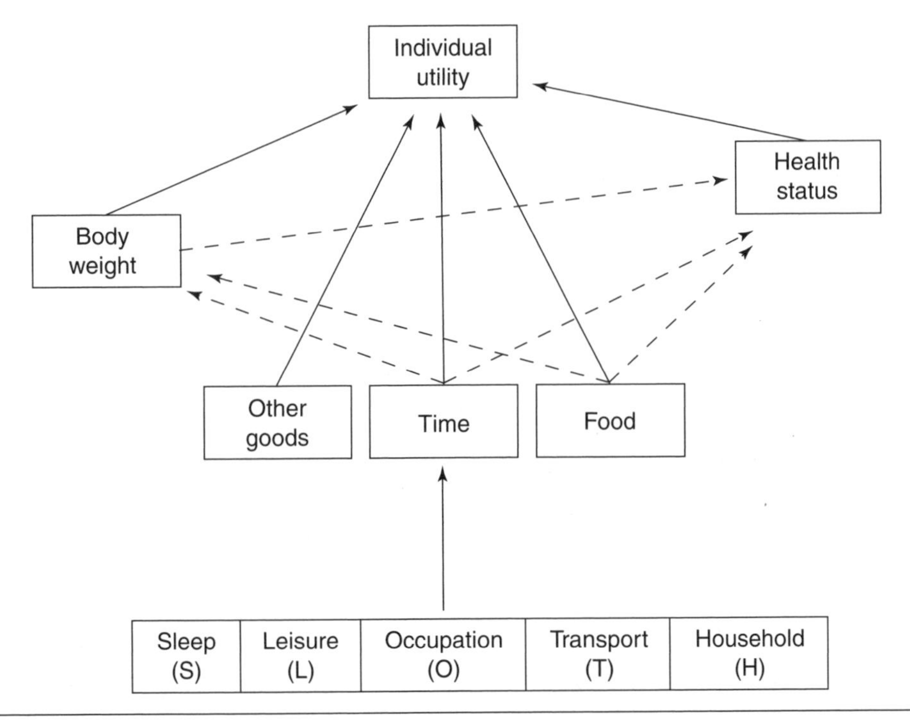

Figure 17.1 The SLOTH model.

labor market earnings, or the "budget constraint." A second constraint reflects the fact that time allocated to sleeping, leisure, occupation or education, travel, and home production cannot exceed the 24 h in a day, while the final constraint defines the biological process governing changes in body weight through time allocation and food consumption decisions.

Of course, individuals make physical activity and time allocation decisions over the course of a lifetime and not just during a single day. The SLOTH model can be modified to incorporate dynamic behaviors across multiple time periods by assuming that individuals maximize their current and future utility subject to a recurring set of constraints in every time period. Future levels of utility are given less weight, or discounted, to reflect individuals' preferences for immediate gratification over future pleasures. Likewise, economic resources in the model's constraints must be appropriately discounted to reflect their value in the current period. The first constraint would ensure consistency between the present discounted value of lifetime consumption and lifetime labor market earnings, while the second and third constraints account for time allocation and weight changes in all periods.

Those who highly discount future utility are said to be myopic in their consumption and time allocation decisions, while those who give it greater weight are more forward looking. When it comes to investments in health, discount rates can vary widely, and children and young adults tend to have much larger discount rates than older adults (Chapman, 1996; Chapman & Elstein, 1995).

The SLOTH model presupposes rationality on the part of the decision maker, which can be problematic when one is modeling the physical activity decisions of children and adolescents. Economists generally define rational actions as those that are in the current and long-term interests of the individual. Often parents intervene in their children's decisions to keep them from adopting habits that threaten their health or doing things that are socially unacceptable. If one assumes that parents have complete control over their children's activities, one can simply include the children's time allocation and goods consumption variables in the parental utility maximization model and proceed as usual. This may be realistic for very young or preschool-age children; but as children mature, parental control tends to diminish. One way to model parental influence is to include an additional constraint in the child's utility maximization framework, specifying that parental utility (defined over child and parental time allocations) cannot fall below a certain level. Over time, children allow parental utility to drop to lower and lower levels as parental approval becomes less important. Mathematically, the constraint on parental utility (U_p) can be expressed as

$$U_p \geq (1 + \theta)^{-a} U_0,$$

where U_0 is the base level of utility that is scaled down to a greater extent as the age of the child increases. θ is then the child's discount rate of parental approval, while $a = 1, \ldots, 18$ indexes the child's age.

Interestingly, the debate on rationality is not confined to decisions made by children or special subpopulations such as persons with mental disabilities. Adults have also been shown to exhibit time-inconsistent or irrational choices in cases of food (Cutler, Glaeser, & Shapiro, 2003) and tobacco consumption (Gruber & Mullainathan, 2002). Although the SLOTH model requires certain simplifications and assumptions, the insights into child and parental physical activity decisions gained through use of the model are applicable across a broad range of different behaviors.

Constrained Optimization of the SLOTH Model

The SLOTH model provides a general framework for understanding how adults and their children make decisions about the time they devote to physical and sedentary activities and the goods they consume. Irrespective of the exact mathematical specification given to each function, the same basic principles govern how all specific forms of the model are solved. In particular, children and adolescents will divide their available time across all possible activities in a utility-maximizing manner, subject to fixed resources and the approval of their parents. Their utility will be at its highest level only if the contribution to utility from the last hour spent in each activity is the same across all activities. Cawley (2004) refers to this as the "last hour rule."

To see why the last hour rule must be satisfied, suppose that a child obtained more utility from the last hour spent walking the dog than from playing basketball with friends. In this case, the child could make him- or herself better off by shifting an hour of the basketball time to dog walking, up to the point where the increments in utility obtained from the two activities were identical. By implication, if something in the environment changes to alter the benefit a child receives from a given activity, the child will adjust his or her allocation of time across all activities until the last hour rule is satisfied. Sturm (2004), for example, conjectures that improvements in technology related to sedentary forms of leisure (cable TV, video games, DVDs) may have increased the utility individuals get from these activities relative to more active forms of leisure (sport, exercise) or household chores, leading to a time reallocation and increase in the average weight of children and adolescents. Likewise, Frank (2004) demonstrates that changes in urban planning and transportation investments have increased the benefit people get from driving relative to walking or biking, making it less likely that parents will walk with their children to events rather than shuttle them by car.

Analogous to the last hour rule, economic theory also implies a "last dollar rule" whereby consumers purchase goods and services at levels that equate the increase in the utility derived from the consumption of an additional unit of each good to the last dollar spent on that good. In the parlance of economics, consumers are said to set the *marginal utility* of consumption equal to the *marginal cost* of the purchase. Therefore, if something alters the attributes or price of a good relative to other goods, consumers will reallocate expenditures across all goods and services. Cutler and colleagues (2003) claim that a relative decrease in the preparatory-time costs of calorie-dense mass-produced foods led to a shift in consumption patterns in their favor, fueling the recent obesity epidemic in the United States. In any case, by allocating their time and money in this self-interested manner, individuals settle upon optimal consumption or "demand" levels for various activities (S^*, L^*, O^*, T^*, H^*) as well as food and other goods and services. (Note that asterisks denote optimal, or equilibrium, levels of demand and prices.)

The sum of individual demands for every time allocation or commodity input in the SLOTH model generates an empirical relationship between the price of the input and its quantity of consumption. In the case of time allocations, the "price" is a measure of the individual's value of time, which for adults or adolescents of working age is generally set equal to the market wage rate (either their actual wage or the wage they would receive upon entering the labor market). Few, if any, economists have attempted to rigorously quantify the value of time for younger children, but it would presumably be based on the discounted present value of their investments in education. This sum of individual demands relating price to quantity across individuals is called a market demand curve. Market demand curves for each time and commodity input in the SLOTH model are decreasing in price, reflecting the principle that people will demand more of a good as its price decreases.

Supply Response and Market Equilibrium

Through the utility maximization process, consumers generate demand for a multitude of goods and services. Those related to physical activity include things like pool and gym memberships, athletic equipment, and sports drinks, as well as shared goods such as playing fields, sidewalks, and clean air. Private enterprise supplies many (but not all) of these goods using a profit-oriented management structure. Like consumers, private firms act to increase their well-being through the optimization of a profit function. This is simply the sale price of a unit of produced output times the number of units sold minus the price of inputs used to manufacture the product (or provide the service) times the level of inputs required. The producer's constraints are technical ones that specify which combinations of inputs can be used to generate the desired output.

Producers follow their own last dollar rule, requiring that the last dollar spent on each input, including labor, must yield an identical increase in the production of output. Otherwise, the mix of inputs could be altered and a larger quantity of output produced at the same input cost. The maximization of profits in this way leads to an optimal level of supply produced by the firm, which typically occurs at a point where the cost of producing an additional unit of the good is increasing. Or, to put it another way, since most production processes require an up-front investment in technologies whose cost is spread over all units produced, the cost of producing the first unit of a good is higher than for subsequent units. Over time, firms will expand production to the point where technological efficiencies are fully exploited and the cost per additional unit begins to rise. This is referred to as "decreasing returns to scale." As a result, the firm must charge more for each unit to justify the higher production level, implying that the relationship between quantity produced and output price is increasing. Analogous to consumer demand, individual supply curves are aggregated across firms to define the market supply curve for a product. The intersection of a market supply and market demand curve, as shown in figure 17.2, is what determines a stable equilibrium level of output, Q*, and price, P*. At this level of output the price consumers pay for the good is equal to the supplier's cost of producing one additional unit of the good, or the *marginal cost*.

When markets are competitive in the sense that there are a large number of consumers and producers and certain other assumptions are met, the determination of equilibrium levels of output and price occurs simultaneously in every product market via the process just described. The assumptions that must be met include perfect and complete information on the part of consumers and producers, product homogeneity such that goods made or services offered by different producers are indistinguishable, equal access to technologies by firms, and the ability of firms to freely enter and exit all markets. In addition to markets for manufactured goods, there are competitive markets for service provision and for people's time (e.g., the labor market). Two key results from economic theory, called the *first and second fundamental theorems of welfare economics*, describe the desirable properties of these markets.

The first theorem states that the allocation of resources (e.g., time, money, raw materials) achieved through competitive markets has the following characteristic: There is no alternative combination of resources such that someone can be made better off without making someone else worse off. This property is called Pareto-efficiency or Pareto-optimality. Pareto-efficiency is the manifestation of Adam Smith's (1937) "invisible hand," whereby competition and the pursuit of selfish interests lead individuals like an invisible hand to do what is best for society as a whole. However, the Pareto-efficiency of an economy doesn't imply anything about the distribution of resources across individuals, which could be very equal or highly unequal. To see this graphically, suppose that the economy has only two people and let the curve in figure 17.3, known

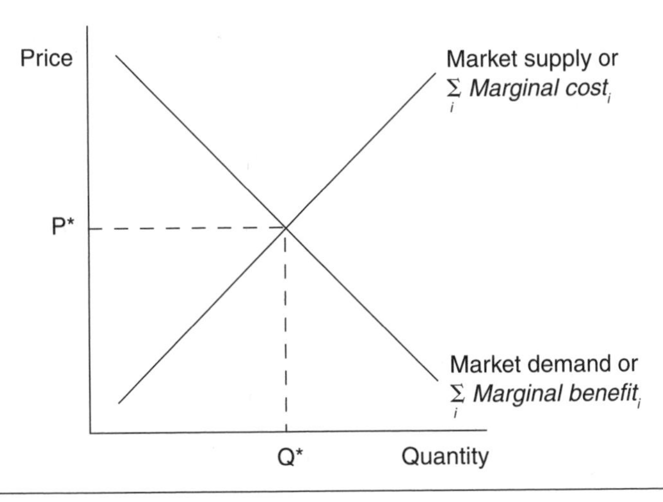

Figure 17.2 Market supply and demand.

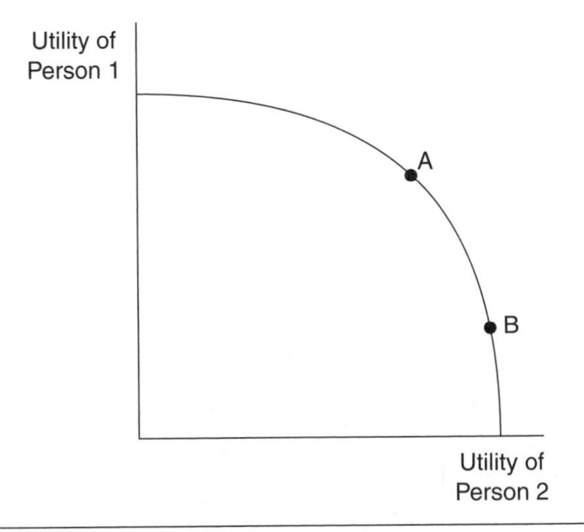

Figure 17.3 Utility possibilities frontier.

as a utility possibilities frontier, represent the maximum level of utility that the first person can have conditional on the second person's utility. Since points A and B lie directly on the curve rather than inside it, they are both Pareto-efficient, but point A represents a much more equitable distribution of utility, or welfare, than point B.

The second fundamental theorem of welfare economics states that any point on the utility possibilities frontier can be achieved through manipulation of the initial distribution of resources across individuals in a competitive economy. Suppose that the economy is at point B but that we would prefer the situation at point A. The second fundamental welfare theorem implies that once the appropriate redistribution of resources is achieved through lump sum transfers, the functioning of competitive markets will take the economy to point A. Therefore, unadulterated competitive market mechanisms lead to an allocation of resources that is Pareto-efficient, and any specific allocation of this type can be achieved through the appropriate redistribution of resources.

Market Failures

The welfare-maximizing implications of free market economics suggest that careful deliberation about others' physical activity choices should precede actions to change them, even if they at first seem misguided or inappropriate. While there are known health risks associated with chronic lack of exercise, some may consider the physical and psychological costs of increasing their current levels of exercise too great to outweigh the increased risk of future diseases. Likewise, overtraining for a sport may increase the risk of injuries, but some people may be willing to accept these risks in order to achieve their competitive goals. Recognition of the benefits of consumer sovereignty makes the promotion of healthy lifestyles more complicated than simply advocating against sedentary activities perceived to compete with more active ones.

This is not to say that interventions to increase physical activity levels and improve health are not justified, rather that they have an economic justification only if there is a failure of one of the assumptions underlying the functioning of competitive markets. In the presence of a market failure, individuals' time and resource allocations will not maximize social welfare, and potential exists for government intervention to improve outcomes. Market failures can be broadly categorized into two areas: (1) public goods and objective information and (2) externalities.

Public Goods and Objective Information

Private markets will underprovide (or not provide at all) certain goods and services that are necessary in order for children and adults to engage in physical activity, despite their clear value to consumers. These are called public goods, and they have two general characteristics. (1) One person's use of the public good in no way diminishes the benefit someone else gets from its use. That is, the marginal cost of providing the good to an additional person is zero. (2) It is either very difficult or impossible to allow some individuals to utilize the public good while excluding others from doing so. Pure public goods, such as clean air, are both nonrival (as in 1) and nonexcludable (as in 2), but many other goods have just one of these two characteristics. For example, people cannot be excluded from taking fish from the ocean, but the more fish they catch, the less there are for others to eat.

In the case of nonexcludability, competitive markets fail to provide Pareto-efficient levels of the public good because of what is known as the "free rider problem." Since there is no way for a private firm to finance the good by charging a price, the only way for a private entity to supply the good is through voluntary contributions by individuals. However, individuals reason that they will benefit from the existence of the public good irrespective of whether they contribute to it, so no contributions are made. To see why competitive markets' under-provision of nonrival public goods leads to lower levels of utility, or a "welfare loss," consider the case of open backyards.

Because parents want a safe and convenient place for their children to play, they purchase, design, and maintain backyards. In doing so, however, they take into account only the needs of their own children, without realizing that other children in the neighborhood could benefit by using the yard. While the collective benefits to these other children may far exceed the cost to the homeowner of purchasing and maintaining additional land, there is no reasonable way for the homeowner to recoup the associated costs, so open backyards are smaller than is optimal from a neighborhood standpoint. The same is true of the decision by homeowners to fence their yards or restrict access to others in the community. In theory, the homeowner could charge neighborhood children a small fee for use of the yard, but others in the community would be so offended by the proposition that it would probably result only in the homeowner's alienation and exclusion from all future community events.

The difficulty in coordinating contributions among private individuals to support open backyards or green space within a community is the primary reason why local and national governments fund public parks and playgrounds. Such facilities play a prominent role in youth physical activity levels, as do publicly funded playing fields and sporting arenas. Just as important as the facilities themselves, or perhaps more so, are the characteristics of neighborhoods and local communities where they are located. For example, neighborhood safety and aesthetics are also public goods and have been shown to affect physical activity levels (Ball et al., 2001; Centers for Disease Control and Prevention [CDC], 1999; Timperio et al., 2004).

Transportation networks are another area in which publicly provided goods are pervasive and the implications for youth physical activity are substantial. For example, historical trends imply that infrastructure investments related to active forms of transportation, such as walking and bicycling, have lagged behind the construction of roads for automobile use, particularly in the United States (Surface Transportation Policy Project, 2002). Although both children and adults rely on sidewalks, bike lanes, and public transportation systems to engage in active forms of transportation, these types of infrastructure investments are arguably more important for children's mobility. Because they cannot drive, or have restrictions placed on their driving privileges, children and adolescents are less likely to travel if they cannot reach a destination by human power.

In addition, studies have documented that the ability of children to travel increases their physical activity levels through several mechanisms. For example, Mackett and colleagues (2005) show that children are least active at home, so the ability to leave home by any means results in higher activity levels. Children burn a large number of calories just by walking to activities, and are more active when they reach their destination. Furthermore, unstructured out-of-home events, which have been shown to result in the largest expenditure of calories, are more strongly associated with arrival by walking than with traveling by car (Mackett et al., 2005). Similarly, Cooper and colleagues (2005) find that children who walk or bicycle to school are more physically active throughout the day compared to those who are driven (although because neither Mackett and colleagues nor Cooper and colleagues control for selection effects, a reason for some of these associations may be that children with greater preferences for physical activity who are already more fit choose to walk to events). Even transportation through the mechanized modes of public transit systems can increase youth physical activity, because it typically requires walking to and from the departure and arrival point (Besser & Dannenberg, 2005).

Given the benefits to society from increased youth physical activity, why are public investments in promoting that activity on the part of children so low? The economic subdiscipline of Public Choice Theory is often used to explain why public investments favor particular groups. As noted by Buchanan and Tullocks (1962), votes are a resource used to "purchase" policy decisions; and because children do not vote, transportation funding is more likely to reflect the preferences of adults for roads and highways. Another likely factor in the failure to invest more in promoting higher levels of physical activity for young people is that unbiased and objective information related to diet and physical activity is another public good that is underprovided by competitive markets.

While the private sector has an incentive to reveal information that is favorable to goods or services for sale, this information is at best incomplete and at worst highly misleading. Recently, the advertising of unhealthy food to children has come under attack because of aggressive marketing tactics by the food and drink product industry. Children take advertising claims at face value without proper consideration of health consequences, and it can be difficult for parents to obtain objective information about how much of this food can be incorporated into a healthy diet (Kunkel et al., 2004). While the risks of unhealthy diets and low physical activity levels are known to researchers, many parents are less aware of the consequences of physical inactivity and overweight in children

(Eckstein et al., 2006). If these risks were known, the utility parents get from their children's sedentary or physical activities might decrease or increase, respectively, and cause a reallocation of children's time in accordance with the mechanisms discussed in connection with the SLOTH model. Although most countries have government agencies that are responsible for disseminating unbiased information on food and other products to consumers, their budgets pale in comparison to those of the private sector (Consumers Union and California Pan-Ethnic Health Network, 2005).

Externalities

Externalities arise from a discrepancy between the private (internal) costs and benefits to individuals of engaging in certain activities and the external costs and benefits that are unrecognized by competitive markets. Activities or products that have unmeasured benefits are underprovided by the market, and those that have unmeasured costs are overprovided. To see why this is the case, consider the example of a power plant that could reduce its emissions by 50% through the adoption of green technologies. However, efficiency losses associated with the adoption would increase operating costs by 20%. While the private cost to the plant of adopting the cleaner technology is substantial, it receives little if any private benefit, making it almost certain that the plant will continue to operate without the technology. However, the collective benefit to people living where air quality is affected by the power plant's emissions may far exceed the 20% increase in its operating cost. For the level of green technology adopted to be Pareto-efficient, the marginal cost of the technology must be compared to the marginal social benefit of equipping the plant. Without government or external intervention, however, competitive market mechanisms will lead the plant to compare its private costs to its private benefit, which is negligible, leading to an inefficient level of technology adoption and a welfare loss.

The power plant example is illustrative of the problem of pollution in general, which almost always leads to uncaptured external costs. Pollution can be a significant impediment to physical activity by both children and adults. It decreases the utility that children get from all types of outdoor physical activities and is a particularly strong disincentive for children with asthma or other respiratory conditions. The presence of pollution essentially increases the marginal utility of sedentary indoor activities relative to more active outdoor activities, causing a shift in children's time allocations toward the former. Another environmental externality that causes individuals to reallocate time away from outdoor physical activities is high traffic congestion. When individuals make driving decisions, they consider their own private costs (gasoline, time costs, automobile depreciation, etc.) without acknowledging that the presence of their vehicle makes the road less safe and less attractive to those traveling by bicycle or walking. An externality exists because this social cost is not incorporated into the private cost of driving.

Externalities are often caused by coordination failures, which are a common problem in the provision of pedestrian infrastructure. For example, suppose a developer is considering putting sidewalks in a housing development. If no sidewalks are already present along the major roads or in adjacent housing developments, then the developer has little incentive to go through with the project. Sidewalks that are not part of a broader pedestrian infrastructure do not increase the value of homes in isolated developments, nor do they promote physical activity. Therefore, lack of coordination among developers leads to negative "network externalities," or the underprovision of pedestrian infrastructure.

Perhaps one of the largest areas of inquiry related to the external costs of physical inactivity has to do with disease incidence and health care expenditures. Lack of physical activity is a risk factor for a variety of diseases, including coronary heart disease, hypertension, type 2 (non-insulin–dependent) diabetes, osteoporosis, colon cancer, anxiety, and depression (Booth & Chakravarthy, 2002). Many of these diseases occur in adults, but even young overweight children can exhibit risk factors such as hyperlipidemia, elevated blood pressure, and increased insulin levels (Freedman et al., 1999). Of course, physically active children who remain active into and through adulthood have a much lower risk of such health complications than those who cease to be active as adults. A number of researchers have documented that physically active children and adolescents are significantly more likely to be active as they grow older, suggesting that important health behaviors are established during childhood (Tammelin et al., 2003; Telama et al., 1997; Yang et al., 1999).

Although individuals bear some of the health care costs associated with their physical activity decisions, a substantial share of health care expenditures is financed by public health insurance systems. This is clearly the case in countries with universal coverage, such as Canada, Australia, and countries in Europe. Even in the United States, where employer-provided health insurance is more prevalent, policies are not risk adjusted for physical activity levels or overweight, so much of the cost of treating medical conditions is borne by others in the same insurance pool. (Ironically, this type of externality is actually perpetuated by government intervention to the extent that health insurance is publicly provided. Nonetheless, risk pooling through insurance is the root cause of externality, so ending government involvement in insurance markets wouldn't necessarily solve the problem.) Studies show that the external costs generated by physical inactivity and sedentary lifestyles can be substantial.

The most direct evidence of negative externalities from sedentary living is from Keeler and colleagues (1989), who estimate that over the course of their lifetime, individuals with a sedentary lifestyle receive a subsidy of $1900 ($3109 in 2006 USD) from others. This is greater than the subsidy to smokers, but less than that to heavy drinkers. This study is exceptional because it measures a comprehensive list of costs over a person's lifetime, including those for medical care, sick leave, group life insurance, nursing home expenses, and retirement pensions. Other studies focusing solely on direct medical costs of sedentary lifestyles estimate that they account for 2.4% of total health care costs in the United States ($24 billion USD; Colditz, 1999), 2.5% of total costs in Canada (2.1 billion CD; Katzmarzyck, Gledhill, & Shepard, 2000), and 1.6 billion CHF per year in Switzerland (Swiss Federal Office of Sports, 2001). Because individuals often pay for some small percentage of their health care expenses out-of-pocket, not all of these costs are external, but a large share of them are.

Accounting for the independent effect of weight status is a methodological challenge for those attempting to estimate medical costs attributable to physical inactivity. Although obesity and physical inactivity are separate risk factors for many of the same diseases, their close correspondence makes it difficult to attribute medical expenditures to one versus the other. While overweight and obesity may result from a lack of physical activity, it is also possible for active children to become overweight through poor diet alone. In addition, overweight individuals with higher physical activity levels will be at lower risk for many diseases than those of the same weight status with lower activity levels. Colditz (1999) attempted to separately estimate the costs of inactivity and obesity in adults and found that inactivity accounted for 2.4% of total U.S. health care costs, while obesity separately accounted for 7% of total costs. More recent estimates of the medical costs due to just obesity by Finkelstein, Fiebelkorn, and Wang (2003) place them at 5.3% of total U.S. medical expenditures (9.1% for both overweight and obesity), roughly half of which are financed by public insurance programs.

Policy Solutions

Under ideal circumstances, unconstrained consumption and time allocation decisions by individuals and production decisions by firms lead to competitive market outcomes that maximize social welfare. However, the existence of public goods, biased and incomplete information, and externalities leads to the failure of competitive markets. Government intervention is justifiable under these circumstances to reduce social welfare losses and achieve a "second-best" solution. Designing and implementing interventions to correct market failures is a nontrivial task and an active area of economic research. Most interventions will also produce some type of welfare loss themselves, so care must be taken to ensure that the intervention does not introduce a distortion into the economy that is worse than the market failure itself. Typical public policy interventions related to youth physical inactivity include government-subsidized provision of public goods, strategies aimed at internalizing externalities, and specific initiatives that operate within the public educational system.

Public Provision and Finance of Public Goods

The most obvious solution to poor provision of public goods is simply to increase government funding for public facilities, infrastructure, and information provision. Nearly all municipalities and tiers of government fund public goods to some extent, and there are many potential public investments that could, in theory, increase physical activity levels of young people. The key to effective public policy is determining which investments will have the largest health and welfare benefits. One prominent trade-off in comparing investment alternatives is between targeting and impact. Investments in sporting stadiums and playing fields may increase the physical activity levels of a small subset of children (those who play particular sports) by a large amount. On the other hand, investments in sidewalks, bicycle paths, traffic-calming devices, and even public transportation will increase activity levels to a smaller extent, but for a much larger number of children. Since physical inactivity is a condition afflicting a large portion of the child and adolescent population, broad-based policy interventions tend to be favored by public health advocates.

Several indications suggest that investments in pedestrian infrastructure and neighborhood connectivity stand to have one of the largest impacts on child and adolescent activity levels. First of all, children burn a substantial number of calories just by walking or bicycling to events (Cooper et al., 2005; Mackett et al., 2005). Furthermore, the ability to walk to events is associated with higher participation in unstructured play, which exhibits greater calorie expenditure than even organized sport (Mackett et al., 2005). Therefore, by increasing opportunities for children to walk to events and meet their friends and playmates, investments in pedestrian infrastructure promote additional physical activity in multiple ways. Facilitating additional unstructured play is particularly important since the decline in this activity is thought to be a contributor to the childhood obesity epidemic (Sturm, 2005).

Other public investments may be necessary to ensure that the benefit children get from improved pedestrian infrastructure and neighborhood connectivity is fully realized. In areas with high crime rates, complementary investments in protective services and community policing are clearly important to children's ability to move freely about the neighborhood. Such improvements in public safety have been shown to have an independent beneficial effect on physical activity levels (Besser & Dannenberg, 2005; CDC, 1999). The proper integration of parks and play areas within communities can also create synergies with pedestrian infrastructure investments. Community planning to preserve *usable* green space either within residential developments or at close walking distance is key to increasing the accessibility and utility of play areas, effectively lowering the time and effort required by children to engage in physical activity.

Deciding which investments to make with public funds is part of the challenge in public goods provision. Funding the investments using available tax instruments requires careful calculation to avoid introducing distortions into the economy that offset the benefits gained from public goods. Economic theory suggests that the only tax that is free of distortions is a lump sum tax, or a head tax, whereby every individual is charged the same fixed amount. Because each person faces the same tax burden regardless of his or her relative income, asset holdings, or family size, head taxes are generally perceived as highly inequitable and as a result are rarely used to generate revenue. All other types of tax instruments introduce some type of distortion into the economy. Although a detailed discussion of optimal tax policy is beyond the scope of this chapter, there are some particularly undesirable features of employing user fees to fund public facilities.

User fees, which are common at public parks, swimming pools, or recreational areas, are popular with policy makers because they are highly targeted toward those who patronize public facilities. While this type of targeting may seem logical, user fees discourage some potential users of the facilities from patronizing them, limiting their opportunities for physical activity. In addition, the argument for charging only those who make use of public parks and recreational areas does not recognize that the positive externalities associated with physical activity accrue to *everyone* through lower health care costs and greater economic productivity. From the standpoint of equity, user fees are regressive in the sense that they are a larger percentage of a poor person's income than of a wealthy person's income, discouraging use to a greater extent among the poor, who in turn have inferior health outcomes.

Therefore, funding public recreational facilities out of general revenues has both a public health and an economic justification in the presence of externalities. Similar arguments have been made to reduce or eliminate user fees for health care in developing countries (Deininger & Mpuga, 2005; Meyerhoefer, Sahn, & Younger, 2007). However, there is an economic justification for charging user fees if the marginal cost of allowing an additional person to use the facilities is not close to zero. This may occur in cases of severe overcrowding, necessitating some type of rationing mechanism. It is still possible to preserve equity in access under these circumstances by issuing vouchers to low-income residents so that they can avoid the user fees levied on wealthier members of the community.

Wide dissemination of objective information about the benefits of physical activity and healthy diets may be another way to affect the behavior of a large number of individuals and improve health outcomes. To raise awareness of the health benefits of regular exercise, some have proposed the airing of public service announcements to children and adolescents during after-school hours and on weekend mornings (Pratt et al., 2004). Of course, the SLOTH model implies that it is also important to target information at parents, mothers in particular, who have substantial influence over their children's time allocation decisions. Such a multipronged approach has been taken by the U.S. Centers for Disease Control and Prevention through an initiative called VERB, which uses paid advertising to reach children and adults separately through a variety of multimedia outlets (Wong et al., 2004). Experimentation with similar initiatives is needed to determine whether the attitudes and behavior of children and adults can be affected in this manner. Experience from anti-tobacco media campaigns suggests that there is scope to engender positive attitudes toward physical activity through the mass media (McVey & Stapleton, 2000; Sly, Heald, & Ray, 2001).

Internalizing Externalities

Public policy approaches to externalities typically involve the development of some mechanism allowing competitive markets to account for, or internalize, external social costs or benefits. This is often done through highly targeted taxes or subsidies on the activities producing externalities. For example, levying a percentage tax on electricity purchased from a power plant is one way to adjust the price of energy to reflect the added social costs of pollution and to generate public revenue for pollution abatement projects. Likewise, gasoline taxes are sometimes justified as a way of transferring the social costs of pollution and traffic congestion to car owners. While pollution control and congestion reduction are clearly important to the promotion of physical activity, their wide-reaching social and economic consequences make it unlikely that major initiatives would garner sufficient support if justified based on youth physical activity alone. Proposals to curb major problems like pollution need to account for the social and environmental costs falling on all parties so that these can be more effectively compared to the economic benefits of the status quo.

Because it can be difficult to develop comprehensive solutions to problems like pollution and traffic congestion among a variety of constituents, more targeted approaches to physical inactivity externalities have been proposed that focus on specific industries. For example, Pratt and colleagues (2004) suggest levying taxes on products related to sedentary activities, such as computers, telecommunications, and home entertainment technologies to discourage their use and fund pedestrian infrastructure. The criticism of this approach is that it will not necessarily lead to an increase in physical activity; children may just substitute other types of sedentary activities such as reading or board games. In addition, home entertainment systems can be part of an active lifestyle if used in moderation, so the taxes are not as well targeted toward physical inactivity externalities as other approaches.

Rather than focusing on the social costs of inactivity and taxation, one could attempt to internalize the social benefits of physical activity through targeted subsidies. Possible interventions include the issuance of vouchers to adolescents for the purchase of memberships in private health clubs or to children's parents to defray the cost of access to recreational facilities. The vouchers could also be used to purchase basic athletic equipment or other products that facilitate physical activity. Communities could subsidize physical activity shuttles to take children to team sporting events, practices, or public parks and playgrounds that are not otherwise accessible. Alternatively, tax laws could be changed to allow parents to deduct the cost of transportation related to children's participation in community- or school-sanctioned sports and recreational events.

Legal remedies are another way to internalize externalities. A well-known result in economics, attributed to Coase (1960), is that if trade of the externality can occur, then bargaining in the context of private markets will lead to an efficient outcome irrespective of which party is initially assigned property rights over the externality. This theoretical result assumes that transaction costs are zero, but the principle can be applied in practice so long as transaction costs are recoupable.

In the power plant example, if a mechanism was established that allowed neighbors to pay the plant to reduce its emissions, trade over the negative externality would lead to an efficient level of pollution. Neighbors would "purchase" lower emissions to the point where their marginal expenditures equaled their marginal gain in utility from less pollution, providing the power plant with the means and incentive to finance adoption of the green technology. This same principle is often applied to the preservation of green space through the purchase of conservation easements by private groups on developable land. More direct examples of legal remedies include the enforcement of zoning laws mandating that new housing developments include sidewalks and other pedestrian infrastructure amenable to active forms of youth transportation, thereby mitigating coordination failures among individual developers.

Private markets can sometimes account for externalities without government intervention if the legal environment is supportive of appropriate private arrangements between groups of individuals. In the case of physical inactivity, this might amount to a transfer from the families of physically inactive children to those of physically active children to compensate them for the higher private insurance premiums and public health care costs they will face in the future due to some children's sedentary lifestyles. While it is difficult to conceptualize how these types of arrangements could be executed in reality, the risk adjustment (or rating) of insurance policies based on body mass index (BMI) is one way of compensating physically active and health-conscious families in the same employer group for the otherwise external costs generated by sedentary families. Indeed, there is some evidence that employers sponsoring health insurance do this implicitly by imposing a cash wage penalty on workers who are obese (Bhattacharya & Bundorf, 2005).

Educational Policy and Recommendations for Practitioners

Since children spend such a large fraction of their time in school, the educational system is often considered a primary conduit for implementing physical activity policies. Due to the substantial positive social benefits associated with an educated population, most countries support schools with public funds, making them more responsive to public policy interventions than private entities. In addition, schools generally have the requisite infrastructure and human capital necessary to support a broad range of sports and physical activities, so they are well suited to carry out youth-oriented interventions. Since exercise is incorporated into the school day differently in various parts of the world, this section focuses on U.S. education policy, with the hope that parallels can easily be drawn to other countries.

Chapter 13 is devoted to discussing physical activity in relation to the U.S. schools as incorporated into (1) recess periods for younger children, (2) physical education classes during the normal school day, and (3) active transport to school. A fourth area that should be considered is that of team sports and intramural activities after school. The level of each of these activities varies across states and local districts, reflecting both financial resources for education and local preferences for physical activity.

At present, the least is known about whether recess periods significantly contribute to physical activity and caloric expenditure over the long run. Most research in this area has focused on quantifying physical activity levels, rather than establishing the connection between recess and the formation of preferences for physical activity that shape behavior throughout life. Nonetheless, the studies discussed in chapter 13 (pp. 327-334) suggest that recess periods held in appropriate environments can significantly raise activity levels. This is important because concerns over safety and preserving instructional time have led some districts to curtail or eliminate recess, despite the fact that it may be an important tool in the fight against child obesity. Recess is also used to facilitate child social development in addition to physical activity, so holding recess periods in areas that are very conducive to exercise and active games may be the best way to achieve complementary goals for younger children.

In the United States, physical education classes are the primary mechanism used to promote physical activity in the educational setting. Debate over the costs and benefits of physical education has heightened in recent years as new education policies such as "No Child Left Behind" have put pressure on schools to reduce physical education in favor of supplemental reading and math education (Cawley et al., 2007). At the same time, physical education is viewed by many as critical to efforts aimed at stemming the rising tide of child obesity. The majority of recent research on physical education has been descriptive of or has focused on the evaluation of specific interventions. These studies, which are discussed at length in chapter 13 (pp. 334-341), have generated some important and policy-relevant findings. For example, opportunities to enroll in physical education are often limited, and students can be relatively inactive for much of the lesson time. The latter is particularly true of secondary school programs that are relatively more focused on teaching specific skills.

Research is lacking on the rigorous evaluation of whether existing physical education programs raise activity levels and reduce body weight. If current programs are found to be effective, then recent proposals to mandate more physical education class time might represent attractive policy interventions. One notable study along these lines showed that physical education instruction at the kindergarten and 1st-grade level reduced the BMI of girls who were overweight or at risk for overweight, but had no effect on overweight or at-risk boys or normal-weight boys and girls (Datar & Sturm, 2004). Another study on high school physical education programs yielded no statistical evidence that physical education lowered BMI or decreased the probability of a student's being overweight. However, it did show that state physical education requirements raised the number of minutes students were physically active in physical education and that additional exposure to physical education increased the number of days per week girls reported having exercised vigorously (Cawley, Meyerhoefer, & Newhouse, 2007). These recent evaluations suggest that the effectiveness of physical education varies significantly by gender; and while there is some justification for increasing physical education time at the elementary school level, initiatives to increase the proportion of time students are physically active in physical education class should precede higher time mandates.

There is no question that children who participate in after-school interscholastic sports are compelled to maintain high levels of physical activity during the sporting season, although it should be noted that these children already have strong preferences for physical activity (Davison & Schmalz, 2006) and are likely to be more active than average regardless of sport offerings. However, there are many children at the margin who would participate in school-sponsored sports but either do not have sufficient skill levels or do not enjoy the pressure associated with interscholastic competition. These children are much more likely to take part in less competitive intramural athletics or club sports. Between 38% and 51% of schools sponsor intramurals, but the relative levels of funding and availability are far below the levels for interscholastic athletics (Wechsler et al., 2000). In a world of limited resources, the question is again which activities generate the largest social benefit per dollar invested. While interscholastic team sports raise the activity levels of a small number of relativity fit children, intramurals have the potential to promote physical activity among a larger number of more sedentary children. There is scope for both types of school-sponsored activities, but it does seem that resources are skewed toward interscholastic competition, in part due to a lack of appreciation for the positive externalities associated with youth physical activity.

Offering after-school intramurals in addition to interscholastic sports is one way of lowering the costs (or increasing the utility) of physical activity in an effort to increase sport participation rates. The costs are both psychological (the pressure of interscholastic competition and fear of inadequacy) and time oriented (required practices, travel, and competitions). In order to induce even more children to engage in physical activity, the costs of participation must be lowered even further. This might include the integration of physical activity into other types of after-school clubs or the provision of additional resources to increase opportunities for children to be physically active. An example of the former might be a fantasy-adventure gaming club in which children physically act out the games; of the latter, providing transportation home following after-school events or opening school athletic facilities during nonschool hours. In either case, practitioners need to determine through experimentation or other means (surveys, etc.) which activities are likely to draw the largest number of children into exercise-related programs at a reasonable cost and promote them in educational settings.

The Field in Perspective

Economics is only beginning to be applied to the analysis of physical activity in children and adults. Although this is a growing area of research, it is still in its infancy, and there is a great potential for future contributions. At its core, economics is a science of how individuals respond to incentives that are present in the marketplace and in their general living environment. Economists have developed sophisticated statistical techniques that can be used to tease out the consequences of changes in the environment on individual behavior, but these have yet to be fully exploited in the area of youth physical activity. At present our understanding of which interventions children and adolescents respond to is very limited, and further research is required. Furthermore, economists and public policy analysts need to encourage the collection of data that are appropriate for answering these questions and to make better use of existing data on child behaviors. Longitudinal time-use data are particularly important to the investigation of how children allocate their time between physical and sedentary activities, and what policy levers can be used to lower the costs of the former relative to the latter. Finally, collaborations between economists and other behavioral and physical scientists are necessary to address the fundamentally multifaceted social problem of youth physical inactivity. Hopefully, this chapter will facilitate future discussion among researchers and practitioners interested in the economic aspects of this important social issue.

Applications for Researchers

Policies aimed at promoting physical activity in children and adolescents will be most effective if they are designed to mitigate the market failures that lead to suboptimal physical activity levels. These include the underprovision of public goods that serve as essential inputs to physical activity, the presence of unmeasured external costs and benefits related to exercise, and the underprovision of objective and unbiased information about activity levels and proper diet. There is often a trade-off in the targeting of physical activity policies between increasing the activity levels of a small, homogenous group of individuals by a large amount and increasing the activity levels of a larger group to a lesser extent. Economic evaluations can help to determine which of these interventions generates the greatest social welfare gains.

APPLICATIONS FOR PROFESSIONALS

The time allocation decisions that children and adolescents make between sedentary and physical activities are not always in their own best interest or those of society as a whole. Interventions to increase exercise and promote healthier lifestyles should encourage participation in activities that children are likely to enjoy at least as much as competing sedentary activities. The school environment is particularly well suited to the implementation of these public health interventions. School-based programs that stand to have the greatest influence on sedentary children are new physical education curricula that include less competitive and more individualized alternatives to team sports and more widely available after-school intramural sport teams and physical activity clubs. The latter need not have exercise as their only focus but should include it as an integral component.

References

Ball, K., Bauman, A., Leslie, E., & Owen, N. (2001). Perceived environmental and social influences on walking for exercise in Australian adults. *Preventive Medicine, 33*, 434-440.

Besser, L., & Dannenberg, A. (2005). Walking to public transit: Steps to help meet physical activity recommendations. *American Journal of Preventive Medicine, 29*, 273-280.

Bhattacharya, J., & Bundorf, M. (2005). *Incidence of health care costs of obesity.* National Bureau of Economic Research (NBER) working paper No. 11303. Cambridge, MA: NBER.

Booth, F., & Chakravarthy, M. (2002). Cost and consequences of sedentary living: New battleground for an old enemy. *President's Council on Physical Fitness and Sports Research Digest, 3*(16), 1-8.

Buchanan, J., & Tullock, G. (1962). *The calculus of consent: Logical foundations of constitutional democracy.* Ann Arbor: University of Michigan Press.

Cawley, J. (2004). An economic framework for understanding physical activity and eating behaviors. *American Journal of Preventive Medicine, 27*(3S), 117-125.

Cawley, J., Meyerhoefer, C., & Newhouse, D. (2007). The impact of state physical education requirements on youth physical activity and overweight. *Health Economics, 16*, 1287-1301.

Centers for Disease Control and Prevention. (1999). Neighborhood safety and the prevalence of physical activity—selected states, 1996. *Morbidity and Mortality Weekly Report, 47*, 143-146.

Chapman, G. (1996). Temporal discounting and utility for health and money. *Journal of Experimental Psychology: Learning, Memory, and Cognition, 22*, 771-791.

Chapman, G., & Elstein, A. (1995). Valuing the future: Temporal discounting of health and money. *Medical Decision Making, 15*, 373-386.

Coase, R. (1960). The problem of social cost. *Journal of Law and Economics, 1*, 1-44.

Colditz, G. (1999). Economic costs of obesity and inactivity. *Medicine and Science in Sports and Exercise, 31*(11, S1), 663-667.

Consumers Union and California Pan-Ethnic Health Network. (2005). *Out of balance: Marketing of soda, candy, snack, and fast foods drowns out healthful messages.* San Francisco: Consumers Union.

Cooper, A., Andersen, L., Wedderkopp, N., Page, A., & Froberg, K. (2005). Physical activity levels of children who walk, cycle, or are driven to school. *American Journal of Preventive Medicine, 29*, 179-184.

Cutler, D., Glaeser, E., & Shapiro, J. (2003). Why have Americans become more obese? *Journal of Economic Perspectives, 17*(3), 93-118.

Datar, A., & Sturm, R. (2004). Physical education in elementary school and body mass index: Evidence from the early childhood longitudinal study. *American Journal of Public Health, 94*, 1501-1506.

Davison, K., & Schmalz, D. (2006). Youth at risk of physical inactivity may benefit more from activity-related support than youth not at risk. *International Journal of Behavioral Nutrition and Physical Activity, 3*, 5.

Deininger, K., & Mpuga, P. (2005). Economic and welfare impact of the abolition of health user fees: Evidence from Uganda. *Journal of African Economies, 14*, 55-91.

Eckstein, K., Mikhail, L., Ariza, A., Thomson, J., Millard, S., & Binns, H. (2006). Parents' perceptions of their child's weight and health. *Pediatrics, 117*, 681-690.

Finkelstein, E., Fiebelkorn, I., & Wang, G. (2003). National medical spending attributable to overweight and obesity: How much, and who's paying. *Health Affairs, W3*, 219-226.

Frank, L. (2004). Economic determinants of urban form: Resulting trade-offs between active and sedentary forms of travel. *American Journal of Preventive Medicine, 27*(3S), 146-153.

Freedman, D., Dietz, W., Srinivasan, S., & Berenson, G. (1999). The relation of overweight to cardiovascular risk factors among children and adolescents: The Bogalusa heart study. *Pediatrics, 103*, 1175-1182.

Gruber, J., & Mullainathan, S. (2002). *Do cigarette taxes make smokers happier?* National Bureau of Economic Research (NBER) working paper No. 8872. Cambridge, MA: NBER.

Katzmarzyck, P., Gledhill, N., & Shepard, R. (2000). The economic burden of physical inactivity in Canada. *Canadian Medical Association Journal, 163*, 1435-1440.

Keeler, E., Manning, W., Newhouse, J., Sloss, E., & Wasserman, J. (1989). The external costs of a sedentary life-style. *American Journal of Public Health, 79*, 975-981.

Kunkel, D., Wilcox, B., Cantor, J., Palmer, E., Linn, S., & Dowrick, P. (2004). Psychological issues in the increasing commercialization of childhood. In *Report of the APA task force on advertising and children.* Washington, DC: American Psychological Association.

Mackett, R., Lucas, L., Paskins, J., & Turbin, J. (2005). The therapeutic value of children's everyday travel. *Transportation Research Part A, 39*, 205-219.

McVey, D., & Stapleton, J. (2000). Can anti-smoking television advertising affect smoking behavior? Controlled trial of the health education authority for England's anti-smoking TV campaign. *Tobacco Control, 9*, 273-282.

Meyerhoefer, C., Sahn, D., & Younger, S. (2007). The joint demand for health care, leisure, and commodities: Implications for health care finance and access in Vietnam. *Journal of Development Studies, 43*, 1475-1500.

Pratt, M., Macera, C., Sallis, J., O'Donnell, M., & Frank, L. (2004). Economic interventions to promote physical activity: Application of the SLOTH model. *American Journal of Preventive Medicine, 27*(3S), 136-145.

Sly, D., Heald, G., & Ray, S. (2001). The Florida "Truth" anti-tobacco media evaluation: Design, first year results, and implications for planning future state media evaluations. *Tobacco Control, 10*, 9-15.

Smith, A. (1937, originally published 1776). *The wealth of nations.* New York: Modern Library.

Sturm, R. (2004). The economics of physical activity: Societal trends and rationales for interventions. *American Journal of Preventive Medicine, 27*(3S), 126-135.

Sturm, R. (2005). Childhood obesity—what society can learn from existing data on societal trends, part 1. *Preventing Chronic Disease: Public Health Research, Practice, and Policy, 2*(1). Retrieved July 31, 2006, from www.cdc.gov/pcd/issues/2005/jan/04_0038.htm.

Surface Transportation Policy Project. (2002). *Mean streets 2002.* Retrieved July 31, 2006, from www.transact.org/report.asp?id=202.

Swiss Federal Office of Sports. (2001). Economic benefits of the health-enhancing effect of physical activity: First estimates for Switzerland. *Schweiz Z Sportmed Sporttraumatol, 49*(3), 131-133.

Tammelin, T., Näyhä, S., Hills, A., & Järvelin, M-R. (2003). Adolescent participation in sports and adult physical activity. *American Journal of Preventive Medicine, 24*, 22-28.

Telama, R., Yang, X., Laakso, L., & Viikari, J. (1997). Physical activity in childhood and adolescence as predictors of physical activity in young adulthood. *American Journal of Preventive Medicine, 13*, 317-323.

Timperio, A., Crawford, D., Telford, A., & Salmon, J. (2004). Perceptions about the local neighborhood and walking and cycling among children. *Preventive Medicine, 38*, 39-47.

Wechsler, H., Devereaux, R., Davis, M., & Collins, J. (2000). Using the school environment to promote physical activity and healthy eating. *Preventive Medicine, 31*, S121-S137.

Wong, F., Human, M., Heitzler, C., Asbury, L., Bretthauer-Mueller, R., McCarthy, S., et al. (2004). VERB™—a social marketing campaign to increase physical activity among youth. *Preventing Chronic Disease: Public Health Research, Practice, and Policy, 1*(3). Retrieved July 31, 2006, from www.cdc.gov/pcd/issues/2004/jul/04_0043.htm.

Yang, X., Telama, R., Leino, M., & Viikari, J. (1999). Factors explaining the physical activity of young adults: The importance of early socialization. *Scandinavian Journal of Medicine and Science in Sports, 9*, 120-127.

CHAPTER

18

Culturally Appropriate Research and Interventions

Suzanna M. Martinez, MS ▪ Elva M. Arredondo, PhD
▪ Guadalupe X. Ayala, PhD ▪ John P. Elder, PhD

Physical activity and inactivity occur around the globe at any given hour; however, the types of physical activity youth participate in and the extent of activity that is achieved are not the same for all populations. With respect to cultural issues, many research questions are still unanswered, particularly given the rise in immigration, the growing numbers of developing countries, and globalization. Research should consider how immigration, sociodemographic factors, and acculturation are associated with current trends in youth physical inactivity. Selecting one or more models or a theoretical framework that is applicable to diverse communities is necessary for examining the cultural, social, and environmental factors contributing to youth inactivity around the world. Obesity is a growing epidemic, and chronic disease affects immigrants and indigenous populations at disproportionate rates (Cossrow & Falkner, 2004; Slattery et al., 2006).

Because of this unfortunate trend, the issue of well-designed research targeting immigrant and ethnic minority groups is increasingly important for public health. Conducting culturally appropriate qualitative and quantitative research is necessary for examining familial, parental, and social pressures that affect youth's physically active and sedentary behaviors. In the present chapter, we review the literature on these issues and discuss how researchers and professionals can develop culturally informed programs that promote physical activity and decrease sedentary behavior. To assist the reader in interpreting central points of this chapter, we provide definitions of some key terms in table 18.1.

We gratefully acknowledge support from the San Diego Prevention Research Center (U48 DP00036-03); Aventuras para Niños, funded by the National, Heart, Lung, and Blood Institute (5R01HL073776); and SDSU Minority Biomedical Research Scientist (MBRS) National Institute of General Medical Sciences (1 R25 GM58906-08).

Table 18.1 Definitions of Key Terms Used in This Chapter

Term	Definition
Acculturation	Process of multidimensional change that occurs when members of one cultural group come into continuous and firsthand contact with members of another more dominant cultural group;[1] associated with the adoption of healthy and unhealthy lifestyle behaviors
Culture	A learned system of beliefs about the manner in which people interact with their social and physical environments, shared among an identifiable segment of a population and transmitted from one generation to the next
Collectivistic	Emphasizing family or group goals rather than individual goals; allocentric; traditional
Individualistic	Emphasizing personal achievement
Traditional	Emphasizing customs or practices handed down from one generation to another
Diverse communities/ populations	Communities represented by a heterogeneous population of different races and ethnicities
Ethnicity	A category used to group individuals by cultural or national-origin characteristics
Familismo	A cultural belief in which there is a strong identification with members of immediate and extended family;[2] cultural variable influencing Latino health behaviors
Sociocultural	Derived from social factors and cultural influences
Socio-environmental	Derived from social factors and environmental exposures
Western world	Originally defined as "Western Europe," but includes countries whose dominant culture is derived from European culture

[1]Berry 2006; [2]Marín and Marín 1991.

Physical Inactivity in Immigrant and Ethnic Minority Populations

A few statistics will serve to illustrate the seriousness of the lower physical activity and higher obesity rates that characterize immigrant and ethnic minority populations worldwide.

- In Canada, the indigenous populations, such as the Northern Arctic remote Aboriginal communities, exhibit the highest prevalence of obesity at 26% (Statistics Canada, 2002).

- Australian studies show a consistent increase in overweight and obesity in the indigenous populations (Bull et al., 2004). In the Aboriginal and Torres Strait Islander populations, data from a 1994 survey showed that 60% of men and 58% of women were obese or overweight. This contrasts with the findings of a 1989 study of the Australian population showing that 48% of men and 34% of women were overweight (Australian Bureau of Statistics, 2001; Bull et al., 2004).
- In the United States, Latino and Asian youth report lower activity levels than any other ethnic group (Wolf et al., 1993).
- Among Aboriginal youth in Canada, declines in physical activity have been attributed to barriers such as lack of awareness about the benefits of physical activity, economic issues, cultural insensitivity of current programs and activities, lack of coaching for traditional Aboriginal activities, and lack of access to resources in remote communities (Government of Canada, 2005).

First-generation immigrants are not expected to achieve socioeconomic parity, which may account for the limited access to resources for physical activity (Gordon-Larsen, McMurray, & Popkin, 1999; Popkin & Udry, 1998). The rise in worldwide immigration, and the resultant increase in mixed ethnic populations, increase the urgency of dealing with such disparities. That rise is highlighted by the following statistics:

- Two percent of the world's population are immigrants.
- Eight percent of the U.S. population are foreign born (Swerdlow, 1998).
- In 2000, one in every five children (13.5 million) in the United States under the age of 18 years was living in an immigrant family (Population Reference Bureau, 2005).

Acculturation, a dual process of cultural and psychological change, occurs when two or more cultural groups or individuals come into continuous and firsthand contact (Berry, 2006). As a consequence of the acculturative process, behavioral adjustment can range from good to poor on the part of individuals in a new society and may explain the adoption of negative health behaviors such as a sedentary lifestyle (National Research Council and Institute of Medicine, 1999). In Australia, a growing number of individuals and families are immigrating from a variety of European countries (Australian Bureau of Statistics, 2005). Fewer acculturated Southern European immigrants are less likely to report being active and are two to three times more likely to be overweight and obese than those who are Australian born (Bull et al., 2004). In the United States, an increase in sedentary behaviors, such as television or video and computer or game use, has been observed among acculturating Mexican and Cuban generations (Gordon-Larsen et al., 2003). Television watching and lower levels of vigorous activity are also more prevalent among non-Hispanic Blacks and Mexican Americans compared to their White counterparts (Andersen et al., 1998). Because watching television more than 2 h daily has been related to overweight and obesity (Hu et al., 2003), it is important to note its impact on acculturating ethnic minority populations.

The rising trends in youth physical inactivity are universal, especially among underserved populations; this trend will exacerbate current global health problems and widen health disparities between ethnic groups. Various factors play a role in the declining trends of physical activity among ethnic minority youth. Barriers to physical activity are ubiquitous and include social, environmental, and socioeconomic factors, but these forces are especially powerful in immigrant populations living in disadvantaged communities. Because research on youth physical activity and sedentary behavior is often undertaken in the context of the dominant culture, research-based interventions are often less effective—or even counterproductive—for minority groups. Thus, it is imperative that theories and models continue to evolve to ensure that they are used or adapted effectively for minority populations.

Adapting Models and Theories

Individually oriented physical activity and sedentary behavior research is based on models and theories such as the health belief model (Becker, 1974; Janz & Becker, 1984), the theory of reasoned action and theory of planned behavior (Ajzen, 1985; Ajzen & Fishbein, 1980; Fishbein, 1980), social cognitive theory (Bandura, 1986, 1989), and Kanfer's parallel theory of self-control (Kanfer, 1975). The transtheoretical model (Prochaska & DiClemente, 1983) comprises a cognitive-behavioral model of change not grounded in a specific health behavior theory. These perspectives share several themes, including intentions to behave, environmental constraints impeding the behavior, skills, outcome expectancies, norms for the behavior, self-standards, affect, and self-confidence with respect to the behavior. In short, in order to change, a person must

- have a strong positive intention to perform a behavior,
- face a minimum of environmental barriers to performing the behavior,
- perceive her- or himself as having the requisite skills for performing the behavior,
- believe that positive reinforcement will follow the activity and that this reinforcement will exceed that for inactivity,
- believe that there is normative pressure to perform and none to avoid the behavior (or no negative consequences for a lack of engagement in non-active behaviors, such as frequent television watching),
- believe that the behavior is consistent with his or her self-perception,
- have positive affect regarding the behavior, and
- encounter cues or enablers to engage in the behavior at the appropriate time and place (Elder et al., 1998).

Photo courtesy of San Diego Prevention Research Center.

Traditional theories of health behavior change emphasize the individual over culture.

The aforementioned leading theories of health behavior change to a large extent emphasize the role of the individual and her or his thought processes in the development, maintenance, or loss of adaptive health-related behaviors. Most of the theorists state that they consider their perspective on health behavior to be universally applicable and that for physical activity promotion to be effective, individual efforts must be made to change those factors that contribute to the adoption of some health-related behaviors and the failure to adopt others.

However, the following issues have been raised about the extent to which these theories apply to people from nations outside of European-rooted cultures or to diverse communities within "Western" countries (e.g., Elder et al., 1998):

- The model of individuals as relatively autonomous beings implies that individuals evaluate potential outcomes and their own "self"-efficacy regarding personal behaviors. This may have little relevance for members of more traditional cultures who more closely adhere to norms set by family or peers (Marín & Marín, 1991).
- Collective identity must be considered in the development of health programs targeting many minority groups. An individual from a traditional, collectivistic culture typically emphasizes the needs of the group over her

or his own needs (Marín & Triandis, 1985). Individuals who emphasize the needs of the group (especially the primary and extended family, but also the community) may be less influenced by messages that emphasize individual well-being. In contrast, modern industrialized countries, especially the United States, are more individualistic, so personal gain and achievement are stressed over the needs of the family or community. Individuals in these societies tend to be credited directly for success, but also may receive personal blame for failure.

- In collectivistic cultures, the individual may approach questions of health and illness with fatalism (Spector, 1991). Disease is often seen as a punishment for past sins, and health interpreted as a gift from God (Neff & Hoppe, 1993; Spector, 1991). Because of the nature of the collectivistic culture in addition to this fatalistic orientation, modification of negative lifestyle behaviors through an appeal to personal autonomy and self-efficacy is less likely to occur (Pepitone, 1995).

Psychosocial models of health behavior must be evaluated in the context of the cultural and societal milieu in which they were developed. In most cases these models are normalized on economically stable, middle-class populations; often college students are used as participants, and instruments are developed to assess essentially White, middle-class norms. Immigrants and indigenous people alike bring with them beliefs and values associated with their country or place of origin. As a result of contact with the dominant culture, they may become disassociated from these beliefs and values or may cling to them as a source of self-support and self-worth. This choice, either freely committed or forced, can be understood only within the broader sociopolitical context. Three salient sociopolitical issues affecting immigrant communities are socioeconomic status, the geographical proximity of the homeland, and the political atmosphere.

Unique Contextual Factors

Selecting a theoretical model to guide culturally tailored research in ethnically diverse populations—those not rooted in European culture—must take into account unique cultural circumstances; "Western" perspectives regarding individual autonomy and related attitudes and behaviors may not be entirely applicable (Elder et al., 1998). Figure 18.1 illustrates how traditional and collectivistic cultural factors may be related to an individual's level of physical activity. In the following sections we detail many of these relationships.

Family Environment

The home environment plays an important role in shaping the health practices of youth (Spence & Lee, 2003). Parents' provision and support of healthy activities can positively influence their children's health (McGuire et al., 2002). For instance, research shows that parents who have rules about smoking are more

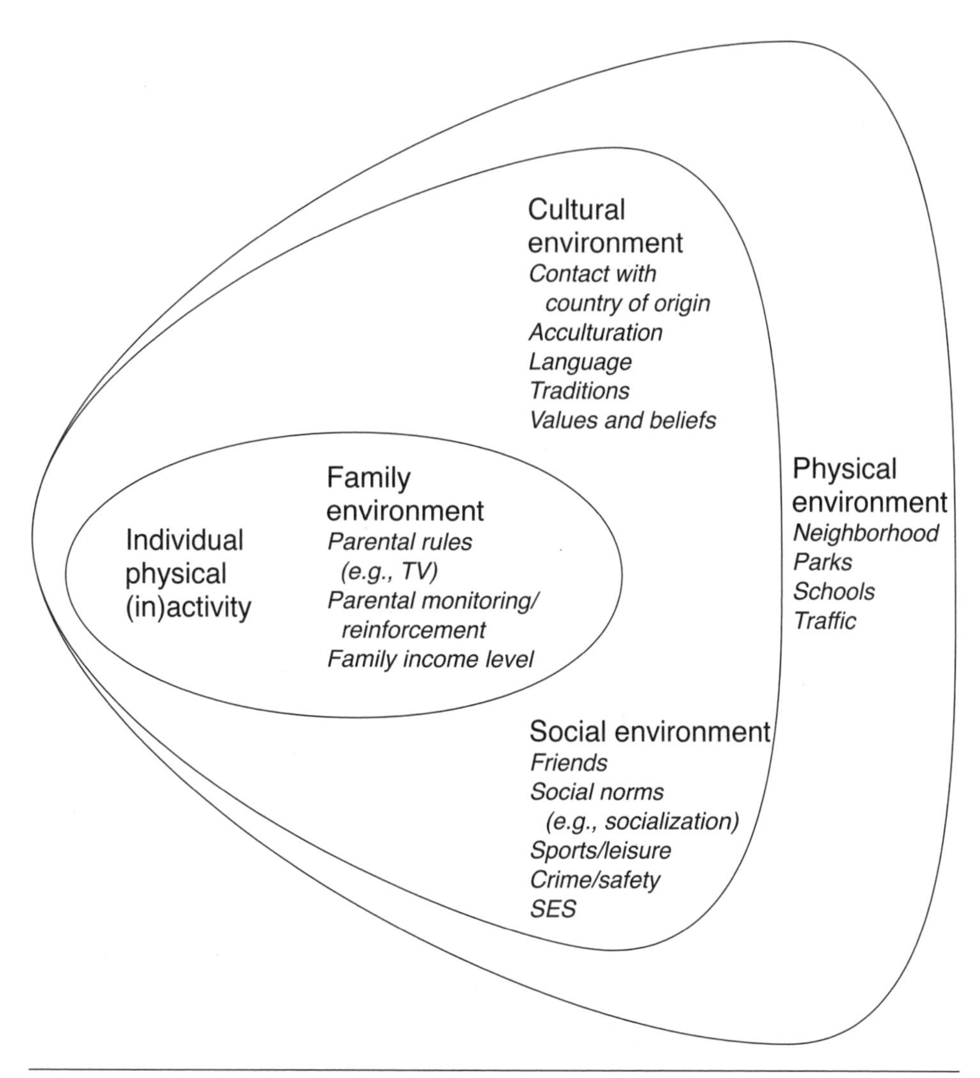

Figure 18.1 Contextual factors influencing individual levels of physical activity.

likely to have children who do not smoke than parents who do not have such rules (Kandel & Wu, 1995). Also, parents may influence the health practices of youth in less direct ways through their own behavior. Research suggests that parents who smoke are likely to have offspring who smoke, even when parents have rules against smoking (Proescholdbell, Chassin, & MacKinnon, 2000). The types of health practices parents model may communicate to their children that parents value or at least accept such behaviors.

Although studies suggest that families influence youth's dietary behaviors by their food preparation and shopping behaviors (Birch & Fisher, 2000), research is less clear about the extent to which families influence their children's activity levels. A comprehensive review involving national and international studies showed mixed evidence of this source of influence (Sallis, Prochaska, & Taylor, 2000). The modest amount of research that includes ethnic minority samples also is relatively inconclusive. McGuire and colleagues (2002) examined parental correlates of youth physical activity in an ethnically diverse sample and found that parents' physical activity was not associated with their children's physical activity but that parental encouragement for physical activity was. Among African American adolescents, Bungum and Vincent (1997) found that paternal support was positively associated with girls' physical activity participation. Also, Sallis and colleagues (1988) reported that the amounts of time mothers and older children spent in leisure-time physical activity were significantly correlated among both Mexican American and Anglo families.

Other studies suggest that the influence of parents on children's physical activity may be moderated by the child's gender (DiLorenzo et al., 1998; Trost et al., 1997). A stronger association between child activity levels and parenting practices regarding physical activity has been reported for girls than for boys (Gottlieb & Chen, 1985; Reynolds et al., 1990). Norman and colleagues (2005) examined psychosocial and environmental correlates of sedentary behaviors in an ethnically diverse adolescent sample of 11- to 15-year-old youth. Results from their multivariate analysis suggest that parental support for physical activity and TV limit setting were negatively associated with the sedentary behavior of girls, but not of boys. The majority of studies examining the influence of parents on minority adolescents' physical activity are quantitative, therefore providing little evidence of the sociocultural processes (due to limited scope of measures) that may explain the mechanism by which parents influence girls' physical activity. The results of a qualitative study involving an ethnically diverse sample of Latino adolescent girls in the United States suggest that adolescents are more likely to embrace behaviors, attitudes, and values similar to those of their same-sex parent compared to their opposite-sex parent (Shakib & Dunbar, 2004). Because the mothers of ethnic minority adolescents are less likely to engage in leisure-time physical activity than non-Hispanic White mothers, this may explain, in part, why minority adolescent girls are involved in less physical activity.

Because parenting practices are influenced by other contextual factors, such as cultural values and beliefs (Bronfenbrenner, 1979), it is important to consider the external forces that may guide parents' socialization of children's physical activity. In the case of many Latino families, research suggests that parents socialize boys and girls according to specific gender roles ascribed in the Latino culture (Diaz-Guerrero, 1993). In general, Latino boys are encouraged to play sports, such as soccer, whereas the girls are discouraged from engaging in leisure-time physical activity. Consistent with ethnic-specific stereotypes, Frenn and colleagues (2005) found that African American and Hispanic 7th-grade girls perceived less support from their parents to engage in physical activity than boys. These parenting practices are in contrast to those found in many White families in which girls are often enrolled in extracurricular activities that involve physical activity. These sociocultural factors and their interaction with parenting practices may be important reasons why we find low physical activity rates in ethnic minority girls compared to boys (Sallis et al., 1996). It is clear that more studies are needed for a better understanding of parenting factors that influence the physical activity and sedentary behavior of young people from various ethnic groups, as well as whether such factors or their degrees of influence differ by stage of youth development. Without these considerations, physical activity interventions developed in majority populations are not likely to be effective with ethnic minority youth.

A cultural variable found to affect Latinos' health practices is *familismo*, or strong identification with members of the immediate and extended family (Marín & Marín, 1991). Latino families have a strong sense of family identification and structure as well as support for extended family. Such attachment promotes loyalty, unity, reciprocity, and solidarity (Marín et al., 1987). Supportive family interactions within the Latino family not only buffer stress but directly affect the personal functioning of each individual family member. An important area for future research is the influence of *familismo* on the physical activity of Latino youth. It may be that parents and extended family who are the primary role models of physical activity play a stronger role in influencing physical activity among Latino youth than among White youth.

Finally, differences in physical activity definitions and measurement (leisure physical activity vs. sedentary behavior) may be among the reasons studies are mixed in showing parental influences on children's physical activity. More studies involving ethnic minority families that use familiar physical activity measures and terminology are needed for a better understanding of the direct and indirect influences of parents on the physical activity of ethnic minority youth. Moreover, further research is needed to better explain the types of dynamics and cultural factors present in ethnic minority families that may influence the physical activity and sedentary behavior of youth.

Socio-Environmental Factors

Other contextual factors that affect levels of physical activity among ethnic and immigrating youth and adolescents are social, cultural, and environmental. As populations diversify, there is a need to understand different cultures and social norms among children and adolescents. In addition, because economics strongly dictates what resources an individual has access to, the socioeconomic impact on physical activity levels and health status is concerning. For immigrating populations who are likely to lack economic resources, it is necessary to address neighborhood safety and crime when promoting physical activity in marginal communities.

- **Social norms.** Social norms appear to influence young people's physical activity levels. The findings of several qualitative studies shed light on individual and social factors that may not motivate ethnic minority girls to engage in physical activity. Mabry and colleagues (2003) conducted six focus groups with 9th- through 12th-grade African American and White adolescent girls. They found that African American participants were more accepting of their body image than were the White participants. These results suggest that desiring to achieve an ideal body weight may not be a motivating factor for physical activity in African American girls as it is for White girls. These reports corroborate the findings of other studies (Adams et al., 2000; Neff et al., 1997). Another factor that may affect ethnic minority adolescents' physical activity is gender role socialization: In some cultures, engaging in sport and physical activities is a male role (Taylor et al., 1999). Moreover, studies have indicated that the lack of physically active female role models may negatively influence ethnic minority girls' physical activity (Garcia et al., 1998). Social norms influence gender role socialization in relation to physical activity and the development of youth.

- **Social-cultural factors.** Research suggests that ethnic minority adolescents' attitudes toward physical activity may change as families acculturate into mainstream culture. Several studies show a positive association between acculturation and youth physical activity. In a sample of over 20,000 children ages 7 to 12 years enrolled in the National Longitudinal Study of Adolescent Health, Gordon-Larsen and colleagues (1999) found that low-intensity physical activity increased with generation of U.S. residence among Hispanic subgroups. Through increased social involvement, acculturating ethnic minority youth are likely to participate in structured physical activity programs; observing their peers engage in physical activity and experiencing the benefits of physical activity may change ethnic minority youths' views of physical activity. Because of the modest amount of research in this area, qualitative and quantitative studies are needed to better delineate the extent to which cultural values toward active living and sedentary behavior change as a function of acculturation.

- **Socioeconomic factors.** Those of ethnic minority status report lower socioeconomic status (SES) and are more likely to live in economically disadvantaged neighborhoods than those in the ethnic majority. Health disparities in underserved communities have been observed, showing a relationship between community income inequality and increased body mass (Robert & Reither, 2004). Low-income groups also report higher levels of physical inactivity, regardless of ethnicity. This puts low-income youth at increased risk for chronic disease (Bauman & Craig, 2005; Centers for Disease Control and Prevention [CDC], 1996; Kavanagh et al., 2005) compared to youth living in higher-SES neighborhoods (Gordon-Larsen, McMurray, & Popkin, 2000). Studies conducted in underserved neighborhoods in Australia and the Netherlands have shown that walking or cycling and vigorous activities are least likely during leisure time; however, limited access to a motor vehicle increased the likelihood of active transportation to shops and work (Giles-Corti & Donovan, 2002). In addition, economic barriers, such as facility and program fees for sport and recreation, as well as transportation costs, limit access to safe leisure-time physical activity and recreational sport opportunities.

- **Crime and safety.** The socio-environment may also lead to concerns about crime and safety for low-income youth living in disadvantaged neighborhoods. Americans living in disadvantaged neighborhoods have worse self-reports of physical functioning, more chronic disease, and impaired health linked to stress mediated through perceived neighborhood disorder and fear than do their more advantaged counterparts (Ross & Mirowsky, 2001). Daily stress is associated with living in neighborhoods that are frequently dangerous and have high crime. Crime in underserved communities heightens concerns about being safe, in turn discouraging individuals from seeking access to facilities and resources. Future physical activity programs should focus on creating accessible safe places for recreation and physical activity for children of marginalized communities.

The Physical Environment

As noted by Biddle and colleagues, levels of physical activity and sedentary behavior are more complex than previously thought and involve more than just the individual (Biddle et al., 2004). Indeed, a young person's physical environment can promote or impede physical activity. From an ecological perspective, multiple factors in the built environment affect the individual, such as city planning, land use for recreation, transportation, and pedestrian safety. These factors have been discussed in chapters 13 and 16. Thus we will simply summarize the main points, citing research not mentioned in those chapters, and add information and conclusions unique to the ethnic minority and immigrant perspectives.

In ethnic communities populated by low-SES groups, access to and availability of facilities and resources play a large role in the amount of vigorous and recreational physical activity that occurs (Sallis et al., 2000). Lack of private transportation among disadvantaged low-income populations is an obstacle, and even more so in neighborhoods with insufficient city transport (Hoehner et al., 2005). Environmental decay (e.g., cracked sidewalks and abandoned structures), common to disadvantaged neighborhoods (Ross & Mirowsky, 2001), presents less walkable environments for residents who live in these neighborhoods.

Low levels of active transportation have been observed in ethnic and disadvantaged neighborhoods in the United States and Australia (Kavanagh et al., 2005). Barriers common to school-aged children include distance to school and traffic-related danger (U.S. Environmental Protection Agency, 2003). These barriers lead to poor access to recreational facilities and fewer opportunities for physical activity (Popkin, Duffey, & Gordon-Larsen, 2005). Furthermore, Giles-Corti and Donovan (2002) performed an environmental scan of recreational facilities of low-SES neighborhoods and found poor use of recreational facilities in disadvantaged communities even though facility access was superior. They concluded that other factors involved in low use of resources were sociocultural influences or low visibility of the activity programs offered.

Saelens and his collaborators (2003) are currently examining neighborhood quality of life in the United States and are finding that inefficient urban planning encourages vehicle transportation, resulting in high traffic and low walkability, especially in low-income neighborhoods. This may explain comparable trends observed in Australia, Portugal, and Belgium. For instance, in Australia, low-income communities have been associated with a decreased likelihood of jogging, walking, and overall outdoor physical activity levels (Kavanagh et al., 2005). In Portugal and Belgium, active transportation and recreational walking have been related to walkability in marginal neighborhoods (De Bourdeaudhuij et al., 2005).

The physical environment clearly presents social and structural barriers to physical activity that are unique to ethnic minority and immigrant youth. Underserved youth populations are at particular risk for being inactive as they are less likely than other populations to perceive that their neighborhood is safe and attractive with respect to active transportation. These established environmental barriers may be fundamental to the health disparities that exist in underserved populations around the world.

Culturally Appropriate Interventions

The U.S. Surgeon General has issued a call to action in the prevention of overweight and obesity in children and adolescents (Committee on School Health, 2004). The United States and Australia are just two of the countries conducting national efforts to promote physical activity. Australia's strategic plan for the prevention of overweight and obesity summarizes reasons for promoting physical activity on a national level (National Health and Medical Research Council, 1997). The *Healthy People 2010* (U.S. Department of Health and Human Services, 2000) guidelines aim to increase the proportion of children and adolescents who walk to school from 31% to 50%. However, there is no single strategy that can be generalized to diverse populations, and the challenge at hand is how to create interventions that can be effective within and across cultures.

General Principles

Developing culturally appropriate interventions, particularly for a new target population, can seem daunting. This is particularly true when one is designing interventions for young people, given that they may not subscribe to the same cultural traditions as their parents (Félix-Ortiz, Fernandez, & Newcomb, 1998). This may explain the dearth of research in this area, particularly as it pertains to physical activity intervention. However, despite the state of the field, there is now a growing body of research on how to develop what are referred to as culturally tailored interventions (Kreuter & McClure, 2004), taking into account that theory has been normalized to mainstream populations as mentioned previously. The spectrum of culturally tailored interventions ranges from those interventions that are simply translated into the target populations' dominant language to those that acknowledge and build on the target populations' cultural beliefs and practices (Kreuter et al., 2003). A parallel line of research approaches the development of culturally sensitive interventions by considering surface and deep-structure elements, with surface elements including such components as "people, places, language, music, foods, brand names, aesthetics, and locations familiar to, and preferred by, the target audience" (p. 339) and deep elements reflecting the social, political, and cultural norms of a target population (Resnicow et al., 2005). Developing and implementing culturally tailored interventions bears cost implications, but the benefits associated with receptivity by the target population and for cohort maintenance (i.e., program effectiveness) far outweigh the costs in time and resources (Elder et al., 2005; Kreuter et al., 2003).

Within the physical activity intervention literature, there are several notable studies that should serve as role models for those interested in developing efficacious and culturally appropriate interventions. The Pathways Obesity Prevention Program is a school-based intervention designed to prevent obesity among American Indians (Davis et al., 2003; Going et al., 2003). In this groundbreaking study, traditional Native American games were adapted to increase energy expenditure among participants through changes in the setup and organization of the games but not their overall objectives. The curriculum was designed to reflect traditional Native American concepts and traditions and included music, stories, artwork, food, and family activities relevant to the target population (Davis et al., 2003). Although Caballero and colleagues (2003) reported no significant dietary differences between control and intervention schools, they did find that

Photo courtesy of San Diego Prevention Research Center.

Physical activity interventions may be more efficacious when they are culturally tailored.

self-reported physical activity levels were higher for the intervention than for the control school students at the end of the intervention.

A related strategy is the recent emphasis on developing physical activity interventions tailored to the needs of girls versus boys. In their innovative Lifestyle Education for Activity Program intervention, Pate and colleagues (Dishman et al., 2005; Pate et al., 2005) developed both gender-specific and gender-separate physical education classes (activities that appeal to girls and activities for homogeneous groups, respectively) that appeal to both African American and Caucasian female students in the 9th grade. At follow-up, regular participation in vigorous physical activity among high school girls was 8% greater in the intervention than in the control schools. This was consistent with results from an intervention implemented in Chile in which high school students were allowed to select from a menu of physical activities; girls selected two noncompetitive activities compared with none among boys (Bonhauser et al., 2005). The school-based program effectively improved physical fitness of low-SES students. Overall, these studies provide evidence that culturally appropriate interventions are relevant to the target population and thus may contribute to observed intervention effects.

In addition to tailoring by integrating main features of existing health behavior theories by integrating cultural circumstances, there are simple strategies one can use to become more adept at developing culturally appropriate interventions:

- First, make sure that intervention activities occur in a location and at a time that are convenient for the target population. This may be particularly important for new immigrant populations living in newer receiving communities, where social networks are smaller and access to resources is limited due to language barriers and low literacy skills (Evenson, Sarmiento, & Ayala, 2004). New immigrants are less likely to travel outside their normal activity space because of risk of experiencing racism and discrimination (Ayala et al., 2005).
- Second, participants are often more receptive to an intervention delivered by someone who looks and sounds like them. For example, we incorporated lay health advisors (or *promotoras*) into our intervention design discussed later in this chapter. There is a wealth of research on the pros and cons of matching the interventionist's ethnicity with the participants' ethnicity, most of which concludes that such matching improves reach and cohort maintenance (Karlsson, 2005).
- Third, well-developed formative research techniques (Krueger, 1994) can be used to identify culturally relevant facilitators and barriers to physical activity and to incorporate the one and minimize the other (Belza et al., 2004). This is part of the multimethod approach that has been developed by the San Diego Prevention Research Center, which we describe later.
- Finally, a fourth strategy involves identifying and integrating culturally appropriate activities. A recent innovation using this theme integrated gender-specific preferences for types of physical activity, as noted earlier (Pate et al., 2005).

Three additional strategies, while less specific to culture, are nonetheless relevant to this topic:

- One is to focus more on developing behavioral skills (Fitzgibbon et al., 2002) and less on imparting knowledge about the health benefits of physical activity and the pathology of inactivity. Developing behavioral skills is more likely to lead to behavior change in other settings (CDC, 1998).
- Another strategy—based on the recognition that many racial or ethnic minorities are more likely to be uninsured, to work in low-wage jobs and often in multiple jobs, and to experience additional impediments to accessing services such as language barriers and discrimination (Rutt & Coleman, 2005)—is to develop intervention approaches that maximize participants' ability to engage in the behaviors once contact with the interventionist has ceased.
- A third strategy is to take different literacy levels into account (Rosal et al., 2005). Currently, close to 18% of the total U.S. population are linguistically isolated (i.e., have difficulty with the English language), and 20% have less than a high school education.

Considering these strategies on a local level, we describe two interventions taking place in San Diego County (California, U.S.) to promote physical activity in Latinos, who make up 60% of the county's population. Researchers are using multimethod approaches that involve environmental assessments and modifications to the environment to promote community nutrition and physical activity with the aim of preventing overweight and obesity.

A Comprehensive Community Intervention

The San Diego Prevention Research Center (SDPRC) is a CDC-funded partnership among a community health center and two local universities. The primary mission of the SDPRC is to promote physical activity among families residing along the U.S.-Mexico border of San Ysidro. Part of its mission included an extensive community-wide needs and assets mapping project to create an environmental picture of programs and community resources for physical activity. A mixed-methods approach used key informant interviews and a community survey to identify cultural and socio-environmental barriers and facilitators related to physical activity among Latinos. Existing data from national, state, and local resources were also used to gather statistics on crime, disease rates, health clinic activities, walking and biking trails, and park and school locations. The mixed-methods approach identified existing health programs, perceptions of physical activity, perceived needs for and barriers to physical activity in the community, and recommendations for physical activity interventions. Recommended activities, such as walking and dancing, were identified as potential intervention activities. A community advisory board was established to build community capacity for sustaining community-wide physical activity programs. As part of the *Familias Sanas y Activas* intervention, the community advisory board and researchers developed a train-the-trainer physical activity kit and trained *promotoras* to promote leisure-time physical activity among families. *Promotoras* are trusted community members who disseminate information about healthy behaviors to the study's participants. They also act as role models for parents, who in turn are role models for their children.

In addition to the intervention, mixed-method findings were used to develop bilingual newsletters and to inform the first annual community conference. The program also initiated the Latino Health Network, a system for researchers who are interested in assessing physical activity in Latino communities.

Using multiple sources to assess community needs and assets creates a clear representation of what is occurring in the environment and increases the chances for a successful community intervention for lifetime physical activity. This unique approach can be used to develop culturally tailored physical activity interventions for underserved low-income communities. The deep elements mentioned earlier can be used to access information from various levels of influence (individual, family, community). The community-driven model demonstrates how different groups (community, academic, government) can collaborate to promote physical activity for an entire community. This strategy can be used nationwide or internationally to promote physical activity that is ethnically and regionally tailored as a means for preventing chronic disease in diverse populations and disadvantaged neighborhoods.

A Home- and School-Based Intervention

Energy balance is the key to a healthy lifestyle, which is the main focus of *Aventuras Para Niños*, a culturally and environmentally centered intervention aiming to prevent childhood overweight among elementary-aged Latino children living in neighborhoods in the southern region of San Diego County (Duerksen et al., in press). This area represents a large immigrant population facing cultural and environmental barriers as well as acculturation. This study, funded by the National Institutes of Health, employs a two-by-two factorial design in which schools are randomly assigned to one of four intervention conditions: micro environment (home), macro environment (school), micro+macro environment (home plus school), or no-treatment control. A key role in the intervention is that of the *promotora*. The study's primary aim is to decrease or maintain the healthy body mass index of kindergarten through 2nd-grade participants. Secondary aims pertain to dietary, physical activity, and sedentary behaviors, specifically, increasing fruit, vegetable, and water consumption; decreasing soda consumption; increasing active play; and decreasing TV watching.

In the micro intervention, *promotoras* (monolingual Spanish speaking and bilingual Spanish-English speaking) counsel parents about how to improve the child's home environment. A series of culturally sensitive home tools are used to modify the home environment through presentation of stories of families similar to the target audience. Home newsletters encourage parents to make home environmental changes to facilitate healthy dietary behavior and physical activity by the child. By focusing on the home environment, it is possible to enforce rules and boundaries to facilitate healthful dietary behaviors, increase child play, and decrease time spent engaging in sedentary behaviors. Diaries are given to parents for observing the child's eating habits and physical activity, and goal-setting sheets help them to monitor the level of comfort in making such changes. Physical activity among children is measured by parental report and, for a subsample of participants, by Actigraph (accelerometry) and home observations. Home assessments include 30 min observations through which a child's behavior (e.g., sitting or playing), and cues for healthy or unhealthy eating and playing or sedentary behavior, are recorded.

In the macro intervention, the *promotoras* provide information and interventions to the entire school community. They also work closely with any interested parents to make changes that affect everyone. The school environment includes aspects of the social environment, such as ways to teach physical education and the type of encouragement children receive in making lunch choices, as well as the physical condition of the playground and cafeteria. Physical education was modified to emphasize increases in the number of children being active at any one time as opposed to activities that required taking turns. California teachers are pressured to increase student academic performance, which decreases physical education time; however, teachers were highly encouraged to teach physical education three to five times a week. Recess was another opportunity to promote physical activity; playgrounds were painted with colorful game designs to keep children moving and occupied. The school environment offered opportunities for different incentive-driven activities, such as walking after lunch and 10 min academically oriented exercise breaks in the classroom. Safety concerns, such as those described as existing in low-income communities, were a barrier to implementing the CDC Kids-Walk-to-School program; however, this evolved into a child and parent after-school walking club. The micro+macro environment was an additive approach in relation to the micro and macro interventions. Assessments for the macro and micro+macro approaches include parental report of child's physical activity including walking to or from school, sport participation, and physical activity at school.

To increase fruit and vegetable consumption, children were encouraged to eat what was available to them in the school cafeteria lunch. Teachers were trained to use nonfood rewards and encouraged to offer healthy food options for class parties and other events. Tips were provided for nonfood fund-raisers and for making healthy options available at school events. Neighborhood restaurants and local grocers also were encouraged to participate by offering healthier items for children and promoting increased fruit and vegetable consumption. Culturally tailored bilingual newsletters were distributed to the entire school to inform all parents, encourage parent participation, and communicate special recognition. This type of community approach engaged parents, teachers, and cafeteria service workers by increasing their knowledge of children's health, and also provided avenues for changing the physical environment toward presenting healthier food options and increasing physical activity in youth. Preliminary assessment of this culturally centered program targeting multiple levels of influence suggests that it is effective in helping children maintain their body mass and engage in healthy eating and activity behaviors.

Conclusion

In sum, it is clear that multiple levels of influence such as family, culture, socio-environmental factors, and the physical environment play a role in the socialization of young people's physical activity. Considering the application of theories when one is developing physical activity interventions is essential. However, given that many of our health behavior theories were developed in nondiverse populations, they often need to be modified to attend to the cultural norms of diverse populations. The many factors that influence physical activity in diverse and disadvantaged populations may render past research outdated. Qualitative and quantitative studies that follow the simple strategies outlined in this chapter are needed to address and better understand the extent to which values toward physical activity and sedentary behavior change as a function of acculturation, in addition to the physical and social environmental barriers to physical activity. Culturally appropriate methods are needed for creating and evaluating interventions appropriate to ethnically diverse and underserved communities. Without use of these methods, physical inactivity will grow into a behavioral pandemic affecting youth of all ethnicities.

Applications for Researchers

A mixed-methods approach to conducting culturally appropriate research helps researchers know and understand their target population. Formative methods (e.g., key informant interviews, needs assessment) improve our understanding of sociocultural and contextual issues that influence physical activity and sedentary behaviors. Encouraging stakeholders to collaborate builds community capacity and support for long-term program implementation. Working with community members (e.g., coaches, teachers, and parents) provides the necessary support for children to become more active and less sedentary. Modified theoretical models (e.g., *promotora* model) can serve as a guide for culturally appropriate physical activity promotion. Promoting environmental change (e.g., neighborhoods and parks) can foster community cohesion and support for physical activity. Comprehensive approaches (e.g., behavioral, social, environmental) can initiate advocacy to address environmental barriers related to youth and community physical activity. Lastly, study findings should be disseminated (e.g., conferences, papers, newsletters, and Internet) as a resource to other researchers and community partners.

APPLICATIONS FOR PROFESSIONALS

As in the pursuit of research, a mixed-methods approach is necessary in order for professionals to know and understand the sociocultural contexts in which physical activity and sedentary behavior occur in the target population. Establishing an advisory board and collaborating with individuals from a variety of community agencies can inform culturally appropriate programs in addition to identifying approaches for communicating and disseminating community-wide physical activity–related information. Partnering with local agencies, state health departments, local schools, and universities can build community empowerment and provide the skills to evaluate program outcomes. Using surveillance data can assist with program planning and evaluation. Finally, working with lay health workers and parents to develop culturally appropriate strategies to modify youth activity and sedentary behaviors can have lasting effects.

References

Adams, K., Sargent, R.G., Thompson, S.H., Richter, D., Corwin, S.J., & Rogan, T.J. (2000). A study of body weight concerns and weight control practices of 4th and 7th grade adolescents. *Ethnicity and Health, 5*, 79-94.

Ajzen, I. (1985). From intentions to actions: A theory of planned behavior. In J. Kuhl & J. Beckmann (Eds.), *Action-control: From cognition to behavior* (pp. 11-39). Heidelberg: Springer.

Ajzen, I., & Fishbein, M. (1980). *Understanding attitudes and predicting social behavior.* Englewood Cliffs, NJ: Prentice Hall.

Andersen, R.E., Crespo, C.J., Bartlett, S.J., Cheskin, L.J., & Pratt, M. (1998). Relationship of physical activity and television watching with body weight and level of fatness among children: Results from the third national health and nutrition examination survey. *Journal of the American Medical Association, 279*, 938-942.

Australian Bureau of Statistics. *Australian social trends 2001.* Retrieved December 21, 2005, from www.abs.gov.au.

Ayala, G.X., Maty, S., Cravey, A., & Webb, L. (2005). Mapping social and environmental influences on health: A community perspective. In B. Israel & E. Eng (Eds.). *Multiple methods for conducting community-based participatory research for health* (pp. 188-209). San Francisco: Jossey-Bass.

Bandura, A. (1986). *Social foundations of thought and action: A social cognitive theory.* Englewood Cliffs, NJ: Prentice Hall.

Bandura, A. (1989). Perceived self-efficacy in the exercise of personal agency. *The Psychologist: Bulletin of the British Psychological Society, 10*, 411-424.

Bauman, A., & Craig, C.L. (2005). The place of physical activity in the WHO global strategy on diet and physical activity. *International Journal of Behavioral Nutrition and Physical Activity*, *2*(10).

Becker, M.H. (1974). The Health Belief Model and personal health behavior. *Health Education Monographs*, *2*, 409-419.

Belanger-Ducharme, F., & Tremblay, A. (2004). National prevalence of obesity. Prevalence of obesity in Canada. *Obesity Reviews*, *6*, 183-186.

Belza, B., Walwick, J., Shiu-Thornton, S., Schwartz, S., Taylor, M., & LoGerfo, J. (2004). Older adult perspectives on physical activity and exercise: Voices from multiple cultures. *Preventing Chronic Disease*, *1*(4), A09.

Berry, J.W. (2006). Acculturation: A conceptual overview. In M.H. Bornstein & L.R. Cote (Eds.), *Acculturation and parent-child relationships: Measurement and development* (pp. 13-32). Mahwah, NJ: Erlbaum.

Biddle, S.J.H., Gorely, T., Marshall, S.J., Murdey, I., & Cameron, N. (2004). Physical activity and sedentary behaviors in youth: Issues and controversies. *Journal of the Royal Society for the Promotion of Health*, *124*, 29-33.

Birch, L.L., & Fisher, J.O. (2000). Mothers' child-feeding practices influence daughters' eating and weight. *American Journal of Clinical Nutrition*, *71*, 1054-1061.

Bonhauser, M., Fernandez, G., Puschel, K., Yanez, F., Montero, J., Thompson, B., et al. (2005). Improving physical fitness and emotional well-being in adolescents of low socioeconomic status in Chile: Results of a school-based controlled trial. *Health Promotion International*, *20*, 113-122.

Bronfenbrenner, U. (1979). *The ecology of human development: Experiments by nature and design*. Cambridge, MA: Harvard University Press.

Bull, F., Bauman, A., Bellew, B., & Brown, W. (2004). *Getting Australia active II: An update of evidence on physical activity for health*. Melbourne, Australia: National Public Health Partnership.

Bungum, T.J., & Vincent, M.L. (1997). Determinants of physical activity among female adolescents. *American Journal of Preventive Medicine*, *13*, 115-122.

Caballero, B., Clay, T., Davis, S.M., Ethelbah, B., Rock, B.H., Lohman, T., et al. (2003). Pathways: A school-based, randomized controlled trial for the prevention of obesity in American Indian schoolchildren. *American Journal of Clinical Nutrition*, *78*, 1030-1038.

Centers for Disease Control and Prevention (CDC). (1996). *Physical activity and health: A report of the Surgeon General*. Atlanta, GA: U.S. Department of Health and Human Services.

Centers for Disease Control and Prevention (CDC). (1998). Guidelines for school and community programs to promote lifelong physical activity among young people. *Morbidity and Mortality Weekly Report*, 46:1-36.

Committee on School Health. (2004). Soft drinks in schools. *Pediatrics*, *113*, 152-154.

Cossrow, N., & Falkner, B. (2004). Race/ethnic issues in obesity and obesity-related comorbidities. *Journal of Clinical Endocrinology and Metabolism*, *89*(6), 2590-2594.

Davis, S.M., Clay, T., Smyth, M., Gittelsohn, J., Arviso, V., Flint-Wagner, H., et al. (2003). Pathways curriculum and family interventions to promote healthful eating and physical activity in American Indian schoolchildren. *Preventive Medicine*, *37*(S1), 24-34.

De Bourdeaudhuij, I., Teixerira, P., Cardon, G., & Deforche, B. (2005). Environmental and psychosocial correlates of physical activity in Portuguese and Belgian adults. *Public Health Nursing, 8*, 886-895.

Diaz-Guerrero, R. (1993). Mexican ethnopsychology. In U. Kim & J.W. Berry (Eds.), *Indigenous psychologies: Research and experience in cultural context* (pp. 44-55). Newbury Park, CA: Sage.

DiLorenzo, T.M., Stucky-Ropp, R.C., Vander Wal, J.S., & Gotham, H.J. (1998). Determinants of exercise among children II. A longitudinal analysis. *Preventive Medicine, 27*, 470-477.

Dishman, R.K., Motl, R.W., Saunders, R., Felton, G., Ward, D.S., Dowda, M., et al. (2005). Enjoyment mediates effects of a school-based physical-activity intervention. *Medicine and Science in Sports and Exercise, 37*, 478-487.

Duerksen, S.C., Elder, J.P., Arredondo, E.M., Ayala, G.X., Slymen, D., Campbell, N., et al. (in press). Family restaurant choices: Associations with body mass index (BMI) in Mexican American children and adults. *Journal of the American Dietetic Association.*

Elder, J.P., Apodaca, J.X., Parra-Medina, D., & de Nuncio, M.L.Z. (1998). Strategies for health education: Theoretical models. In S. Loue (Ed.), *Handbook of immigrant health* (pp. 567-586). New York: Plenum Press.

Elder, J.P., Ayala, G.X., Campbell, N.R., Slymen, D., Lopez-Madurga, E.T., Engelberg, M., et al. (2005). Interpersonal and print nutrition communication for a Spanish-dominant Latino population: Secretos de la buena vida. *Health Psychology, 24*, 49-57.

Evenson, K.R., Sarmiento, O.L., & Ayala, G.X. (2004). Acculturation and physical activity among North Carolina Latina immigrants. *Medicine and Science in Sports and Exercise, 59*, 2509-2522.

Félix-Ortiz, M., Fernandez, A., & Newcomb, M. (1998). The role of intergenerational discrepancy of cultural orientation in drug use among Latina adolescents. *Substance Use and Misuse, 33*, 967-994.

Fishbein, M. (1980). A theory of reasoned action: Some applications and implications. In H.E. Howe (Ed.), *1979 Nebraska symposium on motivation* (pp. 65-116). Lincoln, NE: University of Nebraska Press.

Fitzgibbon, M.L., Stolley, M.R., Dyer, A.R., VanHorn, L., & KauferChristoffel, K. (2002). A community-based obesity prevention program for minority children: Rationale and study design for Hip-Hop to Health Jr. *Preventive Medicine, 34*, 289-297.

Frenn, M., Malin, S., Villarruel, A.M., Slaikeu, K., McCarthy, S., Freeman, J., et al. (2005). Determinants of physical activity and low-fat diet among low income African American and Hispanic middle school students. *Public Health Nursing, 22*, 89-97.

Garcia, A.W., Pender, N.J., Antonakos, C.L., & Ronis, D.L. (1998). Changes in physical activity beliefs and behaviors of boys and girls across the transition to junior high school. *Journal of Adolescent Health, 22*, 394-402.

Giles-Corti, B., & Donovan, R. (2002). Socioeconomic status differences in recreational physical activity levels and real and perceived access to a supportive physical environment. *Preventive Medicine, 35*, 601-611.

Going, S., Thompson, J., Cano, S., Stewart, D., Stone, E., Harnack, L., et al. (2003). The effects of the pathways obesity prevention program on physical activity in American Indian children. *Preventive Medicine, 37*(Suppl. 1), S62-S69.

Gordon-Larsen, P., Harris, K.M., Ward, D.S., & Popkin, B.M. (2003). Acculturation and overweight-related behaviors among Hispanic immigrants to the US: The National Longitudinal Study of Adolescent Health. *Social Science and Medicine, 57*, 2023-2034.

Gordon-Larsen, P., McMurray, R.G., & Popkin, B.M. (1999). Adolescent physical activity and inactivity vary by ethnicity: The National Longitudinal Study of Adolescent Health. *Journal of Pediatrics, 135*, 301-306.

Gordon-Larsen, P., McMurray, R.G., & Popkin, B.M. (2000). Determinants of adolescent physical activity and inactivity patterns. *Pediatrics, 105*, e83.

Gottlieb, N.H., & Chen, M.S. (1985). Sociocultural correlates of childhood sporting activities: Their implications for heart health. *Social Science and Medicine, 21*, 533-539.

Government of Canada. (2005). *Sport Canada's policy on Aboriginal peoples' participation in sport.* Ottawa: Ministry of Supply and Services Canada.

Hoehner, C.M., Brennan Ramirez, L.K., Elliott, M.B., Handy, S.L., & Brownson, R.C. (2005). Perceived and objective environmental measures and physical activity among urban adults. *American Journal of Preventive Medicine, 28*(S2), 105-116.

Hu, F.B., Li, T.Y., Colditz, G.A., Willett, W.C., & Manson, J.E. (2003). Television watching and other sedentary behaviors in relation to risk of obesity and type 2 diabetes mellitus in women. *Journal of the American Medical Association, 289*, 1785-1791.

Janz, N.K., & Becker, M.H. (1984). The Health Belief Model: a decade later. *Health Education Quarterly, 11*, 1-47.

Jolliffe, D. (2004). Extent of overweight among US children and adolescents from 1971 to 2000. *International Journal of Obesity, 28*, 4-9.

Kandel, D.B., & Wu, P. (1995). The contributions of mothers and fathers to the intergenerational transmission of cigarette smoking in adolescence. *Journal of Research on Adolescence, 5*, 225-252.

Kanfer, F.H. (1975). Self-management methods. In F.H. Kanfer & A.P. Goldstein (Eds.), *Helping people change* (pp. 309-355). New York: Pergamon Press.

Karlsson, R. (2005). Ethnic matching between therapist and patient in psychotherapy: An overview of findings, together with methodological and conceptual issues. *Cultural Diversity and Ethnic Minority Psychology, 11*, 113-129.

Kavanagh, A., Goller, J., King, T., Jolley, D., Crawford, D., & Turrell, G. (2005). Urban area disadvantage and physical activity: A multilevel study in Melbourne, Australia. *Journal of Epidemiology and Community Health, 59*, 934-940.

Kreuter, M.W., Lukwago, S.N., Bucholtz, R.D., Clark, E.M., & Sanders-Thompson, V. (2003). Achieving cultural appropriateness in health promotion programs: Targeted and tailored approaches. *Health Education Behavior, 30*, 133-146.

Kreuter, M.W., & McClure, S.M. (2004). The role of culture in health communication. *Annual Review of Public Health, 25*, 439-455.

Krueger, R.A. (1994). *Focus groups: A practical guide for applied research* (2nd ed.). Thousand Oaks, CA: Sage.

Mabry, I.R., Young, D.R., Cooper, L.A., Meyers, T., Joffe, A., & Duggan, A.K. (2003). Physical activity attitudes of African American and White adolescent girls. *Ambulatory Pediatrics, 3*, 312-316.

Marín, G., & Marín, B. (1991). *Research with Hispanic populations.* Newbury Park, CA: Sage.

Marín, G., Sabogal, F., Marín, B., Otero-Sabogal, R., & Pérez-Stable, E. (1987). Development of a short acculturation scale for Hispanics. *Hispanic Journal of Behavioral Sciences, 9,* 183-200.

Marín, G., & Triandis, H.C. (1985). Allocentrism as an important characteristic of the behavior of Latin Americans and Hispanics. In R. Diaz Guerrero (Ed.), *Cross-cultural and national studies in social psychology* (pp. 69-80). Amsterdam: North-Holland.

McGuire, M.T., Hannan, P.J., Neumark-Sztainer, D., Cossrow, N.H., & Story, M. (2002). Parental correlates of physical activity in a racially/ethnically diverse adolescent sample. *Journal of Adolescent Health, 30,* 253-261.

National Health and Medical Research Council. (1997). *Acting on Australia's weight: A strategic plan for the prevention of overweight and obesity.* Canberra, Australia: Australian Government Publishing Service.

National Research Council and Institute of Medicine. (1999). Children of immigrants: Health, adjustment, and public assistance. In D. Hernandez (Ed.), *Committee on the health and adjustment of immigrant children and families.* Washington, DC: National Academies Press.

Neff, J., & Hoppe, S. (1993). Race/ethnicity, acculturation and psychological distress: Fatalism and religiosity as cultural resources. *Journal of Community Psychology, 21,* 3-20.

Neff, L.J., Sargent, R.G., McKeown, R.E., Jackson, K.L., & Valois, R.F. (1997). Black-White differences in body size perceptions and weight management practices among adolescent females. *Journal of Adolescent Health, 20,* 459-465.

Norman, G.J., Schmid, B.A., Sallis, J.F., Calfas, K.J., & Patrick, K. (2005). Psychosocial and environmental correlates of adolescent sedentary behaviors. *Pediatrics, 116,* 908-916.

Pate, R.R., Ward, D.S., Saunders, R.P., Felton, G., Dishman, R.K., & Dowda, M. (2005). Promotion of physical activity among high-school girls: A randomized controlled trial. *American Journal of Public Health, 95,* 1582-1587.

Pepitone, A. (1995). Foreword. In L.L. Adler & B.R. Mukherji (Eds.), *Spirit vs. scalpel: Traditional healing and modern psychotherapy* (pp. xiii-xiv). London: Bergin & Garvey.

Popkin, B.M., Duffey, K., & Gordon-Larsen, P. (2005). Environmental influences on food choice, physical activity and energy balance. *Physiology and Behavior, 86,* 603-613.

Popkin, B.M., & Udry, J.R. (1998). Adolescent obesity increases significantly in second and third generation U.S. immigrants: The National Longitudinal Study of Adolescent Health. *Journal of Nutrition, 128,* 701-706.

Population Reference Bureau. (2005). Retrieved December 21, 2005, from www.prb.org.

Prochaska, J.O., & DiClemente, C.C. (1983). Stages and processes of self-change of smoking: Toward an integrative model of change. *Journal of Consulting and Clinical Psychology, 51,* 390-395.

Proescholdbell, R.J., Chassin, L., & MacKinnon, D.P. (2000). Home smoking restrictions and adolescent smoking. *Nicotine and Tobacco Research, 2,* 159-167.

Resnicow, K., Jackson, A., Blissett, D., Wang, T., McCarty, F., Rahotep, S., et al. (2005). Results of the Healthy Body Healthy Spirit trial. *Health Psychology, 24,* 339-348.

Reynolds, K.D., Killen, J.D., Bryson, S.W., Maron, D.J., Taylor, C.B., Maccoby, N., et al. (1990). Psychosocial predictors of physical activity in adolescents. *Preventive Medicine, 19,* 541-551.

Robert, S., & Reither, E. (2004). A multilevel analysis of race, community disadvantage, and body mass index in adults in the US. *Social Science and Medicine, 59,* 2421-2434.

Rosal, M.C., Olendzki, B., Reed, G.W., Gumieniak, O., Scavron, J., & Ockene, I. (2005). Diabetes self-management among low-income Spanish-speaking patients: A pilot study. *Annals of Behavioral Medicine, 29,* 225-235.

Ross, C.E., & Mirowsky, J. (2001). Neighborhood disadvantage, disorder, and health. *Journal of Health and Social Behavior, 42,* 258-276.

Rutt, C.D., & Coleman, K.J. (2005). Examining the relationships among built environment, physical activity, and body mass index in El Paso, TX. *Preventive Medicine, 40,* 831-841.

Saelens, B.E., Sallis, J.F., Black, J.B., & Chen, D. (2003). Neighborhood-based differences in physical activity: An environment scale evaluation. *American Journal of Public Health, 93,* 1552-1558.

Sallis, J.F., Patterson, T.L., Buono, M.J., Atkins, C.J., & Nader, P.R. (1988). Aggregation of physical activity habits in Mexican-American and Anglo families. *Journal of Behavioral Medicine, 11,* 31-41.

Sallis, J.F., Prochaska, J.J., & Taylor, W.C. (2000). A review of correlates of physical activity of children and adolescents. *Medicine and Science in Sports and Exercise, 32,* 963-975.

Sallis, J.F., Zakarian, J.M., Hovell, M.F., & Hofstetter, C.R. (1996). Ethnic, socioeconomic, and sex differences in physical activity among adolescents. *Journal of Clinical Epidemiology, 49,* 125-134.

Shakib, S., & Dunbar, M.D. (2004). How high school athletes talk about maternal and paternal sporting experiences: Identifying modifiable social processes for gender equity physical activity interventions. *International Review for the Sociology of Sport, 39,* 275-299.

Slattery, M.L., Sweeney, C., Edwards, S., Herrick, J., Murtaugh, M., Baumgartner, K., et al. (2006). Physical activity patterns and obesity in Hispanic and non-Hispanic White women. *Medicine and Science in Sports and Exercise, 38,* 33-41.

Spector, R.E. (1991). *Cultural diversity in health and illness* (3rd ed.). Norwalk, CT: Appleton & Lange.

Spence, J.C., & Lee, R.E. (2003). Toward a comprehensive model of physical activity. *Psychology of Sport and Exercise, 4,* 7-24.

Statistics Canada. (2002). National longitudinal survey of children and youth: Childhood obesity 1994 to 1999. *The Daily,* October 18, 6-7.

Swerdlow, J.L. (1998). Population. *National Geographic, 194*(4), 2-5.

Taylor, W.C., Yancey, A.K., Leslie, J., Murray, N.G., Cummings, S.S., Sharkey, S.A., et al. (1999). Physical activity among African American and Latino middle school girls: Consistent beliefs, expectations, and experiences across two sites. *Women and Health, 30,* 67-82.

Trost, S.G., Pate, R.R., Saunders, R., Ward, D.S., Dowda, M., & Felton, G. (1997). A prospective study of the determinants of physical activity in rural fifth-grade children. *Preventive Medicine*, *26*, 257-263.

U.S. Department of Health and Human Services. (2000). *Healthy people 2010* (2nd ed., Vols. I & II). Washington, DC: U.S. Government Printing Office.

U.S. Environmental Protection Agency. (2003). *Travel and environmental implications of school siting*. Washington, DC: U.S. Environmental Protection Agency.

Wolf, A.M., Gortmaker, S.L., Cheung, L., Gray, H.M., Herzog, D.B., & Colditz, G.A. (1993). Activity, inactivity, and obesity: Racial, ethnic, and age-differences among schoolgirls. *American Journal of Public Health*, *83*, 1625-1627.

Index

Note: The italicized *f* and *t* following page numbers refer to figures and tables, respectively.

J

K

L

M

T

About the Editors

Alan L. Smith, PhD, is associate professor of health and kinesiology at Purdue University. He is recognized internationally for his research in developmental sport and exercise psychology, serves as associate editor of the *Journal of Sport & Exercise Psychology*, and is a consulting editor of *Child Development*. He is a fellow of the Research Consortium of the American Alliance for Health, Physical Education, Recreation and Dance and is a past chair of the Sport Psychology Academy of the National Association for Sport and Physical Education. He earned his PhD in exercise and movement science from the University of Oregon.

Stuart J. H. Biddle, PhD, is professor of exercise and sport psychology at Loughborough University. A recognized leader in the field of physical activity and health for young people, he has worked in the area for nearly 30 years. He is coauthor of the first textbook on exercise psychology and has delivered keynotes and other lectures in more than 20 countries. Dr. Biddle is past president of the European Federation for the Psychology of Sport and Physical Activity and was academic cochair of the Young and Active Project leading to national guidelines for physical activity for young people in the United Kingdom. He earned his PhD in psychology from Keele University.